METHODS OF
LASER SPECTROSCOPY

METHODS OF LASER SPECTROSCOPY

Edited by

Yehiam Prior

The Weizmann Institute of Science
Rehovot, Israel

Abraham Ben-Reuven

Tel-Aviv University
Tel-Aviv, Israel

and

Michael Rosenbluh

Bar-Ilan University
Ramat-Gan, Israel

PLENUM PRESS • NEW YORK AND LONDON

5 3 3 2 7 7 5 5

Library of Congress Cataloging in Publication Data

Fritz Haber International Symposium on Methods of Laser Spectroscopy (1985: Weizmann Institute of Science and En Bokek, Israel)
  Methods of laser spectroscopy.

  "Proceedings of the Fritz Haber International Symposium on Methods of Laser Spectroscopy, held December 16-20, 1985, at the Weizmann Institute of Science Rehovot, Israel, and Ein-Bokek, Dead Sea, Israel"—T.p. verso.
  Includes bibliographies and index.
  1. Laser spectroscopy—Congresses. 2. Nonlinear optics—Congresses. 3. Coherence (Optics)—Congresses. I. Prior, Yehiam. II. Ben-Reuven, Abraham. III. Rosenbluh, Michael. IV. Title.
QC454.L3F75 1985   621.36'6   86-12304
ISBN 0-306-42285-9

Proceedings of the Fritz Haber International Symposium on Methods of Laser Spectroscopy, held December 16-20, 1985, at the Weizmann Institute of Science, Rehovot, and Ein-Bokek, Dead Sea, Israel

© 1986 Plenum Press, New York
A Division of Plenum Publishing Corporation
233 Spring Street, New York, N.Y. 10013

FRITZ HABER SYMPOSIUM
ON
METHODS OF LASER SPECTROSCOPY

December 16-20, 1985

ISRAEL

SPONSORED BY:

- Israel Academy of Sciences
- Weizmann Institute of Science
- Fritz Haber Center for Molecular Dynamics
  of the Hebrew University
- Tel-Aviv University
- Bar Ilan University

ORGANIZING COMMITTEE:

Yehiam Prior (Weizmann Institute of Science)(Chairman)
Abraham Ben-Reuven (Tel-Aviv University)
Yehuda Haas (Hebrew University)
Ron Naaman(Weizmann Institute of Science)
Michael Rosenbluh (Bar Ilan University)
Amnon Yogev (Weizmann Institute of Science)

Coordinator:

Yitzchak Berman (Weizmann Institute of Science)

The Organizing Committee acknowledges the
generous support of the Robert Bosch Stiftung,
Stuttgart, W.Germany for making this meeting
possible.

The Organizing Committee also expresses its
gratitude for the support from the following:

- Maurice & Gabriela Goldschleger Foundation
  at the Weizmann Institute of Science, Rehovot

- Spectra Physics (USA)

- Coherent Inc (USA)

- Lambda Physik (Germany)

v

PREFACE

The Fritz Haber Symposium on Methods of Laser Spectroscopy was held in Ein Bokek, Israel, on the shores of the Dead Sea, on December 16-20, 1985. The location is the lowest place on earth, 392 meters below sea level. It was hoped that 120 active laser scientists, so lowly trapped in such a place, with the nearest entertainment 100 km away, will have no choice but to discuss laser spectroscopy.

On the average, the Dead Sea area receives 3-4 days of rain each year, and this year these days all occurred during the conference. This did not mean the cancellation of the hikes, although the trip to Massada was conducted in the rain. The unexpected rains also caused flash floods in the area, and Ein Bokek was completely cut-off on Thursday night. The archeologist scheduled to speak after dinner, and the belly dancer scheduled to appear afterwards, could not arrive, resulting in the only serious deviation from the original plan.

The scientific program consisted of invited talks and contributed posters. The emphasis in selection of invited speakers and topics was on the methods rather than specific molecular systems, and an attempt was made to allow ample time for discussion after each lecture. The same philosophy guided us in editing this book, and authors were requested to write manuscripts longer than usual for standard conference proceedings.

We wish to thank the other members of the organizing committee, Yehuda Haas, Ron Naaman and Amnon Yogev for their efforts during the organization stages. We are grateful to the Robert Bosch Stiftung, Stuttgart, West Germany, for the generous support which made the meeting possible. We also thank our sponsors, academic and corporate, for their interest and support. Special thanks are due to Yitzchak Berman, the conference coordinator. If one has to organize a conference, one cannot hope for a better person to work with. We are grateful to Jacqueline Biran, Batya Reindorf and Rama Avni who worked endlessly before, during and after the conference to cover all aspects of the organization, and we thank Leo Sapir for his skills in running the projection equipment and the emergency generators during the power failure.

Although organizers are not supposed to enjoy a meeting they organize, we nonetheless managed to do so, and this was possible only because of the pleasant, cooperative atmosphere for which we thank all the participants.

February 1986

Yehiam Prior
Abraham Ben-Reuven
Michael Rosenbluh

CONTENTS

NONLINEAR OPTICS AND COHERENT PHENOMENA

LASER-MOLECULE INTERACTIONS

NEW THEORETICAL TOOLS FOR SINGLE ATOM LASER SPECTROSCOPY

C. Cohen-Tannoudji and J. Dalibard

Collège de France and Laboratoire de Spectroscopie Hertzienne
de l'Ecole Normale Supérieure*
24, rue Lhomond - 75231 Paris Cedex 05 - France

INTRODUCTION

During the last few years, experimental and theoretical developments
in laser spectroscopy have increased our understanding of photon atom interactions. New methods have been invented for controlling the translational
degrees of freedom of atoms or ions[1]. A new type of spectroscopy is arising,
which deals with a single atom or a single ion, confined in a very small volume of space, with a kinetic energy so much reduced by laser cooling that
first and second order Doppler effects become completely negligible. The signal given by a photodetector recording the fluorescence light emitted by
such a single atom looks like a random sequence of pulses. In this conference, devoted to methods of laser spectroscopy, we would like to present
new theoretical tools for extracting the spectroscopic information contained
in this sequence of pulses.

EXAMPLES OF SINGLE ATOM EFFECTS

A first example of single atom effect is the so called photon antibunching[2]. The probability per unit time, $g_2(t,t+\tau)$, if one has detected one
photon at time t, to detect another one at time t+τ, tends to zero when τ
tends to zero[3]. The interpretation of this effect is that the detection of
one photon projects the atom into the ground state, so that we have to wait
that the laser reexcites the atom, before we can detect a second photon[4-6].

Another interesting example of single atom effect is the phenomenon of
"electron shelving" proposed by Dehmelt as a very sentitive double resonance
scheme for detecting very weak transitions on a single trapped ion[7]. Consider
for example the 3 level atom of Fig. 1-a, with two transitions starting from
the ground state $|g>$, one very weak $g \leftrightarrow e_R$, one very intense $g \leftrightarrow e_B$ (which we
will call for convenience the "red" and the "blue" transitions), and suppose
that two lasers drive these two transitions. When the atom absorbs a red
photon, it is "shelved" on $|e_R>$, and this switches off the intense blue
fluorescence for a time of the order of Γ_R^{-1} . We expect therefore in this
case that the sequence of pulses given by the broadband photodetector recording the fluorescence light should exhibit "periods of brightness", with
closely spaced pulses, corresponding to the intense blue resonance fluorescence, alternated with "periods of darkness" corresponding to the periods of
shelving on $|e_R>$ (Fig. 1-b). The absorption of one red photon could thus be

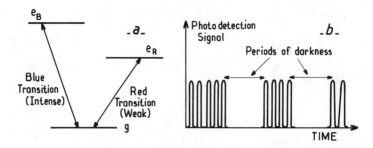

Fig. 1. (a) *3-level atom with two transitions starting from the ground state |g>.*
(b) *Random sequence of pulses given by a photodetector recording the fluorescence of a single atom. The periods of darkness correspond to the shelving of the electron on the metastable level |e_R>.*

detected by the <u>absence</u> of a large number of blue fluorescence photons[8] .
It has been recently pointed out[9] that such a fluorescence signal could pro-
vide a direct observation of "quantum jumps" between |g> and |e_R>, and seve-
ral theoretical models have been presented for this effect, using rate equations and random telegraph signal theory[9], or optical Bloch equations and
correlation functions such as g_2 [10-13].

INTRODUCTION OF NEW STATISTICAL FUNCTIONS

In this lecture, we would like to introduce another statistical function which, in our opinion, is more suitable than g_2 for the analysis of signals such as the one of Fig. 1-b. We define $w_2(\tau)$ as the probability, if
one has detected one photon at time t, to detect <u>the next one</u> at time t+τ
(and not <u>any other one</u>, as it is the case for g_2)[14]. We suppose for the moment that the detection efficiency is equal to 1, so that w_2 and g_2 refer
also to emission processes. The delay function $w_2(\tau)$ is directly related to
the repartition of delays τ between two <u>successive</u> pulses and thus provides
simple evidence for the possible existence of periods of darkness. We would
like also to show in this lecture that $w_2(\tau)$ is very simple to calculate and
is a very convenient tool for extracting all the spectroscopic information
contained in the sequence of pulses of Fig. 1-b.

We first introduce, in parallel with $w_2(\tau)$, a related function $P(\tau)$
defined by :

$$P(\tau) = 1 - \int_0^\tau d\tau' . w_2(\tau') \tag{1}$$

From the definition of w_2 , it is clear that $P(\tau)$ is the probability for not
having any emission of photons between t and t+τ , after the emission of a
photon at time t . $P(\tau)$ starts from 1 at $\tau = 0$ and decreases to zero as τ
tends to infinity. We now make the hypothesis that P and w_2 evolve in time

2

with at least two very different time constants. More precisely, we suppose that $P(\tau)$ can be written as :

$$P(\tau) = P_{short}(\tau) + P_{long}(\tau) \tag{2}$$

where

$$P_{long}(\tau) = p \exp(-\tau/\tau_{long}) \tag{3}$$

and where $P_{short}(\tau)$ tends to zero very rapidly, i.e. with one (or several) time constant(s) τ_{short} much shorter than τ_{long} . We shall see later on that this splitting effectively occurs for the three level system described above.

LOOKING FOR PERIODS OF DARKNESS

Our main point is that this form for $P(\tau)$ proves the existence of bright and dark periods in the photodetection signal, and furthermore allows the calculations of all their characteristics (average duration, repetition rate, ...). Our analysis directly follows the experimental procedure that one would use in order to exhibit such dark and light periods in the signal. We introduce a time delay θ such as :

$$\tau_{short} \ll \theta \ll \tau_{long} \tag{4}$$

and we "store" the intervals Δt between successive pulses in two "channels" : the interval Δt is considered as short if $\Delta t < \theta$, as long if $\Delta t > \theta$. We now evaluate quantities such as the probability Π for having a long interval after a given pulse and the average durations T_{long} and T_{short} of long and short intervals. If none of these three quantities depends (in first approximation) on θ , this clearly demonstrates the existence of bright periods (i.e. : succession of short intervals) and of dark ones (i.e. : occurence of a long interval).

The probability Π for having an interval Δt larger than θ is directly obtained from the function P : $\Pi = P(\theta)$. Using the double inequality (4), we get $P_{short}(\theta) \simeq 0$ and $P_{long}(\theta) \simeq P_{long}(0) = p$, so that :

$$\Pi = p \tag{5}$$

The average durations T_{long} and T_{short} of long and short intervals are given by :

$$T_{long} = \frac{1}{\Pi} \int_{\theta}^{\infty} d\tau . \tau . w_2(\tau)$$
$$T_{short} = \frac{1}{1-\Pi} \int_{0}^{\theta} d\tau . \tau . w_2(\tau) \tag{6}$$

After an integration by parts and using again the double inequality (4), this becomes :

$$\begin{cases} T_{long} = \tau_{long} \\ \\ T_{short} = \frac{1}{1-p} \int_{0}^{\infty} d\tau . P_{short}(\tau) \end{cases} \tag{7}$$

We see that the average length of long intervals is just the long time constant of $P(\tau)$, while the average length of short intervals is related to the rapidly decreasing part of $P(\tau)$. None of the three quantities obtained in (5) and (7) depends on θ, which indicates the intrinsic existence of dark periods and of bright ones. The average duration of a dark period $\mathscr{C}_D$ is just τ_{long}, while the average duration of a bright period $\mathscr{C}_B$ is the product of the duration of a short interval T_{short}, by the average number $\overline{N}$ of consecutive short intervals[15] :

$$\left\{ \begin{array}{ll} \mathscr{C}_D = \tau_{\text{long}} & (8.a) \\ \mathscr{C}_B = T_{\text{short}} \cdot \overline{N} & (8.b) \end{array} \right.$$

This average number $\overline{N}$ can be written $\sum_N N P_N$ where $P_N = (1-p)^N p$ is the probability for having N short intervals followed by a long one. Actually, the notion of "brightness" for a period has a sense only if it contains many pulses. We are then led to suppose $p \ll 1$, so that :

$$\overline{N} = \frac{1-p}{p} \simeq \frac{1}{p} \gg 1 \qquad (9)$$

Using (7) and (8.b), the length of a bright period can finally be written :

$$\mathscr{C}_B = \frac{1}{p} \int_0^\infty d\tau \cdot P_{\text{short}}(\tau) \qquad (10)$$

Note that if the efficiency of the detection ε is not 100%, results (8.a) and (10) are still valid provided certain conditions hold. Remark first that in a bright period, the mean number of pulses is multiplied by ε , and that the interval between two successive pulses is divided by ε . In order to still observe dark and bright periods, one has to detect many pulses in a given bright period, and the average delay between two detected pulses must be much shorter than the length of a dark period :

$$1 \ll \varepsilon N$$
$$(11)$$
$$T_{\text{short}}/\varepsilon \ll T_{\text{long}}$$

Provided these two inequalities are satisfied, it is still possible to detect dark and bright periods, whose lengthes are again given by (8.a) and (10).

METHOD OF CALCULATION OF THE DELAY FUNCTION

We now tackle the problem of the calculation of w_2 and P for the 3-level atom described above, for which we shall use a dressed atom approach. Immediately after the detection of a first fluorescence photon at time t, the system is in the state $|\phi_0\rangle = |g,N_B,N_R\rangle$, i.e. atom in the ground state in presence of N_B blue photons and N_R red photons. Neglecting antiresonant terms, we see that this state is only coupled by the laser-atom interactions to the two other states $|\phi_1\rangle = |e_B,N_B-1,N_R\rangle$ and $|\phi_2\rangle = |e_R,N_B,N_R-1\rangle$ (the atom absorbs a blue or a red photon and jumps from $|g\rangle$ to $|e_B\rangle$ or $|e_R\rangle$). These three states form a nearly degenerate 3-dimensional manifold $\&(N_B,N_R)$ (see Fig. 2), from which the atom can escape only by emitting a second fluorescence photon. The detection of this photon then projects the atom in a lower manifold. Consequently, the probability $P(\tau)$ for not having any emission of photon between t and t+τ after the detection of a photon at time t is simply equal to the population of the manifold $\&(N_B,N_R)$ at time t+τ knowing that the system starts from the state $|\phi_0\rangle$ at time t.

In order to calculate this population, we look for a solution for the total wavefunction of the form :

4

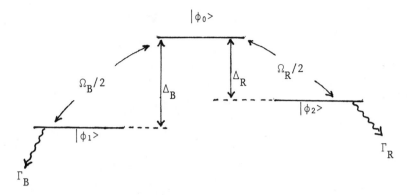

Fig. 2. *Manifold of unperturbed dressed states involved in the calculation of* $P(\tau)$.

$$|\psi(t+\tau)\rangle = \sum_{i=0,1,2} a_i(\tau) |\phi_i\rangle \times |0 \text{ fluorescence photon}\rangle$$

$$+ |\text{states involving fluorescence photons}\rangle \qquad (12)$$

with $a_0(0) = 1$, all other coefficients being equal to zero at time t. From (12), we then extract P :

$$P(\tau) = \sum_i |a_i(\tau)|^2 \qquad (13)$$

The equations of motion for the a_i's read :

$$\begin{cases} i \, \dot{a}_0 = \dfrac{\Omega_B}{2} a_1 + \dfrac{\Omega_R}{2} a_2 \\[2mm] i \, \dot{a}_1 = \dfrac{\Omega_B}{2} a_0 - (\Delta_B + \dfrac{i\Gamma_B}{2}) a_1 \\[2mm] i \, \dot{a}_2 = \dfrac{\Omega_R}{2} a_0 - (\Delta_R + \dfrac{i\Gamma_R}{2}) a_2 \end{cases} \qquad (14)$$

where Ω_B and Ω_R represent the blue and red Rabi frequencies, $\Delta_B(\Delta_R)$ the detuning between the blue (red) laser and the blue (red) atomic transition, and where Γ_B^{-1} and Γ_R^{-1} are the natural lifetimes of levels $|e_B\rangle$ and $|e_R\rangle$. This differential system is easily solved by Laplace transform, and each $a_i(\tau)$ appears as a superposition of 3 (eventually complex) exponentials. The main result is then that, provided Γ_R and Ω_R are small enough compared to Γ_B and Ω_B , $P(\tau)$ can be written as in (2)-(3) : this proves the existence of periods of darkness in the photodetection signal.

We shall not give here the details of the general calculations, and we shall only investigate the two limiting cases of weak and strong blue excitations.

THE LOW INTENSITY LIMIT

This limit corresponds to $\Omega_B \ll \Gamma_B$ (blue transition not saturated). We suppose the blue laser tuned at resonance ($\Delta_B = 0$) and we consider first $\Delta_R = 0$. The system (14) has 3 time constants, two short τ_1 and τ_2 and one long τ_3 :

$$
\begin{cases}
\dfrac{1}{\tau_1} = \dfrac{\Gamma_B}{2} & \text{(15.a)} \\[3mm]
\dfrac{1}{\tau_2} = \dfrac{\Omega_B^{\,2}}{2\Gamma_B} & \text{(15.b)} \\[3mm]
\dfrac{1}{\tau_3} = \dfrac{\Gamma_R}{2} + \dfrac{\Gamma_B}{2}\dfrac{\Omega_R^{\,2}}{\Omega_B^{\,2}} & \text{(15.c)}
\end{cases}
$$

The weight of τ_2 is predominant in $P_{short}(\tau)$ and we find :

$$
T_{short} = \tau_2/2 \tag{16}
$$

Physically, $2/\tau_2$ represents the absorption rate of a blue photon from $|g\rangle$ to $|e_B\rangle$. It can be interpreted as the transition rate given by the Fermi golden rule, with a matrix element $\Omega_B/2$ and a density of final states $2/\pi\Gamma_B$ and corresponds to the width of the ground state induced by the blue laser.

On the other hand, the long time constant in $P(\tau)$ is proportional to τ_3

$$
T_{long} = \tau_3/2 \tag{17}
$$

Physically, $2/\tau_3$ represents the departure rate from $|e_R\rangle$, due to both spontaneous (first term of (15.c)) and stimulated (second term of (15.c)) transitions (Fig. 3). The second term of (15.c) can be written $(\Omega_R/2)^2\tau_2$ and then appears as a Fermi golden rule expression. It gives the stimulated emission rate of a red photon from $|e_R\rangle$ (matrix element $\Omega_R/2$) to the ground state $|g\rangle$ broadened by the blue laser (density of states τ_2/π). Note that the condition $T_{long} \gg T_{short}$ implies :

$$
\Gamma_R \, , \, \Omega_R \ll \frac{\Omega_B^{\,2}}{\Gamma_B} \tag{18}
$$

From now on, we choose Ω_R such that the two spontaneous and stimulated rates of (15.c) are equal, and we calculate from (8.a) and (10) the variation with the red detuning Δ_R of the ratio $\mathcal{C}_D/\mathcal{C}_B$. We find that this ratio exhibits a resonant variation with Δ_R (Fig. 4) :

$$
\frac{\mathcal{C}_D}{\mathcal{C}_B} = \frac{1}{2 + (\tau_2 \, \Delta_R)^2} \tag{19}
$$

This shows that it is possible to detect the $g - e_R$ resonance by studying the ratio between the lengthes of dark and bright periods. Note that this ratio can be as large as $1/2$ (for $\Delta_R = 0$) and that the width of the resonance is determined by the width of the ground state induced by the laser. We have supposed here that $\Delta_B = 0$; if this was not the case, one would get a shift of the resonance given in (19) due to the light shift of $|g\rangle$.

THE HIGH INTENSITY LIMIT

This limit corresponds to $\Omega_B \gg \Gamma_B$ (blue transition saturated). We still suppose $\Delta_B = 0$. The two short time constants τ_1 and τ_2 of (14) are now equal to $4/\Gamma_B$, so that $T_{short} = 2/\Gamma_B$. The corresponding two roots r_1 and r_2 of the characteristic equation of (14) :

$$
\begin{cases}
r_1 = -\dfrac{\Gamma_B}{4} - i\,\dfrac{\Omega_B}{2} & \text{(20.a)} \\[3mm]
r_2 = -\dfrac{\Gamma_B}{4} + i\,\dfrac{\Omega_B}{2} & \text{(20.b)}
\end{cases}
$$

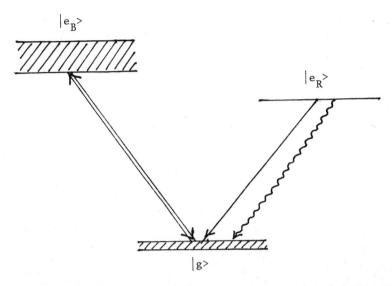

Fig. 3. The two possible desexcitation processes of the shelving state e_R : spontaneous transitions (wavy arrow) from e_R to g and stimulated transitions (full arrow) from e_R to g broadened by the blue laser (double arrow).

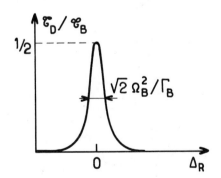

Fig. 4. Variations of $\mathscr{C}_D/\mathscr{C}_B$ with Δ_R in the low intensity limit.

now an imaginary part $\pm i\Omega_B/2$, which describes a removal of degeneracy
:ed in the manifold $\mathcal{E}(N_B , N_R)$ by the atom blue laser coupling : the two
:turbed states $|\phi_0>$ and $|\phi_1>$ of $\mathcal{E}(N_B , N_R)$, which are degenerate for
Ω_B $\mathcal{0}$, are transformed by this coupling into two perturbed dressed states :

$$|\psi_\pm> = \frac{1}{\sqrt{2}} \left(|\phi_0> \pm |\phi_1> \right) \tag{21}$$

having a width $\Gamma_B/4$ and separated by the well known dynamical Stark splitting
Ω_B [16]. The interaction with the red laser couples the third level $|\phi_2>$ to
$|\psi_\pm>$ with matrix elements $\pm \Omega_R/2\sqrt{2}$. This coupling is resonant when $|\phi_2>$ is
degenerate with $|\psi_+>$ or $|\psi_->$, i.e. when $\Delta_R = \pm \Omega_B/2$ (Fig. 5). Such a resonant
behaviour appears on the general expression of the slow time constant τ_3 of
(14) :

$$\frac{1}{\tau_3} = \frac{\Gamma_R}{2} + \frac{\Omega_R^2}{32} \frac{\Omega_B^2 \Gamma_B}{\left(\frac{\Omega_B^2}{4} - \Delta_R^2 \right)^2 + \frac{\Delta_R^2 \Gamma_B^2}{4}} \tag{22}$$

which reaches its maximum value :

$$\frac{1}{\tau_3} = \frac{\Gamma_R}{2} + \frac{\Omega_R^2}{2\Gamma_B} \tag{23}$$

for $\Delta_R = \pm \Omega_B/2$. As in (15.c), the first term of (22) or (23) represents the
effect of spontaneous transitions from $|e_R>$. The second term of (23) can be
written as $(\Omega_R/2\sqrt{2})^2$. $(4/\Gamma_B)$ and appears as a stimulated emission rate of a
red photon from $|e_R>$ to the broad $|\psi_+>$ or $|\psi_->$ states. If, as above, we
choose Ω_R such that the 2 rates of (23) are equal, we get for $\mathcal{C}_D/\mathcal{C}_B$ the
double peaked structure of Fig. 6 . When $\Delta_R = 0$, we find :

$$\frac{\mathcal{C}_D}{\mathcal{C}_B} = \frac{\Gamma_B^4}{2\Omega_B^4} \ll 1 \tag{24}$$

so that the dark periods have a very small weight in this case. On the other
hand, around $\Delta_R = \pm \Omega_B/2$, we get :

$$\frac{\mathcal{C}_D}{\mathcal{C}_B} = \frac{1}{2} \frac{\left(\frac{\Gamma_B}{4} \right)^2}{\left(\Delta_R \mp \frac{\Omega_B}{2} \right)^2 + 2\left(\frac{\Gamma_B}{4} \right)^2} \tag{25}$$

It follows that the two peaks have a maximum value equal to 1/4 and a width
$\Gamma_B/\sqrt{2}$.

Finally, Fig. 6 shows that measuring in this case the ratio between the
lengthes of dark and bright periods gives the possibility to detect, on a
single atom, the Autler-Townes effect induced on the weak red transition by
the intense blue laser excitation.

CONCLUSION

We have introduced in this paper new statistical functions which allow
a simple analysis of the electron shelving scheme proposed by Dehmelt for
detecting very weak transitions on a single trapped ion. We have shown that
there exist, in the sequence of pulses given by the photodetector recording
the fluorescence light, periods of darkness. The average length $\mathcal{C}_D$ of such
dark periods, which is determined by the spontaneous and stimulated lifetimes
of the shelving state, can reach values of the order of the average length
$\mathcal{C}_B$ of the bright periods. They should then be clearly visible on the recording of the fluorescence signal. We have also shown that it is possible to

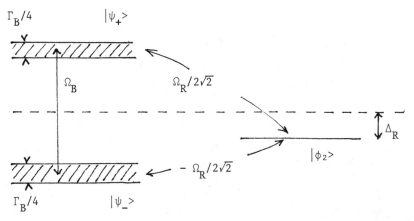

Fig. 5. *Two perturbed dressed states* $|\psi_+\rangle$ *and* $|\psi_-\rangle$ *resulting from the strong blue coupling between* $|\phi_0\rangle$ *and* $|\phi_1\rangle$. *The weak red coupling between* $|\phi_2\rangle$ *and* $|\psi_\pm\rangle$ *is resonant when* $\Delta_R = \pm\, \Omega_B/2$.

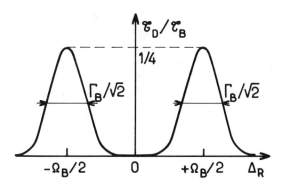

Fig. 6. *Variations of* $\mathscr{C}_D/\mathscr{C}_B$ *with* Δ_R *in the high intensity limit.*

9

get spectroscopic informations by plotting the ratio $\mathscr{C}_D/\mathscr{C}_B$ versus the detuning of the laser driving the weak transition. The smallest width obtained in this way is the width of the ground state due to the intense laser. Note that this width is still large compared to the natural width of the shelving state. It is clear that, in order to get resonances as narrow as possible, the two lasers should be alternated in time.

★ Laboratoire associé au C.N.R.S. (LA 18) et à l'Université Paris VI.

REFERENCES

1. P.E. Toschek, Ecole d'Eté des Houches 1982, in : New Trends in Atomic Physics, eds G. Grynberg and R. Stora (North-Holland, Amsterdam, (1984) p. 381.
   See also the lectures of H. Metcalf and S. Chu in this volume.
2. H.J. Kimble, M. Dagenais and L. Mandel, Phys. Rev. Lett. 39, 691 (1977).
3. We only consider in this lecture stationary random processes, so that all correlation functions such as g_2 only depend on τ .
4. C. Cohen-Tannoudji, Ecole d'Eté des Houches 1975, in : Frontiers in Laser Spectroscopy, eds R. Balian, S. Haroche and S. Liberman (North-Holland, Amsterdam, 1977) p. 1.
5. H.J. Carmichael and D.F. Walls, J. Phys. B8, L-77 (1975) ; J. Phys. B9, L-43 and 1199 (1976).
6. H.J. Kimble and L. Mandel, Phys. Rev. A13, 2123 (1976).
7. H.G. Dehmelt, IEEE Trans. Instrum. Meas., Vol. IM-31, n° 2, 83 (1982) Bull. Amer. Phys. Soc. 20, 60 (1975).
8. A similar amplification scheme has recently been used in microwave – optical double resonance experiment :
   D.J. Wineland, J.C. Bergquist, W.M. Itano, R.E. Drullinger, Opt. Lett. 5, 245 (1980).
9. R.J. Cook and H.J. Kimble, Phys. Rev. Lett. 54, 1023 (1985).
10. J. Javanainen, to be published.
11. D.T. Pegg, R. Loudon and P.L. Knight, to be published.
12. F.T. Arecchi, A. Schenzle, R.G. de Voe, K. Jungmann and R.G. Brewer, to be published.
13. A. Schenzle, R.G. de Voe and R.G. Brewer, to be published.
14. Similar functions have been introduced in S. Reynaud, Ann. Phys. Paris 8, 315 (1983).
15. We treat here the durations of intervals between pulses as independant variables. This is correct since two successive intervals are uncorrelated. At the end of a given interval, the detection of a photon projects the atom in the ground state so that any information concerning the length of this interval is lost. This is to be contrasted with the fact that two successive pulses are correlated (anti-bunching effect for example).
16. For a detailed analysis, see [14] and C. Cohen-Tannoudji and S. Reynaud, in : Multiphoton Processes, eds J.H. Eberly and P. Lambropoulos (Wiley, New York, 1978) p. 103.

THE ONE-ATOM MASER AND CAVITY QUANTUM ELECTRODYNAMICS

G. Rempe and H. Walther

Max-Planck-Institut für Quantenoptik und
Sektion Physik der Universität München
D-8046 Garching, Fed. Rep. of Germany

INTRODUCTION

When a valence electron of an atom is excited in an orbit with high
principal quantum number and therefore far from the ionic core, the
energy levels of the atom can simply be described by the Rydberg
formula. This is the reason why atoms in these highly excited states are
often called Rydberg atoms.

The existence of hydrogen Rydberg atoms has been known from radio
astronomy for many years. In space, they are formed when protons capture
electrons in high orbits; radio waves are then emitted when the electron
jumps down to lower lying levels. Until recently, however, it was impossible to generate such atoms in the laboratory. In particular, it is not
possible to populate highly excited states in discharges, owing to their
large collisional cross sections. It is the availability of tunable
lasers of sufficient intensity and narrow bandwidth that first made the
selective excitation of these states possible. Because of their high
principal quantum numbers, and also because the relative energy difference between neighbouring levels is small, Rydberg atoms are expected
to exhibit a number of classical properties. In particular, according to
Bohr's correspondence principle, the transition frequency between neighbouring levels approaches the orbital frequency of the electron.

Rydberg atoms represent an ideal testing ground for some of the
most fundamental models and predictions of low energy quantum electrodynamics (QED). The reasons are the following:
a) The matrix elements for electric dipole transitions between neighbouring Rydberg states scale as n^2, where n is the principal quantum number.
For high enough n, stimulated effects overcome spontaneous emission already for very small photon numbers. Therefore Rydberg atoms are very
sensitive e.g. to blackbody radiation (see Ref. [1] and [2] for recent
reviews).
b) The transitions to neighbouring levels are in the region of millimeter waves, therefore it is possible to physically modify the nature of
the environment into which they decay using for example conducting
walls. Introducing conductors imposes boundary conditions on the electromagnetic field and leads back to a descrete spectrum in the case of a
finite volume enclosed in a cavity. In priciple, there are essentially

two distinct cases to be discussed. First, the situation of an atom in close proximity to a conducting plate [3-8]. The induced image charges give rise to extra contributions of a van der Waals type force to the inner atomic forces, thus leading to position dependent level shifts. Second, there are effects from a discrete mode structure of the electromagnetic field inside a cavity due to its geometry. Of course, it is not possible to consider one of these phenomena without the other, but in most cases only one of the two produces the major influence. Consequences of the discrete mode structure of a cavity for Rydberg atoms are: the rate of the spontaneous emission is enhanced or diminished depending upon the cavity being tuned on or off resonance with a transition frequency [9-13,27], as well as modifying the Lamb shift of Rydberg levels [14].

c) For cavities with high quality factors, the photon emitted by an atom in a Rydberg state remains stored inside the resonator long enough to be reabsorbed by the same atom with a finite probability. In this way, it is possible to realize a single-atom maser [15]. A single Rydberg atom inside a low loss, single mode resonator is an experimental realization of the Jaynes-Cummings model [16], describing the interaction between a single two-level atom and a single mode of the electromagnetic field. This model has been the object of considerable attention in the past, and a number of purely quantum mechanical predictions on the dynamics of this system have been made. These include the collapses and revivals in the dynamics of the atomic population. Rydberg atoms will for the first time offer the possibility to test these predictions [16-18].

In the following we will briefly review the most important properties of Rydberg atoms and then discuss the QED in a cavity and in the last chapter we will describe the present status of our one-atom maser experiment.

PROPERTIES OF RYDBERG ATOMS

The scaling laws for the properties of Rydberg atoms are compiled in Table I. The energy can be simply described by the Rydberg formula with n replaced by an effective principal quantum number n^*. In general, n^* depends on the phenomenological quantum defect δ_ℓ of the state of angular momentum ℓ. For low-ℓ states, where the orbits of the classical Bohr-Sommerfeld theory are ellipses of high eccentricity, the penetration and polarization of the electron core by the valence electron lead to large quantum defects and strong departures from hydrogenic behaviour. As ℓ increases, the orbits become more circular and the atom more hydrogenic: as a result δ_ℓ scales as ℓ^5. Table 1 compiles the scaling laws for further properties of Rydberg atoms: The radius of the charge distribution of the valence electron scales as n^{*2}, and for $n^* = 50$ the linear dimension of the atom is already comparable with the wavelength of light in the visible region and competes with the size of large biomolecules.

Table I. Scaling laws for properties of Rydberg atoms

Energy:	$E_n = R/(n-\delta_\ell)^2 = R/n^{*2}$ δ_ℓ quantum defect, n^* effective quantum number R Rydberg constant
Radius:	$\langle r \rangle \sim n^{*2}$
Lifetimes:	$\tau \sim n^{*3}$ (low angular momentum states) $\tau \sim n^{*5}$ (high angular momentum states)
Fine-structure interval:	$\Delta E \sim 1/n^{*3}$

The electric polarizability for the quadratic Stark effect increases as $n*^7$ and the diamagnetic interaction as $n*^4$. This allows one to perform experiments at field strengths high enough to make the interaction energy in the external electric or magnetic field comparable with or larger than the Coulomb energy of the atom. For practical reasons the corresponding field strengths for ground state atoms cannot be reached in the laboratory. The study of highly excited atoms in external electric and magnetic fields is therefore interesting in itself. (For reviews see Refererences [19-22]). The sensitivity of Rydberg atoms to external electric fields also means that the atoms already ionize in rather weak fields. This opens the possibility of a very effective detection, as will be discussed later.

The large Rydberg atom orbitals are characterized by natural life-times much longer than the ones of less excited atoms. In the case of hydrogen Rydberg states, the dependence of the lifetime on n can be obtained by fully quantum mechanical radiation rate calculations involving hydrogenic coulombic wave-functions. For Rydberg states of other species the lifetimes (and the other radiative parameters) scale not exactly as a power of n but rather as a power of n*. The n* scaling law can be determined using Bates and Damgaard type of calculations [23]. The lifetimes scale either with $n*^3$ (when ℓ is small) or with $n*^5$ (when $\ell \cong n$).

The rate of spontaneous emission of radiation for a transition from a state n to n' is given by the Einstein A coefficient:

$$A_{n \to n'} = 16\pi^3 \upsilon^3 \; <r_{nn'}>^2 / 3\varepsilon_o hc^3,$$

where υ is the transition frequency and $<r_{nn'}>$ the matrix element of the electric dipole operator between the initial n and the final state n'. For the case $n' << n$ one has a small matrix element for dipole transitions $<r_{nn'}>$ owing to the small overlap of the radial wave functions for n and n'. $A_{n \to n'} \sim n^{-3}$. If n' is close to n, the energy difference $E_n - E_{n'} \sim n^{-3}$ and $<r_{nn'}>^2 \sim n^4$, and so $A_{n \to n'}$ becomes proportional to n^{-5}.

Since the matrix element $<r_{nn'}>$ ($n \cong n'$) scales with n^2, this leads to rather high transition probabilities for induced transitions. Rydberg atoms therefore strongly absorb microwave or far-infrared radiation (For a Review of these effects see [1], [2] and [24]). As a consequence, blackbody radiation may cause strong mixing of the states. This is especially the case for states with high angular momenta since the spontaneous lifetimes for these are much larger than for low ℓ states and the induced transitions can therefore be saturated more easily. The corresponding saturating power fluxes are proportional to n^{-10} for low and to n^{-14} for high angular momentum states. A very vivid way of describing the behaviour is to express the saturation power flux in terms of number of photons per surface of the size λ^2 and per lifetime, (the size λ^2 corresponds to the resonant cross-section). For $n \cong 30$ one obtains 10^2 and 1 for low and high angular momentum states, respectively. This means that for high angular momentum states a single photon is required (in the chosen units) to saturate the transition to a neighbouring Rydberg level [24].

CAVITY QUANTUM ELECTRODYNAMICS

It is well-known that the spontaneous lifetime of an excited atom is proportional to the density of modes of the electromagnetic field $\rho(\omega_k)$ about the atomic transition frequency ω_o. Specifically, the Weisskopf-Wigner spontaneous lifetime is given by [25]

$$\gamma = 2\pi \int d\Omega_k \ [\ /V_{12}{}^k/\ ^2\ \rho(\omega_k)\ ]\ _{\omega k\ =\ \omega o} \tag{1}$$

where $\rho(\omega_k)d\omega_k\ d\Omega_k$ represents the number of field modes in the frequency range ω_k to $\omega_k + d\omega_k$, $d\Omega_k$ is the differential solid angle, and $V_{12}{}^k$ is the dipole matrix element between the states "atom in excited state and no photon present" and "atom in ground state and one photon present in mode k". The vacuum density of modes per unit volume is

$$\rho_f\ (\omega_k)\ d\omega_k = \frac{1}{2}\ (\frac{1}{4\pi})\ \omega_k^2\ d\omega_k/\pi^2 c^3 \tag{2}$$

Thus the spontaneous emission rate for a two-level system is increased if the atom is surrounded by a cavity tuned to the transition frequency. This was already noted years ago by Purcell [9]. Conversely, the decay rate decreases when the cavity is mistuned [10]. In the case of an ideal cavity far off the atomic resoncance, no mode is available for the photon, and spontaneous emission cannot occur. In order to eliminate spontaneous emission completely, every propagating mode must be suppressed. The advent of Rydberg atoms has rendered the realisation of such experiments possible.

Consider a cubic cavity of length L and quality factor Q. Owing to the finite Q, the cavity exhibits some losses which yield a cavity linewidth $\Delta\omega = \omega/Q$. For a Lorentzian lineshape, the density of modes around a cavity mode of frequency ω_c and mode volume V_c is given by

$$\rho_c(\omega) = \frac{1}{2\pi V_c Q}\ \cdot\ \frac{\omega}{(\omega-\omega_c)^2 + (\omega_c/2Q)^2} \tag{3}$$

At resonance $\omega_o = \omega_c$, this gives

$$\gamma_c = \gamma_f\ (Q/4\pi^2)\cdot(\lambda_o^3/V_c), \tag{4}$$

where γ_c and γ_f are the spontaneous emission rates in the resonator and in vacuum, respectively.

From this equation, we draw two important conclusions: At optical frequencies, the size of the resonator is large compared to the wavelength, resulting in a small ratio λ_0^3/V_c. Since even the best available resonators, of $Q\sim10^5$, are unable to compensate for this factor, one cannot hope to observe cavity-enhanced spontaneous decay in this regime. However, for cavities operating near their fundamental frequencies, the ratio $\lambda_0^3/V_c \sim 1$, which results typically in high enhancements in the case of high Q resonators. We conclude that when an atom is placed inside a cavity with a single mode at its transition frequency, it radiates about Q/π^2 faster than in free space. (Here, we have taken $V_c \sim \lambda_o^3$.) This enhancement of the radiation rate was first pointed out by Bloembergen and Pound [26] and is the rationale for using resonant cavities in lasers and masers.

Since a resonant cavity can enhance spontaneous emission, it is not surprising that a nonresonant cavity can depress it. Consider for instance a cavity whose fundamental frequency is at twice the resonant frequency of the atomic transition. In this case, the radiation rate becomes

$$\gamma_c = \gamma_f\ \cdot\ \rho_c(\omega_o=2\omega)/\rho_f(\omega) = \gamma_f/4\pi^2 Q \tag{5}$$

In principle, γ_c can be made arbitrarily small by making Q sufficiently large.

It is important to realize that a cavity is not absolutely necessary to modify the spontaneous decay rate of an atom. Any conducting
surface placed near it will affect the mode density and hence its decay
rate. For instance, parallel conducting plates can somewhat alter the
emission rate, but at most reduce it by a factor of 2 because of the
existence of TEM modes, which are independent of the separation. The
effect of conducting surfaces on the radiation rate has been studied
theoretically in a number of investigations (for details see Reference
[1]).

To demonstrate experimentally the modification of the spontaneous
decay rate, it is not always necessary to go to single-atom densities.
The experiments where the spontaneous emission is inhibited can also be
performed with higher densities. However, in the opposite case where an
increase of the spontaneous rate is observed, a large number of excited
atoms increases the field strength in the cavity and the induced transitions disturb the experiment.

The first experimental work on inhibited spontaneous emission is
due to Drexhage [6]. The fluorescence of a thin dye film near a mirror
was investigated. Drexhage observed an alteration of the fluorescence
lifetime arising from the interference of the molecular radiation with
its surface image. An experiment with Rydberg atoms was recently performed by Vaidyanathan,Spencer and Kleppner [11]. They observed a wavelength dependent cutoff in the absorption of blackbody radiation by
Rydberg atoms arising from a discontinuity in the density of modes
between parallel-conducting plates. Absorption at a wavelength of 2/3 cm
by atoms between planes 1/3 cm apart was measured at a temperature of
180 K. The discontinuity in the absorption rate occurred when the absorption wavelength was varied across the cutoff of the parallel-plate
modes. The experiment was performed with Na atoms and the transition
employed was 29d → 30p. For the tuning of the atomic resonance across
the cutoff frequency a small electric field was applied to the parallel
plates.

Inhibited spontaneous emission was recently also observed in a very
clear experiment by Gabrielse and Dehmelt [12]. In this experiment with
a single electron stored in a Penning trap, they observed that the
cyclotron excitation shows a lifetime up to ten times larger than that
calculated for a cyclotron orbit in free space. The electrodes of the
trap form a cavity which decouples the cyclotron motion from the vacuum
radiation field, leading to a longer lifetime. Recently Hulet and Kleppner also observed an increased lifetime for Rydberg atoms in circular
states (large principal quantum number n, and magnetic quantum number
/m/ = n-1 [27]).

The first observation of enhanced atomic spontaneous emission of
Rydberg atoms in a resonant cavity was published by Goy, Raimond, Gross
and Haroche [13]. Their experiment was performed with Rydberg atoms of
Na excited in the 23s state in a niobium superconducting cavity resonant
at 340 GHz. Cavity-tuning dependent shortening of the lifetime was
observed taking advantage of the very strong electric dipole of these
atoms and of the high Q value of the superconducting resonator. This
cooling, necessary for superconducting operation, also had the advantage
of totally suppressing the blackbody field contributions, a necessary
requirement to test purely spontaneous emission effects in the cavity.

Of course, the cavity affects only the partial decay rate from the
23S to the 22P state, which in vacuum is $(\gamma_{23S \to 22P})_f = 150$ sec^{-1}, this

being only a small fraction of the total decay rate $(\gamma_{23S})_f \sim 6.9 \cdot 10^4$ sec^{-1}. To make the effect observable, the probability of a cavity-enhanced spontaneous transition during the transit time $\Delta t \sim 2 \cdot 10^{-6}$ sec of the atom through the cavity has to be near one, thus calling for a quality factor of about 10^6 in view of $(\gamma_{23S \rightarrow 22P})_c \cdot \Delta t \sim 0.22 \cdot 10^{-6}$ Q. This has actually almost been achieved and an enhanced spontaneous emission rate of $(\gamma_{23S \rightarrow 22P})_c \sim 8 \cdot 10^4$ sec^{-1} has been observed.

In spite of this enhancement of spontaneous emission, the cavity damping rate of the radiation was still faster by a factor of 35 in this experiment. The probability of reabsorbing the emitted photon either by the same atom, or by a new atom entering the cavity could be neglected. Improving further on the cavity Q, however, leads to a regime where these effects become important. This allows in particular the experimental realization of a single-atom maser [15], and brings us for the first time to experimental conditions close to those required to test the Jaynes-Cummings model [16], which describes the simplest and at the same time most fundamental interaction between a single mode of the electromagnetic field and a single two-level atom. Experiments of this type will be described in the next section.

Now we are going to discuss the question to what extent radiation corrections like the Lamb shift can be altered if they are calculated under the restriction of a certain mode structure due to a cavity [14]. For the calculations we assume that the cavity only affects the propagation of real and virtual photons leaving the atomic system unchanged. This can be justified by comparing the extension of the atomic wave function ($\cong$ 1 µm even for highly excited Rydberg states) to the size of the cavity. Under this assumption the QED corrections have to be calculated as usual but the conventional propagation function of the photon has to be replaced by that of the photon in the cavity. It is obvious that this procedure causes tiny "apparatus-dependent" deviations from the high-precision predictions of QED, as recently pointed out e.g. by Brown et al. [28] for the anomalous magnetic moment of the electron.

These deviations are expected to depend on both the shape and size of the surrounding cavity. Is has been shown for rectangular cavities [29] that the photon propagator in ordinary space has the same form as for an unaltered vacuum; however, changes appear in the momentum space representation because of the discrete structure of the modes. Essentially, the integral over the wave vector is replaced by a sum over the mode wave vectors. For a certain high-frequency cutoff Λ, the cavity is assumed to become transparent so that the mode structure only enters for $\omega < \Lambda$, whereas beyond Λ the usual integration is performed. In the case of the Lamb shift, this argument excludes cavity effects on the vacuum polarization and on the high-frequency part of the self-energy diagram for one virtual photon emission.

In the calculation of the low-frequency part of the Lamb shift only those intermediate states will contribute that accompany the emission of a virtual photon with momentum appropriate to the cavity. Since there are broadening mechanisms both for the atomic levels of the atomic system due to the natural line width and for the cavity modes due to a finite quality factor Q, the coincidences between atomic transition frequencies and eigenfrequencies of the cavity are appreciably larger than it might be expected. A special case is the waveguide for which the mode density approaches that of the undisturbed vacuum just above its fundamental cutoff. Therefore the calculation of the modification of the Lamb shift is especially simple for this case and our calculation was restricted to this situation. That means that in the Lamb shift calculation those terms were omitted corresponding to energies $\hbar\omega$ of the virtual photons below the cutoff frequency ω_c of the waveguide. It should be noted that the mass renormalization counterterm is also changed owing to the cavity, however, this correction is the same for all levels with the same main quantum number.

The results for the change of the Lamb shift of s-states $\delta E_{n,o}$ due to the model waveguide outlined above are plotted in Fig. 1 versus n and the cutoff wavelength $\lambda_c = 2\pi c/\omega_c$. Some points are worth mentioning. The overall effect is extremely small, even for very short cutoff wavelengths. For a fixed λ_c, the largest $\delta E_{n,o}$ occurs for the state with the smallest quantum number n^* that is just above the threshold for a suppressed transition to the $(n^*+1)p$ state. As λ_c is decreased, also n^* gets smaller. For values $n > n^*$, two facts tend to decrease $\delta E_{n,o}$: First, the general n^{-3} behaviour of the Lamb shift itself; second, there may be virtual cancellation of contributions to $\delta E_{n,o}$ resulting from transition to neighbouring states as e.g. $n^*s \rightarrow (n^*+1)p$ and $n^*s \rightarrow (n^*-1)p$ since they enter with opposite signs.

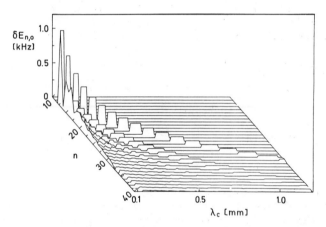

Fig. 1. Absolute change of the Lamb shift of hydrogen s-states as a function of principal quantum number n and cut-off wavelength λ_c [14].

Since the change of the Lamb shift of an hydrogen atom in a waveguide shows a strong variation with the main quantum number n, it is not necessary to perform a measurement in the optical region in order to demonstrate the phenomenon experimentally. It has been shown in Ref. [14] that the change of the Lamb shift can be measured by comparing transitions of the kind $n^*s \rightarrow (n^*+1)p$ with $(n^*+2)s \rightarrow (n^*+1)p$. To achieve the necessary precision the Ramsey method with two spatially separated microwave fields [30] has to be used. In the space between the microwave fields the atoms are moving between the conducting plates of a waveguide (for details see ref. [14]),

As mentioned above Rydberg atom experiments in high Q superconducting cavities approach the conditions required to test the Jaynes-Cummings [16-18, 31-33] model of interaction between a single two-level atom and a single mode of the electromagnetic field. This model is amenable to an exact analytical solution, and has been the object of extensive theoretical analysis. A number of purely quantum-mechanical

17

effects, whose experimental verification would be of considerable interest, have been predicted. These include the so-called Cummings collapse in the evolution of the population inversion [16,17,31], as well as its partial revival for longer interaction times [18].

The Jaynes-Cummings hamiltonian is

$$H = \hbar\omega_o\sigma_{22} + \hbar\omega a^+ a + g(\sigma_{21}a + a^+\sigma_{12}),$$ (6)

where $\sigma_{21} = |2\rangle\langle 1|$, etc..., $|1\rangle$ and $|2\rangle$ being the lower and upper atomic levels, and a, a^+ are the usual boson operators of the radiation field, $[a,a^+] = 1$. At resonance $\omega = \omega_o$, the eigenstates of this system are of the form

$$|n,\pm\rangle = (1/\sqrt{2})\{|n+1,1\rangle \pm |n,2\rangle\},$$ (7)

with corresponding eigenenergies

$$E_{n,\pm} = \hbar\omega n \pm g\sqrt{n}.$$ (8)

Supposing the atom to be initially in its ground state $|1\rangle$ and the field to be given by the initial density matrix $\rho_f(0) = \sum_n p_n |n\rangle\langle n|$, the probability $p_1(t)$ for the atom to be in state $|1\rangle$ at time t is

$$p_1(t) = (1/2) \sum_n p_n \{1 + \cos(2g\sqrt{n}t/\hbar)\}.$$ (9)

For a coherent field of average photon number $\langle n\rangle$,

$$p_n = e^{\langle n\rangle} \langle n\rangle^n/n!,$$ (10)

and for a chaotic field

$$p_n = \langle n\rangle^n/(1 + \langle n\rangle)^{n+1}.$$ (11)

As first shown by Cummings [16], $p_1(t)$ first exhibits a collapse, i.e., there is a range of interaction times for which $p_1(t)$ becomes essentially independent of time, and this, in the absence of decay mechanisms in the model. This collapse is described in the case of resonant excitation by the approximate envelope $\exp(g^2 t^2/2)$, independent of the photon number $\langle n\rangle$. (This independence does not hold for nonresonant excitation.) For longer times, $p_1(t)$ exhibits then recorrelations (revivals) and starts oscillating again in a very complex way. The recorrelations occur at times of the order [18]

$$t = kT_R \quad (k = 1,2,3,\ldots),$$

where

$$T_R \approx 2\pi\sqrt{\langle n\rangle}/g.$$ (12)

Eberly et al. have shown [18] that the physical reason for the collapse is the Poisson photon statistics of coherent states, while the revivals find their origin in the discrete nature of the photon number distribution, leading to a kind of "granularity" of the quantized radiation field present even at high average photon numbers.

THE SINGLE-ATOM MASER

For the Rydberg maser experiment [15] an atomic beam of highly excited Rydberg atoms was used. The beam oven is carefully heat-shielded from the cavity by copper plates cooled by water, liquid nitrogen and liquid helium; the atoms pass through small apertures into the liquid

18

helium cooled part of the apparatus. There they are excited to the upper maser level and enter the cavity. Behind the cavity the atoms in Rydberg states were monitored by field ionization (see Fig. 2).

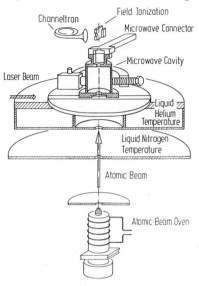

Fig. 2.   Vacuum chamber with the atomic-beam arrangement and the micro-wave cavity. The upper part is cooled to liquid-helium temperature (see also Reference [15]).

The Rydberg states were populated using the frequency doubled radiation of a continuous wave ring dye laser. The constant stream of Rydberg atoms is ionized in an inhomogeneous dc electric field of a plate capacitor. The atoms attain the point of maximum field strength in front of a hole in the anode through which the ejected electrons pass and reach a channeltron multiplier. If the field strength is adjusted properly mainly the atoms in the upper maser level are monitored. Transition from the initially prepared state to the lower maser level are thus detected by a reduction in the electron count rate. The cylindrical cavity (diameter 24.8 mm, length 24 mm) was manufactured from pure niobium rods. It is enclosed in a cryoperm shield to reduce the influence of ambient magnetic fields on the Q value due to frozen-in flux. The temperature of the cavity could be varied from 4.3 to 2.0 K, corresponding to Q factors of 1.7×10^7 and 8×10^8, respectively. The atomic beam passes through the cylindrical cavity along its axis, where only the TE_{1np} and TM_{1np} modes possess a nonvanishing transversal electric field. For our experiment the TE_{121} mode was used. This mode has a plane field distribution and is doubly degenerate in an ideal cylindrical cavity. The degeneracy is removed by a slight deformation of the circular cross section into an oval shape, which then determines the direction of polarization of the field mode. The deformation is achieved by squeezing the cylinder with a screw and a piezoelectric transducer for fine tuning (0.5 MHz/1500V). The upper maser level was the $63p_{3/2}$ level of ^{85}Rb. The fine structure splitting between $63p_{3/2}$ and $63p_{1/2}$ amounts to 396 MHz (see Fig. 3). It is, therefore no problem to excite a single fine structure level with the narrow-band ultraviolet radiation ($\Delta\upsilon \cong 2$ MHz).

To demonstrate maser operation, the cavity was tuned over the $63p_{3/2}$ - $61d_{3/2}$ transition by changing the voltage of the piezoelectric transducer; the field ionization was recorded simultaneously. Transitions from the initially prepared $63p_{3/2}$ state to the $61d_{3/2}$ level (21.50658 GHz) are detected by reduction of the electron count rate.

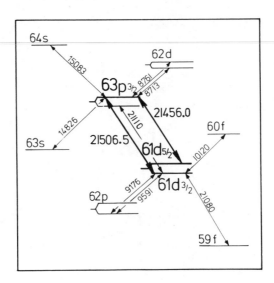

Fig. 3.   Rubidium level scheme with the maser transition.

In the case of measurements at a cavity temperature of 2K [15], a reduction of the $63p_{3/2}$ signal could be clearly seen for atomic fluxes as small as 800 atoms/s. An increase in flux caused power broadening and finally an asymmetry and a small shift (Fig. 4). This shift is attributed to the ac Stark effect, caused predominantly by virtual transitions to the $61d_{5/2}$ level, which is only 50MHz away from the maser transition (Fig. 3). The fact that the field ionization signal at resonance is independent of the particle flux (between 800 and 22 x 10³ atoms/s) indicates that the transition is saturated. This, and the observed power broadening show that there is a multiple exchange of photons between Rydberg atoms and the cavity field.

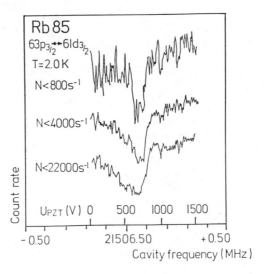

Fig. 4.   Maser resonance at a cavity temperature of 2 K (see also Reference [15]).

With an average transit time of the Rydberg atoms through the cavity of 80µs one calculates for a flux of 800 atoms/s a probability of

0.06 of finding a Rydberg atom in the cavity. According to Poisson statistics this implies that more than 99% of the events are single atom. This clearly demonstrates that single atoms are able to maintain continuous oscillation of the cavity.

Since the transition is saturated, half of the atoms initially excited in the $63p_{3/2}$ state leave the cavity in the lower $61d_{3/2}$ maser level. The decay to other levels can be neglected for the average transit time of 80 µs. The energy radiated by those atoms is stored in the cavity for its decay time, increasing the average field strength. The average number of photons left in the cavity by the Rydberg atoms is given by $n = \tau_d N/2$ where τ_d is the characteristic decay time of the cavity and N the number of Rydberg atoms entering the cavity in the upper maser level per unit time. For the highest particle flux used in our experiment $N = 22 \times 10^3$ atoms/s, one finds $n \cong 55$ photons at 2 K ($\tau_d \cong 5$ ms) and $n \cong 1.4$ photons at 4.3 K ($\tau_d \cong 0.13$ ms). This last value is smaller than the average number of blackbody photons $n_{b\ell} \cong 4$ at 4.3 K. For $N \cong 800$ atoms/s one obtains $n \cong 2$ at 2 K, which means that the radiation generated by the Rydberg atoms in the cavity has about the same energy as the blackbody radiation ($n_{b\ell} \cong 1.5$).

When the squares of the halfwidth $\Delta\upsilon$ of the signal curves are plotted versus the Rydberg atom flux, a straight line is obtained as expected. This line intersects the $(\Delta\upsilon)^2$ axis at a finite value, from which the number of blackbody photons originally in the cavity can be evaluated. The result (3 ± 1) is in reasonable agreement with the value given above. It follows that as the atomic flux decreases the thermal radiation becomes the dominant part of the field [15].

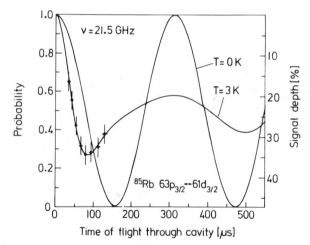

Fig. 5. Rabi-nutation with and without a thermal field. The crosses on the T = 3 K curve have been obtained with a velocity selected beam. The velocity distribution of the atomic beam only allows to measure for interaction times in the cavity between 30 and 140 µs.

The experimental setup described above is suitable to test the Jaynes-Cummings model describing the dynamics of the interaction of a single atom with a single cavity mode. An important requirement is, however, that the atoms of the beam have a homogeneous velocity so that it is possible to observe the Rabi nutation in the cavity directly. In a

modified setup a Fizeau type velocity selector is inserted between
atomic beam oven and cavity, so that a fixed atom-field interaction time
is obtained. Changing the selected velocity leads to a different interaction time and leaves the atom in another phase of the Rabi cycle when
it reaches the detector.

Since the cavity contains thermal photons the transient behaviour
has to be described by a sum of elementary Rabi oscillations in a field
in which the number of photons is a random quantity following the Bose-
Einstein statistics. The distribution of Rabi frequencies results in an
apparent random oscillation which for larger n values collapses very
quickly and then revives again. This behaviour is typical of a chaotic
quantum field; a semi-classical description of a random field does not
give this result. Fig. 5 shows the Rabi nutation for a single atom
entering the cavity in the excited state for T = 0 K and T = 3 K. The
single photon Rabi frequency assumed is 10 kHz. The cavity contains zero
thermal photons at T = 0 K and 2.5 thermal photons on the average at T =
3 K and 21.5 GHz. Plotted is the probability P for the population of the
upper (P = 1) and the lower maser level (P = 0). In a thermal field the
time dependence of P is given by

$$P(t) = \frac{1}{2}(1-\exp(-h\upsilon/kT))\sum_{n} \exp\left(-\frac{h\upsilon n}{kT}\right)(1+\cos(2\Omega\sqrt{n+1}\,t))$$

The crosses on the T = 3 K curve are obtained in the experiment.
For these measurements the cavity temperature was raised to 3 K in order
to have more thermal photons (2.5) in the cavity. The agreement between
theory and experiment is excellent.

For the transition $63p_{3/2}$ - $61d_{5/2}$ which has a larger transition
matrix element as $63p_{3/2}$ - $61d_{3/2}$ and therefore a larger Rabi frequency
also a revival could be observed so that it has been demonstrated that
the described apparatus is suitable to test the QED in a cavity in
detail [35] (Fig 6).

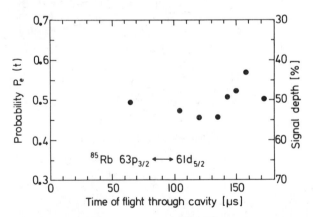

Fig. 6.   Revival of the Rabi-nutation for the transition $63p_{3/2}$-$61d_{5/2}$.
For this measurement the cavity contained about 2 photons of
blackbody radiation and about 3 photons generated by the Rydberg atoms.

REFERENCES

1. S. Haroche, J.M. Raimond, Advances Atomic and Molecular Physics, Vol. 20, Eds. D. Bates and B. Bederson, pp. 350-411, Academic Press New York (1985).
2. J.A. Gallas, G. Leuchs, H. Walther, H. Figger, Advances in Atomic and Molecular Physics, Vol. 20, Eds. D. Bates and B. Bederson, pp.413-466, Academic Press New York 1985.
3. V.B. Berestetskii, E.M. Lifshitz, and L.P. Pitaevskii, "Quantum Electrodynamics" (Pergamon Press, Oxford 1982)
4. G. Barton, Proc. Roy. Soc. London A320, 251 (1970)
5. P. Stehle, Phys. Rev. A2, 102 (1970)
6. K.H. Drexhage, Progress in Optics, Vol. 12, Ed. by E. Wolf (North Holland, Amsterdam 1974)
7. P.W. Milonni, and P.L. Knight, Opt. Commun. 9 : 119 (1973)
8. E.A. Power, and T. Thirunamachandran, Phys. Rev. A25 : 2473 (1982)
9. E.M. Purcell, Phys. Rev. 69 : 681 (1946)
10. D. Kleppner, Phys. Rev. Lett. 47 : 233 (1981)
11. A.G. Vaidyanathan, W.P. Spencer, and D. Kleppner, Phys. Rev. Lett. 47 : 1592 (1981)
12. G. Gabrielse, H. Dehmelt, Phys. Rev. Lett. 55 : 67 (1985)
    G. Gabrielse, R. van Dyck, Jr., J. Schwinberg, H. Dehmelt, Bull. Am. Phys. Soc. 29 : 926 (1984)
13. P. Goy, J.D. Raimond, M. Gross, S. Haroche, Phys. Rev. Lett. 50 : 1903 (1983)
14. P. Dobiasch, and H. Walther: Annales No6 - Alfred Kastler Symposium (Editions de Physique 1985) in print
15. D. Meschede, H. Walther, G. Müller, Phys. Rev. Lett. 54 : 551 (1985)
16. E.T. Jaynes, and F.W. Cummings, Proc. IEEE 51 : 89 (1963)
17. P. Meystre, PhD Thesis, Ecole Polytechnique Fédéderale Lausanne (1974)
    P. Meystre, E. Geneux, A. Quattropani, A. Faist, Nuovo Cimento 25B : 521 (1975)
18. J.H. Eberly, N.B. Narozhny, J.J. Sanchez-Mondragon, Phys. Rev. Lett. 44 : 1323 (1980), and references therein
19. S. Feneuille, P. Jacquinot, Advances Atom. Mol. Phys. 17 : 99 (1981)
20. D. Kleppner, in "Laser-Plasma Interactions", Les Houches XXXVI, ed. by R. Balian pp. 733-784, North-Holland, Amsterdam 1982.
21. D. Kleppner, M.G. Littman, M.L. Zimmerman, in "Rydberg States of Atoms and Molecules" edited by R.F. Stebbings and F.B. Dunning, pp. 73-116, Cambridge University Press, Cambridge 1983.
22. D. Delade, J.C. Gay, Comments At. Mol. Phys. 13 : 275 (1983)
23. D.R. Bates, A. Damgaard, Philos. Trans. Roy.Soc. London, 242 : 101 (1949)
24. S. Haroche, in "Atomic Physics 7", edited by D. Kleppner, and F.M. Pipkin, Plenum Press, New York, London 1981, p. 141
25. V. Weisskopf and E. Wigner, Z. für Phys. 63 : 54 (1930)
26. N. Bloembergen and R.V. Pound, Phys. Rev. 95 : 8 (1954)
27. R.G. Hulet, E.S. Hilfer, D. Kleppner, Phys. Rev. Lett. 55 : 2137 (1985)
28. L.S. Brown, G. Gabrielse, K. Helmerson, J. Tan, Phys. Rev. Lett. 55 : 44 (1985)
29. E. Ledinegg, Acta Phys. Austr. 51 : 85 (1979)
30. N.F. Ramsey: Phys. Rev. 76 : 996 (1949)
31. T. von Foerster: J. Phys. A8 : 95 (1975)
32. S. Stenholm: Phys. Rep. 6 : 1 (1975)
33. P.L. Knight, and P.W. Milonni: Phys. Rev. C66 : 21 (1980)
34. P. Filipowicz, J. Javanainen, P. Meystre, to be published
35. G. Rempe, H. Walther, N. Klein, publication in preparation

RYDBERG ATOMS AND RADIATION

S. Haroche

Ecole Normale Supérieure, Paris and Yale University
Laboratoire de Spectroscopie Hertzienne
24, rue Lhomond - 75231 Paris Cedex 05 - France

INTRODUCTION

The interaction of very excited atomic systems (Rydberg atoms) with electromagnetic fields has been the object of a large number of theoretical and experimental investigations during the last five years[1-2-3]. Rydberg atoms have indeed remarkable radiative properties. Due to the very large size of the valence electron orbit, these atoms are very strongly coupled to radiation (the electric dipole matrix elements between Rydberg levels scale as n^2 where n is the principal quantum number of the Rydberg state). Moreover, this coupling is resonant with transitions between nearby Rydberg states whose wavelength scales as n^3 and falls in the millimeter and microwave range for $n \gtrsim 20$. Consequently these atoms exhibit quite spectacular radiative behaviour. The rate of spontaneous emission between Rydberg levels, although small in absolute value, is very large when compared to those of ordinary atoms or molecules radiating in the same frequency range. Very small external fields acting on these atoms induce unusually large transition rates. In particular, blackbody radiation, even at low temperature, has dramatic effects on Rydberg states lifetimes[4] and energies[5]. Of particular interest are the effects observed when the Rydberg atoms are prepared in a resonator of very high Q (superconducting microwave cavity) resonant with a transition between two adjacing Rydberg levels. One then realizes the quasi ideal model of Quantum Electronics corresponding to the coupling of a two-level atom system with a single mode of the electromagnetic field[6]. Isolated Rydberg atoms prepared in the cavity are sensitive to the vacuum field in the mode and the one Rydberg atom-cavity system provides a way of studying for the first time spontaneous effects induced by a single mode of the field, which are quite different from the corresponding effects in free space, where all the modes of the field interact with the atom.

Ensembles of N Rydberg atoms prepared in the cavity also exhibit interesting effects. The atoms couple symmetrically to the field mode and collective radiative effects are observable even if the atomic sample is made of a few radiators only. Tests of superradiance theory and related radiative effects have been studied on these systems which—when N goes down to unity—constitute the smallest maser systems ever achieved.

The statistical properties of the field radiated by these masers have been analyzed and it has been shown that they should exhibit large amount of "squeezing". Most of the Rydberg atom radiation experiments so far have been performed on low ℓ Rydberg states (corresponding to ℓ ≪ n where ℓ is the level orbital momentum), these states being readily prepared by laser excitation from an atomic ground state. The possibility of preparing Rydberg levels with large ℓ's , and in particular the "circular" states[7] of maximum ℓ and m (ℓ = m = n - 1 ; m is the projection of the orbital momentum on the quantization axis), has recently opened the way to new Rydberg atom radiative studies. Circular Rydberg states have indeed special properties which make them particularly suitable for metrological applications and a new determination of the Rydberg constant should be obtained by measuring the transition frequency between two adjacing "circular" Rydberg states. In this short presentation, we will restrict ourselves to a qualitative description of this expanding field and refer the reader looking for more details to recent articles or review papers where theoretical calculations are carried out and more extensive descriptions of experiments are presented[1-2-3].

SINGLE ATOM CAVITY EFFECTS

In order to discuss the coupling of an isolated atom to the mode of a resonant cavity it is convenient to introduce the elementary Rabi frequency $\Omega_0 = \dfrac{d_{ef}}{\hbar} \sqrt{\dfrac{\hbar \omega_{ef}}{2\varepsilon_0 v}}$ which measures the rate at which the atom precesses between the two levels e and f in the field of a single photon stored in a cavity of volume v .resonant at the atomic frequency ω_{ef} (d_{ef} is the electric dipole matrix element between these levels). For usual atomic systems Ω_0 is quite small compared to the reciprocal of all relevant experimental times so that single atom cavity effects are negligible. It is only when a large number n_ϕ of photons is stored in the cavity that the Rabi frequency in the field, $\Omega_0 \sqrt{n_\phi}$ becomes large enough to yield observable effects. In other words, coherent atom-field coupling effects with ordinary atoms require macroscopic fields, which explains why most of laser physics can be dealt with in the semi-classical formalism where the field is described as a classical system.

With Rydberg atoms, this situation is drastically changed, due to the large value of the electric dipole matrix element d_{ef} between adjacing Rydberg levels. Coherent atom-field effects become observable for a field initially in its vacuum state, the atom evolving in the field of the single photon it itself radiates in the cavity. The theory of the atom-field evolution is then very simple and has been developed in several articles[1-2]. We restrict here ourselves to the case where the atom is initially in the upper level e of the Rydberg transition with no photon present. The system evolution is then readily analyzed in the subspace spanned by the three states $|e,0\rangle$, $|f,1\rangle$ and $|f,0\rangle$ corresponding respectively to the upper level without photon and the lower level with one and zero photon. In the absence of field damping, the system oscillates at frequency Ω_0 between $|e,0\rangle$ and $|f,1\rangle$. In other words, spontaneous emission in a loss-less cavity, contrary to the corresponding effect in free space, is a reversible process and one has a non zero probability of finding the atom excited at arbitrary large times.

Field damping (finite cavity Q) results in the photon loss at a rate ω/Q (irreversible decay of the system towards the $|f,0\rangle$ state). Two cases have thus to be distinguished when this irreversible process is taken into account :

(i) when $\Omega_0 > \omega/Q$, several atom-field oscillations occur before the photon disappear and the reversible behaviour described above remains observable.

(ii) when $\Omega_0 \lesssim \omega/Q$, the atomic system undergoes an irreversible over-damped evolution and the probability of finding the system in state $|e,0\rangle$ decays exponentially[8-1-2]. It is easy to show in this case that the rate of this decay process is $\Gamma_{cav} = \Omega_0^2/(\omega/Q)$ which can also be written as $\Gamma_{cav} = \eta \, \Gamma_0$ where Γ_0 is the spontaneous emission rate of the same transition in free space and $\eta = \frac{3}{4\pi^2} \frac{Q\lambda^3}{v}$ is a dimentionless enhancement factor proportional to the finesse of the cavity mode ("enhanced" spontaneous emission).

Both regimes (i) and (ii) have been recently investigated : the enhanced spontaneous emission effect which requires cavities of relatively moderate Q ($\sim 10^6$) has been observed by Goy et al. in 1983[9]. The oscillatory exchange of photon between the atom and the cavity requires larger Q's ($\sim 10^8$) and is just now in the process of being directly observed by Walther and Rempe[10].

The experimental scheme used in these experiments is quite simple in its principle (see Figure 1). The Rydberg atoms are prepared by laser excitation of an atomic beam. The laser radiation is attenuated until one makes sure that no more than one atom at a time is prepared in the cavity. The optical selection rules result in a preparation of low angular momentum Rydberg levels (practically s , p or d states with $\ell = 0$, 1 or 2 depending upon the number of photons involved in the optical transition).

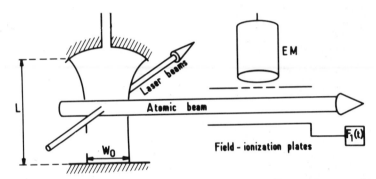

Fig. 1. Scheme of experimental set-up used for Rydberg atom cavity experiments.

The Rydberg atoms then cross a resonant superconducting cavity in which they interact with the field mode. The atomic evolution is analyzed (after the atom has emerged from the cavity) by a Rydberg detector which measures the quantum state of the atomic final state (e or f). This detector operates on the well known field ionization method[1]. An electric field $F_1(t)$ is applied to the atoms between the plates shown on the right part of Figure 1. This field ionizes the Rydberg levels and the resulting electron are counted by the electron multiplier (EM). The threshold ionizing field being slightly different for levels e and f , the method allows us to discriminate between the levels and to selectively measure the final population of levels e and f . Figure 1 corresponds to the scheme of the Goy and al. experiment which makes use of an open cavity of the Fabry-Perot type. The resonator used in the Walther experiment is a closed cylindrical cavity.

This kind of set up clearly recalls a typical maser systems with its population inversion device (here the laser excitation bringing the atoms in level e), and its resonant cavity. It corresponds in fact to the limit of a maser when the number of radiators in the cavity is reduced to unity. This point was clearly exhibited by the Meschede, Walther and Muller experiment in which a continuous maser operation has been obtained with such a device[10-11] : in this experiment, the Rydberg atoms are prepared one at a time by a low intensity CW laser beam ; each atom interacts with the mode, leaves the cavity at some point of the coherent atom-field evolution with a finite probability of depositing a photon in the mode and the maser field builds up as a result of successive such processes (see Walther's paper in these proceedings).

We should remark that, in order to observe genuine spontaneous effects in these single atom cavity experiments it is important to control the blackbody field and to reduce the number of thermal photons in the mode well below unity ($k_B T/\hbar \; \omega_{ef} < 1$), which requires very low temperatures. If this condition is not fulfilled, one observes the oscillations of the atomic system in the random thermal field, which also present interesting features. A discussion of these effects, along with the effect of quantum collapse and revivals of Rabi nutation in an applied coherent field can be found in reference[10] .

We have discussed so far the radiation of a single atom in a cavity resonant with the atomic transition. The opposite situation, namely the inhibition of spontaneous decay for Rydberg atoms in off-resonant cavities has also been predicted[12] and recently observed[13] : the atom is then prevented from radiating at all by the absence of suitable field modes in the cavity. Observation of this effect requires the excitation of the atom in circular Rydberg levels, which can decay only along microwave transitions (low angular momentum Rydberg levels are also coupled to ground states or low excited states via optical transitions which cannot be inhibited by microwave size cavity structures). The technique of preparation of these high angular momentum Rydberg levels is described in[7] and the spontaneous emission inhibition experiment in[13]. We discuss more about circular Rydberg levels and radiation in the last section of this paper.

COLLECTIVE RADIATIVE EFFECTS IN A RESONANT CAVITY

The experimental arrangement shown on Figure 1 is also well adapted to the study of collective radiative effects of Rydberg atoms in resonant cavities[1-2]. By increasing the laser intensity, one can easily prepare samples of N Rydberg atoms in the cavity ($1 \leqslant N \lesssim 10^6$). The system is prepared in a pulsed regime in the inverted state (all atoms in level e at $t = 0$) and one studies the evolution of the atomic population in the cavity following this initial preparation. This system thus constitutes the transient version of a Rydberg atom maser. It is important to remark that the sample is initially in a state invariant by atom permutation and that it is coupled to the field by an interaction which basically exhibits the same invariance[1-2]. As a result, the atomic system remains -at least until it leaves the antinode position where it has been prepared- in a state fully symmetrical by atomic exchange. This state which presents strong interatomic correlations can be described as a macroscopic angular momentum $J = N/2$ and the atomic system coupled to the cavity behaves as a single quantum object. In fact, the collective emission of the N atom system appears as a mere generalization to the N atom case of the single atom effects described above which corresponded to $J = 1/2$. The theory of this collective emission has been developed in several references[1-2].

It predicts, as in the $N = 1$ case, a coherent exchange of energy between atoms and field at the frequency $\Omega_0 \sqrt{N}$ and one has again to distinguish two regimes depending upon the magnitudes of $\Omega_0 \sqrt{N}$ and ω/Q . The overdamped regime ($\Omega_0 \sqrt{N} < \omega/Q$) is associated to the irreversible escape of electromagnetic energy from the system and corresponds to the superradiance phenomenon of Dicke[14]. The oscillatory regime ($\Omega_0 \sqrt{N} > \omega/Q$) in which the atoms and the field reversibly exchange their excitation correspond to the so called "ringing regime" of superradiance[15].

Both these regimes have been experimentally observed in experiments which constitute precise checks of the superradiance theory[15-16].

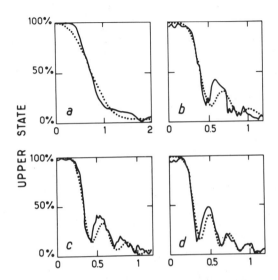

Fig. 2 . *Collective emission of Rydberg atoms in cavity : averaged atomic energy versus time ($36S_{1/2} \rightarrow 35P_{1/2}$ transition in Na). Recordings a to d correspond to increasing atom number [N=2000 (a), 19000 (b), 27000 (c) and 40.000 (d)]. Full line (experiment) and dotted line (theory). Time unit : 1 microsecond.*

Figure 2 shows the atomic energy, averaged over a large number of identical pulses, as it is observed with the atomic field ionization detector as a function of the delay t following the initial atomic inversion in the cavity. Traces a, b, c, d correspond to increasing atom numbers, i.e. to increasing exchange frequencies $\Omega_0\sqrt{N}$. Figure 2a shows the overdamped monotonous case whereas figures 2b, c, d exhibit the ringing cases. (Full line curves are experimental and dotted lines theoretical). The way the system is probed at time t in these experiments is discussed in details in reference[15].

We have also studied the fluctuations of the atomic energy around its average[1-2]. These fluctuations come from the quantum and thermal radiation noise acting on the atomic system around t = 0 . In order to analyze these fluctuations, it is convenient to describe the J = N/2 pseudo-angular momentum of the system as a "Bloch vector" evolving in an abstract space. The system evolution then appears as a pendulum motion, in which the potential and kinetic energies correspond respectively to the atom and field energy in the cavity. The pendulum starts from its unstable inverted position and is triggered away from this position by random fluctuations of its initial tipping angle and velocity[1-2]. It is possible to analyze the process as a Brownian motion of the tip of the Bloch vector, followed by the irreversible fall of the pendulum. We have experimentally studied the pulse to pulse fluctuations of the atomic energy as a function of time and we have been able to interpret all the results of the experiment in terms of the falling pendulum model[1-2-16]. In this study, the N-atom coupled to the cavity constitute an interesting example of a system exhibiting large macroscopic fluctuations originating from a microscopic quantum noise at an earlier stage of the system evolution. Other examples of such situations do exist (ferromagnetism, etc...).

In the experiments described so far, the atoms are initially prepared in the upper state |e> of the transition resonant with the cavity (collective emission process). We have also carried out collective absorption experiments in resonant cavities with the atoms starting from level |f> . The atoms then absorb the thermal radiation in the cavity mode in a collective process. The experiment can also be analyzed as a Brownian motion of the Bloch vector around its stable equilibrium position[1-2-17].

The Rydberg atom experiments described above are well adapted to the study of the atomic observables via the very sensitive field ionization method. The observation of the field itself and its fluctuations would also be very interesting. (In the Bloch vector model, the field variables are associated to the pendulum velocity whereas the atomic ones are related to its position). It has recently been shown[18] either by full quantum mechanical calculations or by the Bloch vector semi-classical approach that if the system is initially triggered by a small external field impinging on the cavity, the fluctuations on one phase of the field become at some time smaller than in the vacuum field. This is a case of radiation "squeezing" which would be very interesting to study on Rydberg atom maser systems. Such a study would require the use of low noise quantum limited microwave detectors.

CIRCULAR RYDBERG ATOMS AND RADIATION

As mentioned above, radiative studies of atoms in circular Rydberg states (such that $\ell = |m| = n-1$) open the way to new promising applications. The valence electron distribution in these states has the shape of a thin torus of radius $n^2 a_0$ and width $n\, a_0$ (a_0 : Bohr radius) perpendicular to the quantization axis and centered around the atomic core. When compared to low ℓ and $|m|$ Rydberg levels, circular Rydberg levels present the following advantages :

(i) Because the Rydberg electron practically does not penetrate the atomic core, the so-called quantum defects of these circular levels are very small and their energies are extremely close to those of the corresponding levels in hydrogen, differing from them by tiny and well calculable corrections.

(ii) Since the valence electron moves in a quasi two dimension orbital, these atoms do not have any dipole along the orbit axis and are thus relatively insensitive to electric field components perpendicular to the orbit plane.

(iii) All circular states radiatively decay only through $\Delta n = -1$, $\Delta m = -1$ radiofrequency transitions (and not at all through $|\Delta n| \gg 1$ optical channels). This gives to circular states very long radiative lifetimes. (We have already mentionned above the importance of (iii) in connexion with the spontaneous emission inhibition experiment).

The above characteristics make these systems very promising for radiation effect studies and in particular for metrological applications such as the measurement of the Rydberg constant directly in frequency units. One can indeed expect very narrow resonances between circular states, with spectral lines only quadratically sensitive to stray electric fields and frequencies depending only slightly upon the atomic ion core properties and being easily related to the hydrogen frequencies via the determination of very small quantum defects corrections.

The preparation of circular Rydberg levels require, in addition to the laser excitation, a stage of circularly polarized microwave photon absorptions performed by adiabatic fast passage in a directing electric field[7]. We have adapted this method to our Rydberg atom-cavity set-up and we have been able to prepare these circular states and to induce a millimeter·wave transition between circular states of Lithium. The details of the experiment can be found in reference[19].

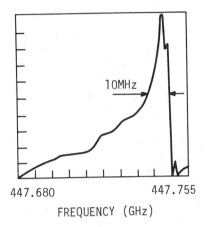

FREQUENCY (GHz)

Fig. 3. Circular to circular Rydberg resonance observed at 447.7 GHz between the $n = 25$, $m = 24$ and $n = 24$, $m = 23$ states of ^7L_i.

We present on Figure 3 the resonance line we have observed in this preliminary work. This signal corresponds to the $n = 25 \rightarrow n = 24$ circular to circular Rydberg transition. The resonance is found to be strongly asymmetrical, with a sharp edge on the high frequency side. This shape —characteristic of a quadratic Stark effect inhomogeneous broadening— is due to the fact that the resonance was recorded in a large residual electric field (several Volt/cm) leaking from the region where the circular Rydberg levels were prepared. Elimination of this field, under progress, should considerably decrease the linewidth. The position of the sharp frequency edge of the line provides a measure of the unperturbed Rydberg frequency. We find ν_{25-24} = 447 749.5 (5) MHz, in good agreement with the Rydberg formula for this transition in ^7L_i .

We are planning to improve very soon on these very preliminary results by shielding the millimeter wave interaction region against electric fields and reducing the background thermal field. Use of the Ramsey method of separated oscillatory fields with stray fields in the 100 mV/cm range should allow us to reach a resolution of a few part in 10^{11}. The Rydberg constant measured in this vay will then have to be compared with the one obtained from optical measurements in hydrogen (see T. Hänsch's contribution in these proceedings).

I hope to have shown in this brief review that Rydberg atom radiation experiments are of central interest in a variety of studies not only in fundamental Quantum Optics, but also in the technology of new radiation sources and detectors and in metrology.

The work of the Ecole Normale Supérieure laboratory reported here has been carried out in collaboration with Drs Fabre, Goy, Gross, Raimond, Kaluzny and Liang.

REFERENCES

1. S. Haroche in "New Trends in Atomic Physics", Proceedings of Les Houches Summer School, Session XXXVIII, G. Grymberg and R. Stora editors, North Holland, Amsterdam (1982).
2. S. Haroche and J.M. Raimond in "Advances in Atomic and Molecular Physics 20, 347 (1985).
3. J.A.C. Gallas, G. Leuchs, H. Walther and M. Figger in "Advances in Atomic and Molecular Physics 20, 413 (1985).
4. T.F. Gallagher and W.E. Cooke, Phys. Rev. Lett. 42, 835 (1979).
5. L. Hollberg and J.L. Hall, Phys. Rev. Lett. 53, 230 (1984).
6. E.T. Jaynes and F.W. Cummings, Proc. IEEE 51, 89 (1963).
7. R.G. Hulet and D. Kleppner, Phys. Rev. Lett. 51, 1430 (1983).
8. E.M. Purcell, Phys. Rev. 69, 681 (1946).
9. P. Goy, J.M. Raimond, M. Gross and S. Haroche, Phys. Rev. Lett. 50, 1903 (1983).
10. H. Walther, in these proceedings.
11. D. Meschede, H. Walther and G. Müller, Phys. Rev. Lett. 54, 551 (1985).
12. D. Kleppner, Phys. Rev. Lett. 47, 233 (1981).
13. R. Hulet, E. Hilfer and D. Kleppner, Phys. Rev. Lett. 55, 2137 (1985).
14. R.H. Dicke, Phys. Rev. 93, 99 (1954).
15. Y. Kaluzny, P. Goy, M. Gross, J.M. Raimond and S. Haroche, Phys. Rev. Lett. 51, 1175 (1983).
16. J.M. Raimond, P. Goy, M. Gross, C. Fabre and S. Haroche, Phys. Rev. Lett. 49, 1924 (1982).
17. J.M. Raimond, P. Goy, M. Gross, C. Fabre and S. Haroche, Phys. Rev. Lett. 49, 117 (1982).
18. A. Heidmann, J.M. Raimond and S. Reynaud, Phys. Rev. Lett. 54, 326 (1985) ; A. Heidmann, J.M. Raimond, S. Reynaud and N. Zagury, Opt. Comm. 54, 54 (1985) and 54, 189 (1985).
19. J. Liang, M. Gross, P. Goy and S. Haroche (to be published).

LASER COOLING AND MAGNETIC TRAPPING OF NEUTRAL ATOMS

Harold Metcalf

Physics Department
State University of New York
Stony Brook, NY 11794   USA

Laser cooling is the compression of the velocity distribution of an atomic sample (e.g., atomic beam) by momentum exchange with laser light [1,2]. It can be achieved by exploiting the Doppler shift to assure that fast atoms are decelerated more than slow ones [3,4]. It has been developed in part to ameliorate the effects of two distinct limits to ultrahigh resolution spectroscopy and atomic clocks caused by the thermal motion of atoms. One of these arises from the linewidth produced by the finite interaction time between the measuring equipment and rapidly moving atoms. At a typical thermal speed of 1000 m/s, an experimenter has only a few milliseconds to interact with free atoms in an apparatus of reasonable size (i.e., few meters). The other limit arises from the relativistic time differences between reference frames in relative motion (second order Doppler effect). Although atoms in a thermal beam have a velocity distribution characterized by the temperature of their source, and even though this can be calculated, the details of the distribution at the low velocity end depend very sensitively on the details of the source, and sometimes cannot be adequately known.   A sample of slowly moving or monovelocity atoms produced by laser cooling can therefore provide a substantial improvement in spectroscopic resolution [5].

Another application for such a sample is in the currently expanding area of electromagnetic traps for neutral atoms. These traps are so shallow that atoms must be cooled considerably in order to be confined. Still   a third major application for a collection of slowly moving and/or monovelocity atoms would be as an ideal target or beam for a variety of low energy and/or high energy-resolution scattering experiments, chemical reaction studies, or surface studies.   There   are   several   other applications that require monoenergetic atomic beams, including sensitive beam deflection experiments for the study of quantum statistics in optical absorption and emission.

The radiation force from scattering light of frequency ν at or near the   resonance frequencies of atomic transitions derives from the optical momentum $h\nu/c$. When an atom of mass M absorbs light it undergoes a velocity change $h\nu/Mc$, which is very small compared with thermal velocity (only 3 cm/s in our experiment), but repeated scattering can be used to produce a large total velocity change. By directing a laser beam against an atomic beam, each atom can absorb light very many times along its path through the apparatus.   Of course, excited state atoms can not absorb light efficiently from the laser that excited them, so they must return to the ground state between absorptions. If they return by spontaneous decay, accompanied by emission of fluorescent light, there is also a change of the momentum of the atoms. But the spatial symmetry of the fluorescence results in an average of zero net momentum transfer after many such scattering events (stimulated emission results in no net momentum change). The net deceleration of the atoms is therefore in the direction of the laser beam.

The Doppler shifted laser frequency in the moving atom's reference frame should match that of the atomic transition to maximize the light absorption and scattering rate. Of course, as the atoms in the beam slow down, their changing Doppler shift will take them out of resonance with the (spectrally narrow) laser, and they will eventually cease deceleration after their Doppler shift has been decreased by a few times the natural width of the optical transition (10 MHz corresponding to 6 m/s for sodium). Although this 6 m/s is considerably larger than $h\nu/Mc = 3$ cm/s (for the sodium D-line), it is still only a small fraction of the atoms' average thermal velocity, so that significant cooling or deceleration cannot be accomplished.

The two methods that have succeeded in overcoming this problem are sweeping the laser frequency to keep it in resonance with the decelerating atoms, and spatially varying the atomic resonance frequency to keep the decelerating atoms in resonance with the fixed frequency laser. The first method, called chirping, has succeded in three laboratories, including our own [6,7,8]. The second method, using a spatially varying magnetic field to tune the atomic levels along the beam path, is favored by us because it produces a continuous beam of laser cooled atoms. It works because an external magnetic field shifts the energy levels of an atom (Zeeman effect), and as long as the shift of the ground and excited states is different, the energy difference between them and hence resonant frequency of the atomic transition induced by the laser will also be shifted. The resonant frequency for an atom in moderate fields can be readily found, and the field tailored to provide the appropriate Doppler shift along the moving atom's path. For uniform acceleration a the field should decrease from B_o along z according to $B = B_o\sqrt{(1 - x)}$ where $x=2az/v_o^2$. We constructed a solenoid to produce such a spatially varying field by winding layers of decreasing lengths on a cylindrical form, and running the same current through each layer.

Measurement of the velocity distribution and the laser-induced changes to it is most easily done by observing the fluorescence from atoms excited by a second, very weak laser propagating nearly parallel to the atomic beam in a region where there is zero magnetic field. Because of the Doppler shift, the absorption and hence the intensity of this fluorescence will depend on the atomic velocity, and a slow scan of this laser's frequency will result in a fluorescence signal that reflects the velocity distribution. This second, probe laser should be sufficiently weak that it doesn't perturb the atomic velocity distribution, and should cross the atomic beam sufficiently far away from the end of the solenoid that there is no residual magnetic field from it.

Our apparatus is shown schematically in Fig. 1. A 450 °C oven produces a slightly supersonic atomic beam through a 0.12 mm aperture. After being collimated to about 0.01 rad, the beam enters a 1.1 m long solenoid that has windings to produce the inhomogeneous or tapered field to compensate for the changing Doppler shift. The circularly polarized cooling laser opposes the atomic beam and is chopped by a spinning slotted wheel to allow convenient observation of fluorescence induced by the probe laser while the main cooling laser beam is blocked. The atomic

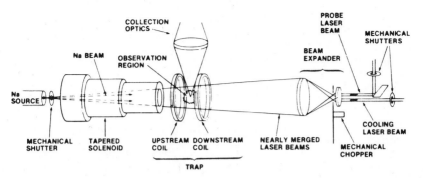

Fig. 1. Schematic of the apparatus. The solenoid is 1.1 m long and the trap is 40 cm from its end. The combination of shutters and chopping wheel allows the cooling and probe laser beams to be turned on and off rapidly and independently.

velocity distribution is measured by observing the fluorescence induced by a very weak probe that crosses the beam at a small angle about 40 cm beyond the end of the 8 cm diameter solenoid. The purpose of the 40 cm drift space is to allow observation in a region free of the fringing magnetic field from the solenoid end. The velocity resolution of this measurement scheme is dictated by the natural width of the atomic transition and corresponds to about 6 m/s.

Some atoms in the thermal velocity distribution are moving too fast to be decelerated at all because the laser frequency is Doppler shifted too far out of resonance for them to absorb light, even where the magnetic field is strongest at the solenoid entrance. Others have speed that Doppler shifts the laser frequency to match the magnetic field shift and begin slowing down as soon as they enter the solenoid. Still others are moving so slowly that they do not absorb light until they have travelled to a point where the static but spatially varying magnetic field has decreased to the appropriate value to match their smaller Doppler shift and produce resonance. Thus all atoms with velocity slower than some maximum can be decelerated by the laser beam to some lower velocity at the end of the solenoid. This final velocity is determined by the atomic resonance condition in the field at the end of the solenoid at the chosen laser frequency, and the atoms are swept into a narrow velocity group around this final velocity. The result is that most of the originally wide thermal velocity distribution is compressed and shifted to lower velocities.

The maximum attainable deceleration is obtained for very high light intensities, and is limited because the atom must then divide its time equally between ground and excited states. High intensity light can produce faster absorption, but it also causes considerable stimulated emission. This produces neither deceleration nor cooling because the momentum transfer is in the opposite direction to what it was in absorption. The deceleration therefore saturates at the value a(max) = $h\nu/2Mc\tau$ where τ is the excited state lifetime.

The choice of an atom to use for deceleration and cooling experiments depends on the availability of suitable lasers and the ease of making an atomic beam. The most suitable lasers for cooling available today are tunable dye lasers that operate in the yellow-green region of the spectrum, and of several atoms with resonance transitions in this spectral region, sodium is the most suitable. It can readily produce a strong atomic beam, it is easily handled, and its spontaneous fluoresence rate is sufficiently high to obtain deceleration from thermal to zero velocity in a reasonable size apparatus. For these reasons, all the experiments reported to date have been done with this atom. For sodium, the maximimum deceleration is about 10^6 m/s^2, and it takes about 33,000 scattering events to decelerate thermal atoms (v =1000 m/s) to rest. The process occurs along a beam at least 0.5 m long and requires at least 1 ms.

Like other alkalis, sodium has a ground state hyperfine splitting (hfs) caused by the interaction of the atomic electrons with the spin of the nucleus. It is considerably larger than the natural width of the optical transition (hfs = 1772 MHz), resulting in two distinct zero field ground state levels. If the cooling laser is tuned for excitation out of one of these hfs levels and the atom decays from its excited state into the other hfs level, it will be out of resonance with the laser, unable to absorb and scatter any more light, and lost from the deceleration and cooling process. This is called optical pumping.

There are a number of methods to overcome this loss of atoms by optical pumping. For example, there could be two laser beams tuned for excitation out of each hfs state, the laser spectrum could be sufficiently broad to excite both hfs states, there could be an rf or microwave field that would induce hfs transitions to return atoms to the appropriate hfs state, or optical pumping could be inhibited by careful choice of experimental conditions. For a variety of carefully considered technical reasons, we employ the last of these alternatives. We use circularly polarized light and the axis provided by the magnetic tuning field to allow only excitations to a particular sublevel of the atomic excited state. The only strongly allowed decay process returns the atom to the original state.

Our measurements of the atomic velocity distribution are made some time after the cooling laser beam has been shut off by the chopping wheel. Even though the magnetic field falls rapidly at the end of the solenoid and the light scattering rate is very low, deceleration does not stop completely when the atoms leave the solenoid. During the long time required for the slowest atoms to reach the observation region 40 cm away, many may be stopped by the laser light before they reach the observation region. In order to avoid this difficulty, the observation is delayed after the beam is blocked by the chopper allowing the slowest atoms to drift into the observation region in the dark. Longer delays allow the observation of slower atoms.

Even with this delayed observation technique it is somewhat difficult to observe atoms much slower than about 30 m/s because of spreading of the beam of slow atoms. We have therefore made a slight variation to the procedure [9]. First we decelerate and cool atoms to about 50 m/s as described above. Then we allow them to drift for about 8 ms in the dark, and then illuminate them with a short (100 - 400 μs) pulse of light from the cooling laser while they are in the observation region. We found that the cooled distribution, about 12 m/s fwhm and centered near 50 m/s, was slightly broadened and decelerated to near zero m/s. The gas of atoms thus produced is at rest in the laboratory, and has an effective temperature of less than 0.1 Kelvin.

With the production of such very slow, cold atoms, neutral atom deceleration has been taken from the status of an interesting demonstration to a practical tool. For example, the width of the energy distribution of these atoms is comparable to the well depth of a magnetic neutral atom trap.

Electromagnetic trapping of neutral atoms is a new branch of applied physics that has potential for use in very many areas. We will restrict our attention to the proven technique of trapping by inhomogeneous dc magnetic fields [10], although quasi-trapping with suitable configurations of laser beams has also been demonstrated [11], and real optical trapping is possible [12]. Such confinement depends on the interaction between an inhomogeneous magnetic field and the magnetic moment of the atomic ground state (in this paper we restrict our attention to the magnetic dipole moment of the sodium atom). Although Earnshaw's theorem prohibits an electrostatic field from stably trapping a charged particle (similarly for magnetic fields and monopoles), dipoles may be trapped by a local field minimum (local field maxima are forbidden [13]).

In order to confine any object it is necessary to exchange kinetic for potential energy, and in neutral atom traps, the potential energy must be stored in internal atomic degrees of freedom. There are two immediate and extremely important consequences of this requirement. First, the atomic energy levels will necessarily shift as the atoms move in the trap, and these shifts will affect the precision of spectroscopic measurements, perhaps severely. Since one of the potential applications of trapped atoms is in high resolution spectroscopy, such inevitable shifts must be carefully considered. Second, practical traps for ground state neutral atoms are necessarily very shallow compared with thermal energy because the energy level shifts that result from convenient size fields are typically much less than 1 K. Neutral trapping therefore depends on substantial cooling of a thermal atomic sample, and is inextricably connected with the cooling process.

The small depth of neutral atom traps also dictates stringent vacuum requirements because a trapped atom can not survive a collision with a thermal energy background gas molecule. Since these atoms are "sitting ducks" for any such molecules remaining in the necessarily imperfect vacuum of any real apparatus, the mean free time between collisions MUST exceed the desired trapping time. The liklihood of such destructive collisions is quite large because even a gentle collision can impart enough energy to eject the atom from the shallow trap.

We have built a spherical quadrupole trap originally suggested by W. Paul [14] comprised of two identical coils carrying opposite currents. This trap clearly has a single center where the field is zero, and is the simplest of all possible magnetic traps (see Fig. 2). The trapping force derives from the interaction $F = -\nabla|\vec{\mu} \cdot \vec{B}|$ and is constant along a line through the trap center (not harmonic). The magnetic potential is then given by $U \propto (x^2 + y^2 + 4z^2)^{\frac{1}{2}}$. The trap has equal depth in the radial (x-y plane) and longitudinal (z-axis) directions when the coils are separated by 1.25 times their radius. Its experimental simplicity makes it most attractive, because of ease of construction and of optical access to the interior. Such a trap has been used in our experiments [10] to trap laser cooled Na atoms for times exceeding one second, and that time is limited only by background gas pressure.

There is a simple expression for the field change of a magnetic trap required to confine atoms decelerated by laser light using Zeeman compensation for the changing Doppler shift. The Doppler shift to be compensated by the the change of magnetic field ΔB is $\Delta v / \lambda = \mu B / h$, where μ is the atomic magnetic moment and Δv is the velocity change of the atoms. Also, the randomness of the spontaneous emission in the deceleration process leaves the atoms with an rms velocity of the order of $\delta v = \Delta v / \sqrt{n}$ where $n = M \Delta v \lambda / h$ is the number of scattered quanta (this is approximate because the component along the laser direction may be

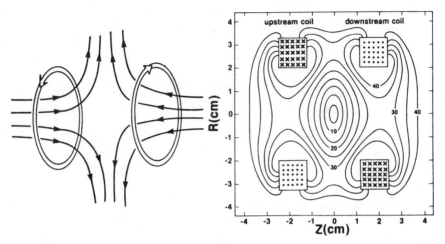

Fig. 2. Part (a) shows the arrangement of the two coils of the magnetic quadrupole trap and part (b) shows the equipotentials. The force is not central (angular momentum not conserved), and is not proportional to the distance from the origin so that the motion is not harmonic.

partially cooled by the laser). The kinetic energy associated with this random motion is $M(\delta v)^2/2 = \Delta vh/2\lambda = \mu B/2$. The magnetic field change required to trap atoms having this kinetic energy is simply $B/2$.

In order for this kind of trap to work, the atomic magnetic moments must be oriented so that they are repelled from regions of strong field. This is certainly the orientation produced by the optical pumping process that occurs during Zeeman compensated laser deceleration and cooling of atoms, but it must be preserved while the atoms are in the trap even though the trap fields change directions in a very complicated way. As long as the atoms move slowly enough, the magnetic field changes sufficiently slowly that the magnetic moments precessing about the field at the frequency ω_L can follow it adiabatically.

This requires velocities slow enough to assure that the interaction between the moment and the field remains adiabatic, especially if the atom's path passes through a region where the energy separation between the trapping orientation and some non-trapping orientation is sufficiently small (e.g., if the field magnitude is very small). For the case of circular orbits, this requires that $\omega_L \gg (dB/dt)/B = \omega_T$, the frequency of orbital motion in the trap. (More general orbits must satisfy a similar condition. Therefore energy considerations are not sufficient to determine the stability of such a magnetic neutral atom trap: orbit calculations and their consequences must also be considered.

The adiabatic condition can be easily calculated for Na atoms in a circular orbit in the x-y plane (z = 0) of a two coil trap with field gradient 1 T/m. We find that it requires $r \gg 0.3$ µm as well as $v \gg 0.8$ cm/s, corresponding to a kinetic temperature $\gg 0.1$ µK. For $v = 2.5$ m/s (typical of our experiment) we find $r = 2$ cm, and this condition is clearly satisfied. The adiabatic condition is most severely violated by the fastest atoms (3.5 m/s in our experiments) moving in very non-circular orbits that take them nearest the trap center where the field is smallest and changes direction most rapidly. Since the smallest field gradient is 1.25 T/m, the most stringent requirement for adiabaticity is $r \gg 6$ µm. As long as the orbiting atom stays several µ m away from the trap center where the field vanishes, the adiabatic condition is satisfied. Since the non-adiabatic region of the trap is so small (less than 10^{-10} of the trap volume), nearly all the orbiting atoms behave adiabatically and we expect that most atoms will stay trapped for many orbits.

The present results of our experiments have shown trapping times longer than one second, and are limited only by scattering from background gas molecules. We are presently working on improving the vacuum apparatus to reduce the number of these residual molecules, and thereby observe much longer trapping times. We do not know if we will reach the adiabaticity limit with the presently planned vacuum apparatus but we expect to achieve trapping times longer than one minute.

Since the force confining the atoms in the trap is neither harmonic nor central, it is impossible to characterize a general orbit. Circular orbits of radius r in the z = 0 plane have a period of $T = 2\pi\sqrt{r/a}$ where a is the centripetal acceleration supplied by the field gradient. For our trap a = 300 m/s^2 for z = 0, and the fastest atoms in circular orbits have v = 2.5 m/s at r = 2 cm resulting in T = 51 ms. Atoms that oscillate linearly through the trap center have periods of $T = 4v_{max}/a$, and the fastest atoms (v_{max} = 3.5 m/s) have oscillation periods of 47 ms in the x-y plane and 23 ms along the z-axis where the acceleration is twice as large. In all cases, slower atoms have shorter periods. Since the non-adiabatic region of the trap is so small (less than 10^{-10} of the trap volume) nearly all the orbiting atoms behave adiabatically permitting trapping stability for very many orbits and suggesting trapping times of many minutes, or possibly longer.

The equations of motion in this very unusual, conical potential have not been solved in closed form. The potential conserves the z-component of angular momentum but not the total angular momentum, much like a molecule. Detailed numerical studies of classical atomic orbits in this quadrupole trap have shown that some orbits are nearly closed while others seem chaotic. The trap cannot have its equilibrium position displaced from the origin by gravity as can a harmonic trap. In addition to studies of the classical orbits, there have also been studies of the quantum states that would be occupied by extremely cold atoms in the quadrupole trap.

We have been studying the cooling of trapped atoms. We have demonstrated cooling of atoms in a beam; in fact such cooling is necessary in order to trap atoms. Cooling trapped atoms is somewhat easier because we have more time than the one millisecond flight time of thermal velocity atoms, but somewhat more difficult because we are trying to approach closer to absolute zero. Although we have succeeded in trapping thousands of atoms at a temperature of 0.05 K, this has been accomplished only by selecting the slowest ones from our cooled atom sample. A great many more such atoms could be contained at this or lower temperature if we could learn to cool them in the trap. The principle difficulty is the presence of the strongly inhomogeneous trap fields which would always permit optical pumping by the cooling laser light in some part of a quadrupole trap. Excitation of the trapped atoms may very well result in their decay to states that can not be trapped. It may be possible to dodge this problem in a trap with an odd number of coils (such as a harmonic hexapole trap).

It is interesting to consider the behavior of extremely slow (cold) atoms in this trap. For this situation we make a heuristic quantization of the orbital angular momentum using $mvr = n\hbar$ and readily find $\omega_L = n\omega_T$. More careful study of the quantized states of very cold atoms in this trap is necessary because of the coupling of magnetic transitions and orbital motion at small n. As long as n is large, the two phenomena are readily separated because of their different magnitudes, but when n is small, the classical idea of orbital motion is no longer rigorous. The adiabatic condition is satisfied only for $n \gg 1$; for $v = 3$ m/s, $n = 5 \times 10^7$.

The future for trapped neutral atoms is very difficult to predict. Of course there will be opportunities for precision spectroscopy in parts of a trap where the field is zero or very uniform. Maybe the trap will serve as a place for further refrigerating of atoms before they are released into free fall where precision work can be done. Perhaps the zero-gravity environment of space can be exploited for further work. There will also be collision studies, perhaps leading to observation of collective effects. The long-sought Bose condensation may very well be accomplished with trapped atoms. Also, study of the traps themselves will provide new applications of both classical and quantum mechanics. Calculations of the orbits on a classical mechanical scale have shown signs of an interesting form of chaos. Orbits on a quantum scale exhibit very close relations with the non-adiabatic condition discussed above because the atoms must necessarily be close to the origin where the field is zero. Finally, studies of the thermodynamics and statistical mechanics of laser cooling, either in or out of traps, may provide new insights into non-equilibrium processes that can be tested in the laboratory.

This work would not have been possible without the insight and ideas of Bill Phillips. I also wish to acknowledge the help of J. Prodan, A. Migdall, T. Bergeman, I. So, and J. Dalibard. This work was supported by the Office of Naval Research (USA) and the National Bureau of Standards (USA).

REFERENCES

1. W. Phillips and H. Metcalf, Phys. Rev. Lett. 48, 596(1982)
2. J. Prodan, W. Phillips, and H. Metcalf, Phys. Rev. Lett. 49, 1149(1982).
3. T. Hansch and A. Schawlow, Opt. Commun. 13, 68 (1975).
4. D. Wineland and H. Dehmelt, Bull. Am. Phys. Soc. 20, 637 (1975).
5. W. Phillips, J. Prodan, H. Metcalf, J. Opt. Soc. Am. B 2, 1751(1985). This review article (in a special issue of JOSAB) contains detailed description of many aspects of neutral atom cooling and trapping.
6. S. V. Andreev et al., JETP Lett. 34, 442 (1981).
7. W. Phillips and J. Prodan in Coherence and Quantum Optics V, ed. by L. Mandel and E. Wolf, Plenum, 1984, p. 15.
8. W. Ertmer et al., Phys. Rev. Lett. 54, 996 (1985).
9. J. Prodan et al., Phys. Rev. Lett. 54, 992 (1985).
10. A. Migdall et al., Phys. Rev. Lett. 54, 2596 (1985).
11. Chu et al., Phys. Rev. Lett. 55, 48 (1985).
12. J. Dalibard and W. Phillips, Bull. Am. Phys. Soc. 30, 748(1985)
13. W. Wing, Prog. Quant. Elect. 8, 181 (1984).
14. W. Paul, private communication. Some of this is described in W. Paul and U. Trinks, Inst. Phys. Conf. Ser. 42, 18 (1978).

COOLING AND TRAPPING OF ATOMS WITH

LASER LIGHT

Steven Chu, J. E. Bjorkholm
A. Ashkin, L. Hollberg*, and Alex Cable

AT&T Bell Laboratories
Holmdel, New Jersey 07733

Abstract:

Recent experiments have shown that it is now possible to cool and trap neutral atoms. We review the work done at Bell Laboratories to confine and cool sodium atoms to $T = 2.4 \times 10^{-4}K$ in a viscous molasses of photons tuned to the yellow "D-line" resonance of sodium. Potential optical traps are presented, and several ideas for cooling atoms to temperatures of $10^{-6}-10^{-10}K$ are suggested. Possible uses for ultra cold atoms are also mentioned.

The manipulation of atoms with laser light has seen major advances in recent years. Deflection [1] and focusing [2] of atomic beams was followed by the slowing and finally stopping [3] of a beam of sodium atoms by a laser beam directed opposite the atomic velocity. In the laser stopping experiments, the residual velocity spread corresponded to a temperature of 50-100mK, and such a sample of atoms led to the magnetic trapping of sodium atoms [4]. The confinement of sodium atoms in an "optical molasses" of light has also been demonstrated [5]. In addition to fairly long containment times, the optical confinement scheme simultaneously cools the atoms to temperatures of a fraction of a mK in a time much less than a millisecond. Because of the relative ease in obtaining a high density sample of cold atoms, optical molasses seems to be a good starting point for a variety of experiments.

After a brief review of our work on optical molasses, we will present schemes for laser trapping of atoms that we are trying. We also present several schemes for laser cooling atoms to temperatures of $10^{-6}K$ and possibly to $10^{-10}K$. Finally, the realization of ultra-cold atoms has opened up the potential for new experiments. We will briefly mention a few areas of research that can be addressed with these atoms.

This paper is not intended to be a review of the field of laser manipulation of atoms, but rather a snap-shot of the work currently underway at AT&T Bell Laboratories. In particular, we will not discuss the elegant magnetic trapping experiments of Phillips and his co-workers.[4] For a review of the earlier work in the field, the reader is referred to the earlier review articles of Ashkin [6] and Letokhov and Minogin [7]. More recent collections of articles in the field can be found in dedicated journal issues edited by Phillips [8] and Meystre and Stenholm [9].

---

* Present address: National Bureau of Standards, 325 Broadway, Boulder, Colorado 80303

I. OPTICAL MOLASSES

The basic physics of the optical molasses scheme is briefly outlined. Consider an atom irradiated by a laser tuned near a resonant absorption line. For each photon absorbed, the atom receives a net change of velocity $\Delta v = {}^{h}/m\lambda$, where m is the mass of the atom and λ is the wavelength of the absorbed photon. Since the subsequent reemission of the photon has no preferred direction, an average over many scattering events gives a net scattering force along the direction of the light [10]. Hänsch and Schawlow [11] noted that if counterpropagating beams were tuned to the low-frequency side of the absorption line, a moving atom will Doppler shift the opposing laser beam closer to resonance and shift the laser beam co-propagating with the atom away from resonance. After averaging over many photon absorptions, the atom will see a scattering force opposite its velocity. Thus, an atom imbedded in six counterpropagating beams along three orthogonal axes will always see a damping force $\vec{F} = -\alpha \vec{v}$ opposing its motion. The force is extremely viscous giving damping times on the order of 10^{-4} sec, hence the name "optical molasses." In order to calculate the minimum temperature achievable by this technique, statistical fluctuations in the optical force must be considered. These fluctuations lead to heating and are balanced by the cooling effect of the damping force. In the absence of damping, the continual absorption and emission of photons will cause the atom to execute a random walk in velocity. Although $<v> = 0$, $<v^2>$ increases linearly with the total number of scattered photons. The heating effect of the scattering force was first observed by Bjorkholm, et al.[12] A detailed analysis of quantum fluctuations, including stimulated processes [13], predicts a minimum temperature based on a balance of heating and cooling of $kT_{min} = h\gamma/2$ where γ is the width (fwhm) of the absorption line. For sodium, $\gamma = 10$ MHz and $T_{min} = 240\mu K$.

An estimate of the confinement time is made by noting that the damped motion of atoms in the photon bath is analogous to Brownian motion. The time t necessary for a displacement $<x^2>$ is given by $t = \alpha<x^2>/2kT$. This result assumes an infinite medium, but for a random walk in a medium surrounded by a spherical escape boundary [14], the storage is reduced by a factor of 3.1 for our experimental conditions.

The experiment arrangement is shown in Fig. 1. An intense pulsed atomic beam is produced by irradiating a pellet of sodium metal with a 10 nsec long, 30 mj pulse at 532 nm focused to $\sim 5 \times 10^{-2}\ cm^2$. Atoms with initial velocities of about 2×10^4 cm/sec are slowed down (pre-cooled) in less than 6 cm to velocities of 2×10^3 cm/sec before they are allowed to drift into the optical molasses region.

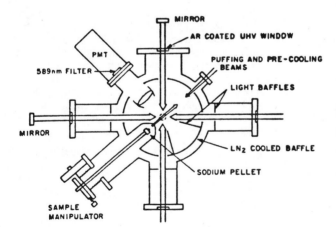

Figure 1 - Schematic of the vacuum chamber and intersecting laser beams and atomic beam. The vertical confining beam is indicated by the dotted circle. The "puffing" beam is from the pulsed YAG laser.

The optical schematic is shown in Fig. 2. The single frequency ring dye laser is typically tuned to the crossover lines of the Doppler-free saturation cell containing sodium vapor. A wideband electro-optic modulator (e.o.) similar to the ones made by Ertmer, et al, [3] and Devoe and Brewer [15] is used to generate sidebands at 856.2 MHz, which corresponds to one half of the frequency difference between $3S_{1/2}$, $F = 2$ to $3P_{3/2}$, $F = 3$ transition and the $3S_{1/2}$, $F = 1$ to $3P_{3/2}$, $F = 2$ transitions in sodium. Approximately two-thirds of the laser power is contained in the two sidebands. The two frequencies prevent the optical pumping of the atoms into states not accessible to laser excitation. The laser beam passes through an acoustic-optic (a.o.) modulator that is used to switch out a pre-cooling laser beam. The pre-cooling beam passes through a second electro-optic modulator that generates another series of sidebands on each of the frequencies generated by the first modulator. Following Ertmer, *et al.* [3], this modulator is driven by an rf sweep oscillator which allows us to sweep a sideband to stay in resonance with the changing Doppler-shift of the atomic resonance frequency as the atoms slow down. After slowing for 0.7 msec, the pre-cooling beam is shut off by turning off

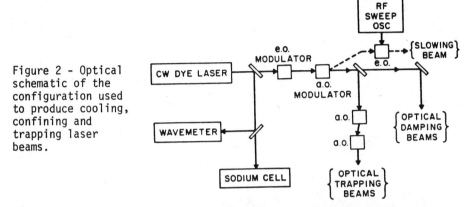

Figure 2 - Optical schematic of the configuration used to produce cooling, confining and trapping laser beams.

the a.o. modulator. The undeflected beam is split into three optical molasses beams and three "trap beams". The trap-beam line passes through two additional a.o. modulators that give us the option of constructing a gradient beam trap [16] with a de-tunning of up to >250 MHz from the resonance line, or a scattering force ac trap [17] which requires the laser beams to be tuned close to the center of the resonant line. Not shown in the figure are a myriad of mirrors, telescopes, spatial filters, attenuators, beam splitters, polarizers, e.o. modulators, etc. used to manipulate the laser beams through various molasses and trap lines. Also not shown are the optical and rf spectrum analyzers used to monitor the a.o. and e.o. modulators. With ~600mw out of the dye laser, we easily have enough power in 1 cm diameter beams to saturate the resonance line.

When the pre-cooled atoms are properly introduced into the optical molasses (if the atoms hit the molasses with too little velocity, they don't penetrate into the center of the confinement region, but mushroom out on the side nearest the atomic beam) a monotonically decreasing fluorescent signal is seen by the phototube. (See Fig. 3.) The decay of the signal can be then fit to a model of atoms executing a random walk in a spherical region of space defined by the intersecting laser beams. In Fig. 3, we also plot the number of atoms remaining in the observation region as a function of Dt/R^2, where D is the diffusion constant, t is the time and R is the effective radius to the spherical boundary. If we take $R = 0.4$ cm, the decay curve gives us an effective diffusion constant $D \sim 4.8 \times 10^{-2}$ cm^2/sec. Using a computed maximum value of the damping constant $\alpha_{max} = 5.8 \times 10^{18}$ gm/sec, we obtain an upper limit on $T < T_{max} = D\alpha_{max}/k = 1.9$ mK.

A more direct measurement of temperature is shown in Fig. 4. After cooling and confining the atoms for 10-15 msec, all six beams were turned off in about 0.1 msec. The atoms then stop their random walk motion and begin to ballistically leave the interaction region with their instantaneous velocities. When the light is turned back on, only the slower moving atoms will be observed in the confinement region. The fraction of atoms remaining as a function of the light-off time is compared to a calculation based on an initial uniform distribution of atoms with a Maxwell-Boltzman distribution of velocities. After many scattering events, the velocity distribution has been explicitly shown to be a Gaussian distribution in the absence of damping. [18] Our experimental results are consistant with a Gaussian distribution in the presence of damping. Given a radius of 0.4 cm, we obtain a

measured temperature of $T = 240$ μK $\begin{pmatrix} +200\mu K \\ -60\mu K \end{pmatrix}$ where the error is largely due to the uncertainty in assigning an effective confinement radius.

43

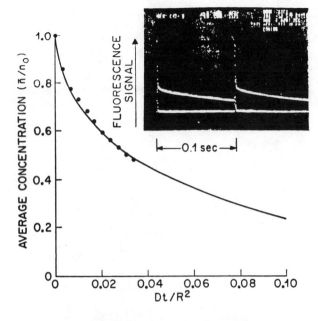

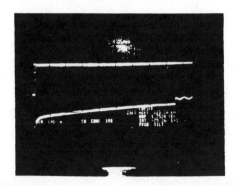

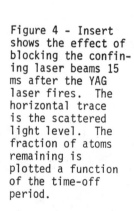

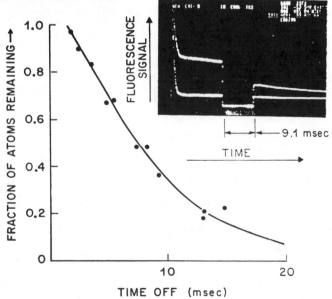

Figure 3a - Fluorescence signal as a function of time is shown in the insert. The confinement region is loaded every 0.1 sec governed by the repetition rate of the YAG laser.

Figure 3b - Fluorescent decay in a larger confinement volume. (~1.2 cm diameter) In 0.2 sec, the atom density decreases by ~20%.

Figure 4 - Insert shows the effect of blocking the confining laser beams 15 ms after the YAG laser fires. The horizontal trace is the scattered light level. The fraction of atoms remaining is plotted a function of the time-off period.

II. LASER TRAPPING

We are currently working on two types of laser traps: an alternating light beam trap [17], with modifications first mentioned by Dalibard and Phillips [19], and a six-beam gradient trap [16] in the optical molasses geometry.

The ac light trap uses the scattering force to confine the atoms. Because of the Optical Earnshaw Theorem [20], cw laser beams using only the scattering force can not be configured to make a stable trap. An ac light trap, in direct analogy to electromagnetic ac traps [21], has been proposed by Ashkin[17]. Ashkin's trap, however, does not exploit the powerful damping forces of optical molasses. A more stable trap is indicated schematically in Fig. 5. A spherical distribution of atoms is first cooled by molasses. Next, pushing forces are applied along the x-axis with diverging laser beams tuned exactly to the atomic resonance. The outward forces in the $y-z$ plane are defeated by optical molasses beams tuned so that the sphere will be pushed into a pancake as shown in Fig. 5a. Next, the x-axis is switched over to a damping mode while the pushing force is applied to the y-axis. The pancake should then deform into a cigar shaped distribution. (Fig. 5b). Finally, the pushing forces are applied to the z-axis while the damping

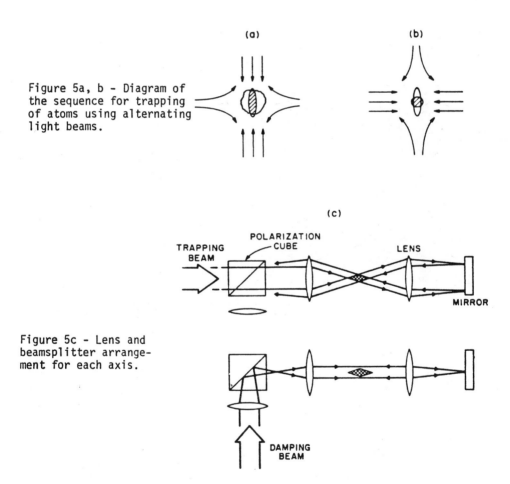

Figure 5a, b - Diagram of the sequence for trapping of atoms using alternating light beams.

Figure 5c - Lens and beamsplitter arrangement for each axis.

present in the $x-y$ plane. The optical "patty-cake" arrangement can be realized by the lens arrangement shown in Fig. 5c, where only one dimension is shown. A computer simulation of the trap is shown in Fig. 6a. The advantage of this trap is that the capture of atoms is possible over large volumes with laser beam intensities on the order of 10 mW/cm². The disadvantages are that small f-number lenses and a complex array of light shutters are necessary.

The other type of trap we are exploring is a gradient trap. Gradient traps generally require fairly high intensity laser beams focused to a small diameter.[16] One can both maximize the depth of the trap and minimize heating [22] by tuning far below resonance. The use of circular polarization [23] can further improve the trap. The most important advance, however, is the realization that a trap of a very modest depths is sufficient if the atoms can be cooled to temperatures of $h\gamma/2$. For example, in a gradient trap in the molasses geometry, (6 nearly parallel beams propagating along $\pm x, y, z$) the scattering forces are canceled by the retro-reflected beams. Alternate cycling between the gradient beams tuned hundreds of MHz below the resonance, and the molasses beams tuned half a linewidth below the resonance will create a stable trap. For laser powers of 30 mW, a beam focused to a 0.5 mm dia and tuned 400 MHz below resonance is expected to create a trap with a depth of 20 $h\gamma$, or approximated 40 times deeper than temperatures obtained with optical molasses. A simulation of the gradient trap for the above parameters is shown in Fig. 6b. The advantages of this configuration over an ac trap are the simplicity and self-correcting nature of the optical arrangement. However, greater laser power is required for smaller trapping volumes.

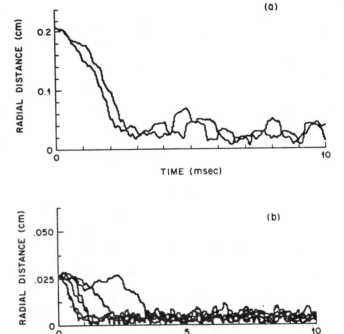

Figure 6 - Computer simulation of (a) ar ac trap and (b) a gradient trap with alternate cooling cycles. Heating due to photon scattering has been included. Induced dipole heating is not included.

III. FURTHER COOLING

The temperature given by $T_{min} = h\gamma/2$ is not the minimum temperature achievable with laser cooling. One can begin cooling on a broad transition before switching to a more forbidden transition. Thus, atoms can be initially cooled quickly when it is important to scatter many photons/sec. Once cooled to less than 1 mk, a more leisurely cooling process can be used to reach temperatures below 10^{-5}K. At these temperatures, the recoil due to a single photon momentum becomes a factor.

In combination with magnetic or optical traps, further cooling can be shown. Pritchard [24] has suggested a combination of $+$rf and visible photon transitions that can cool atoms in a magnetic trap to temperatures on the order of 10^{-5}K. If the number of atoms in the trap is large enough to establish collisional temperature equilibrium (in molasses, the atoms are in thermal contact with the photon bath and need not be in contact with each other) and the high energy tail of the Boltzman distribution is allowed to escape out of the trap, the average energy of the remaining atoms decreases. This "evaporative cooling" scheme has been proposed for the cooling of magnetically confined spin aligned hydrogen experiments.[25] Adiabatic cooling, achieved by the slow weakening of an atom trap, is another method for achieving further cooling. We have suggested a cooling technique particularly well suited to light traps.[26] In this scheme, the atoms are first localized and cooled by a combination of optical molasses and a tightly confining gradient trap. Next, the deep, small trap is turned off and a broad harmonic trap is substituted in its place. The atoms, initially near the center of the trap will begin to oscillate. The trap light is turned off after approximately one quarter of a harmonic oscillator period - the time when most of the atoms will have reached their turning points in the oscillation. The reduction in atomic velocity is determined by Liouville's Theorem $\Delta v_i \Delta x_i = \Delta v_f \Delta x_f$, how closely a gradient trap can approximate a harmonic potential, and the probability of heating due to the trap. For realistic gradient traps and an initial temperature of $T \sim h\gamma/2 \sim 2.4 \times 10^{-4}K$, we have shown that temperatures below $10^{-6}K$ are possible.[26]

The above discussion treats the atoms as classical particles. If we consider sodium atoms in a trap with a potential depth $U \sim 4 \times 10^2 h\gamma$, 30 μm in diameter at a temperature $T = h\gamma/2$, the quantum mechanical harmonic oscillator energy $\hbar\omega$ corresponds to approximately $10^{-2}kT$. For lithium atoms, the lighter mass and narrower D-line resonance (2s -> 2p at 671 nm) will mean that only ~ 15 quantum states will be occupied. If the 2s -> 4p (323 nm) transition is used to cool the Li atoms, temperatures on the order of $10^{-5}K$ should be achieved. At this temperature, kT is comparable to $\hbar\omega$ If the atoms can be cooled into the ground state of the harmonic oscillator potential, the zero point "temperature" can be decreased by adiabatically enlarging the size or decreasing the depth of the trap. Since the oscillator frequency will initially be on the order of $\omega/2\pi \sim 1$ MHz, the adiabatic cooling time is short compared to the trap heating time. The limitation to atom cooling in this situation seems to be the size of the localization volume before the trap cannot support the weight mg of the atom. Atoms with harmonic oscillator wavefunctions on the order of $\lambda = 100$ μm should be accessible. The corresponding localization temperature which we define by $\lambda = \dfrac{h}{\sqrt{3KTm}}$, is $T \sim 10^{-10}K$.

IV. APPLICATIONS

Laser cooled and confined atoms open the possibility for a variety of novel experiments. Atoms cooled and confined in a mostly uniform magnetic field [27], or in free fall as in an atomic fountain [28], will have the long coherence times needed for precision spectroscopy. In addition to improved frequency standards, small frequency shifts due to feeble physics effects can be explored. For example, an electron dipole moment d_e will induce a linear Stark shift in a nondegenerate state of an atom. The present limit of $d_e \leq 2 \times 10^{-24} e - cm$ [29] corresponds to frequency shifts in strong electric fields in the millihertz range. It is possible that a two to four order of magnitude improvement can be obtained by using an intense beam or fountain of slow atoms.

Micro-Kelvin temperatures will allow us to examine atom-atom and atom-surface collisions in a new regime. In the gas phase, pure s-wave scattering, low energy resonances, and spin flip collisions can be studied.[30] In surface collisions where both the atom and surface temperatures approach zero, it is uncertain whether the atom will stick or bounce on the de-Broglie wavelength λ of the atom is significantly larger than the extent of the attractive potential. The results from ^4He atoms scattering from a liquid ^4He surface at grazing incidence agree surprisingly well with a model that neglects inelastic processes.[31] In these experiments, the atom momentum components perpendicular to the surface correspond to de-Broglie wavelengths as long as 150 Å. If the reflection probability goes to unity, cold atom scattering from surfaces would have important applications. One can imagine a box that could store the atoms in much the same way that ultra-cold neutrons are stored. In the limit of very long de Broglie wavelengths, the atom wavefunction should have a 180° phase shift in an elastic collision. The internal degrees of freedom, however may not be affected. Thus, the coherence in atomic transitions will be maintained, and the confinement box would be the ideal confinement volume for ultra-precise metrology.

The study of quantum collective effects such as Bose Condensation are also possible. For the case of ^7Li, temperatures of $10^{-5}K$ require densities of $\sim 10^{15}$ atoms/cm^3 before Bose condensation is expected. At these densities, three body recombination will create hot atoms, and molecules that can rapidly heat or condense out the remaining atoms. Rough estimates of the three body rate constant are $10^{-30} - 10^{-32}$ $cm^6 sec^{-1}$[30]. In the event that the three body rate is too high, adiabatic expansion immediately after condensation will quickly turn off the recombination and give the experimenter an extended period to examine the behavior of the dilute Bose gas.

In summary, we have shown how to confine and cool atoms with laser light to temperatures of a fraction of a millidegree. An outline of possible candidates for an ac scattering force trap and a dc gradient trap are also given. Proposals for cooling atoms to temperatures in the sub-microkelvin range are suggested. Finally, we briefly mention the possible experiments one can imagine with atoms properly prepared by laser light.

REFERENCES

1. O. R. Frisch, F. Phys. *86*, 42 (1933).

2. J. E. Bjorkholm, R. R. Freeman, A. Ashkin and D. B. Pearson, Phys. Rev. Lett. *41*, 1361, (1968).

3. J. Prodan, A. Migdall, W. D. Phillips, I. So, H. Metcalf and J. Dalibard, Phys. Rev. Lett. *54*, 992 (1985); W. Ertmer, R. Blatt, J. L. Hall, and M. Zhu, Phys. Rev. Lett. *54*, 996 (1985).

4. A. L. Migdall, J. V. Prodan, W. D. Phillips, T. H. Bergeman, and H. J. Metcalf, Phys. Rev. Lett. *54*, 2596 (1985).

5. S. Chu, L. Hollberg, J. E. Bjorkholm, A. Cable, and A. Ashkin, Phys. Rev. Lett. *55*, 48 (1985).

6. A. Ashkin, Science *410*, 1081 (1980).

7. V. S. Letokhov and V. G. Minogin, Phys. Rep. *73*, 1 (1981).

8. "Laser Cooled and Trapped Atoms," ed. W. D. Phillips, Prog. in Quant. Elec. *8*, 204 (1984).

9. "The Mechanical Effects of Light," eds. P. Meystre, S. Stenholmm, J. of Opt. Soc. of Am. B, *2*, 1706, (1985).

10. A. Ashkin, Phys. Rev. Lett. *24*, 156 (1970); ibid, *25*, 1321 (1970).

11. T. W. Hansch and A. L. Schawlow, Opt. Comm. *13*, 68, (1975).

12. J. E. Bjorkholm, R. R. Freeman, A. Ashkin, and D. B. Pearson, Optics Lett. *5*, 111 (1980).

13. J. P. Gordon and A. Ashkin, Phys. Rev. A, *21*, 1606 (1980); R. J. Cook, Phys. Rev. A, *20*, 224 (1979); R. J. Cook, Phys. Rev. Lett., *22*, 976 (1980).

14. N. A. Fuchs, *The Mechanics of Aerosols*, (Pergamon Press, Oxford, 1964) pp. 193-200.

15. R. G. DeVoe and R. G. Brewer, Phys. Rev. A, *20*, 2449 (1979).

16. A. Ashkin, Phys. Rev. Lett., *40*, 729 (1978).

17. A. Ashkin, Opt. Lett. *9* 454 (1984).

18. J. L. Picque, Phys. Rev. A., *19*, 1622 (1979).

19. J. Dalibard and W. D. Phillips, Bull. Am. Phys. Soc.

20. A. Ashkin and J. P. Gordon, Opt. Lett. *8*, 511 (1983).

21. R. F. Wuerker, H. Shelton, and R. V. Langmuir, J. Appl. Phys. *30* 342 (1959).

22. J. P. Gordon and A. Ashkin, Phys. Rev. A *21*, 1606, (1980).

23. J. Dalibard, S. Reynard and C. Cohen-Tannoudji, Opt. Comm., *47*, 395 (1983); J. Phys. B *17*, 4577 (1984).

24. D. E. Pritchard, Phys. Rev. Lett. *51*, 1336 (1983).

25. H. F. Hess, Bull. Am. Phys. Soc. *30*, 854 (1985); R. V. E. Lovelace, C. Mehanian, J. J. Tommila, and D. M. Lee, to be published, Nature (1985).

26. S. Chu, J. E. Bjorkholm, A. Ashkin, J. P. Gordon and L. W. Hollberg, to be published, Optics Letters; also in Laser Spec. VII, ed. T. W. Hänsch and Y. R. Shen (Springer-Verlag) 1985.

27. D. Pritchard, private communication.

28. R. G. Beausoleil and T. W. Hansch, Optics Lett. *10*, 547, (1985).

29. M. A. Player and P. G. H. Sandars, J. Phys. B*3*, 1620 (1970).

30. D. E. Pritchard, in XIV Int. Conf. on Phys. of Electron and Atomic Collisions, eds. D. C. Lorents, W. E. Meyerhof and J. R. Petersen, (1985).

31. V. U. Nayak, D. O. Edwards, and N. Masuhara Phys. Rev. Lett. *50*, 990 (1983).

INVESTIGATION OF NONTHERMAL POPULATION DISTRIBUTIONS
WITH FEMTOSECOND OPTICAL PULSES

Charles V. Shank
AT&T Bell Laboratories
Holmdel, New Jersey USA

ABSTRACT

Advances in femtosecond optical pulse techniques have provided a unique opportunity to excite and probe nonthermal population distributions in semiconductors and complex molecular systems. In this paper I will describe a recent application of high resolution femtosecond optical pulse techniques to the dynamics of nonthermal excitations. Using high resolution time resolved absorption spectra we have been able to observe the time evolution of a nearly monoenergetic population of carriers in a semiconductor, excited with a short optical pulse, to a carrier distribution in which a temperature can be defined.

Ultrashort optical pulse techniques have been used previously to excite hot carriers in semiconductors[1,2] and to probe the process of carrier cooling to the lattice temperature. In the experiments reported here we excite a non-equilibrium carrier population in such a short time that the temperature of the distribution cannot yet be defined. Using femtosecond spectroscopic techniques, we are able to follow the evolution of the non-thermal carrier distribution to a thermalized distribution within 200 femtoseconds at room temperature in a GaAs Multiple Quantum Well Structure (MQWS).[3] In addition, by monitoring the time development of the bleaching of the exciton resonance we are able to determine the relative effectiveness of Coulomb screening and near band edge state filling on excitonic absorption.

Optical transitions in a semiconductor are strongly influenced by the distribution of carriers in the conduction and valence bands. For direct transitions away from exciton resonances the absorption coefficient α can be expressed as

$$\alpha = \alpha_o(1-f_e-f_h)$$

where $f_{e,h}$ are the distribution functions for electrons and holes and α_o is the absorption coefficient in the absence of state filling. With a sufficiently short optical pulse a narrow band of states can be excited creating a non-thermal distribution. As time progresses, carrier-carrier interactions lead to thermalization of the carrier distribution.

For optical transitions near an exciton resonance the influence of carriers is more complex.[4] In quasi-two-dimensional semiconductors, exciton resonances are well resolved at room temperature and have been the subject of extensive experimental[5] and theoretical work.[6,7]

In previously reported work [8] femtosecond optical pulses were used to excite excitons resonantly in a room temperature GaAs MQWS and the exciton ionization time was measured in a room temperature phonon bath by monitoring the bleaching of the exciton resonance. In a recent theoretical study, Schmitt-Rink et al.[7] have determined the influence of carriers excited above the band edge on the exciton resonance absorption. In their study they attempt to assess the relative effectiveness of excited carriers on exciton absorbance through direct Coulomb screening and through Pauli exclusion, i.e., the effect of occupation of near band edge states, which they refer to as phase-space filling and exchange screening. In the experiments reported here, we also monitor the change in exciton absorbance early in time induced by the non-thermal population distribution and compare it to the change at later times when the carriers have occupied near band edge states. This experiment allows us to determine for the first time the relative importance of these two processes on bleaching the exciton resonance. All these measurements were performed at room temperature.

Using a high repetition rate (8 kHz) data acquisition system, we have been able to measure time resolved spectra of induced transmittance changes to an accuracy of one part in 10^4 with better than 100 femtosecond time resolution.

Optical pulses of 50 femtosecond duration were generated in a colliding pulse mode-locked laser[9,10] and amplified to ~3 microjoules energies at 8 kHz repetition rate using a copper-vapor-laser-pumped amplifier system.[11] These pulses are used to generate white light continuum pulses. About 5% of the continuum beam was used to probe the absorption and the remainder of the beam was passed through an interference filter with a 10 nm bandwidth. The pump pulse was broadened to approximately 100 femtoseconds as a result of the spectral narrowing upon passage through the filter. The pump and probe beam were focused to a ~20 micron diameter spot. A stepping motor was used to provide the relative time delay and the spectra were measured in the differential transmittance mode of an optical multichannel analyzer (OMA III) with a reticon detector. In such a mode, the data acquisition system directly measures the transmittance difference between the excited and the unexcited sample normalized to that of the unperturbed sample, i.e., $(T_0 - T)/T_0$, which for small changes of the transmission is proportional to $(\alpha_0 - \alpha)$. In the following, we refer to these spectra as Differential Transmittance Spectra (DTS).

The sample is a MQWS that consists of 65 periods of 96 Å GaAs quantum wells and 98 Å $Al_{0.3}Ga_{0.7}As$ barrier layers. This sample has been described and analyzed in Reference [12].

The temporal evolution of the nonthermal population is seen in the DTS presented in Figure 1. DTS at intervals of 50 fs for delays varying between -100 fs and 200 fs are shown. The point of zero time delay is estimated with a precision of approximately 50 fs. The pump energy has a distribution 20 meV wide and is centered at 1.509 eV as shown by the spectrum at the bottom of the figure.

Early in time, a narrow band of increased transmission is observed which is broader than the pump spectrum and slightly shifted to lower energies. As time evolves, the region of higher transmission appears to

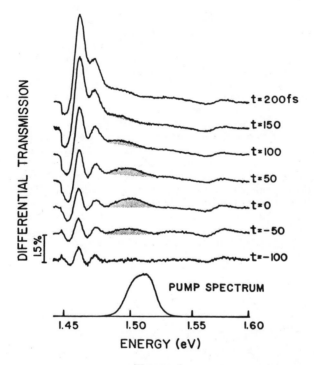

Figure 1

Differential transmission spectra for a pump centered
at 1.509 eV and observed at 50 fs intervals between
t=-100 fs and t=+200 fs. The spectra have been
displaced vertically for clarity. The pump spectrum
is shown at the bottom of the figure. Various
features are discussed in the text.

shift to lower energy and broadens. Also observed is an increase in transmission near both the n=1 and the n=2 excitons. The DTS profile close to the excitons is characteristic of broadening of resonances and reduction of oscillator strength.[5] AFter 200 femtoseconds, the spectra remain practically unchanged suggesting that the carriers have thermalized. Using the data in the figure, we estimate that average carrier energy remains essentially constant during the thermalization process indicating very little interaction with the lattice and showing that carrier-carrier interaction is the dominant thermalization mechanism.

The increase in transmission near the n=1 excitons is delayed in time compared to the absorbance changes near the pump energy, showing that the exciton bleaching does not reach its peak value until carriers thermalize and fill the near band edge states. Just following excition, the carriers occupy states well above the band edge. At that point in time, the primary mechanism for changes in the exciton absorption resonance is long range Coulomb screening. However, as the carriers begin to be thermalized, near band edge states become occupied and phase-space filling and exchange begin to bleach the exciton absorption more strongly. From the data we estimate that bleaching due to screening is at least six times smaller than that due to phase-space filling and exchange in accord with the recent predictions.[7] It is also interesting to note that the bleaching of the n=2 exciton does not seem to show a strong variation as the carriers thermalize because for these resonances the only bleaching mechanism is Coulomb screening. The initial evolution of the continuum-state bleaching indicates that the carriers leave their initial states in less than 100 femtoseconds with delayed exciton bleaching occurring in approximately 150 femtoseconds.

REFERENCES

1. C. V. Shank, R. L. Fork, B. I. Greene, C. Weisbuch, and A. C. Gossard, Surface Sci., 113, 108 (1982).
2. C. V. Shank, R. L. Fork, R. Yen, C. Weisbuch, and J. Shah, Solid State Comm., 47, 981 (1983).
3. W. H. Knox, C. Hirlimann, D. A. B. Miller, J. Shah, D. S. Chemla, and C. V. Shank, (to be published).
4. For a recent review see: H. Haug, S. Schmitt-Rink, Progress in Quantum Electronics, Vol. 9, 1 (1985).
5. For a recent review of excitonic effects in MQWS, see: D. S. Chemla and D. A. B. Miller, J. Opt. Soc. Am., B2, 1144 (1985) and references therein.
6. S. Schmitt-Rink, and C. Ell, J. of Lumin., 30, 585 (1985).
7. S. Schmitt-Rink, D. S. Chemla, and D. A. B. Miller, (to be published in Phys. Rev. B32 (1985)).
8. W. H. Knox, R. L. Fork, M. C. Downer, D. A. B. Miller, D. S. Chemla, C. V. Shank, A. C. Gossard, and W. Wiegmann, Phys. Rev. Lett., 54, 1306 (1985).
9. R. L. Fork, B. I. Greene, and C. V. Shank, Appl. Phys. Lett., 38, 671 (1981).
10. J. A. Valdmanis, R. L. Fork, and J. P. Gordon, Opt. Lett., 10, 131 (1985).
11. W. H. Knox, M. C. Downer, R. L. Fork, and C. V. Shank, Opt. Lett., 9, 552 (1984).
12. D. S. Chemla, D. A. B. Miller, P. W. Smith, A. C. Gossard, and W. Wiegmann, IEEE J. Quan. Elec., QE-20, 265 (1984).

INTRA- AND INTERMOLECULAR ENERGY TRANSFER OF LARGE MOLECULES IN SOLUTION AFTER PICOSECOND EXCITATION

W. Kaiser and A. Seilmeier

Physik Department
Technische Universitaet Muenchen
Arcisstrasse 21, D-8000 Muenchen, Germany

In small molecules the vibrational energy flows from well defined initial modes to known final states. Fermi resonances in absorption and Raman spectra tell us which modes interact by anharmonic coupling. The situation is different in large molecules. The number of vibrational quantum states rises strongly with the number of atoms and with the vibrational energy. The energy supplied to a specific vibrational state is transferred to numerous overtones and combination modes. Evidence is presented on the rapid intramolecular redistribution time of ≤ 1 ps and on the transient vibrational distribution of the energetic molecules. On a somewhat longer time scale of tens of picoseconds the excess energy is transferred from the excited molecules to the liquid surrounding, restoring the thermal equilibrium.[1] The "cooling" of the molecules is inferred from the decay of the excess absorption (or fluorescence) at the long-wavelength tail of the absorption spectrum.

Energy is supplied to the molecules either by tunable infrared photons (~ 3000 cm^{-1}) exciting NH- or CH-stretching modes or by visible photons (~ 19000 cm^{-1}) making transitions to the first electronic state with subsequent rapid internal conversion to the vibrational manifold of

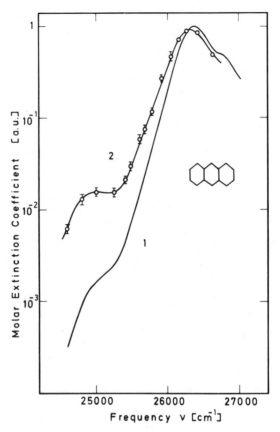

Fig. 1. Standard absorption edge of anthracene at room temperature (1). Transient absorption taken 7 ps after IR excitation by photons of 3050 cm⁻¹ (2). Note the enhancement of the hot bands of frequencies of 1400 cm⁻¹.

the electronic ground state. Our experiments indicate that the excitation energy is rapidly redistributed over the many degrees of freedom of the molecule as if the molecule has acquired a higher internal temperature. This transient internal temperature may be estimated from the excitation energy and from the known vibrational modes of the molecule.

As an example of intramolecular vibrational redistribution experimental data of anthracene are presented.[2] In Fig.1, the absorption

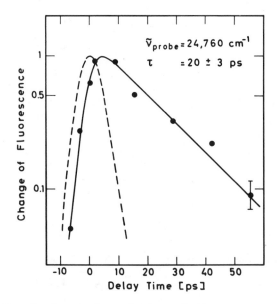

Fig. 2. Temporal evolution of the excess
fluorescence of anthracene after IR
excitation. The signal rises rapidly
during the excitation and decays by
intermolecular interaction. The broken
curve gives the correlation curve of
excitation and probe pulse.

tail of anthracene at room temperature (1) is compared with the transient

absorption (2) taken 7 ps after excitation with infrared photons of

3050 cm^{-1}. The small shoulder seen at the room temperature curve results

from three hot bands around 1400 cm^{-1}. After IR-excitation of the

CH-stretching modes, the energy is rapidly redistributed and the

absorption of the hot bands increases drastically. The experimental data

suggest a rise of internal temperature of 170 K. The same increase of

temperature is calculated when the energy of 3050 cm^{-1} is distributed over

all the vibrational modes of anthracene. The temporal evolution of the

excess fluorescence is presented in Fig.2 measured for a probing frequency

at the absorption tail of 24760 cm^{-1}. With this frequency one monitors the

population of states around 1600 cm^{-1} since the 00-transition of

anthracene is at 26380 cm^{-1}. The figure shows a rapidly growing signal

which results from the fast intramolecular redistribution of energy. The rise time of ≤ 2 ps is given by the time resolution of the system (see the broken curve, the cross correlation between excitation and probing pulse). The relatively slower decay with a time constant of 20 ps is due to intermolecular energy transfer to the molecular surrounding.

Similar findings were made for a number of other molecules. In many large dye molecules without symmetry the absorption at long wavelengths falls exponentially over many orders of ten as a true Boltzmann edge. At room temperature we find exactly 300 K from the absorption tail. The momentary internal temperature can be directly deduced from the transient absorption edge. The latter is determined by the population of numerous vibrational modes in the ground state. As an example we point to investigations of the laser dye coumarin 7, where a NH-stretching mode was excited by infrared pulses of 3400 cm^{-1}.[3] The absorption tail measured 8.5 ps after excitation indicates an internal temperature of 400 K. The same internal temperature is estimated when the excess energy of 3400 cm^{-1} is distributed over the many vibrational modes of the coumarin 7 molecule.

Internal conversion represents the most common radiationless process in large molecules. Large amounts of energy (ten thousands of wavenumbers) are transferred from electronic excited states to the vibrational manifold of the ground state. According to the preceding findings, very high internal temperatures are anticipated especially for medium size molecules with a limited number of vibrational modes. As an example we present data of azulene,[4] which is known for its very fast internal conversion with a S_1 lifetime of approximately 2 ps. The molecule is excited to the first excited electronic state with visible pulses of ~ 19000 cm^{-1} and the long wavelength absorption is measured with a delay time of 10 ps. In Fig.3 the absorption edge at room temperature (1) is depicted together with the

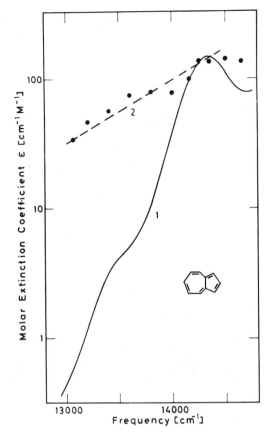

Fig. 3. Absorption edge of azulene at
300 K (1) and after excitation by
visible pulses of 19000 cm^{-1} (2).
The broken line corresponds to an
internal temperature of 1200 K.

transient data of the excited molecules (2). The drastic change of the

absorption edge suggests a large temperature rise of the azulene

molecules. The broken line through the experimental data corresponds to a

Boltzmann slope of 1200 K. The same internal temperature is calculated

when the supplied energy of 19000 cm^{-1} is distributed over all vibrational

degrees of freedom of azulene.

How long does the high internal temperature remain in the excited

azulene molecules? In Fig.4 the change of absorption is plotted versus

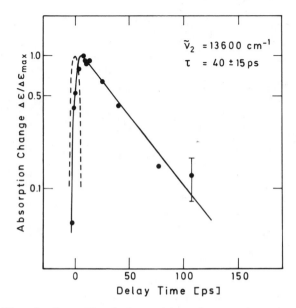

Fig. 4. Normalized absorption change at a
probe pulse frequency of 13600 cm⁻¹
A first rise of absorption − due to
very rapid vibrational redistribution
(< 2 ps) − is followed by a cooling of
the molecules with a time constant of
40 ps.

time. The data are taken with a probing frequency of 13600 cm⁻¹ (see
Fig.3). The excess absorption rises rapidly with the excitation pulse.
Considering the time resolution of the experimental system one estimates
an upper limit of 2 ps for the internal conversion and for the
redistribution of vibrational energy. The data suggest that the
redistribution process is faster than 1 ps. According to Fig.4 the
intermolecular energy transfer to the solvent, i.e. the cooling of the
molecule, proceeds considerably slower. From the figure one deduces a time
constant of 40 ps for the solvent CCl_4. Obviously, molecules of very high
internal temperature can exist in the liquid phase for times of several
10^{-11} s.

References

1. A. Seilmeier, P.O.J. Scherer, and W. Kaiser, Ultrafast energy dissipation in solutions measured by a molecular thermometer, Chem. Phys. Lett. 105:140 (1984); P.O.J. Scherer, A. Seilmeier, and W. Kaiser, Ultrafast intra- and intermolecular energy transfer in solutions after selective infrared excitation, J. Chem. Phys. 83:3948 (1985).
2. N.H. Gottfried, A. Seilmeier, and W. Kaiser, Transient internal temperature of anthracene after picosecond infrared excitation, Chem. Phys. Lett. 111:326 (1984).
3. F. Wondrazek, A. Seilmeier, and W. Kaiser, Ultrafast intramolecular redistribution and intermolecular relaxation of vibrational energy in large molecules, Chem. Phys. Lett. 104:121 (1984).
4. W. Wild, A. Seilmeier, N.H. Gottfried, and W. Kaiser, Ultrafast investigations of vibrationally hot molecules after internal conversion in solution, Chem. Phys. Lett. 119:259 (1985).

NEW TECHNIQUES OF TIME-RESOLVED INFRARED AND RAMAN SPECTROSCOPY USING
ULTRASHORT LASER PULSES

Alfred Laubereau

Physical Institute
University of Bayreuth
Bayreuth, W. Germany

Considerable progress has been made in recent years in the field of spectroscopic applications of ultrashort laser pulses. In vibrational spectroscopy numerous investigations have been performed studying ultrafast relaxation processes in condensed matter, mostly using Raman techniques.[1] These measurements benefit from the extreme time resolution and high intensity achieved by short light pulses and are also facilitated by the large number density of liquids and solids. In this short article two different approaches will be discussed: (i) an IR technique is presented which complements coherent Raman scattering; (ii) a Fourier Raman method is described with high frequency resolution.

The advent of sensitive, ultrafast light gates in the IR[2] and recent progress in the generation of widely-tunable ultrashort pulses in the long-wavelength range[3] have opened up the field of ultrafast IR spectroscopy. A first technique of this kind has been recently demonstrated studying the transient propagation of resonant low-intensity pulses.[4] Drastic pulse reshaping was observed for IR pulses of several ps tuned to the resonance frequency of a vibration-rotation transition. Working in gases at medium pressure level and with optically thick samples, quantitative information was, for the first time, substracted from the coherent propagation effect.

Studying moderately thin samples, the molecular time constants can be directly obtained from the experimental data without the need of computer simulations.[5] This situation is termed nearly-free induction decay (NFID) which is related to the case of free induction decay in very thin samples. Some numerical results are depicted in Fig. 1.[5] A specific experimental situation is considered with resonant, weak input pulses of Gaussian shape and duration t_p (FWHM) and two values of the normalized propagation length, $\alpha \ell = 0.2$ and $\alpha \ell = 1$. Here α denotes the conventional absorption coefficient at the maximum of the absorption band. The intensity of the transmitted pulse is evaluated from Maxwell-Bloch equations for homogeneous line broadening. Fig. 1a refers to a single molecular transition frequency. Corresponding results for two neighbouring transitions are shown in Fig. 1b.

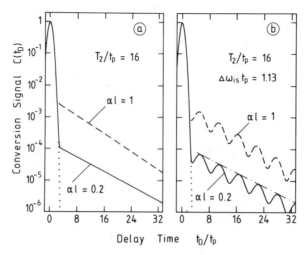

Fig. 1. Calculated conversion signal $C(t_D)$ of the transmitted pulse vs delay time for T_2/t_p = 16 and
two values of absorption length, αl = 0.2 and
1.0. (a) Single transition frequency. (b) Two
transition frequencies leading to a beating
phenomenon. Dotted curves: Gaussian input pulse.

Experimentally, the short pulse intensity is detected by [2]
means of an ultrafast light gate using parametric up-conversion.
This technique provides the convolution of the IR pulse of interest with the probing pulse operating the parametric device.
Assuming Gaussian shape (duration t_p) for the probing pulse,
the conversion signal $C(t_D)$ is plotted in Fig. 1 on a logarithmic scale vs the delay time between the investigated and probing
pulse. For values of t_D/t_p < 2 the shape and peak amplitude of
the transmitted pulse (solid line) are practically unchanged
compared to the signal curve of the input pulse (dotted curve).
For larger values of t_D, the $C(t_D)$ curve in Fig. 1a displays an
extended, weak wing with approximately exponential decay, which
represents NFID of the radiating transition dipoles. The effective decay time τ of the signal is found to differ from $T_2/2$ by
several per cent for the example αl = 0.2, while a larger deviation of $\simeq$ 25 % occurs for αl = 1 (broken line). A simple analytic relation has been shown to exist between the time constants τ and T_2.[5]

Similar results are presented in Fig. 1b for two species
with frequency difference $\Delta\omega_{is} t_p$ = 1.13. Equal dephasing time
T_2 and a ratio of α_1/α_2 = 3 are considered ($\alpha = \alpha_1 + \alpha_2$). Most
important is the oscillatory time behaviour of the decaying
pulse wing. A novel beating phenomenon occurs which is superimposed on an approximately exponential slope. The beating results from the coherent superposition of the adjacent molecular
transitions.

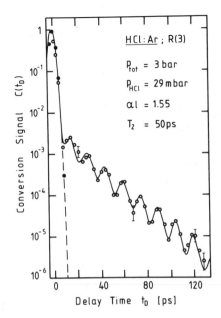

Fig. 2. Time-resolved IR spectroscopy of
HCl:Ar (R(3) line) at P_{tot} = 3 bar;
conversion signal of the transmitted
pulse vs delay time (o-o, P_{HCl} = 29 mbar).
The input pulse (●---●) refers to
the same ordinate scale; theoretical
curves. A novel beating effect and exponential dephasing are observed.

Our time domain IR spectroscopy has been applied to various
vibration-rotation transitions of pure HCl gas and in mixtures
with Ar buffer gas. Working at a total pressure of several bar,
homogeneous collision broadening is adjusted, while Doppler
broadening is completely negligible. The measuring system is
described as follows: Single picosecond pulses, generated by a
Nd:glass laser system enter a double-pass parametric generator-
amplifier system. High quality tunable infrared pulses (dura-
tion t_p ≃ 2 ps and frequency width $t_p \times \Delta\nu$ ≃ 0.7) are efficiently produced at the "signal" and "idler" frequency positions by
stimulated parametric amplification. The "idler" pulse frequency
is tuned to the molecular transition of interest and serves for
the propagation phenomenon. The second pulse at the "signal"
frequency is used as probing pulse and operates the parametric
light gate. Measuring the up-conversion emission of the light
gate with a photomultiplier versus delay time t_D between infrared and probing pulse, the IR intensity is measured as a function of time. Our experimental system achieves femtosecond time
resolution (170 fs) and a dynamic range of 10^7.

An example for the observed signal transients is presented in Fig. 2. The system HCl:Ar with natural isotope abundance of ^{35}Cl and ^{37}Cl is studied. The infrared pulse is resonantly tuned to the R(3) transition frequency of $H^{35}Cl$ at 2963.3 cm^{-1}.[5] The corresponding transition of $H^{37}Cl$ is shifted by approximately 2 cm^{-1} to smaller frequency and also interacts with the incident pulse. The time-integrated conversion signal $C(t_D)$ representing the intensity of the infrared pulse is plotted on a logarithmic scale versus delay time t_D of the probing pulse. $C(o) = 1$ marks the maximum signal measured with evacuated sample cell (transmission = 1). A total pressure of 3 bar and a partial pressure of HCl of 29 mbar are adjusted. Sample length is ℓ = 1.3 cm. Around t_D = 0 the conversion signal of the transmitted pulse (open circles, solid line) follows the rapid build-up and decay of the input pulse (full points, broken line). For larger time values, $t_D \gtrsim 10$ ps, the signal transient changes drastically. Three and more orders of magnitude below the maximum, the pulse wing displays a beating phenomenon, which is superimposed on an (approximately) exponential slope. From the oscillations the beating time T_{is} = 15.0 + 0.3 ps is directly measured corresponding to an isotopic line splitting of $\Delta\omega/2\pi c$ = 2.22 + 0.04 cm^{-1}. This value favourably compares with spectroscopic data and theoretical estimates.[6]

The envelope of the beating maxima yields the time constant τ = 30 ps. This value represents the decay time of NFID. Applying an analytic correction formula or fitting the theory of coherent propagation to the data (solid line) we obtain the de-

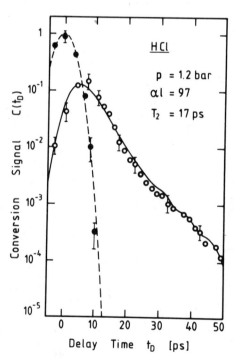

Fig. 3. Same as Fig. 2 for neat HCl at 1.2 bar and large absorption ($\alpha\ell$ = 97). The drastic change of the pulse wing allows the determination of T_2 = 17 + 1 ps.

phasing time of the R(3) line, T_2 = 50 $\pm$ 5 ps. We also note that the observed intensity level of the pulse wing compares favourably with the calculated curve for the estimated value of αl = 1.55; i.e. quantitative agreement between theory and experiment is found.

Results on coherent propagation in a thick sample (αl = 97) are presented in Fig. 3. The R(3) transition of pure HCl is investigated at p = 1.2 bar and 10 cm sample length (open circles). The drastic change of pulse shape compared to the input pulse (full points) and the large peak transmission of $\simeq$ 0.15 compared with Beer's law should be noted. For large αl, repetitive absorption and coherent re-emission of the interacting molecules in addition to the beating of the two isotopic species leads to a complex time dependence of the pulse wing; the resulting rapid fluctuations are smeared out by the finite duration of the probing pulse. Quantitative agreement between theory and experimental points is obtained. The result for the dephasing time is T_2 = 17 ps.

The dephasing process originates from binary HCl:HCl, and for mixtures with Ar, also from HCl:Ar collisions. Varying the pressure values of HCl and buffer gas, a linear dependence of $1/T_2$ on pressure has been observed as expected theoretically. For example, for the R(3) transition of neat HCl we find $T_2 \times p$ = 21.5 $\pm$ 1 ps bar representing self-broadening via HCl:HCl collisions.

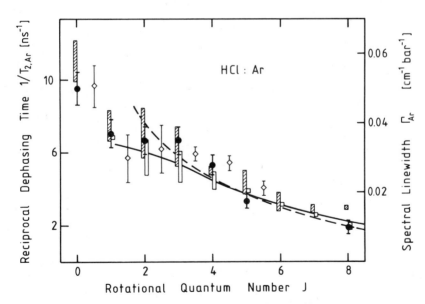

Fig. 4. Foreign gas broadening of HCl. Measured values
of the reciprocal dephasing time $1/T_{2,Ar}$ versus
rotational quantum number of the vibration-rotation transition ($\bullet$); corresponding linewidth
scale for Γ_{Ar}; IR-linewidth data (hatched rectangles) and rotational Raman linewidth ($\Diamond$), are
shown for comparison; theoretical curves (see
text).

Our investigations verify the exponential decay predicted by impact theory for gases. The observed time dependence is equivalent to a spectral line of Lorentzian shape with a half-width (HWHH in units of cm^{-1}) of $\Gamma = 1/(2\pi c T_2)$. Data for the foreign gas broadening of various R-transitions by Ar collisions are compiled in Fig. 4.[7] The experimental T_2-numbers were normalized to $p_{tot} = 1$ bar and $p_{HCl} \to 0$, yielding the dephasing time $T_{2,Ar} = (2\pi c \Gamma_{Ar})^{-1}$ due to foreign gas interaction. In the Fig. the reciprocal time constant $1/T_{2,Ar}$ is plotted (see left ordinate scale). The corresponding linewidth Γ_{Ar} is indicated by the right hand scale. The hatched and open rectangles in Fig. 4 respectively represent the range of reported IR and FIR linewidth data (see literature quoted in Ref. 7); the experimental error of these measurements has been omitted. The results of rotational Raman scattering (J→J+2 transitions) are also shown;[8] because of the different selecting rule the experimental points (◊) are displayed with an arbitrary shift of 0.5 on the abscissa scale. The agreement with our time-resolved data is noteworthy. The solid and broken curves in Fig. 4 represent the theoretical results of Ref. 9, considering rotational relaxation mechanisms for two different interaction potentials.

We recall that our time-resolved data and also the IR results contain a possible contribution of vibrational relaxation, while FIR spectroscopy and rotational Raman scattering do not. It is concluded from Fig. 4 that vibrational dephasing contributes little to the measured relaxation rates. Comparison of the transient IR data ($\Delta J=1$ transitions) with picosecond CARS results ($\Delta J=0,2$ transitions) gives evidence that apart from rotational population decay also pure rotational dephasing is an important relaxation mechanism.[5]

In the past considerable progress has been achieved in high resolution Raman spectroscopy.[10] Using several versions of the stimulated Raman scattering process and narrow band laser sources the limitations of conventional (spontaneous) Raman spectroscopy could be avoided. These techniques necessitate a considerable technical effort towards frequency stabilisation at the high power level required for nonlinear Raman processes. In the following a novel alternative approach is demonstrated applying Fourier transform coherent Raman spectroscopy (FT CARS) with picosecond laser pulses. The advantage of our time domain measurements instead of frequency spectroscopy is readily visualized recalling that a frequency resolution of 10^{-3} cm^{-1} corresponds to time observations over 10^{-8} s, which are readily feasible.

Our method [11,12] consists of two steps: coherent anti-Stokes Raman scattering of delayed probing pulses is measured after short pulse stimulated Raman excitation; subsequent numerical Fourier transformation of the signal transient provides the described spectroscopic information. In a generalized two level approach with a distribution of neighbouring molecular transitions of relative weight f_j and frequency position ω_j, the time integrated scattering signal of the probing pulse is given by

$$S^{coh}(t_D) \propto \Phi^2(t_D)\left[\sum_j f_j^2 + 2\sum_{i<j} f_i f_j \cos(\omega_i - \omega_j)t_D\right] \qquad (1)$$

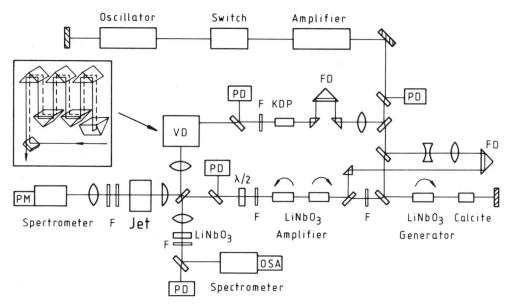

Fig. 5. Schematic of experimental system for Fourier transform
coherent Raman spectroscopy in gases with collinear
beam geometry

The probe delay is denoted by t_D. $\Phi(t_D)$ represents the
phase auto-correlation function, which accounts for the damping
of the molecular excitation. Eq. 1 describes a modulation
of the scattering signal with the frequency differences $(\omega_i-\omega_j)$
and a decay proportional to Φ^2. Measuring $S^{coh}(t_D)$ the frequency
differences can be determined by numerical Fourier transformation. The resolution of the difference frequency spectrum is
dominantly given by the total delay time interval during which
the signal transient is observed.

As a first demonstration of the FT-Raman technique we present experimental data on the Q-branch of the ν_1-vibrational
mode of methane. Using the supersonic expansion the temperature
of CH_4 is effectively lowered leading to a reduced number and
width of lines in the Q-band. In our measurements a 1:2 mixture
of CH_4 with argon buffer gas [11,13] was expanded at approximately
10 bar pressure and room temperature through a pulsed nozzle
into vacuum (nozzle diameter 250 µm, background pressure 10^{-5}
bar). The schematic of our experimental system for FT-CARS is
depicted in Fig. 5. Single pulses of a modelocked Nd-YAG laser
system ($\omega_L/2\pi c$ = 9398 cm^{-1}, duration 23 ps) and of a parametric
generator-amplifier device ($\omega_S/2\pi c$ = 6482 cm^{-1}, duration 12 ps)
are directed into the sample and serve for the excitation of
the Q-lines via stimulated Raman amplification. The second harmonic of the laser pulse ($\omega_{2L}/2\pi c$ = 18796 cm^{-1}) is used for the
coherent Raman probing process with collinear optical beams,
perpendicular to the supersonic stream. A multiple-pass delay
set-up is incorporated in the experimental apparatus which
achieves a maximum delay of the green probe pulse of t_{obs} =
12 ns with sub-picosecond precision.

69

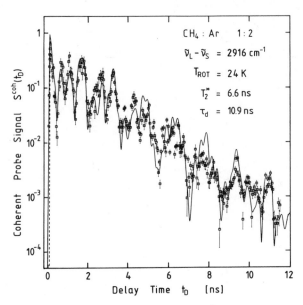

Fig. 6. Coherent probe signal $S^{coh}(t_D)$ versus delay
time for the ν_1-Q band of CH_4 ($T_{ROT} \simeq 24$ K);
experimental points; theoretical (solid)
curve. Broken line: instrumental response
with picosecond decay.

An example for the experimental data obtained approximately
2 mm behind the nozzle is presented in Fig. 6. The coherent
anti-Stokes Raman signal $S^{coh}(t_D)$ is plotted versus delay time.
The data points extend over 4 orders of magnitude and represent
an average of 20 individual measurements. A complicated beating
structure and the decay of the signal envelope are readily seen.

The solid line in the Figure is a theoretical curve which
agrees well with the experimental points; it represents a computer simulation using the data of Ref. 14 for the frequency
positions ω_j and level degeneracies effecting the weight factors f_j. A Boltzmann distribution with same rotational temperature T_{rot} is assumed for the rotational levels of the three
nuclear spin subensembles of CH_4.[10] The value of T_{rot} governs
the line strength parameters f_j and is treated as a fitting
parameter. From the data of Fig. 6 we obtain $T_{rot} \simeq 24$ K in good
agreement with earlier results.[11] As decay function Φ we use

$$\Phi = \exp\left[-(t'-t)/T_2^* - (t'-t)^2/\tau_D^2\right] \qquad (2)$$

with the fitted parameter values $T_2^* = 6.6$ ns and $\tau_D = 10.9$ ns.

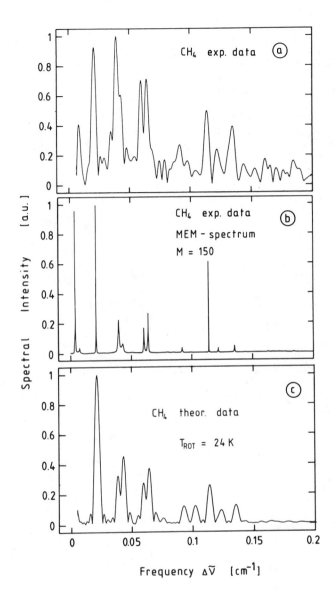

Fig. 7. High resolution Raman spectra of CH_4:
   a) FT spectrum of experimental data of Fig. 6;
   b) MEM spectrum of experimental data of Fig. 6;
   c) theoretical FT Raman spectrum

The desired spectroscopic information is obtained by numerical Fourier transformation of the experimental points of Fig.6. The result is shown in Fig. 7a where the computed spectral intensity distribution is plotted. A number of spectral lines is readily seen. Fig. 7a represents experimental information without use of the available spectroscopic data of CH_4. For improved frequency resolution we also applied the maximum entropy method (MEM). The spectrum obtained for this nonlinear transformation is depicted in Fig. 7b.

For comparison a computer simulation of the difference frequency spectrum is presented in Fig. 7c; i.e. the Fourier transform of the theoretical curve of Fig. 6 implying our knowledge of the CH_4 molecule. The good agreement of the line positions in Figs. 7a and b compared to 7c is noteworthy. Of particular interest is the doublet structure of the two bands around 0.04 cm^{-1} and 0.06 cm^{-1} which verifies the theoretically predicted tensor splitting of the $J = 2$ transitions of the Q-branch. The doublet structure is observed by our high-resolution FT-technique for the first time.

From the data of Fig. 7, the frequency differences of four transition frequencies have been determined with an accuracy of $(2-5) \times 10^{-4}$ cm^{-1}; the result for the tensor splitting of the $J = 2$ transitions is $(3.4 \pm 1.5) \times 10^{-3}$ cm^{-1}.[12]

In summary it is pointed out that coherent propagation of low-intensity picosecond pulses has been applied for a time-resolved IR spectroscopy that provides data on frequency differences of adjacent transitions and molecular time constants. Working in the regime of nearly-free induction decay with moderately short samples the spectroscopic information can be readily derived from the decay of the trailing pulse wing. The application of the phenomenon is facilitated by its linear intensity dependence; i.e. the signal transients are intensity independent apart from a scaling factor. Because of the different selection rules the transient IR method complements coherent Raman techniques yielding additional information on molecular dynamics. The technique is applicable to gases and condensed matter.[16]

Ultrashort pulses may be also used for vibrational spectroscopy with high frequency resolution. As a first example we have demonstrated FT-CARS of a supersonic expansion. Several advantages of the technique should be noted.[11,12] The effect of transit time broadening can be eliminated. Artifacts via the nonresonant part of the third order susceptibility are negligible. A possible dynamic Stark effect during the excitation process does not influence the ns signal transient. Precise spectroscopic information is provided without narrow-band laser sources.

REFERENCES

1. See, for example "Ultrafast Phenomena", Vo. IV, eds. D.H. Auston and K.B. Eisenthal (Springer, Berlin, 1984); A. Laubereau and W. Kaiser, Rev. Mod. Phys. 50:607 (1978).
2. H.-J. Hartmann and A. Laubereau, Appl. Opt. 20:4259 (1981).
3. A. Laubereau, L. Greiter and W. Kaiser, Appl. Phys. Lett 25: 87 (1974); A. Fendt, W. Kranitzky, A. Laubereau and W. Kaiser, Opt. Commun. 28:142 (1979).
4. H.-J. Hartmann and A. Laubereau, Opt. Commun 47:117 (1983).
5. H.-J. Hartmann and A. Laubereau, J. Chem. Phys. 80:4663 (1984); H.-J. Hartmann, K. Bratengeier and A. Laubereau, Chem. Phys. Lett. 108:555 (1984).
6. G. Herzberg, Molecular Spectra and Molecular Structure, Vol. I Van Nostrand, Princeton, N.J. (1950).
7. H.-J. Hartmann, H. Schleicher and A. Laubereau, Chem. Phys. Lett. 116:392 (1985).
8. G.J.Q. van der Peijl, D. Frenkel and J. van der Elsken, Chem. Phys. Lett. 56:602 (1978).
9. E.W. Smith and M. Girard, J. Chem. Phys. 66:1762 (1977).

10. P. Esherik and A. Owyoung, in: Advances in Infrared and Raman Spectroscopy, Vol. 9, eds. R.J.H. Clark and R.E. Hester (Heyden and Sons, Ltd., London, 1982) p. 130-187, and references cited therein.

11. H. Graener, A. Laubereau and J.W. Nibler, Optics Lett. 9: 165 (1984).

12. H. Graener and A. Laubereau, Optics Comm. 54:141 (1985).

13. P. Huber-Wälchli and J.W. Nibler, J. Chem. Phys. 76:273 (1982).

14. A. Owyoung, C.W. Patterson and R.S. McDowell, Chem. Phys. Lett. 59:156 (1978).

15. S. Haykin and S. Kesler, in: Nonlinear Methods of Spectral Analysis, eds. S. Haykin (Springer-Verlag, Berlin, Heidelberg, New York 1979) p. 9-72.

16. K. Bratengeier and A. Laubereau, to be published.

NONLINEAR TRANSIENT SPECTROSCOPY WITH ULTRAHIGH TIME-RESOLUTION USING

LIGHT SOURCES WITH CONTROLLED COHERENCE

Tatsuo Yajima and Norio Morita

Institute for Solid State Physics
University of Tokyo
Roppongi, Minato-ku, Tokyo 106, Japan

1. INTRODUCTION

The study of ultrafast phenomena is one of the major subjects of laser spectroscopy, for which numerous methods have been developed. The progress of ultrafast transient spectroscopy by means of picosecond and femtosecond pulse technology undoubtedly brought about revolutionary development in this field of research.[1] There still remain, however, difficulties in the generation of practical light pulses and their applications in the time region far below 100 fs. The pulse broadening by material dispersion is one of fundamental limitations in all aspects of the technology associated with extremely short pulses. As another approach of studying ultrafast phenomena, various nonlinear spectroscopic methods in the frequency domain have early been developed.[2-4] Although they served as complementary means to the time domain methods, they are still indirect and require careful interpretation.

In conventional transient spectroscopy, the time-resolution is usually limited by the pulse width of light sources or equivalent perturbers. A new approach of transient spectroscopy to overcome this difficulty is recently developed by several groups including ours, where the time-resolution is governed by the correlation time τ_c instead of the time duration of light sources.[5-11] A temporally incoherent light has short τ_c corresponding to the reciprocal of its broad spectral width, and is expected to play the same role as a short pulse with the duration τ_c in a correlation-type measurement. Several experimental demonstrations to confirm this expectation have been reported in the observations of dynamical behaviors of Na vapours,[6,8] Nd^{3+}- doped glasses,[5,10] and dye solutions[11] by means of transient four-wave mixing or equivalently photon echo technique. The incoherent light to be used in this new method cannot be supplied from thermal sources, but must be based on laser technique, because it must have good spatial coherence and sufficient intensity.

Phase-modulated coherent pulse such as due to self-phase modulation has a common feature with incoherent light in that it has broad band-width whose reciprocal is much shorter than the light duration. It, therefore, is expected to play the same role as incoherent light in some nonlinear transient spectroscopy. This expectation has also been confirmed by preliminary theoretical and experimental study.

Although these previous studies revealed that strongly non-transform-limited broad-band light sources, both coherent and incoherent, can play basically the same role as transform-limited short pulses, there, of course, exist differences among the results caused by different types of light sources. The purpose of the present report is to show and compare various nonlinear material responses for a variety of light sources as comprehensive as possible mainly by theoretical consideration, and to clarify what type of light source is really suitable for ultrahigh time-resolution spectroscopy in the extremely short time region. Among many possible nonlinear optical processes, we will concentrate our attention here on the transient degenerate four-wave mixing with two input light beams, which contains basic processes of coherent transient, such as photon echoes and free induction decay,[12,13] and serves mainly to determine phase relaxation time (T_2) of resonant materials. Other types of nonlinear spectroscopic problems will also be mentioned briefly.

2. DEGENERATE FOUR-WAVE MIXING WITH A GAUSSIAN STOCHASTIC FIELD

As a starting point of discussion, we will first review briefly essential points of our earlier analytical study[9] for an idealized model.

Consider an inhomogeneously broadened two-level system whose relaxation properties are described only by two phenomenological decay times T_1 (longitudinal) and T_2 (transverse). When two noncollinear light beams with wave-vectors $\vec{k}_1$, $\vec{k}_2$ and a common center frequency ω_c are incident on the resonant two-level system, output light beams with wave-vectors $\vec{k}_3 = 2\vec{k}_2 - \vec{k}_1$ and $\vec{k}_4 = 2\vec{k}_1 - \vec{k}_2$ are produced through the third-order nonlinear polarization (Fig. 1). The integrated or averaged energy of one output beam as a function of the delay time τ between two incident beams (called as correlation trace) provides information on relaxation times and other properties of the material system. With the incident optical field of the form

$$E(\vec{r},t) = \tilde{E}(t+\tau)\exp\{-i\omega_c(t+\tau)+i\vec{k}_1\cdot\vec{r}\}+\tilde{E}(t)\exp\{-i\omega_c t+i\vec{k}_2\cdot\vec{r}\} + \text{c.c.}, \quad (1)$$

the averaged energy J of the output light with $\vec{k}_3$ is expressed as

$$J \propto <|\tilde{P}^{(3)}(t,\tau)|^2>$$

$$\tilde{P}^{(3)} = M\int_0^\infty g(\omega_0)d\omega_0\int_{-\infty}^t dt_1\int_{-\infty}^{t_1} dt_2\int_{-\infty}^{t_2} dt_3 \exp\{-\gamma_1(t_1-t_2)-\gamma_2(t-t_1+t_2-t_3)\}$$

$$\times [\tilde{E}(t_1)\tilde{E}(t_2)\tilde{E}^*(t_3+\tau)\exp\{-i(\omega_0-\omega)(t-t_1-t_2+t_3)\}$$

$$+\tilde{E}(t_1)\tilde{E}^*(t_2+\tau)\tilde{E}(t_3)\exp\{-i(\omega_0-\omega)(t-t_1+t_2-t_3)\}], \quad (2)$$

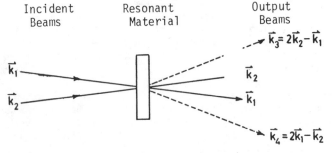

Fig. 1 Configuration of the four-wave mixing
with two incident light beams

where $\gamma_1 = T_1^{-1}$, $\gamma_2 = T_2^{-1}$, M is a factor independent of time variables, $g(\omega_0)$ is the normalized distribution function of resonant frequency ω_0, and $\tilde{P}^{(3)}$ represents the complex amplitude of the third-order nonlinear polarization.

In general, explicit analytical form of $J(\tau)$ is difficult to be given because $J(\tau)$ involves sixth-order moment of $\tilde{E}(t)$. For a Gaussian stochastic field, this moment can be expressed as a sum of the products of three field-correlation-functions of the type $f(\tau) = \langle \tilde{E}^*(t)\tilde{E}(t+\tau)\rangle$. In particular, for $f(\tau)$ with a δ-function shape (corresponding to the limit $\tau_c \to 0$), the form of $J(\tau)$ can easily be obtained and is already given.[9] Figure 2 shows its graphical representation for typical cases in comparison with the results for short incident pulses with a limiting zero width. The figure manifests well the specific features of the problem.

In the case of $T_1 \gg T_2$, the results for incoherent light are entirely the same as those for short pulse except that the T_2 decay characteristic in the correlation trace appears symmetrically with respect to τ for homogeneously broadened transition. In the case of $T_1 \simeq T_2$, however, the appearance of T_1-dependent background and the modification of relaxation decay due to the T_1 effect occur for incoherent light, while the results are independent of T_1 for short pulse input. These features complicate the interpretation of the results for incoherent light. But they still serve for the determination of T_2, because the decay characteristics are dominantly governed by T_2. All these behaviors can well be explained physically by

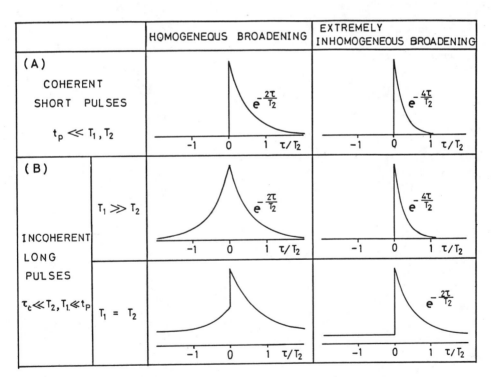

Fig. 2 Typical correlation traces of transient degenerate four-wave mixing calculated by analytical equations for idealized light sources : (A) coherent short pulses with zero pulse width ($t_p \to 0$), and (B) incoherent long pulses with zero correlation time ($\tau_c \to 0$). Here, τ is the delay time, and T_1 and T_2 denote the longitudinal (energy) and transverse (phase) relaxation times, respectively.

regarding the incoherent light as a series of random ultrashort pulses and by considering the integration of numerous transient four-wave mixing processes caused by various combinations of these pulses.

It has been found later that the same results still hold for other stochastic fields with less restrictions than for the Gaussian process.

3. DEGENERATE FOUR-WAVE MIXING WITH A CHIRPED COHERENT LIGHT PULSE

Incoherent light is a kind of strongly non-transform-limited (NTFL) light, whose reciprocal spectral width is much shorter than the light duration. The reciprocal spectral width corresponds to the width of the field correlation function of arbitrary optical field according to the Wiener-Khintchine theorem, and therefore can be defined as a correlation time τ_c for any kind of light. Another kind of broad-band NTFL light such as self-phase-modulated coherent pulse has also short τ_c in this sense, and is therefore expected to provide short resolution time close to τ_c in the transient four-wave mixing spectroscopy.

The narrowing of correlation trace in the four-wave mixing for a chirped pluse has been pointed out, and used to investigate the coherence property of light pulses.[14] We have examined[7] the possibility of high time-resolution measurement of phase relaxation time T_2 with a chirped coherent pulse in the same configuration as described in Sec. 2 by means of computer simulation, which will be summarized below. Recently, a similar attempt has also been reported by other workers.[15]

In order to calculate the correlation trace $J(\tau)$ by eq. (2), we must generally deal with a multiple integral of high degree. For a broadly inhomogeneous transition satisfying $\Delta\omega_i$ (inhomogeneous width) $\gg T_2$, τ_c^{-1}, the multiplicity is reduced by two. This makes the calculation much easier. The result is expressed, using a factor K similar to M, as[16]

$$J \propto \int_{-\infty}^{\infty} |\tilde{P}^{(3)}(t,\tau)| dt$$

$$\tilde{P}^{(3)} = K \int_{-\infty}^{t} \int_{-\infty}^{t'} \tilde{E}(t'-\tau)\tilde{E}(t''-\tau)\tilde{E}^*(t'+t''-t)$$
$$\times \exp\{-\gamma_1(t'-t'')-2\gamma_2(t-t')\}dt''dt' \quad . \tag{3}$$

As a simple model for a self-phase-modulated pulse, we consider a linearly chirped pulse with a fixed envelope function. The corresponding complex field amplitude is given by

$$\tilde{E}(t) = \hat{E}(t)\exp(ibt^2). \tag{4}$$

A typical result of numerical calculation for a Gaussian envelope function $\hat{E}(t)$ is shown in Fig. 3 in comparison with the case for a transform-limited (unmodulated) pulse with the same envelope function.

It is to be noted that with the chirped pulse the decay characteristics due to T_2 clearly appears with a resolution time much shorter than the pulse width t_p even when $T_2 \ll t_p$. While, with the unmodulated pulse having the same pulse envelope, relaxation behavior can hardly be seen except the peak shift of the correlation trace.

The spectral width of a chirped pulse increases with the chirping parameter $\phi_p = bt_p^2$. Detailed examination of the relation between the

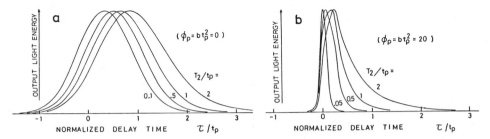

Fig. 3 Calculated correlation traces of transient degenerate four-wave
mixing for two kinds of incident light pulses of the same Gaussian
pulse profile : (a) without phase-modulation (transform-limited)
and (b) with linear chirping (quadratic phase-modulation). Here,
t_p, and T_2 are the pulse width and the transverse relaxation time,
respectively. The condition $T_1 \gg t_p$, T_2 has been assumed.

spectrum and the correlation trace with $T_2 \to 0$ revealed that the correlation width becomes roughly equal to the reciprocal spectral width. This
means that the basic behavior of the four-wave mixing for a chirped pulse is
also governed roughly by the power spectrum of the pulse, and the resolution
time close to τ_c (defined as the reciprocal spectral width) can be achieved.
It should be noted, however, that the result in this case cannot be determined "strictly" only by the power spectrum or the field correlation function of the pulse in contrast to the case with a Gaussian stochastic field.
This point will be clarified by further investigation described in the next
section.

The main advantage of the phase-modulated pulse is its capability of
having very broad spectrum in a controlled manner. A preliminary experiment
of four-wave mixing in dye solutions has been tried in our group using a
broad-band chirped pulse produced in an optical fiber, and confirmed the
achievement of subpicosecond resolution time much shorter than the pulse
width, though T_2 is too short to be observed.

4. DEGENERATE FOUR-WAVE MIXING WITH LIGHT SOURCES OF ARBITRARY INCOHERENCE AND REGULAR PHASE-MODULATION

Even a light source based on laser technique generally involves
various degrees of incoherence and regular phase-modulation due to linear
dispersion and nonlinear effects. An effective way of representing a
variety of light of these kinds in a unified manner is to adopt a multi-mode
model of light. In this model, any kind of light is represented as a
superposition of many frequency components with a fixed power spectrum and

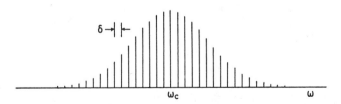

Fig. 4 Power spectrum of light sources
expressed by a multi-mode model

arbitrary relative phases. From only a single model we can represent a
variety of light including transform-limited short pulse, Gaussian incoherent light, partially coherent light, phase-modulated coherent pulse and
also more complicated light with combined coherent and incoherent modu-
lations, by giving various regular and/or irregular phase relations among
the frequency components. An important point of this model is that we can
investigate the differences of specific features associated with various
kinds of light as described above under the condition of common correlation
time τ_c resulting from fixed power spectrum. With this model and condition
we have investigated and compared the correlation traces of degenerate four-
wave mixing for a variety of light fields by means of computer simulation.

The basic equation used is the same as eq. (3), but with different
expression of $\tilde{E}(t)$. The complex amplitude of the light field in a multi-
mode model as shown in Fig. 4 is expressed as

$$\tilde{E}(t) = \sum_n E_n \exp[i\{n\delta t + \phi_n - (n\delta)^2 w\}], \tag{5}$$

where ϕ_n is the random phase representing the incoherence, and the term
$-(n\delta)^2 w$ represents the regular phase-modulation due to material linear
dispersion. A simple way of representing the arbitrary incoherence is to
let $\phi_n = 2\pi P R_n$, where R_n's are a set of random numbers distributed uniformly
between 0-1, and P is defined as the coherence parameter taking a value
between 0-1. The dispersion parameter W can also be defined as
$W = w\tau_c^{-2} = \ell(2\pi c^2 \tau_c^2)^{-1}(\lambda^3 d^2\eta/d\lambda^2)$, where ℓ is the propagation length,
λ is the wavelength, and η is the refractive index. The correlation time
τ_c has been defined as the reciprocal spectral width so as to be equal to
the FWHM width in the case of transform-limited pulse.

By giving a set of arbitrary values for two parameters P and W, we can
express a variety of light fields with incoherent and/or coherent modulation
for computer simulation. Figure 5 shows the intensity waveforms of typical
four kinds of light fields with the "same" power spectrum of Gaussian shape,
(a) transform-limted short pulse, (b) Gaussian incoherent light without

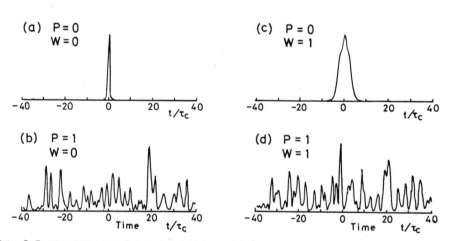

Fig. 5 Intensity waveforms of various light sources having the same power
spectrum with a mode number of 401. (P: coherence parameter, W:
dispersion parameter, see text.)
(a) transform-limited short pulse
(b) Gaussian incoherent light without coherent phase-modulation
(c) phase-modulated coherent pulse
(d) Gaussian incoherent light with coherent phase-modulation

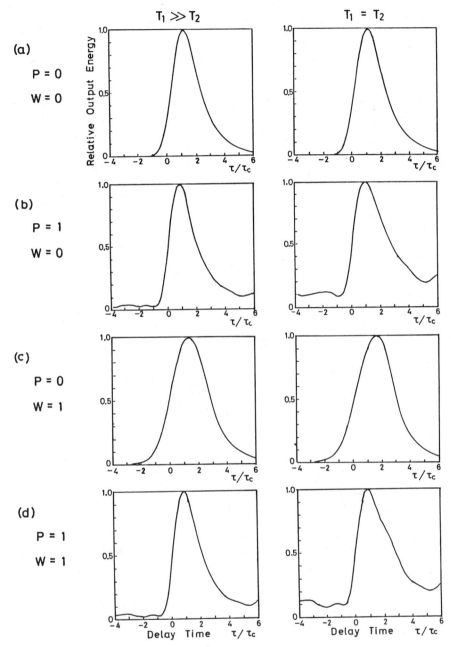

Fig. 6 Calculated correlation traces of degenerate four-wave mixing with various light sources (a), (b), (c) and (d) as expressed in Fig. 5 for a broadly inhomogeneous transition under two conditions of the relaxation parameter T_2/T_1. (T_1: longitudinal relaxation time, T_2: transvers relaxation time) The value $T_2/\tau_c = 5$ has been assumed, where τ_c is the correlation time (reciprocal spectral width) of light sources. The coherence parameter P represents the extent of random phase distribution, and the dispersion parameter W represents the degree of regular phase-modulation due to material dispersion (see text for definition). The cross relaxation effect has been neglected.

coherent phase-modulation, (c) phase-modulated coherent pulse, and (d) Gaussian incoherent light with coherent phase-modulation. The light field (c) has a linear chirping and manifests essentially the same features as those of the field treated in Sec. 3. The difference lies in that the spectral profile is fixed and the pulse profile is broadened for the pulse (c) while for the pulse in Sec. 3 the pulse profile is fixed and the spectral profile is broadened. The value of $W = 1$ corresponds to about five-times pulse broadening. It is to be noted that the two kinds of light fields (b) and (d) has essentially the same statistical properties. This means that the essential feature of the Gaussian incoherent field is not affected by material dispersion.

In Fig. 6 are shown typical calculated correlation traces of the degenerate four-wave mixing with the four different types of incident optical fields shown in Fig. 5 under two conditions of relaxation parameters. The results for the cases (a) and (b) well reproduce the essential features indicated in Fig. 2 for inhomogeneously broadened transition with idealized light sources except that a finite value of τ_c has been involved in Fig. 6. The traces for incoherent light show clear relaxation decay with the same risetime (corresponding to the resolution time) as that for transform-limited short pulse, although complete agreement between the traces (a) and (b) for $T_1 \gg T_2$ has not been achieved due to incomplete averaging with incoherent light. For phase-modulated coherent pulse (case (c)), the risetime of the trace becomes much shorter than the pulse width as expected, but considerably longer than those for transform-limited pulse and incoherent light. The result indicates that the resolution time with phase-modulated coherent pulse approaches to the correlation time τ_c, but is not strictly determined by τ_c or the power spectrum of the light. One should note, however, the case (d) with incoherent light subject to material dispersion. The results are essentially the same as those in the case (b), and the same time-resolution as in the case (a) can be held. Namely, with incoherent light the time-resolution in the transient four-wave mixing spectroscopy is not deteriorated by material dispersion. This is very important feature in the application of incoherent light in the extremely short time region.

In actual materials, there exists more or less the cross relaxation effect within an inhomogeneous broadening, which has been neglected in the previous consideration. Its effect in the transient four-wave mixing will be examined next. The model and basic equations representing the cross

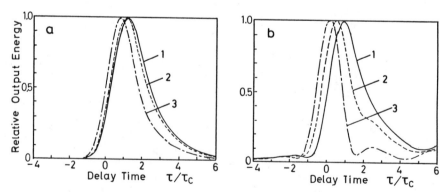

Fig. 7 Effect of cross relaxation time (T_3) on the correlation traces of transient degenerate four-wave mixing with light sources of (a) transform-limited short pulse and (b) Gaussian incoherent light as given in Fig. 5. Broadly inhomogeneous transition and the conditions $T_1 \gg T_2$ and $T_2/\tau_c = 5$ have been assumed. Curves 1, 2, 3 correspond to $T_2/T_3 = 0$, 0.2, 1.0, respectively.

relaxation effect is the same as those used previously.[2,17] The third-order nonlinear polarization with this effect is given by the sum of two terms as

$$\tilde{P}(3) = \tilde{P}_1(3) + \tilde{P}_2(3) \quad .$$ (6)

The first term $\tilde{P}_1(3)$ is of the same form as eq. (3) except that γ_1 is replaced by $\gamma_1 + \gamma_3$, where $\gamma_3 = T_3^{-1}$ is the cross relaxation rate. The second term arises from inverse spectral diffusion, and is expressed as

$$\tilde{P}_2(3) = 4\pi\gamma_3 g(\omega_c) K \int_{-\infty}^{t} \int_{-\infty}^{t'} \tilde{E}(t-\tau)\tilde{E}(t''-\tau)\tilde{E}^*(t'')$$
$$\times \exp\{-\gamma_1(t'-t'')-(\gamma_1+\gamma_3)(t-t')\}dt''dt' \quad .$$ (7)

Typical calculated results based on these equations are shown in Fig. 7 for the case $T_1 \gg T_2$. The presence of cross relaxation generally brings about a change in correlation traces such that the central peak is enhanced relative to the component representing T_2 decay. This tendency appears already in the case with short input pulses, but more prominently in the case with incoherent light. It is found further that the correlation traces are appreciably affected by the time dulation D of incoherent light as illustrated in Fig. 8. Detailed examination revealed that under the condition $T_3 \ll D \ll T_1$, the correlation trace is dominated by the central peak and the T_2 decay component can hardly be seen. This is physically understandable by an interpretation that in this situation a portion of the inhomogeneous width covering the spectral width of incident light behaves as homogeneous broadening. The cross relaxation effect is not prominent in the case of $T_1 \simeq T_2$, because the effect only lasts within a time comparable to T_2 in view of the relation $T_3 \gtrsim T_2$.

These results show that in materials with very fast cross relaxation, the measurement of T_2 with incoherent light becomes difficult and therefore not suitable. From another point of view, however, a new method of investigating cross relaxation phenomena is possible by using incoherent light with variable duration or by comparing the results with coherent and incoherent light sources.

In the case with phase-modulated coherent pulse, the correlation traces

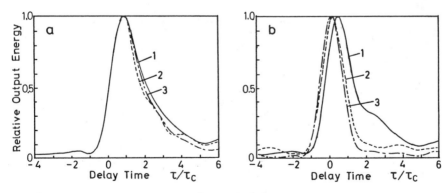

Fig. 8 Effect of time dulation (D) of incoherent incident light on the correlation traces of transient degenerate four-wave mixing (a) without ($T_2/T_3 = 0$) and (b) with ($T_2/T_3 = 0.2$) cross relaxation effect. Other conditions are the same as in Fig. 7. Curves 1, 2, 3 correspond to $D/\tau_c = 19, 38, 76$, respectively.

are found to be not so affected by the cross relaxation effect, but to approach to those with transform-limited pulse at high rate of γ_3.

It has been pointed out[18,19] that, in a multi-level system where one energy level is optically connected with a set of closely spaced many levels, observation of T_2 becomes essentially difficult with transient spectroscopy. The cross relaxation effect is considered to manifest a part of the specific feature of this type of multi-level system.

5. PROBLEM OF ENERGY RELAXATION TIME MEASUREMENT

The measurement of ultrafast energy relaxation times (represented by T_1) in various materials is also important in ultrafast laser spectroscopy. Examination of using non-transform-limited incoherent or coherent light for this purpose is an attractive subject as well as for the purpose of determining T_2. For this purpose, however, other types of optical processes than those dealt with in the previous sections are required to be considered.

Preliminary studies, both theoretical and experimental, have been carried out for the pump-probe method[20] and the four-wave mixing with three incident light fields[21], and revealed that the T_1 decay characteristics can also be observed with incoherent light.

There exists a common problem in these methods with incoherent light, that is, the appearance of large background and central coherence peak (or artifact) in addition to the T_1 decay in the output signal as a function of delay time. In order that the methods be practical, effective ways of eliminating or minimizing these obstacles must be considered. Phase-modulated coherent pulses are ineffective in these methods to provide high time-resolution, because the relevant optical processes are dominantly governed by the intensity correlation function of optical fields.

6. SUMMARY AND CONCLUSION

The results of our investigation are summarized as follows.

Incoherent light sources can effectively be used to determine ultrafast relaxation times, in particular T_2, by means of transient four-wave mixing in relatively simple, inhomogeneously broadened two-level systems. The most important feature is that material nonlinear responses with incoherent light are dispersion-free, and are essentially suitable to achieve ultrahigh time-resolution in the extremely short time region. For optical transitions with very fast cross-relaxation or with complicated multi-level structure, incoherent sources are not so effective for the above purpose.

Phase-modulated coherent pulses are promising for observing phase relaxation effects in view of the possibility of having very broad-band in a controlled manner, though ultimate time-resolution does not strictly reach to the correlation time (reciprocal spectral width). This type of light source is, however, ineffective for the purpose of observing energy relaxation effects.

Although strongly non-transform-limited broad-band light sources of coherent or incoherent nature are certainly effective to provide ultrahigh time-resolution, nonlinear material responses to these sources are more complicated compared with those to transform-limited short pulses, and therefore careful interpretations are required for the results. Anyway, all types of light sources considered above will contribute complementarily in ultrahigh time-resolution spectroscopy through the use of specific features associated with each light source.

The conclusions derivable here are of course limited to a certain extent because they are still based on a restricted model of calculation, and more refined theoretical and corresponding experimental studies must be promoted to establish this type of spectroscopic method. The present results are, however, hoped to provide a useful guide to the future trend of ultrafast laser spectroscopy.

This work was supported in part by the Grant-in-Aid for Scientific Research from the Ministry of Education, Science and Culture, Japan.

REFERENCES

1. See, for example, D.H. Auston and K.B. Eisenthal ed., "Ultrafast Phenomena IV", Springer-Verlag, Berlin (1984).
2. T. Yajima and H. Souma, Phys. Rev. A 17, 324 (1978).
3. T. Yajima, H. Souma and Y. Ishida, Phys. Rev. A 17, 1439 (1978).
4. J.J. Song, J.H. Lee and M.D. Levenson, Phys. Rev. A 17, 1439 (1978).
5. S. Asaka, H. Nakatsuka, M. Fujiwara and M. Matsuoka, Phys. Rev. A 29, 2286 (1984).
6. N. Morita, T. Yajima and Y. Ishida, in "Ultrafast Phenomena IV", D.H. Auston and K.B. Eisenthal ed., Springer-Verlag, Berlin (1984) P.239.
7. T. Yajima, N. Morita and Y. Ishida, J. Opt. Soc. Am. B 1, 526 (1984).
8. R. Beach and S.R. Hartmann, Phys. Rev. Lett. 53, 663 (1984).
9. N.Morita and T. Yajima, Phys. Rev. A 30, 2525 (1984).
10. H. Nakatsuka, M. Tomita, M. Fujiwara and S. Asaka, Opt. Commun. 52, 150 (1984).
11. M. Fujiwara, R. Kuroda and H. Nakatsuka, J. Opt. Soc. Am. B 2, 1634 (1985)
12. T. Yajima and Y. Taira, J. Phys. Soc. Japan, 47, 1620 (1979).
13. P. Ye and Y.R. Shen, Phys. Rev. A 25, 2183 (1982).
14. H.J. Eichler, U. Klein and D. Langhaus, Appl. Phys. 21, 215 (1980).
15. M.A. Vasil'eva, J. Vischakas, V. Kabelka and A.V. Masalov, Opt. Commun. 53, 412 (1985).
16. T. Yajima, Y. Ishida and Y. Taira, in "Picosecond Phenomena II", R.M. Hochstrasser, W. Kaiser and C.V. Shank ed., Springer-Verlag, Berlin (1980) P.190.
17. Y. Taira and T. Yajima, J. Phys. Soc. Japan, 50, 3459 (1981).
18. J.J. Yeh and J.H. Eberly, Phys. Rev. A 22, 1124 (1980).
19. A.M. Weiner, S. De Silvestri and E.P. Ippen, J. Opt. Soc. Am. B 2, 654 (1985).
20. M. Matsuoka, presented at the fall meeting of the Physical Society of Japan (1984).
21. N. Morita and T. Yajima, presented at the Gordon Research Conference, in USA (1985).

PICOSECOND MODULATION SPECTROSCOPY IN SODIUM VAPOR

R.Beach, D.DeBeer, L.G.Van Wagenen, and S.R.Hartmann

Columbia Radiation Laboratory and Department of Physics
Columbia University, New York, New York 10027
U.S.A.

ABSTRACT

Time delayed four wave mixing (TDFWM) using two-frequency light beams, separately resonant with each of the Na D lines, is performed in dilute Na vapor. The time integrated four wave mixed signal modulates with increasing excitation delay with a period of 1.9 psec corresponding to the 6Å splitting of the D lines. The modulation persists well into the region where the 7 nsec excitation pulses no longer overlap and far beyond the 100 psec correlation time of either laser.

INTRODUCTION

The notion of exploiting the short correlation time associated with noise spikes in broadband (incoherent) laser light in order to study Ultra Fast phenomena becomes viable via Time Delayed Four Wave Mixing (TDFWM). This technique offers the opportunity of performing picosecond and femtosecond experiments with conventional laser sources which are neither spectrally very narrow nor designed to provide short excitation pulses.[1-4] A second feature of this technique is that it simultaneously provides spectral information contained within the bandwidth of the excitation fields. Recent work in this laboratory using 12Å wide 7 nsec long excitation pulses covering both Na D line transitions showed that the integrated TDFWM signal measured as a function of the relative delay of the excitation pulses was deeply modulated with the 1.9 psec period corresponding to the inverse of the 3P fine structure splitting.[5] As these pulses were greater in length than the period of the modulation they induced, we call them "long". The noise components in the light however give rise to noise spikes which are short. These have a correlation time of the order of a picosecond and allow the measurement of relaxation times of this order. On the other hand one only expects light which is on or near resonance to effectively interact with any long-lived atomic state. We therefore modified our source so that it peaked separately at each of the Na D lines and repeated our experiment. We again saw a modulated signal. The 1.9 psec modulation we observe in fact persists well beyond the regime where the excitation pulses overlap. Previous photon echo work using 7 nsec excitation pulses has shown that echoes may be observed twenty excited state lifetimes and longer after the initial excitation pulse.[6] The same should obtain here leading to the

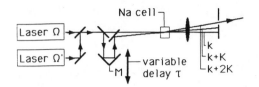

FIG. 1   Schematic diagram of the experimental apparatus used to
generate TDFWM. The outputs of lasers at frequencies Ω and
Ω' are combined and then split into two double wavelength
beams. Time delay is achieved by moving mirror M. The Na D
lines are close, $\lambda-\lambda'=6\text{Å}$, and the angle between prompt and
delayed beams is small, so that to within the accuracy of the
beam divergence when we set $\mathbf{k}\ ||\ \mathbf{k}'$ we have $\mathbf{k}+\mathbf{K}\ ||\ \mathbf{k}'+\mathbf{K}'$.
The pinhole only allows the TDFWM signal at $\mathbf{k}+2\mathbf{K}$ and $\mathbf{k}'+2\mathbf{K}'$
to pass.

possibility for an expanded use of Photon Echo Modulation Spectroscopy (PEMS)
to perform high resolution spectroscopic studies.

EXPERIMENT

     The simultaneous outputs of two pulsed-YAG-pumped dye lasers were combined
into a single beam and used to generate a time delayed four wave mixing signal
in atomic sodium vapor. The laser pulses were 7 nsec long and centered at $\lambda =
2\pi c/\Omega = 5896\text{Å}$ and $\lambda' = 2\pi c/\Omega' = 5890\text{Å}$, the wavelengths corresponding to the
$3S_{1/2} - 3P_{1/2}$ and the $3S_{1/2} - 3P_{3/2}$ transitions respectively. Each laser ran in
4 to 5 longitudinal modes yielding an overall bandwidth of 5 GHz.

     The double frequency pulse was split and recombined to provide two double
frequency pulses directed along $\mathbf{k}$ and $\mathbf{k}+\mathbf{K}$ and angled at $\Theta \approx 2$ mrad to each
other. Mirror M was mounted on a translation mount to provide a variable delay.
These pulses overlapped spatially throughout a 10 cm long sodium cell. Signal
intensity was measured, away from the excitation pulses, along the phased
matched $\mathbf{k}+2\mathbf{K}$ direction. A 100μm pinhole was placed in the focal plane of a 45
cm focal length lens to isolate the signal. Large offset delays were obtained
by inserting additional mirrors in the path of the beam with the variable
delay. Measurements were made using either an EG&G FND-100 photodiode or an RCA
C31034 PMT. In all cases the signal was integrated over its full 7 nsec
duration.

     In the regime where there is considerable temporal excitation pulse
overlap we obtain the results shown in Fig. 2. For three different offset
delays, stepped displacements of M generated delays of several tens of
picoseconds around 0, 230, and -300 picoseconds. The three traces are
essentially the same: any differences can be ascribed to laser fluctuations and
other instabilities. The common important feature is the 1.9 psec modulation
corresponding to the inverse of the 3P fine structure splitting. These results

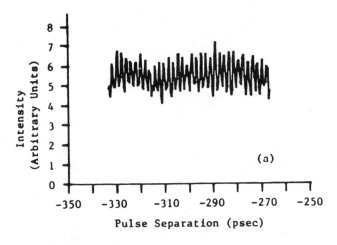

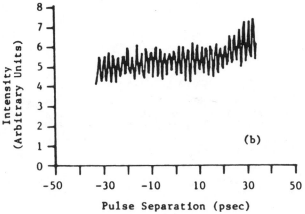

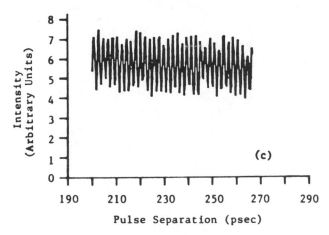

FIG 2.   TDFWM signal intensity vs. pulse delay at
         several delay offsets: (a) –300 psec, (b) 0
         psec, and (c) 230 psec.

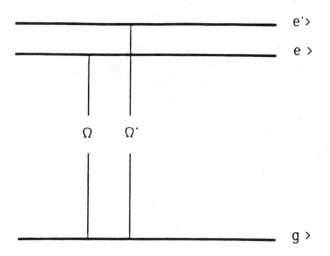

FIG 3. Energy level diagram of a simple three
level system.

were unchanged when we used a spectrometer to look separately at each of the
5896Å and 5890Å transitions instead of looking at them together. Our results
imply that the modulation comes from a modulation of the atomic state
amplitudes rather than from a beating of the dipole moments. This obtains when
two states are coherently excited through a third. An idealized case which
illustrates this occurs when a two-frequency step function excitation
containing prompt and delayed components along $\mathbf{k} \parallel \mathbf{k}'$ and $\mathbf{k}+\mathbf{K} \parallel \mathbf{k}'+\mathbf{K}'$ is used
to resonantly excite a three level system via it's common ground state. Such an
excitation field is written as:

$$\mathcal{E}=\mathcal{E}'=0$$

for t<0 and

$$\mathcal{E}=(E_0/2)\{[\exp i(\Omega t-\mathbf{k}\cdot\mathbf{r})][1+\exp-i(\Omega\tau+\mathbf{K}\cdot\mathbf{r})] + \text{c.c.}\}$$
$$=E\exp(i\Omega t) + \text{c.c.},$$

$$\mathcal{E}'=(E'_0/2)\{[\exp i(\Omega' t-\mathbf{k}'\cdot\mathbf{r})][1+\exp-i(\Omega'\tau+\mathbf{K}'\cdot\mathbf{r})] + \text{c.c.}\}$$
$$=E'\exp(i\Omega' t) + \text{c.c.}$$

for t>0. In the approximation that the interaction Hamiltonian H can be written
as

$$H = \begin{pmatrix} 0 & p^*E & p'^*E' \\ pE^* & 0 & 0 \\ p'E'^* & 0 & 0 \end{pmatrix}$$

the induced dipole moment is given exactly by

$$<P> = (4\pi/h)\,\text{Re}-i(|p|^2E\exp i\Omega t + |p'|^2E'\exp i\Omega' t)[\sin(2ft)]/2f\}.$$

where

$$f = \{ (\Omega_R^2/2) [1+\cos(\Omega\tau+\mathbf{K}\cdot\mathbf{r})] + (\Omega_R'^2/2) [1+\cos(\Omega'\tau+\mathbf{K}'\cdot\mathbf{r})] \}^{1/2}$$

where $\Omega_R=(2\pi|p|E_0/h)$ and $\Omega_R'=(2\pi|p'|E_0'/h)$. The quantity f has Fourier components at all multiples of $(\Omega\tau+\mathbf{K}\cdot\mathbf{r})$ and $(\Omega'\tau+\mathbf{K}'\cdot\mathbf{r})$. For small t we can use

$$[\sin(2ft)]/2f = t - 2f^2t^3/3$$

To this order the induced moment at frequency Ω is phased to radiate along $\mathbf{k}, \mathbf{k}\pm\mathbf{K}, \mathbf{k}\pm\mathbf{K}', \mathbf{k}+2\mathbf{K}$, and $\mathbf{k}+\mathbf{K}+\mathbf{K}'$. The component of P which oscillates at Ω along the last two directions is given by

$$P=[it^3/12]p\Omega_R\{\exp i[\Omega(t-2\tau)-(\mathbf{k}+2\mathbf{K})\cdot\mathbf{r}]\}$$
$$\cdot\{\Omega_R^2+\Omega_R'^2\exp-i[(\Omega'-\Omega)\tau+(\mathbf{K}'-\mathbf{K})\cdot\mathbf{r}]\} + c.c.$$

This shows the modulation at $\Omega-\Omega'$ as long as $\mathbf{K}-\mathbf{K}'$ is small. A similar expression holds for P'. For frequencies which are not close the signals along $\mathbf{k}+2\mathbf{K}$ and $\mathbf{k}'+2\mathbf{K}'$ will spatially separate when holding $\mathbf{K}-\mathbf{K}'=0$ and continue to be modulated.. For $\Omega_R=\Omega_R'$ and $\mathbf{K}=\mathbf{K}'$ which is effectively where we work, the time averaged intensity is

$$I_{av}\sim 1+\cos(\Omega-\Omega')\tau$$

in good agreement with what we observe.

The 1.9 psec modulation is thus understood as originating from a modulation of the dipole density at a single transition and not from an interference of atomic radiators working on different transitions. All our results are accounted for except the depth of modulation. We believe this is due to jitter in the temporal delay we mechanically introduce.

In addition to looking at a signal in the vicinity of zero delay where both excitation pulses overlap in time we also investigated the regime where the excitation pulses were well separated. Now we are in the photon echo regime. For a pulse separation of τ= 20 nsec the integrated signal intensity behavior is similiar to that shown in fig 2. The 1.9 psec modulation is still

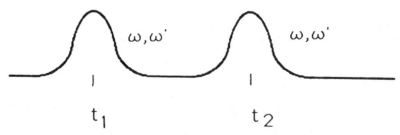

FIG. 4. Laser pulse profile exciting an atom moving along $x=x_i+v_it$. Stationary atoms experience a pulse separation of $t_2-t_1=\tau$. The frequencies ω and ω' are resonant with the Doppler shifted atomic frequencies at Ω and Ω'.

present but the depth of modulation has decreased.

For simplicity we analyse these modulated echoes in the parallel pulse configuration setting $\mathbf{k} \parallel \mathbf{k}'$ and $K=K'=0$. The lasers are tuned at

$$\omega = \Omega(1+v_s/c)$$
$$\omega' = \Omega'(1+v_s/c)$$

to select a particular velocity group which we characterize by v_s. As long as the difference frequency of the two lasers relates to the atomic frequency splitting acccording to

$$\Delta\omega - \Delta\Omega = \Delta\Omega v_s/c$$

it is possible to simultaneously excite a single atom on both transitions at Ω and Ω'. Having explicitly noted which atoms are selected we can consider the excitation pulses to be delta functions acting on a Doppler distribution shifted according to the laser detuning and having a width corresponding to the actual length (bandwidth) of the excitation pulses. The wave function of an atom which experiences two-frequency excitations at t_1 and t_2 develops in the following fashion:

$$|t_{1-}>=|g>$$

$$|t_{1+}>=|g>+|e>+|e'>$$

$$|t>t_{1+}>=|g>+\exp[-i\Omega(t-t_1)]|e>+\exp[-i\Omega'(t-t_1)]|e'>$$

$$|t_{2-}>=|g>+\exp[-i\Omega(t_2-t_1)]|e>+\exp[-i\Omega'(t_2-t_1)]|e'>$$

$$|t_{2+}>=|e>+|e'>+\{\exp[-i\Omega(t_2-t_1)]|g>+\exp[-i\Omega'(t_2-t_1)]|g>\}$$

$$|t>t_{2+}>=\exp[-i\Omega(t-t_2)]|e>+\exp[-i\Omega'(t-t_2)]|e'>$$
$$+\{\exp[-i\Omega(t_2-t_1)]+\exp[-i\Omega'(t_2-t_1)]\}|g>$$

we write $|t_{n-}>$ and $|t_{n+}>$ to mean the wave function immediately before and after, respectively, the atom experiences the excitation at t_n. For simplicity we are setting all coefficients equal to unity. The Ω component of the dipole moment for the atom being followed, labeled i, is obtained from the last equation above as:

$$P_i=\{\exp i\Omega[(t-t_2)-(t_2-t_1)]\}\cdot[1+\exp-i(\Omega'-\Omega)(t_2-t_1)] + c.c.$$

A similar expression exists for the Ω' component. The ith atom moves along the trajectory $x=x_i+v_i t$ where x_i and v_i are the components of position and velocity along the direction of the laser pulses. The dipole moment density is obtained by summing over all atoms and is proportional to

$$P(x,t)=\sum_i P_i(t)\delta(x-x_i-v_i t).$$

This summation will be converted to an integral over an effective line shape according to $\sum_i \to \int dv g(v)$ where $g(v)=(1/\sqrt{\pi})(1/v_p)\exp-(v-v_s)^2/v_p^2$, $v_p=(\tau_D/\tau_p)v_D$, $v_D = \sqrt{(2k_B T/m)}$, k_B is Boltzman's constant, T is the temperature and m is the atomic mass. Here $\tau_D=\lambda/2\pi v_D$ is the normal Doppler dephasing time and τ_p is the excitation pulse width and we have assumed $\tau_p \gg \tau_D$.[7] But first we

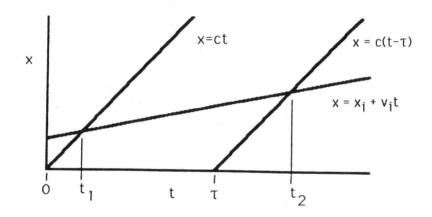

FIG. 5. Timing diagram showing trajectories of excitation pulses
traveling at c and resonant atom traveling at v_i.

calculate the contribution to the radiated field from those atoms moving at v_i.
This is obtained by integrating the dipole density evaluated at the retarded
time $t+x/c$:

$$E \sim \int dx\, P(x, t+x/c)$$
$$\sim \Sigma_i\, P_i[t+(x_i+v_it)/(c-v_i)].$$

The relationship between the various parameters t_1, t_2, τ, x_i, and v_i are
displayed in fig. 5 and used to evaluate P_i to order v_i/c. We obtain

$$E \sim \Sigma_i\, \{\exp[i\Omega(1+v_i/c)(t-2\tau)]\}\cdot\{1+\exp[-i(\Omega'-\Omega)(1+v_i/c)\tau]\}+ \text{c.c.}$$

Here the summation over i is replaced by an integration over the effective
lineshape to give us the radiated field and it's corresponding intensity. The
integrated intensity which we detect has the form

$$<I_{echo}> = \int dt\, I_{echo}$$
$$= 1+\{\exp-[\tau v_p(\Omega'-\Omega)/c]^2/8\}\cdot\cos[(\Omega-\Omega')(1+v_s/c)\tau].$$

The modulation amplitude degrades according to the Gaussian term. The origin of
this degradation can be seen from the recoil diagram of fig. 6. In the language
of the Billiard Ball Model the states at $|e>$ and $|e'>$, starting at time $t=0$,
have recoiled from the ground state at different velocities so that τ later they
are separated by $[h(\Omega-\Omega')/2\pi mc]\tau$. In restricting our analysis to the echo at Ω
the frequency components at both ω and ω' are required in forming $|e>$ and $|e'>$
in the first pulse and subsequently reforming $|g>$ in the second pulse. On the
other hand, only the frequency component at ω in the second pulse which
generates an excited state from $|g>$ is required. The other component at ω'
would produce echoes on transitions whose unshifted resonance is Ω'. Thus only
the $|e>$ state which recoils from $|g>$ at the second pulse need be considered. It
collides sequentially with two reconstituted ground states and forms an echo
with each one. Each echo has a duration of the order $\tau_p=(\lambda/2\pi)/v_p$. The time
between the two collisions is $(1+\Omega'/\Omega)\tau-2\tau=[(\Omega'-\Omega)/\Omega]\tau$. This quantity divided
by τ_p (aside from a numerical factor) is just the argument in the Gaussian

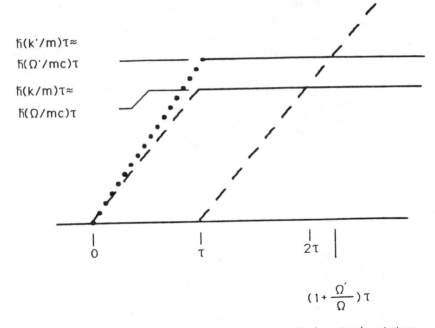

$\hbar(k'/m)\tau \approx$
$\hbar(\Omega'/mc)\tau$

$\hbar(k/m)\tau \approx$
$\hbar(\Omega/mc)\tau$

| | | |
0   τ 2τ

$$(1+\frac{\Omega'}{\Omega})\tau$$

FIG 6. Recoil diagram showing trajectories of the atomic states.
The solid line, dashed line, and dotted line show the
trajectories of the $|g\rangle$, $|e\rangle$, and $|e'\rangle$ state components
respectively. Time increases to the right and recoil
position is given in the vertical dimension.

degradation factor. This term which we rewrite as
$\exp-(1/2)\{(1/2)[(\Omega'-\Omega)/\Omega](\tau/\tau_p)]\}^2$ is of no consequence in the regime in which
we work since $2(\sqrt{2})\tau_p\Omega/(\Omega-\Omega')$ is of order 30 μsec.

We noted earlier that the depth of modulation we found in the echo
experiment was not as great as we had observed in the four wave mixing
experiments where the excitation pulses overlapped. In subsequent experiments
performed with parallel excitation pulses delayed by 30 nsec we found the
modulation was again weaker. As the tables on which we work are not
acoustically isolated we checked to determine if vibrations posed a problem by
repeating our initial four wave mixing experiments, again with small delay but
with extended optical paths. By appropriately moving optical components we
extended the optical paths so that the effective length of the beam travel was
comparable to that in the echo experiments. The result was that the depth of
modulation was reduced to that observed in the associated echo experiment. We
believe that with care many orders of improvement are possible.

Our final major comment centers on the fact that our analysis assumes
narrow band excitation and the modulation frequency vs. pulse separation is
therefore not Doppler free. The error however is of order $(\Omega-\Omega')v_s/c$ and not
$\Omega v_s/c$. Suppose then that two broadband lasers at Ω and Ω' are used with a

bandwidth that covers the Doppler width. The pulses can be decomposed into modes of width $\Delta\omega=\pi/\tau$ each of which excites that bandwidth.[1] Since both lasers are broadband it is certain that all atoms in each bandwidth $\Delta\omega$ throughout the line centered at Ω and their Doppler shifted counterparts at Ω' will be coherently excited in the manner of the calculation just presented above. As all the modes are incoherently related the radiated echo intensities from all the modes add. Each mode is characterized by a particular value of v_s and these are distributed according to $g(v_s)=(1/\sqrt{\pi})(1/v_D)\exp(-v_s^2/v_D^2)$. Averaging over v_s we obtain

$$\langle\langle I_{echo}\rangle\rangle = \int dv_s\ g(v_s) \langle I_{echo}\rangle$$

$$= 1 + \{\exp-[\tau v_p(\Omega'-\Omega)/c]^2/8\}\cdot\{\exp-[\tau v_D(\Omega'-\Omega)/c]^2/4\}$$
$$\cdot\cos[(\Omega-\Omega')\tau]$$

This result gives the modulation in a Doppler free form. The requirement is that the line be excited uniformly. The averaging over the velocity groups has introduced an additional Gaussian degradation term which can be rewritten in the form $\exp-\{(1/2)[(\Omega'-\Omega)/\Omega](\tau/\tau_D)]\}^2$ is of no consequence in the regime in which we work since $2\tau_D\Omega/(\Omega-\Omega')$ is of order 400 nsec. For a pulse separation of $\tau=400$ nsec the echo which appears $2\tau=800$ nsec or 50 fluorescence lifetimes later would suffer a diminished modulation amplitude of only e^{-1}. It seems then that this technique can effectively utilize the abiltiy of the photon echo technique to observe excited atoms which live many times their natural lifetime.

CONCLUSION

In this paper PEMS is extended into a new regime to study transitions between states that are widely separated in frequency compared to the bandwidth of the exciting laser pulses. In the Doppler free limit, the accuracy to which a splitting can be measured using this technique depends on how many beats can be observed. The spectroscopic advantage here arises because the beats are long lived; they last much longer than the coherence time of the lasers. If we block the entrance window of the sodium cell with a white card the two overlapping excitation pulses form an interference pattern on this card only for pulse delays of the order of a hundred picoseconds or less. Echo beats on the other hand last well into the nanosecond regime. This technique does not require stable frequency lasers.

ACKNOWLEDGMENTS

We thank E. Giacobino for valuable discussions. This work was supported by the U. S. Office of Naval Research and by the Joint Services Electronics Program (U.S. Army, U.S. Navy, U.S. Air Force) under contract No. DAAG29-85-K-0049.

REFERENCES

[1]R. Beach and S.R. Hartmann, Phys. Rev. Lett. 53, 663 (1984).
[2]S. Asaka, H. Nakatsuka, M. Fujiwara, and M. Matsuoka, Phys. Rev. A 29, 2286 (1984).
[3]N. Morita and T. Yajima, Phys Rev. A 30, 2525 (1984).
[4]H. Nakatsuka, M. Tomita, M. Fujiwara, and S. Asaka, Opt, Comm. 52, 150 (1984).
[5]R. Beach, D. DeBeer, and S.R. Hartmann, Phys. Rev. A 32, 3467, (1985).
[6]R. Beach, B. Brody, and S.R. Hartmann, J. Opt. Soc. Am. B 1, 189 (1984).
[7]R. Beach, B. Brody, and S.R. Hartmann, Phys. Rev. A 27, 2537 (1983).

MODULATED PUMPING IN CS WITH PICOSECOND PULSE TRAINS

H. Lehmitz, W. Kattau, and H. Harde

Universität der Bundeswehr Hamburg, Fachbereich Elektrotechnik

Holstenhofweg 85, D-2000 Hamburg 70, FRG

Already in the early days of optical pumping spectroscopy it has been shown that modulated optical light can create atomic coherence between close lying atomic substates and therefore can be used to measure the energy splitting between these states [1]. In this context optical pumping with a pulse train has often been mentioned but only scarcely applied for quantitative measurements.

Recently we could demonstrate the feasibility of high-resolution coherence spectroscopy by means of periodic excitation with ultrashort light pulses to measure Zeeman and hyperfine level splittings in the Na ground state [2,3]. In this contribution we report for the first time on the applicability of this technique for measurement of frequency splittings in the 10 GHz-range with an accuracy comparable with rf-experiments. For the well-known hyperfine splitting of the Cs ground state we observe resonance widths down to 30 Hz and small frequency shifts due to buffer gases can sensitively be detected.

The applied method of periodic excitation takes advantage of the fact that a train of short light pulses can create an enhanced coherent superposition of nearly degenerate atomic states or substates. Each pulse of an infinite train resonantly interacts with the atoms and induces an atomic coherence which is freely precessing with the splitting frequency of the coherently superposed states. Constructive interference of the various contributions induced by different pulses gives rise to very sharp resonances in the resulting amplitude of the coherence, whenever the pulse rate or a higher harmonic coincides with the corresponding atomic splitting frequency. Due to the excitation process and the dominating coherence phenomenon we have termed this method as optical pulse train interference spectroscopy (OPTIS).

For investigation of long lived states as ground or metastable states and sufficiently long observation times higher order resonances in atomic coherence can successfully be used to measure GHz splitting frequencies with practicable pulse rates in the MHz-range. Necessary condition for the coherent preparation of the atoms is that the width of the optical pulses is small compared to the reciprocal of the splitting. For measurement of GHz-frequency splittings therefore ps pulses are required.

The actual coherent superposition of adjacent substates which shall be used for a measurement depends on the polarization and propagation of the incident light pulses with respect to an applied magnetic field. While in a transverse field and excitation with circularly polarized light, coherence can be created between Zeeman-states, which differ in their magnetic quantum number m_F by unity, in the longitudinal case with the field parallel to the propagation direction of light only coherent superpositions of levels with identical m_F-values will be generated. For a high precision measurement of the Cs hyperfine ground state splitting especially the coherence between Zeeman-levels $(F, m_F | F', m_F') = (4,0 | 3,0)$ called 0-0 coherence, is of interest, since the splitting between these states depends only in second order on the magnetic field and therefore is less sensitive to stray fields.

We used two different experimental arrangements for periodic excitation and detection of coherence. They are shown in Fig. 1. Cs vapor in a gas cell is resonantly excited on the D_2 line by a train of ultrashort light pulses of circular polarization. To reduce transit-time broadening, additional buffer gas is contained in the cell and the light beam from a laser is expanded to a cross section of about 1.5 cm². The resulting atomic coherence amplitude which is due to the periodic excitation of atoms can be measured by different means. The experimental set-up in Fig. 1a takes advantage of the fact that atomic coherence gives rise to an oscillating optical anisotropy in the sample, as well known from optical pumping experiments [4]. The anisotropy can be detected as a function of the pulse repetition rate by placing the sample between crossed polarizers and monitoring the transmitted intensity of probe pulses which are derived from a low intensity fraction of the laser output [5]. In this arrangement a synchronously pumped mode locked dye laser has been used which generates pulses of 10 ps duration. The pulse rate (83 MHz) is accurately controlled and can slightly be shifted to tune through an atomic resonance.

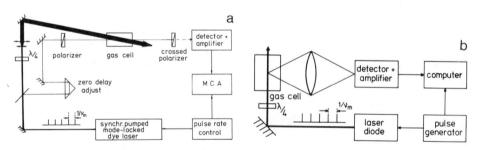

Fig. 1: Experimental set-up for OPTIS with polarization sensitive
detection (a) and fluorescence detection (b)

A second detection scheme for observation of periodic excitation resonances is shown in Fig. 1b. The fluorescence radiation of the atoms which are optically excited by the pulse train, is monitored as a function of the pulse rate. At resonant coherent excitation of the atomic sample the incoming light is less absorbed and reduced fluorescence radiation is observed. In this case a cheap Ga(Al)As injection laser has been applied which is directly modulated and supplied with electrical pulses from a comb generator. The laser generates optical pulses of about 40 ps duration at a pulse rate of 1 GHz.

Measurements of the hyperfine ground state splitting taken with these arrangements are shown in Fig. 2. In a small magnetic field subjected to the atoms, the periodic excitation resonance resulting from the 0-0 coherence, can easily be separated from other resonances and then exclusively be measured as a function of the pulse rate. In the case of polarization selective detection (see Fig. 2a) the signal appears on a constant background which is due to a static polarization originating from population differences between sublevels. The coherence is driven and can be tuned through resonance by the 110th harmonic of the pulse rate. A very narrow resonance with a line width of only 30 Hz related to the hyperfine splitting of 9.1926 GHz, can be observed and seems mainly to be determined by coherence destroying collisions between Cs atoms among themselves and with the buffer gas. The solid line in Fig. 2 is a fit to the measured curve, based on a theoretical signal form.

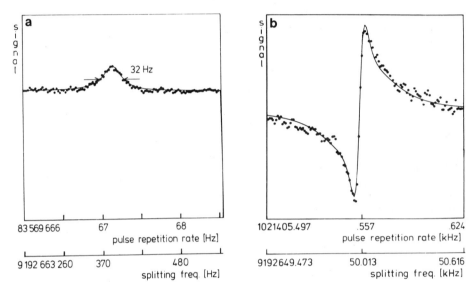

Fig. 2: Measurement of the Cs hyperfine splitting. (a) Polarization selective detection with Ne buffer gas at 70.3 hPa. (b) Fluorescence detection with Ne at 35.2 hPa.

For measurement with fluorescence sensitive detection the magnetic field is slightly modulated, effecting a small periodic shift of the 0-0 resonance. Then with lock-in detection technique the derivative of a resonance signal is observed and a large background in fluorescence radiation and scattered light is suppressed. Fig. 2b shows a measurment of this type. The atomic splitting is measured by the 9th harmonic of the injection laser pulse rate with a width of less than 50 Hz.

These experiments demonstrate the ultrahigh frequency resolution possible with OPTIS and therefore allow to sensitively detect small pressure shifts in the hyperfine frequency caused by buffer gases in the gas cell. With the polarization sensitive detection we have measured the frequency shift at room temperature as a function of pressure for the rare gases He, Ne and Ar. Our preliminary results for the pressure shifts in the Cs hyperfine ground state splitting are shown in Fig. 3. Our measurements give considerably increased accuracy for the frequency shift than attained in earlier experiments [6]. From extrapolation of the straight lines to zero buffer gas pressure a hyperfine splitting of 9 192 631 784 ± 31 Hz is found which is in excellent agreement with the atomic clock frequency. By far the largest uncertainty in our results is determined by measurements of the pressure.

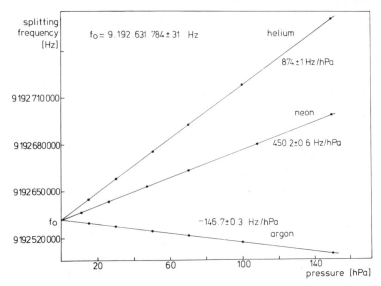

Fig. 3: Pressure shift in the Cs hyperfine splitting for the
buffer gases He, Ne and Ar.

The investigations presented in this contribution demonstrate, as we believe, the potentiality of OPTIS for measurement of large frequency splittings with extremely high accuracy. Higher harmonics of the pulse rate
can successfully be used to measure these splittings with practicable
pulse frequencies. Synchronously pumped dye lasers and semiconductor lasers are appropriate to generate picosecond pulse trains with very low
jitter and high stability in the pulse rate. With polarization and fluorescence sensitive detection of the periodic excitation resonances line
widths comparable with the best rf-experiments are obtained. Thus we expect that the applied method of modulated pumping with ultrashort light
pulses should prove usefully in different branches of high resolution laser spectroscopy for measurement of large energy splittings.

The work was supported by the Deutsche Forschungsgemeinschaft.

References

[1] W.E. Bell, A.L. Bloom, Phys. Rev. Lett. 6, 280 (1961)
[2] J. Mlynek, W. Lange, H. Harde, H. Burggraf, Phys. Rev. A24, 1099(1981)
[3] H. Harde, H. Burggraf, Optics Comm. 40, 441 (1982)
[4] J. Manuel, C. Cohen-Tannoudji, C.R.Ac.Sci. 257, 413 (1963)
[5] For measurement of atomic coherence see also: "Polarization Selective
    Detection of Hyperfine Quantum-Beats in Cs", by H. Lehmitz and H. Har-
    de, this symposium
[6] C.W. Beer, R.A. Bernheim, Phys. Rev. A 13, 1052 (1976)

POLARIZATION SELECTIVE DETECTION OF HYPERFINE QUANTUM BEATS IN CS

H. Lehmitz and H. Harde

Universität der Bundeswehr Hamburg, Fachbereich Elektrotechnik

Holstenhofweg 85, D-2000 Hamburg 70, FRG

Recently we could demonstrate that time-resolved coherence spectroscopy with polarization selective detection is a powerful tool for Doppler-free measurements of atomic level splittings in the GHz-range [1]. In this contribution we report on the observation of fast coherent transients in Cs. Quantum-beats in forward scattering have been detected, which result from the hyperfine splitting of the ground and first excited 6p fine structure states. These investigations have been performed to test the method for measurements of splitting frequencies in the 10 GHz range and to verify a theroretical model which has been derived for description of atomic coherence in alkaline atoms.

The method takes advantage of an excite and probe technique with polarization selective detection. A circularly polarized pump pulse of ps duration generates a coherent superposition of atomic substates. The coherence induces an optical anisotropy in the atomic sample, which oscillates exactly with the splitting frequency of the respective substates. The optical anisotropy can be detected via changes of the optical polarization of linearly polarized probe pulses, which are delayed with respect to the pump pulses. Variation of the delay-time allows to sample the time evolution of the optical anisotropy and by this the atomic coherence.

The light pulses are assumed to satisfy the impact excitation condition. Then the hyperfine interaction can be neglected during the time of excitation. Therefore within this time the atoms can reasonably be described by a system of basic states with quantum numbers J, I, m_J, and m_I. The corresponding level scheme for the $6s^2S_{1/2}$ ground state and the excited $6p^2P_{1/2}$ state for cesium with nuclear spin $I={}^7/_2$ is shown in Fig. 1. The sublevels of a common fine structure state are degenerate in energy.

In the level scheme of cesium, shown in Fig. 1, pump pulses with σ^+-polarization induce optical transitions with $\Delta m_J=1$ and by this create population differences between sublevels with quantum numbers $m_J=-{}^1/_2$ and $m_J={}^1/_2$ in the ground and excited state.

After excitation the hyperfine interaction cannot further be neglected. It can be considered as a perturbation and the time evolution of the atomic system is calculated by the Liouville-equation. This hyperfine interaction gives rise to oscillating population differences between sublevels with identical quantum values $m_F=m_J+m_I$. These oscillations appear on a

101

background resulting from static population differences. The oscillation frequency is exactly given by the hyperfine splitting frequency.

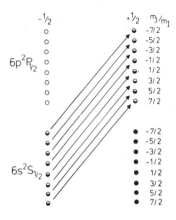

Fig. 1: Level scheme of cesium

The linearly polarized probe pulse can be described as a superposition of σ^+- and σ^--light. Therefore the pulses, acting with the atoms, measure the actual population differences between left ($m_J=-1/2$) and right ($m_J=1/2$) sublevels in ground and excited state. The optical polarization of the probe pulse is changed due to dichroism and can sensitively be detected behind a crossed polarizer. The calculated signal for observation on the D_1-line gives a transmitted intensity for the probe pulses behind the crossed polarizer:

$$I(t) \sim \left\{ w^o/4 \left[e^{-\gamma_e x t}(11+21\cos\omega_{HF}^e t) + e^{-\gamma_g x t}(11+21\cos\omega_{HF}^g t) \right] \right\}^2 \ .$$

The relaxations for the ground and excited state are included by γ_g and γ_e. w^o is the population difference generated by the pump pulse and ω_{HF}^g and ω_{HF}^e are the hyperfine splitting frequencies of the respective states.

Fig. 2 shows the applied experimental set-up. A synchronously pumped mode-locked dye laser generates light pulses of 20 ps duration. The pulse rate is reduced to 40 kHz by a cavity-dumper in order to realize a single pulse experiment. The dye laser is resonantly tuned to the D_1-line (894.4 nm) or the D_2-line (852.1 nm) of cesium. The laser light is split into a circularly polarized pump pulse and a weaker linearly polarized probe pulse, which can be delayed up to 10 ns. Both beams are focussed and directed into a Cs cell, where they overlap within a few cm. At room temperature the atomic density of cesium is 10^{10} cm^{-3}. Behind a crossed polarizer the transmitted light is detected and averaged over many pulses by a photomultiplier and is stored by a computer as a function of the delay-time.

Fig. 3a shows a measurement of the beat structure in the D_1-line of cesium. The two hyperfine splitting frequencies of the ground and excited state clearly show up. The fast oscillation corresponds to the splitting of the ground state, while the envelope with the smaller frequency results from the splitting in the excited state. The difference between the measured and calculated oscillating depths can be explained by the finite pulse duration.

The Fourier transformed signal in Fig. 3b clearly shows the hyperfine splitting frequencies of the excited state with 1.2 GHz and the ground state with 9.2 GHz. Since the beat structure can be described as an amplitude modulation of the fast oscillation with the excited state split-

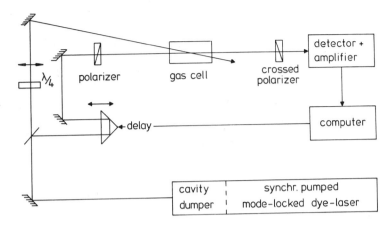

Fig. 2: Experimental set-up for quantum-beats in forward scattering

ting side bands can be observed. While the time resolution is only limited by the pulse length, the accuracy for measurement of frequencies by this technique in principle is limited by the finite delay-time and leads to line-width of typically 100 MHz. This is already distinctly smaller than the Doppler-width for an optical transition and could still further be reduced by increasing the delay-time.

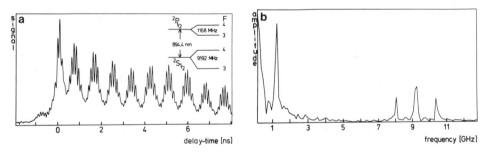

Fig. 3: a) Measurement of the hyperfine beat structure in the Cs D_1-line. b) Fourier spectrum

The measurement on the D_2-line in Fig. 4 shows fast beats at 9.2 GHz and only one period of slow oscillation at 200 MHz, which is due to the hyperfine splitting frequencies of the $6p^2P_{3/2}$ -state.

Both measurements show a fast rising beat structure at negative delay-times. The origin of this signal results from optical coherence generated by the linearly polarized probe pulse and corresponds to a first-order Free-Induction-Decay. An interpretation of this signal is found in ref. [2].

The presented results demonstrate that polarization selective, time-resolved coherence spectroscopy can successfully be applied for Doppler-free measurements of large frequency splittings in the 10 GHz range. Fast coherent transients with excellent signal-to-noise ratio have been detected. The investigations show a very satisfactory agreement with the theoreti-

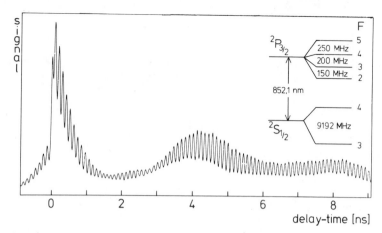

Fig. 4: Measurement of the hyperfine beat structure in the
Cs D$_2$-line

cally calculated signal form and can be explained in terms of oscillating
population differences of atomic substates.

This work was supported by the Deutsche Forschungsgemeinschaft.

References

[1] H. Harde, H. Burggraf, J. Mlynek, W. Lange, Optics Lett. 6, 290 (1981)
[2] see Fritz Haber International Symposium on Methods of Laser Spectroscopy: "Measurement of First-Order Free-Induction-Decay in Cs" by
H. Lehmitz, H. Harde.

OBSERVATION OF 517 GHZ FINE STRUCTURE QUANTUM-BEATS IN NA

H. Burggraf, M. Kuckartz, and H. Harde

Universität der Bundeswehr Hamburg, Fachbereich Elektrotechnik

Holstenhofweg 85, D-2000 Hamburg 70, FRG

The feasibility of time-resolved coherence spectroscopy for measurement of atomic level splittings with frequencies up to the THz-range is demonstrated in this contribution. For the first time quantum-beats, which are due to the fine structure splitting of the first excited electronic state in Na, have been measured. With a beat frequency of 517 GHz this is the fastest quantum-beat signal observed so far.

The applied method is an extension of a previously described technique [1] of time-resolved polarization spectroscopy into the femtosecond range. It relies on the creation of a coherent superposition of adjacent states or substates by an optical pulse, which is short compared to the reciprocal of the frequency splitting of the respective states. Such an atomic coherence causes an optical anisotropy in the sample, oscillating exactly with the splitting frequency of the coherently excited states.

These oscillations can sensitively be detected via changes in the polarization of delayed probe pulses. Therefore, the coherently excited sample is placed between crossed polarizers and the transmitted intensity of the probe beam is monitored. By varying the delay time, one can sample the time evolution of the coherent transients.

For a more quantitative description and explanation of the fine structure beats we consider an excitation of Na atoms by ultrashort light pulses in a level system as shown in Fig. 1 for the ground and first excited state in Na. The width of the optical pulses may be so short that the impact excitation condition is satisfied. Then the fine structure during the excitation cannot be resolved. Within this time an atomic system can reasonably be represented by a fundamental scheme of eigenfunctions labeled by the quantum numbers L, S, m_l, and m_s.

In this description pump pulses with σ^+ polarization create population differences between sublevels with $m_l=1$ and $m_l=0$ in the 3p state. After a preparation the spin-orbit interaction has to be taken into account. The time evolution of the atomic system can be calculated by the Liouville equation. It is found that population oscillations, represented by the diagonal elements of the density matrix, show up between states with identical values of $m_j = m_l + m_s$. Their oscillation frequencies are exactly given by the fine structure splitting. In our case, due to the excitation only coherence between the substates $(m_l|m_s) = (1|-\frac{1}{2})$ and $(0|\frac{1}{2})$ is expected. Its time evolution is sampled by a linearly polarized probe pulse, which measures the population difference between $m_l = \pm 1$ sub-states and according to dichroism is changed in its polarization. Therefore, from simple considerations we find that behind a crossed polarizer a beat signal on a constant background is expected as a function of delay time.

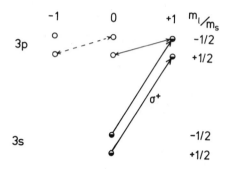

Fig. 1: Level scheme of Na for ground and first excited state in Paschen-Back notation.

The experimental set-up we used is shown in Fig. 2.

For excitation and detection of fine structure beats with a frequency of 517 GHz subpicosecond light pulses are necessary. For this purpose we used a synchronously pumped mode-locked dye laser with saturable absorber in the dye solution. The dye laser generates light pulses of about 400 fs duration at a pulse rate of 84 MHz. It is pumped by a frequency doubled actively mode-locked Nd:YAG laser and tuned to a wavelength of 589.3 nm for resonant excitation of the Na atoms into the 3p fine structure states.

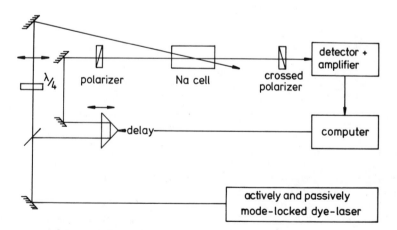

Fig. 2: Experimental set-up for measurement of Na fine structure quantum-beats in forward scattering.

The laser output is split into a strong pump beam with circular polarization and a weak probe beam with linear polarization. Both beams are focussed into a common interaction region where they overlap within a few cm. The pump pulses act on the Na atoms with an input pulse area $\Theta \approx \pi / 100$ rad and induce coherence in the vapor which is contained in a heated cell with a number density of approximately 10^{12} cm³.

The probe pulses can be delayed by a variable optical delay line and sample the time evolution of the resulting anisotropy within the interaction region. Since the time scale is transformed to an optical path length by the speed of light, time resolution is not restricted by an optoelectronic detection system but only by the pulse length. The probe beam intensity passing the crossed polarizer is detected by a photomultiplier, which integrates over many pulses. To suppress scattered light from the pump beam at small crossing angles, the probe beam is modulated and standard lock-in technique is applied. The output of the lock-in is stored by a computer as a function of the delay time.

A beat structure observed with this experimental arrangement is shown in Fig. 3. Within a delay time of 16 ps clearly resolved oscillations with a period of 1.9 ps are monitored. These beats correspond to the 517 GHz fine structure splitting in the 3p state and represent the fastest quantum-beat signal observed so far.

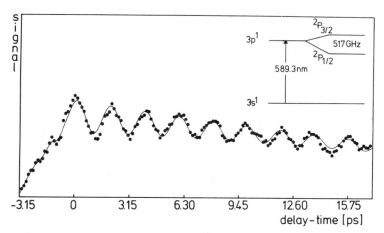

Fig. 3: Measurement of Na fine structure quantum-beats.

The signal appears on a larger background essentially resulting from natural birefringence of the entrance and exit windows of the cell. It has been subtracted in Fig. 3.

Up to now the quality of our measurements still suffers from the thermal drift of this birefringence as well as the amplitude and wavelength stability of the laser. Therefore, the measuring time for one run was restricted to a few minutes and by this the maximum delay time limited to 20 ps.

In principle, however, quantum-beats can be expected over considerably longer times. They decay with the homogenous transverse relaxation rate

and under proper conditions should be detectable over the full mean lifetime of the excited state.

Recently Rothenberg and Grischkowsky [2] observed polarization beats also
originating from the 3p fine structure splitting of Na. However, in their
experiments the beat phenomenon has to be explained as an interference
between the macroscopic polarizations simultaneously excited in the sodium
vapor on the D_1- and D_2-resonance line.

Since the macroscopic polarizations are determined by optical coherences
induced in the sample, their observation is restricted by Doppler dephasing, and the spectral resolution of the polarization beats is limited to
the Doppler linewidths. In comparison with that quantum-beats are nearly
unaffected by inhomogeneous broadening and therefore they are appropriate
for high resolution level splitting measurements.

In summary, the preliminary results presented in this contribution already demonstrate that time resolving polarization spectroscopy offers a
number of favourable and new features for direct observation of fast
evolving events on a femtosecond time scale and detection of oscillations
up to the THz-range. The described technique can be applied to free atoms,
liquids and solids to measure coherent transients in ground and excited
states. Since the observed beats result from an atomic interference effect, narrow structures which may be hidden by inhomogeneous broadening
mechanisms can still be resolved.

The theoretical description of the quantum-beat structure in terms of oscillating population differences between Zeeman substates of different
fine structure levels gives very satisfactory explanation for the appearance and form of the observed signal.

References

[1] H. Harde, H. Burggraf, J. Mlynek, W. Lange, Optics Lett.6, 290 (1981)
[2] J.E. Rothenberg, D. Grischkowsky, Optics Lett. 10, 22 (1985)

MEASUREMENT OF FIRST-ORDER FREE-INDUCTION-DECAY IN CS

H. Lehmitz and H. Harde

Universität der Bundeswehr Hamburg, Fachbereich Elektrotechnik

Holstenhofweg 85, D-2000 Hamburg 70, FRG

We present a new approach for observation of first-order free-induction-decay (FID), which can sensitively be detected by means of polarization selective devices. FID in Cs vapor is measured with picosecond time resolution, and for the first time beats due to the hyperfine splitting of the ground and first excited state with 9.2 GHz and 1.2 GHz can be observed on the fast decay curve.

The experimental set-up is quite similar to an arrangement as used for quantum-beat measurements in forward scattering [1], but with inverted time axis. A linearly polarized pump pulse induces an optical coherence yielding an FID, which is radiated in the direction of the exciting pulse. Then a second delayed pulse with circular polarization, propagating under small crossing angle to the first pulse, interacts with the coherently excited atoms and by this changes the optical polarization of the FID. Then the fast decaying signal can sensitively be detected behind a crossed polarizer, while the exciting light is strongly suppressed. Measurement of this signal with an integrating detector as a function of the pulse separation then allows to deduce the FID from the measured decay curve.

For a theoretical explanation of the signal we consider a simplified level scheme with four sublevels in the ground and four in the excited state (see Fig. 1). This corresponds to an atomic system with angular momentum $J = \frac{1}{2}$ and a nuclear spin $I = \frac{1}{2}$. A complete treatment of the Cs case would require to calculate with 32 sublevels in minimum. However, we can expect that the reduced level system already includes the essential coherence phenomena and in principle shows the same structure only with slightly modified amplitudes for the oscillations.

The atoms may resonantly be excited on the D_1 or D_2 line and the width of the optical pulses τ may be short enough to satisfy the impact excitation condition. Then the hyperfine and magnetic field interactions, spontaneous emission and collisions during the excitation can be neglected. Within this time atoms can well be represented by Paschen-Back states $|JI\ m_J m_I\rangle$, which are characterized by the magnetic quantum numbers m_J and m_I. All sublevels of a common fine structure state are degenerate in their term energy. Excitation with linearly polarized light pulses can be described as a simultaneous excitation of atoms with σ^+ and σ^- polarization with $\Delta m_J = \pm 1$ transtitions. These pulses change the population of the states involved and in particular induce optical coherences between states which are connected by arrows in Fig. 1.

If the atomic system is expressed by the density matrix ρ, the effect of an optical pulse on the atoms can quantitatively be calculated by the Liouville equation

$$\dot{\rho} = \frac{i}{\hbar} [\rho,H] + \text{relaxation terms} \qquad (1)$$

with $H = H_O+H'$. Here H_O is the basic Hamiltonian, which determines the eigenfunction system and H' is the electric dipole interaction term, causing a perturbation of the level system during the pulse length. After such a preparation of the atoms all optical coherences, which are expressed by the off-diagonal elements of the density matrix ρ_{18}, ρ_{27}, ρ_{36}, and ρ_{45} are freely oscillating in time with the same optical splitting frequency. However, for times larger than the reciprocal of the hyperfine splitting the hyperfine interaction cannot further be neglected and has to be included in the Liouville equation for the free evolution phase. In the chosen system of eigenstates this interaction generates a coherent evolution between sublevels with identical values of $m_F = m_J + m_I$ (these levels are connected by dashed lines in Fig. 1) and gives rise to an oscillating population exchange between these states with the hyperfine frequency ω_g of the ground state and ω_e of the excited state respectively. This process also influences the optical coherences, and the amplitude of ρ_{18} and ρ_{27} are modulated with $\omega_g/2$, while ρ_{36} and ρ_{45} are modulated with $\omega_e/2$.

The optical coherences exhibit a macroscopic polarization in the sample which is the origin of a new emitted light wave. Due to phase matching this wave is radiated in the direction of the initial wave with wave vector $\vec{k_1}$, and its intensity is modulated with ω_g and ω_e. In Cs vapor we can expect that the signal decays in a time mainly determined by the Doppler dephasing time T_2^*, while the homogeneous dipole dephasing time T_2 is much larger. Therefore, such transient is dominated by the first-order FID [2], which is described by a Gaussian $\exp[-(t/T_2^*)^2]$. Since the atoms emit σ^+ and σ^- light with the same relative phase and amplitude, the resulting electric field vector of the light wave is parallel to the polarization of the excitation pulse. Therefore, this pulse as well as the signal is blocked by a crossed polarizer and no light detected in forward direction.

If a second delayed pulse with circular polarization and wave vector $\vec{k_2}$ acts on the atoms at time t_2, it generates new optical coherences, e.g. ρ_{18} and ρ_{36} (with σ^+ light), while the free evolution of the previously induced coherences between these states is perturbed. For a quantitative description again the Liouville equation has to be solved for the excitation and free evolution phase, however, with initial conditions at time t_2, which result from the evolution after the first excitation.

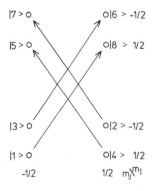

Fig. 1: Level scheme

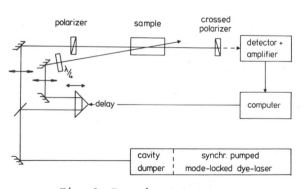

Fig. 2: Experimental set-up

Detailed analysis shows that after the second pulse three waves are emitted by the coherently excited atoms. One of them is radiated in $\vec{k}_2$-direction and only originates from the second excitation. Of particular interest is that wave, which propagates in the direction of the first pulse, however, now with modified polarization and therefore can be monitored behind a crossed polarizer. Its intensity can be calculated to be proportional to:

$$I(t) \sim \sin^2\theta_1(1-\cos\theta_2)^2 \left[\cos\frac{\omega_g}{2}(t-t'_2)\cdot\cos\frac{\omega_g}{2}(t_2-t'_1) + \right.$$
$$\left. + \cos\frac{\omega_e}{2}(t-t'_2)\cdot\cos\frac{\omega_e}{2}(t_2-t'_1)\right]^2 e^{-2(t-t'_1)/T_2} e^{-2((t-t'_1)/T_2^*)^2} \qquad (2)$$

where θ_i is the pulse area of the i-th pulse. For derivation of this equation we have assumed that square shaped pulses of width τ act on the atoms at times t_i till $t'_i = t_i+\tau$. Eq. (2) directly describes the first-order FID, which is modulated by the hyperfine beats from the ground and excited state. This decay already starts at time t'_1, but the signal appears not before the second pulse has interacted with the atoms.

With a very fast optical detector this still remaining signal can directly be observed in real-time, while the first exciting pulse, which is the origin of this signal, can be suppressed. The amplitude of the signal strongly depends on the pulse areas and on the delay-time between first and second pulse. It shows exactly the same temporal behaviour as the free evolution of the signal. Therefore, the FID can also be monitored by measuring the signal amplitude as a function of the delay-time.

Instead of direct observation or measurement of the amplitude, for detection of very fast transients it is more convenient to observe the time integrated signal as a function of the delay-time. At the output of a bandwidth limited detector we then expect a signal of the form

$$S(t'_2) = \int_{t'_2}^{\infty} I(t)dt \qquad (3)$$

which can be measured with high sensitivity and time resolution, only limited by the width of the optical pulses.

A third wave with identical amplitude and temporal behaviour, also described by Eq. (2), is reflected under a direction with wave vector $\vec{k}_e=2\vec{k}_2-\vec{k}_1$ and therefore in principle can also be used for observation of the FID.

The experimental set-up we used is shown in Fig. 2. A synchronously pumped mode-locked and cavity-dumped dye laser, which can be tuned to the D_1 or D_2 line of Cs, generates pulses of about 20 ps duration at a pulse rate of 4 MHz and peak powers of several hundred watt. They are split into linearly polarized pump pulses and stronger circularly polarized probe pulses, which pass an optical delay line. Both beams are focussed into a common interaction region where they act on the Cs vapor, which is contained in a cell at room temperature. The radiated wave propagating in pump pulse direction is detected by a photomultiplier, which measures the transmitted average intensity behind a crossed polarizer as a function of the delay time.

A measurement on the D_2 line, taken with this technique is shown in Fig. 3a (dotted curve). The solid line is the theoretically predicted decay curve as given by equation (2) and (3). It has been calculated with values for $T_2 = 60$ ns, $T_2^* = 1.4$ ns (at room temperature), $\omega_g = 9.2$ GHz and $\omega_e = 200$ MHz. On this time scale the different hyperfine splittings in the excited state (150 - 250 MHz) cannot be distinguished and have been approximated by a unique frequency ω_e.

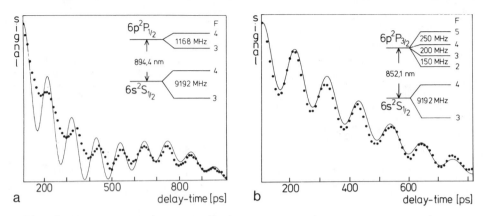

Fig. 3: Measurement (dotted line) and theoretical curve of FID in Cs:
a) D_1-line, b) D_2-line

Comparison of measured and calculated signal shows that a very satisfactory explanation of the observed signal form is achieved. The beat structure can be understood in terms of atomic coherence between substates yielding an amplitude modulation of optical coherence. The fast decay is mainly determined by Doppler dephasing, however, is also slightly influenced
by the excited state splitting frequency.

Measurement of the FID on the D_1 line (see Fig. 3b) shows that within the
observation time now also a first beat originating from the hyperfine
splitting of the excited state with $\omega_e = 1.2$ GHz appears in outlines.
The theoretical curve (solid line) in this case gives not such good conformity with the measured signal as for the D_2 line and especially differs
in the modulation depth. The principal structure, however, again can well
be represented and deviations explained due to the restricted number of
levels enclosed in the calculations.

In summary we can say that the investigations and results presented in
this contribution demonstrate the high sensitivity and time resolution of
the applied technique for measurement of first-order FID. Clearly resolved
hyperfine beats can be observed on the fast decay and the appearance of
the signal as well as the measured signal form can satisfactorily be explained in an eight-level system.

Part of the work is supported by the Deutsche Forschungsgemeinschaft.

References

[1] H. Harde, H. Burggraf, J. Mlynek, W. Lange, Optics Lett. 6, 290 (1981)
[2] R.G. DeVoe, R.G. Brewer, Phys. Rev. A 20, 2449 (1979)

DYNAMICS OF GEMINATE RECOMBINATION IN EXCITED STATE

PROTON TRANSFER REACTIONS

Dan Huppert and Ehud Pines

School of Chemistry, Raymond and Beverly Sackler Faculty of
Exact Sciences, Tel-Aviv University, 69 978 Tel-Aviv, Israel

INTRODUCTION

Ion pairs are produced when aqueous solutions of certain hydroxy aromatic compounds are irradiated by light. The excited hydrated molecules
dissociate in their first excited singlet state according to Scheme 1.(1-3).

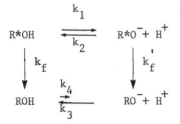

where $R*OH$ is the electronically excited molecule in its acidic form, $R*O^-$
is the excited anion. k_1 and k_2 are the dissociation and recombination
rate constants in the excited state and k_3 , k_4 are the recombination and
dissociation rate constants in the ground state. k_f and k_f' are the natural radiative rate constants of the excited molecule and the excited anion,
respectively. Since k_4 is very small compared to all other constants the
decay of the anion to its ground state is followed by a fast neutralization
reaction. Thermodynamic calculations, known as the "Forster Cycle" (3), can
be applied to determine the pK* values of these molecules. Another method
to obtain the pK* values is by steady state titration of the excited
state (3).

RESULTS AND DISCUSSION

As a generation source for the ion pairs we used the 8-hydroxy 1,3,6
trisulfonate pyrene molecule (HPTS) which has a pK* value of 0.5 (4) - a
7 unit change from its ground state value. The excited state proton dissociation reaction of HPTS occurs within 100 psec. The yield for free ions
generation ($R*O^-$ and H^+) was found to be 70% (5). pH-jump experiments (6)
confirmed that the generated protons are free to react with any basic species present in the solution. The very high yield of free ion generation
was explained by assuming that the RO^- is not reactive in the excited

state (5). Furthermore, it has been shown that most of the reductions in the free ions production is due to a ground state geminate recombination which follow the relaxation of the excited anion to its ground state (5).

However, steady state (7) as well as direct kinetic measurements have shown that recombination in the excited states do occur with an almost diffusion controlled efficiency.

The very high yield of separation of the ionic pairs and the very high reactivity of the excited state thus seem to contradict each other. The well known escape probability of Onsager (8) which measures the very long time yield of separation of the ion pair is given by Eq. 1.

$$\Omega_{(\infty)} = \exp(-R_D/r_0) \tag{1}$$

where R_D is the Debye radius, $R_D = |Z_1 Z_2| e^2 / \epsilon k_B T$ and r_0 is the initial separation between the ionic pair. Here Z_1 and Z_2 are the charge numbers of the reactants, e the electron charge, ϵ the static dielectric constant of the medium, k_B the Boltzmann constant and T the absolute temperature. From Eq. 1 and putting $R_D = 28$ A $\Omega_\infty = 70\%$, r_0 is 80 A. This distance is far greater than the distance where the proton is expected to thermalize after dissociation.

Examination of the fluorescence decay curve of R*OH shown in Fig. 1a reveals that the decay is not purely exponential. The decay rate is found to decrease with time and hence the fluorescence time profile exhibits a decay tail.

We propose a kinetic model which explains the nonexponential behavior as well as all the previous observations in terms of bidirectional excited state dissociation-geminate recombination mechanism (Eq. 2). However for each cycle of dissociation-recombination there is a finite chance for some ions to escape neutralization (Eq. 1). As time passes, more ions escape to the bulk and the dissociation of the R*OH becomes almost complete. The dynamic behaviour is described by time dependent equilibrium kinetics in the form of Eq. 2.

$$R*OH \; \underset{k_2(t)}{\overset{k_1}{\rightleftarrows}} \; R*O^- \;\ldots\; H^+ \tag{2}$$

where k_1 is the rate constant for generating the ion pair and $k_2(t)$ is a time dependent rate constant for the back geminate recombination process.

In order to describe $k_2(t)$, the rate of the geminate recombination, we used the expression given by Hong and Noolandi (9) for the long time behaviour of the ion pair recombination rate after it was generated by a δ function distribution at a mutual separation distance of r_0.

For the diffusion-controlled limit and with $R_D \gg a$, the contact radius, their expression takes the form:

$$k_2(t) = k_2^o t^{-3/2} \tag{3}$$

k_2^o is given by:

$$k_2^o = R_D \exp(-R_D/r_0)/2(\pi D_{AB})^{\frac{1}{2}} \tag{4}$$

D_{AB} is the relative diffusion coefficient between the ionic pair. The solution of the kinetic scheme (Eq. 2) is given by:

$$\frac{[\text{ROH}]_t}{[\text{ROH}]_0} = 1 - \frac{k_1 \int_0^t \exp(k_1 t' - 2k_2 t'^{-\frac{1}{2}})\,dt'}{\exp(k_1 t - 2k_2 t^{-\frac{1}{2}})} \tag{5}$$

Eq. 5 has the same characteristics as the observed decay, namely a fast and almost exponential decay which is followed by a long and nonexponential "tail". A convoluted numerical solution of Eq. 5 is compared with the experimental decay. The best fit is achieved with $k_1 = 1.5 \times 10^{10} \text{s}^{-1}$ and $k_2 = 3.5 \times 10^{-6} \text{s}^{1/2}$. With $R_D = 28 \text{Å}$ and $D_{AB} = 10^{-4} \text{cm}^2 \text{s}^{-1}$, r_0 is calculated according to Eq. 4 to be 35 Å.

The large magnitude of r_0 shows that it represents some average distance of a broad distribution function rather than a δ function value. Since in our model dissociation and recombination occur repeatedly, this conclusion seems natural.

Furthermore, according to the Hong and Noolandi model, the initial distribution function for protons in aqueous solution, characterized by a diffusion coefficient $D \simeq 10^{-4} \text{cm}^2 \text{sec}^{-1}$, is diffusing within 10 psec to form a distribution function with a half width of roughly $R_D/2$, or 14 Å for the given reaction parameters. This period of time is shorter by a factor of 2 than our experimental time resolution. It follows that the real physical situation as well as our time resolution allow us to detect only a diffused distribution of ion pairs distances rather than a sharp distribution.

Our model also predicts that in many cases the observed rate constants for ionic dissociation in solutions are considerably smaller than the molecular dissociation rates due to a very fast geminate recombination. Thus, the dissociation rate for the excited HPTS molecule is found, in our time resolution, to be at least 50% larger than the observed one.

In order to prove that geminate recombination is indeed responsible for the nonexponential luminescence decay of R*OH, we introduced proton scavengers to the aqueous solution. According to our model, it is expected that the contribution to the fluorescence intensity due to geminate recombination will be reduced. In the case of HPTS, the luminescence after few hundred picoseconds is mainly due to geminate recombination. The obvious scavenger for HPTS is its ground state anionic form RO^- with a proton recombination rate of $2 \times 10^{11} \text{M}^{-1} \text{s}^{-1}$.

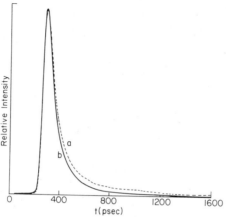

Fig. 1. Time resolved luminescence of HPTS;
a) in pure water, b) in aqueous solution
containing 0.1M Pyridine (see text).

With a ground state pK_a value of 7.7, 90% of the HPTS molecules are in the anionic form at pH = 8.7. At this pH and at concentrations greater than 10^{-3} M the pseudo first order proton capture rate exceeds $2 \times 10^8 \mathrm{s}^{-1}$, large enough to be detected as an intensity reduction of the luminescence tail. In addition to RO^-, we used acetate anion (pK_a = 4.7) and pyridine (pK_a = 5.2) as proton scavengers at neutral pH solutions. Figure 1b portrays the ROH emission as in Figure 1a but the aqueous solution contains 10^{-1} M of pyridine. The pyridine-proton recombination is diffusion controlled with a rate of $5 \times 10^{10} \mathrm{M}^{-1} \mathrm{s}^{-1}$. As expected, the tail intensity is reduced appreciably when pyridine is added. Another prediction of our model is that the isotopic D to H substitution tends to increase the geminate recombination rate by a factor of at least $(D_H^+/D_D^+)^{1/2}$ because of the smaller diffusion coefficient of the deuteron as compared to the proton. Thus the true molecular isotopic effect of dissociation will be smaller than the observed one. Indeed, our preliminary results show that $k_2^D(t)$ values for HPTS in D_2O are found to be larger than their corresponding $k_2^H(t)$ values in H_2O. Consequently a much smaller isotopic effect in k_1 (k_1^H/k_1^D) is found to be 2.5 ± 0.3 as compared to the apparent 3.3 ± 0.2.

In conclusion, the concept of quasi equilibrium between dissociation and geminate recombination is leading to a new description of the excited-states kinetics of molecules like HPTS.

The model explains the very high yield of proton separation in the excited states, in spite of the geminate recombination, in terms of repetitive cycles of dissociation and association. The overall recapture of the protons (the "cage effect") in the excited state is thus negligible as was predicted by Haar et al. (5), although the recombination rate is very high. The model accurately describes the nonexponential nature of the HPTS fluorescence decay and predicts that in many ionic dissociation reactions the apparent dissociation rates are considerably smaller than the molecular ones. The nonexponential decay is not unique to HPTS molecule but is observed by us in other HPTS-like molecules.

Further investigations will probably reveal that this behaviour is common to all dissociation-recombination reactions where recombination does not quench the dissociating molecule.

References

1. E. Pines and D. Huppert, J. Phys. Chem. 87:4471 (1983).
2. J.F. Ireland and P.A.H. Wyatt, Adv. Phys. Org. Chem. 12:131 (1976).
3. A. Weller, Progress in Reaction Kinetics 1:1 (1961).
4. K.K. Smith, K.J. Kaufmann, D. Huppert and M. Gutman, Chem. Phys. Lett. 64:522 (1979).
5. H.P. Haar, U.K.A. Klein and M. Hauser, Chem. Phys. Lett. 58:525 (1978).
6. M. Gutman and D. Huppert, J. Biochem. Biophys. Methods 1:9 (1979).
   M. Gutman, D. Huppert and E. Pines, J. Am. Chem. Soc. 103:3709 (1981).
7. A. Weller, Z. Physics Chem. 17:14 (1958).
8. L. Onsager, J. Chem. Phys. 2:599 (1934).
9. K.M. Hong and J. Noolandi, J. Chem. Phys. 68:5163 (1978).

OPTICAL PROPERTIES AND APPLICATIONS OF REVERSE SATURABLE ABSORBERS

Yehuda B. Band

Department of Chemistry
Ben-Gurion University
Beer-Sheva, Israel

1. INTRODUCTION

A reverse saturable absorber (RSA) is a material whose excited state absorption cross section at wavelength λ, $\sigma_{ex}(\lambda)$, is larger than the ground state absorption cross section, $\sigma_{gr}(\lambda)$, and which possesses four other necessary properties detailed below. Recently, applications of reverse saturable absorbers (RSA's) to a number of areas in optical pulse processing and laser physics have been described.[1-4] RSA's can be used as optical pulse shorteners and optical energy limiters for laser pulses whose temporal widths are small compared with the decay time of the first excited state of the reverse saturable absorber.[1] RSA's in conjunction with saturable absorbers can be used to mode-lock lasers whose gain media have high saturation energies (the saturable absorber cuts the leading edge of the pulse and the RSA cuts the trailing edge of the pulse) and stabilize high gain mode-locked CW lasers against the onset of relaxation oscillations.[2] A combination of RSA and saturable absorber can also be used as an extracavity optical pulse compressor to eliminate the wings of short optical pulses which may have been produced by the unwanted process of superfluorescence.[5] Such a pulse compressor can be inserted into a laser amplifier chain. RSA's can also be used as power limiters and pulse smoothers for pulses whose rate of change of intensity is slow compared with the decay time of the first excited state of the RSA.[1] There are many examples of molecules and crystals which behave as RSA's at specific wavelengths (e.g. the molecular dye rhodamine 6G behaves as a RSA at 435 nm wavelength). However, a suitable RSA may not be readily available, at least not readily known, for a given wavelength of interest. This is due in part to the paucity of information available regarding excited state absorption cross sections as compared with ground state absorption cross sections. Thus, choosing appropriate candidates for RSA's at a given wavelength can prove difficult.

Here we review the properties of RSA's,[1-4] analyze the propagation of optical pulses through RSA's, and describe how to design RSA's for a desired wavelength.[6] In an accompanying paper we shall describe some of the applications of RSA's mentioned above.[7]

2. Criteria for Successful RSA's

The criteria for molecules(crystals, solids or liquids) to be successful reverse saturable absorbers are: (1) $\sigma_{ex}(\lambda)$ should be larger than $\sigma_{gr}(\lambda)$ at the desired wavelength. (2) $\sigma_{gr}(\lambda)$ should be sufficiently large to insure that the incident pulse saturates the ground state transition (at least somewhat). Otherwise excited state absorption can not occur. The pulse may be focused into the RSA sample so this is not too severe a criteria. (3) Neither the first nor the second excited states should decay to other levels thereby trapping the excitation since this would diminish the presence of molecules which can exhibit excited state absorption. (4) the emission cross section $\sigma_{em}(\lambda)$ should be small at the desired wavelength. Recall that in molecules and in crystals the emission is shifted to the red relative to the absorption so it is possible to meet this criteria. (5) the decay time, τ, of the first excited state back down to the ground state should be large compared to the pulse width, τ_p, for pulse smoothing or power limiting applications, and should be small compared to the pulse width for pulse shortening or energy limiting applications. This criterion will be explained more fully below.

3. Analysis of Propagation of a Pulse through a RSA

Consider a plane wave pulse with time dependence of the incident intensity $I_p(t)$ and with central wavelength λ. Let us describe the dynamics of the propagation of the pulse through a reverse saturable absorbing medium of width L. The rate equation for the excited state population density of the RSA at position x and time t, $N(x, t)$, and the wave equation for the intensity of the pulse of light within the RSA at position x at time t, $I(x, t)$ take the form

$$\partial N_r(x, t)/\partial t = \{N_r^0 - N_r\} \sigma_r - N_r \sigma_r^{em}\} I/h\nu - \tau_r^{-1} N_r. \tag{1}$$

$$\partial I(x, t)/\partial t + c\, \partial I(x, t)/\partial x = -c\{N_r^0 - N_r)\sigma_r - N_r \sigma_r^{em} + \sigma_r^* N_r\} I \tag{2}$$

Here, N_r^0 is the concentration of RSA molecules, σ_r is the RSA ground state absorption cross sections at wavelength λ, τ_r is the excited lifetime of the RSA, σ_r^{em} is emission cross sections at wavelength λ, σ_r^* is the excited state absorption cross section at wavelength λ, $h\nu$ is the photon energy, and c is the speed of light in the medium. In Eq.(1) we assumed that the doubly excited state of the molecule decays back down to the excited state on a time scale fast compared to the width of the pulse and therefore no term involving σ_r^* appears. All the molecules are in the ground state initially before pulse, so $N_r(x, 0) = 0$. Also the initial intensity inside the RSA is zero, $I(x, 0) = 0$ and the boundary condition is $(0, t) = I_p(t)$. In what follows we shall assume that the emission cross section σ_r^{em} is is negligible according to criterion (4) above.

There are three dimensionless parameters which aid us in analyzing the Eqs. (1-2): (a) τ_p/τ_r, (b) σ_r^*/σ_r, (c) I_p/I_{sat}, where the saturation energy is defined as $I_{sat} = h\nu/\tau_r \sigma_r$ (or E_p/E_{sat} where E_p is the pulse energy and $E_{sat} = \tau_r I_{sat}$ is the saturation energy and $E_p = \int dt\, I_p(t)$). When $\tau_p/\tau_r \gg 1$ then the steady state approximation can be made in Eqs. (1-2). Eq. (1) is solved to yield

$$N_r(x) = \sigma_r N_r^0 \, I(x)/\{ I_{sat} [1 + I(x)/I_{sat}]\}. \tag{3}$$

Substitution of this result into Eq.(2) yields the differential equation

$$dI/dx = \sigma_r N_r^0 I(x)\{ 1-(1-\sigma_r^*/\sigma_r)I/(I(x) + I_{sat})\}. \tag{4}$$

In the limit $I_p \ll I_{sat}$ Eq. (4) gives $I(x) = I_p \exp(-\sigma_r N_r^0 x)$ whereas if $I_p \gg I_{sat}$, $I(x) = I_p \exp(-\sigma_r^* N_r^0 x)$. Thus the output $I(x = L)$ varies as a function of I_p. Since $\sigma_r^* > \sigma_r$ the relative output $I(L)/I_p$ decreases with increasing I_p and the RSA behaves as a power limiter. Moreover, if there are fluctuations in I_p the RSA will behave as a pulse smoother, transmitting relatively more (less) output when I_p is small (large).

When the pulse width is small, $\tau_p/\tau_r < 1$, and the medium is not optically thin, no analytical solutions to Eqs.(1-2) are known. Numerical solution of the equations show[1] that for $\sigma_r^*/\sigma_r > 1$ the medium behaves as an energy limiter and a pulse shortener if the pulse energy is sufficient to begin saturation, $E_p/E_{sat} > 1$. The leading edge of the pulse begins to saturate the medium, and upon saturation the transmission of the remaining part of the pulse decreases. The larger the pulse energy, the greater the reduction of the trailing edge of the pulse. Since the trailing edge of the pulse is eaten away, the pulse is shortened. However, to shorten the pulse from both the leading and trailing edges, the combination of a saturable absorber and a RSA is required.[3,7] Thus, the RSA behaves as an energy limiter and a pulse shortener for short pulses.

4. Engineering RSA's for Desired Wavelengths

Since RSA's for arbitrary wavelengths are not known, it is important to be able to design molecules which behave as a RSA at a desired wavelength.[6] Briefly, the idea is to use a dimer molecule formed from monomers whose ground state absorption peaks near the wavelength of interest. If synthesis of the dimer with fixed orientation of the monomers at desired distances from one another is possible, and the dimer is sufficiently stable to withstand dissociation upon absorption of two quanta of radiation, and a number of other criteria are satisfied, the dimer may behave as a RSA at the desired wavelength. The dimer ground state absorption is to a state in which the excitation is spread over both monomeric units and the dimer excited state absorption commences from this state to the doubly excited electronic state in which both monomeric units are excited. To zeroth order in the interactions between monomers, the excitation energy from ground to first excited and from first excited to doubly excited states are equal. Therefore, it is expected that if a sufficient density of first excited state is optically excited, excited state absorption should proceed. We now discuss details of the interactions which give rise to this phenomenon.

Consider a dimer molecule synthesized with the monomers tied together so that their relative geometry is fixed. The shift of the dimer ground state absorption spectrum relative to that of the monomer is due mainly to (a) the splitting of the first excited state of the dimer resulting from the dipole-dipole interaction between the monomers, and (b) the stabilization energy of the ground state via its interaction with the doubly excited state of the dimer resulting from the dipole-dipole interactions between the monomers. The sign of the shift of the dimer ground state absorption spectrum relative to the monomer is controlled by the positioning of the orientation of the monomers within the dimer. The first excited dimer states $\psi_{1g}\psi_{2e}$ and $\psi_{1e}\psi_{2g}$ undergo first-order interaction to give two states, $\psi_{\pm} = (\psi_{1g}\psi_{2e} \pm \psi_{1e}\psi_{2g})/\sqrt{2}$, whose energies are separated by a dipole-dipole interaction matrix element. Here g and e indicate the ground and excited monomer states and 1 and 2 indicate the monomer units. The energy separation between the two states $\psi_{\pm}$, $\Delta E_{\pm}$, depends on the relative orientation of the ground state transition dipole moment of the monomers as well as the distance between monomers. By controlling the distance r, and the angle between the dipole moments of the two monomers composing the dimer (i.e. the relative geometrical orientation of the monomers), the energy differences $E_{+} - E_{g}$ and $E_{-} - E_{g}$, where E_{g} is the energy of the ground state of the dimer $\psi_{g} = \psi_{1g}\psi_{2g}$, can be controlled. When the transition dipole moments of the monomers are parallel, the optical transition from the ground state ψ_{g} to the state ψ_{+} is allowed and and the transition moment is equal to $\sqrt{2} < \psi_{e}|\mu|\psi_{g}>$, whereas the transition to the state ψ_{-} is forbidden. It follows that the intensity of the allowed band is twice the intensity of the monomer and its position is controllable via adjusting the distance r and the angle between the dipole moments of the monomers within the dimer. Note that the allowed component of the state ψ_{+} carries oscillator strength to the higher lying state $\psi_{e} = \psi_{2e}$. Reversing the orientation of one of the monomers reverses the sign of the shifts ΔE_{+} and ΔE_{-} (i.e. ψ_{+} will be lower in energy than ψ_{-}), and the state which carries the oscillator strength will then be ψ_{-}. A geometry corresponding to this case is shown in Fig.1. Note that the arrows in Fig.1 depict the transition dipole moments of the monomers, not the permanent electric dipole moments. The transition dipole moments are not necessarily parallel to the permanent dipole moments, although for the special cases shown in Fig.1. this is the case.

The direction of the shift in the energy of the transition from the "first" excited state of the dimer ψ_{+} or ψ_{-} (whichever carries the oscillator strength) to the higher excited state ψ_{e} is difficult to predict. It depends upon whether the shift in the energy E_{e} of state ψ_{e} is larger or smaller than the shift in the energy of state ψ_{+}(or ψ_{-}). Nevertheless, since the dimer is designed so that the interaction between the monomers is not terribly strong, we expect that the position of the excited dimer transition will be near (although somewhat shifted from) the energy of the ground state absorption of the monomer.

Fig.2a schematically shows the absorption and emission spectra of the monomer. If the shift in the ground state absorption is to the red upon dimerization, as shown in Fig.2b for the dimer geometry depicted in Fig.1, successful RSA may be obtained at wavelength λ^{*}. In Fig.2b we draw the red shifted ground state dimer absorption (solid line) and, as a dashed curve, we show a dimer excited state absorption which is shifted to the blue of the monomer ground state absorption. At λ^{*}, $\sigma_{ex}(\lambda^{*}) > \sigma_{gr}(\lambda^{*})$. Thus, the first criterion is met at λ^{*}. In this case state ψ_{-} is lower in energy than state ψ_{+} and therefore at finite temperature ψ_{-} will be populated more prevalently than ψ_+. The excitation will tend to remain within the ψ_- manifold even better than in the blue shifted case discussed above. The nonradiative decay from ψ_e will result in efficient transfer of population to the bottom of the ψ_-, according to Kasha's rule which states that high lying singlet states decay to the first excited singlet, and because of the fast nonradiative decay within the vibrational manifold of the ψ_- state.

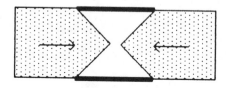

Fig.1. Geometry of orientation of dimers for a red shift in absorption spectrum relative to the monomer.

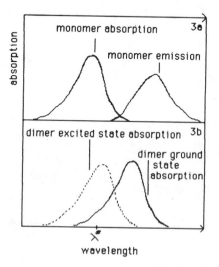

Fig.2.(a) Schematic drawing of absorption and emission spectrum of monomer. (b) Ground and excited state absorption spectrum dimer shown in Fig.1

References

1. D.J. Harter, M.L.Shand, and Y.B. Band, J.Appl.Phys. 56, 865 (1984).
2. D.J. Harter, Y.B. Band, and E.P. Ippen, IEEE J. Quant. Electron, QE-21. 1219 (1985).
3. Y.B. Band, D.J. Harter, and R. Bavli, Chem.Phys.Lett. "Optical Compressor Composed of Saturable and REverse Saturable Absorbers", to be published.
4. Y.B. Band, J. Chem. Phys.,"Reverse Saturable Absorption of Short Polarized Light Pulses", in press.
5. W. Koechner, Solid-State Laser Engineering, (Springer-Verlag,N.Y.,)
6. Y.B. Band and B. Scharf, "Engineering Reverse Saturable Absorbers for Desired Wavelengths", to published.
7. Y.B. Band and R. Bavli, this volume.

EXTRACAVITY PULSE COMPRESSOR AND INTRACAVITY MODE-LOCKING WITH SATURABLE AND REVERSE SATURABLE ABSORBERS

Yehuda B. Band and Raanan Bavli

Department of Chemistry
Ben-Gurion University
Beer-Sheva, Israel

1. INTRODUCTION

We show that an optical pulse compressor, a device to remove the leading and trailing edges of a subnanosecond pulse, can be made of a first stage composed of a saturable absorber (SA) whose function is to reduce the leading edge of the pulse, and a second stage composed of a reverse saturable absorber (RSA) which functions to reduce the trailing edge of the pulse. An RSA is a substance with a larger excited state absorption cross section, $\sigma^*(\lambda)$, than ground state cross section, $\sigma(\lambda)$, at a particular wavelength λ of interest, and which fulfills four other criteria.[1,2]   A numerical example of pulse compression of a ps pulse whose central wavelength is 266 nm using a Coumerine dye as the RSA and quaterphenyl as the SA is presented.   We then consider the effects of introducing SA and RSA within a laser cavity to mode-lock lasers. The introduction of SA and RSA into the laser cavity is shown to be particularly useful for passive mode-locking of lasers with gain media with high gain saturation energies wherein individual pulses cannot saturate the gain. Moreover, introduction of a RSA into the laser cavity stabilizes the laser against the onset of relaxation oscillations.

2. Pulse Compressors

Interest in producing short intense pulses of light is motivated by the hope of measuring fast processes occurring in chemistry, physics, and biology. Often, such pulses are produced in oscillator-amplifier chains wherein the oscillator prepares a short stable pulse which passes through a succession of amplifier stages.   However, there are a number of deleterious processes (e.g. superfluorescence) which degrade the quality of the short pulse produced by the oscillator upon traversal through the amplifier chain.[1]   Here we describe an optical compressor which can be used between amplifier stages or after an amplifier stage to remove the temporally wide noise surrounding a central pulse.   If an RSA is used in conjunction with a SA, both the leading and trailing edges of a pulse can be reduced in intensity relative to the central portion of the pulse.   Thus, the temporal width of a light pulse with central wavelength λ can be compressed by selective elimination of the tails of the pulse.   This compression reduces the noisy tails produced via superfluorescence. We demonstrate that for effective extra-cavity pulse compression and pulse shortening, the RSA and SA should be separated into distinct compartments, with the SA upstream relative to the RSA.

The intensity of the leading edge of the pulse is reduced using a SA with appropriate absorption cross section, concentration, optical thickness, etc. Very little absorption by the SA occurs after the leading edge of the pulse has passed since the SA transition is already saturated and few molecules are left in the ground state to absorb the trailing edge photons. The correct choice of the above mentioned parameters insures that only the desired fraction of the leading edge of the pulse is reduced in intensity. Using a RSA with the correct ground state absorption cross section, excited state cross section, concentration, optical thickness, etc., the intensity of the trailing edge of the pulse is selectively reduced by excited state absorption. The leading edge and central portions of the pulse are reduced somewhat by ground state absorption and consequently the excited state population density of th RSA begins to build up. By the time the desired portion of the trailing edge of the pulse has passed through the RSA, the excited state is sufficiently populated to insure that the absorption is dramatically increased due to excited state absorption.

Consider a medium containing both RSA and SA molecules and an incident plane wave pulse with central wavelength λ impinging on the medium from the left. As the incident intensity $I_0(t)$ at the front surface of the medium propagates through the medium, its intensity is reduced due to absorption by the RSA and SA. The concentration of RSA and SA are denoted by N_r^0 and N_s^0 respectively. The excited state population density of the RSA and SA at time t and position x, $N_r(x, t)$ and $N_s(x, t)$, will begin to increase as a result of absorption. The rate equations for $N_r(x, t)$ and $N_s(x, t)$ and the equation for the intensity of the pulse, $I(x, t)$, are given by

$$\partial_t N_r(x, t) = \{(N_r^0 - N_r)\sigma_r - N_r\sigma_r^{em}\} I/h\nu - \tau_r^{-1}N_r , \tag{1}$$

$$\partial_t N_s(x, t) = \{(N_s^0 - N_s)\sigma_s - N_s\sigma_s^{em}\} I/h\nu - \tau_s^{-1}N_s , \tag{2}$$

$$\partial_t I(x,t) + c\,\partial_x I(x,t) = -c\{N_r^0 - N_r)\sigma_r - N_r\sigma_r^{em} + \sigma_r^* N_r + (N_s^0 - N_s)\sigma_s - N_s\sigma_s^{em}\}I \tag{3}$$

Here σ_r and σ_s are the RSA and SA ground state absorption cross sections at wavelength λ, τ_r and τ_s are the excited state lifetimes of the RSA and SA, σ_r^{em} and σ_s^{em} are the RSA and SA emission cross sections at wavelength λ (assumed small in what follows), σ_r^* is the RSA excited state absorption cross section at wavelength λ, $h\nu$ is the photon energy, and c is the speed of light in the medium. Eq.(3) is presently written for the case when both SA and RSA are located in the same cell. For this case where the RSA and SA are mixed, the leading edge of the pulse populates the RSA as well as the SA. We separate the SA and RSA into separate compartments of length L_s and L_r, with the SA upstream relative to the RSA. This separation is desirable so that the leading edge of the pulse populates the SA excited state and not the RSA excited state, for then the absorption decreases with increasing time and the central portion of the pulse experiences minimum absorption. Only after saturation of the SA and passage of the central portion of the pulse through the RSA will the RSA excited state be populated and cause increased absorption. The separation of SA and RSA can be represented in Eqs.(1-3) by including functions multiplying the density of the SA and RSA which are unity for values of x where the material is present and zero otherwise. It is possible to optimize the choice for parameters $N_r^0, \sigma_r, \sigma_r^*, \tau_r, N_s^0, \sigma_s, \tau_s, L_r$, and L_s, for effective pulse compression.[4]

As a numerical example, we choose an initial pulse of the form $I_0(t) = E_p/2\tau_p \operatorname{sech}^2(t/\tau_p) + E_n/2\tau_n \operatorname{sech}^2(t/\tau_n)$, with central wavelength $\lambda=266$ nm and with energies per unit area of $E_p=0.0245$ J/cm^2 and $E_n=0.0735$ J/cm^2 and widths $\tau_p = 0.6$ ps and $\tau_n = 6.0$ ps. We choose Coumerine dye

490 (C-151) as the RSA [σ_r and σ_r^* are taken from Ref. 5] with $L_s = 0.098$ cm, quaterphenyl as the SA [parameters obtained from Ref.6] with $L_r = 0.119$ cm and ethanol as the solvent for both. Fig. 1 shows the initial pulse shape and the pulse shape after emerging from the SA-RSA pulse compressor. Both pulses are normalized relative to their own maximum intensity, i.e. for the initial pulse we plot $I_0(t)/I_0(0)$ and for the output pulse, $I_{out}(t)/I_{out.max}$. The ratio of the total output energy to total input energy is 43% (most of the reduction is in the noise which accounts for 75% of the input pulse). The intensity reduction at the maximum intensity is about 34% (i.e. $I_{out\ max}/I_0(0)$ is 0.66), whereas the reduction of the leading and trailing edges of the pulse are about 86% and 60% respectively. The leading edge of the noise is almost completely eliminated and the trailing edge is drastically reduced. We have shown that pulse compression can be successfully performed without significantly reducing the central pulse.

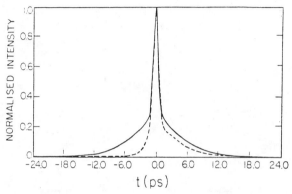

Fig. 1.   Incident pulse shape vs. time (solid line) and exit pulse shape after passage through the medium (dashed line), each normalized to its maximum.

III. Passive Mode-Locking

In passively mode-locked lasers containing gain media with saturation energy too high to be saturated by individual mode-locked pulses, the trailing edge of the pulse continues to see gain and therefore the pulse is not terminated. There is no mechanism to shorten the the pulse to less than the SA relaxation time. Many examples of such gain media exist, including Nd:YAG, ND:glass, ruby, and alexandrite. We shall show that inclusion of a RSA as an additional passive element will succeed in turning off the gain of the laser during the trailing edge of the pulse due to the increased absorption by the excited state of the RSA. Thus, for a gain medium with high gain saturation, the pulse width can be limited to the band gainwidth of the gain medium and not the relaxation time of the SA.[7] Another problem of passively mode-locking high gain medium CW lasers is their instability against relaxation oscillations.[8] The addition of a RSA into the laser cavity suppresses relaxation oscillations. We therefore slay two birds with one stone in the CW passive mode-locking case.

An analytic solution to steady-state passive mode-locking with a RSA can be obtained and a closed form solution for the pulse shape is determined. We consider a homogeneously broadened laser medium with small enough cross section σ_L and long enough lifetime τ_L so that a single pulse does not drive down the gain.

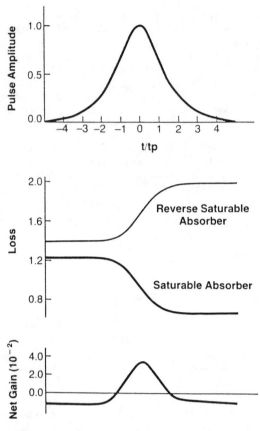

Fig. 2. Pulse amplitude, loss from SA and RSA, and net gain vs time
for mode-locked pulse.

The bandwidth of the laser is limited by an additional intracavity
element which is narrow compared with the laser medium bandwidth. The
SA and RSA are taken sufficiently thin so that longitudinal variations
of the population and intensities can be neglected. The effect of each
element of the laser is assumed small enough so that the exponentials
can be expanded and truncated to the first two or three orders. With
these assumptions an integrodifferential equation for the temporal dependence of the amplitude of the electric field, A(t), can be obtained and
the solution to this nonlinear operator equation can be obtained. It
takes the form[7]

$$A(t) = (E_0/2\tau_p)^{1/2} \operatorname{sech}(t/\tau_p) \tag{4}$$

and the values of the pulse energy E_0 and the pulse width τ_p are
determined in terms of the parameters of the laser medium, the SA and
the RSA.   It would take us too far afield to discuss the details of
the solution here, so we shall content ourselves with presenting a
numerical example.

We take the gain medium to be alexandrite, the SA as
cryptocyanime, and the RSA as a hypothetical molecule which has the
same lifetime and the same ground and excited state cross sections at
750 nm as rhodamine 6G has at 435 nm (where it is a good RSA,

$\sigma_r{}^*/\sigma_r = 40$). Fig.2 shows the pulse amplitude, the loss due to the SA and RSA, and the net gain plotted vs t/t_p. The pulse time is analytically determined to be 0.5 ps. The SA saturates first, and this leads to positive net gain for the duration of the pulse. The RSA then begins to saturate and the net gain becomes negative. The temporal change of the loss of the SA and RSA are responsible for the variation of the net gain.

References

1. Y.B. Band, this volume.
2. D.J. Harter, M.L. Shand, and Y.B. Band, J.Appl.Phys.56, 865 (1984).
3. W. Koechner, Solid-State Laser Engineering, (Springer-Verlag),N.Y., 1976), pp. 162-168.
4. Y.B. Band, D.J. Harter, and R. Bavli, "Optical Pulse Compressors Composed of Saturable and Reverse Saturable Absorbers",to be published.
5. S. Spieser and N. Shakkur, Appl. Phys., in press.
6. I.B. Berlman, Handbook of Fluorescence Spectra of Aromatic Molecules, (Academic Press, N.Y., 1965).
7. D.J. Harter, Y.B. Band, and E.O. Ippen IEEE J. Quant,Electron. QE-21, 1219 (1985).
8. H.A. Haus, IEEE J. Quant. Electron. QE-11, 169 (1976).

PICOSECOND PHOTODYNAMICS OF THE EXCITED STATES OF ORGANIC COMPOUNDS IN SOLUTION USING TIME RESOLVED SPECTROSCOPY

A. Declémy, C. Rullière, and Ph. Kottis

Centre de Physique Moléculaire Optique et Hertzienne
Université de Bordeaux I
33405 Talence Cedex, France

INTRODUCTION

Investigation of excited states of polyatomic molecules in solution provides a good framework in order to study many classes of interactions. But their lifetime is very short (10^{-13} s – 10^{-8} s). So, in order to study such excited states we have built a picosecond absorption spectrometer which allowed us to study many types of molecular systems in solution. We present here two such studies. In the first one (1-4-diphenyl-butadiene) we have shown a solvent assisted electronic level inversion. In the second one we have studied interactions between an intramolecular charge transfer state and polar environment. Before discussing these results, we present the apparatus we have built in our lab.

I) PICOSECOND ABSORPTION SPECTROMETER

Our system[1] is shown on Figure 1. A mode locked YAG/Nd^{3+} laser delivers 25 ps pulses at 1.06 µm. After amplification one part of the pulse is transmitted by the mirror (M1) and passes through KDP crystals generating third (0.353 µm) and fourth (0.265 µm) harmonics and provides the excitation beam. The other part of the pulse which is reflected by the mirror (M1) is once more amplified and focused in D_2O cell generating a picosecond light continuum probe beam and is delayed in time in respect to the excitation beam by moving a reflecting cube corner. After crossing the sample this probe beam is analyzed through a monochromator by means of a linear photodiode array associated to a multichannel analyzer. At the first laser shot the excitation beam is stopped by a disk (D) and we measure the spectral distribution of the probe beam as the reference signal $I_0(\lambda)$. At the second laser shot distribution of the probe beam becomes $I(\lambda)$ due to absorption or emission of excited species. By accumulation of the two signals $I_0(\lambda)$ and $I(\lambda)$ over a great number of laser shots (typically 25) we are able to measure the total absorption (or gain) spectrum of excited species at any time t after excitation (between 0 and 500 ps continuously) in the spectral range 400 nm to 800 nm with a minimum detectable optical density DO {DO = log $(I_0(\lambda)/I(\lambda))$} of about 0.03. Moreover the microprocessor integrated to the multichannel analyzer entirely controls the procedure : Firing laser, accumulation of $I_0(\lambda)$ and $I(\lambda)$, rotation of the disk (D) and optical density calculation. On the other hand, in order to prevent the effects of laser fluctuations two "sample and hold" measure the intensity of the two beams at each laser shot. As a result of this measure the microprocessor takes into account or not this

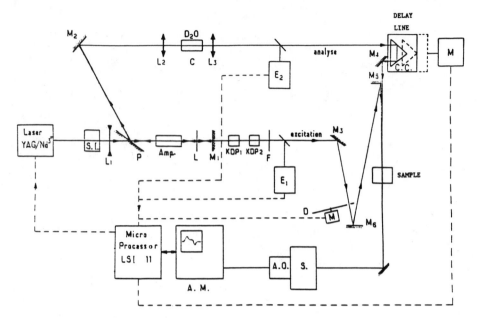

Fig.1. Experimental set-up S.I. : Pulse selector. P : Dielectric polarizer.
L : Quarterwave plate at 1.06 μm. S : Spectrograph. A.O. : Optical
analyzer (linear photodiode array). A.M. : Multichannel analyzer.

laser shot. At last, the microprocessor drives the delay line. So, with this
entirely automatized picosecond spectrometer we are able to obtain the entire
absorption (or gain) spectrum of excited species at any time t after excitation in a few minutes.

II) EVIDENCE FOR EXCITED STATE CONFORMATION CHANGE IN 1-4-DIPHENYL-BUTADIENE (DPB)

This compound has been extensively studied[2,3] but the exact symmetry of
the first excited singlet state S_1^* of this compound was not well established :
Absorption experiments reveal an A_g symmetry (the same as the ground state)
while emission studies show a B_u symmetry. On the other hand this compound
has a number of internal degrees of freedom (twisting of double or single
bonds in the polyenic chain, twisting of phenyl groups) as revealed in Figure 2. So in order to reconcile these apparently contradictory experimental results, and due to flexible nature of this compound, some authors[4]
suggested that the excited molecules of DPB undergo fast conformational reorganization leading to an inversion of the first excited singlet energy levels
with different symmetry. In order to verify this hypothesis we have made a
full analysis of the absorption shape of excited DPB molecules in solution,
with various values of temperature and viscosity[2]. This study led us to conclude unambiguously in favour of the scheme shown on Figure 3, which recon-
ciles all the experimental results : Just after absorption of one photon, the
molecule has the ground state conformation and the lower S_1^* state has the A_g
symmetry. Then the molecule evolves to the relaxed conformation, the two
lower singlet states cross and the lower S_1^* state has now B_u symmetry, this
process being hindered at low temperature. At last, the short reorganization
time of about 10 ps we estimated for the excited DPB at room temperature[2]
is similar to the observed twisting time of phenyl groups in some similar
compounds at room temperature and we attributed the reorganization process
involved in the relaxation of the excited DPB to the phenyl rings twisting.
Moreover, in a subsequent theoretical study, we have shown[3] that the pre -
pared excited state with A_g symmetry has a phenyl rings twisted conformation

GROUND   EXCITED
STATE   STATE
CONFORMATION   CONFORMATION

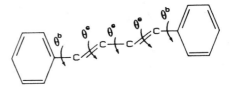

Fig. 2. 1-4 diphenyl-butadiene with   Fig. 3. Illustrative diagram showing
   angles characterizing confor-   the level ordering dependence upon con-
   mational changes.   formational change of 1-4-diphenyl-
                                         butadiene.

while the relaxed excited state with B_u symmetry is more planar. So this picosecond study of DPB allowed us to conclude in a complete fashion on an open
problem, due to the quantity of information available from it.

III) INTERNAL CHARGE TRANSFER (I.C.T.) EXCITED STATE IN POLAR SOLVENTS

In this last part we present our up to date work relative to high interactions between a polar compound in polar solvents. As shown on Figure 4 the
first excited singlet state of this compound (7-amino-3-methyl-1-4-benzoxa-
zine-2-one : AMBO) is characterized by an important electron transfer from
the amino group (NH_2) to the moitie containing carboxyl part. As a consequence the interactions between this excited state and a polar environment
are very strong. So in order to study the different interactions between this
molecule and its environment we have considered the time behaviour
of the gain spectrum of this compound in different classes of solvents[5].
In pure polar solvents with no OH group such as acetone or dimethyl-formami-
de (DMF), a red shift of the gain spectrum is observed, revealing a larger
interaction of the excited state with polar environment than for the ground
state, due to the higher dipole moment of the excited state ; But the time
dependence of the relaxation part of this shift is not observable inside our
temporal resolution ($\sim$ 10 ps) and this fact corresponds to a rapid reorientation of the permanent dipoles of the solvent molecules at room temperature.
In alcoholic solvents giving hydrogen bonds the total amplitude of the shift
is more important than in equivalent pure polar solvents without hydrogen
bonds but with identical dielectric constant showing that in alcoholic
solvents the hydrogen bonds between excited solute and solvent molecules are
largely strengthened. But the point to focus is the different kinetics observed in different alcohols (Figure 5). Before discussing this point we must
keep in mind two essential particularities of these solvents : i) Through
their intermolecular hydrogen bonds they form aggregates whose cohesion and
geometrical extension vary from one alcohol to another ; ii) The dielectric
measurements on these solvents exhibit three dispersion regions with associated relaxation times and the greater of these times is attributed to the
breaking of hydrogen bonds in the solvent[6]. In our case the equilibrium configurations of solvent molecules around the solute in its ground state are
roughly perturbed by picosecond excitation of the solute, and at this first
time they are bent configurations. So assuming a damped harmonic oscillator
model to describe the return to equilibrium around the excited solute, we

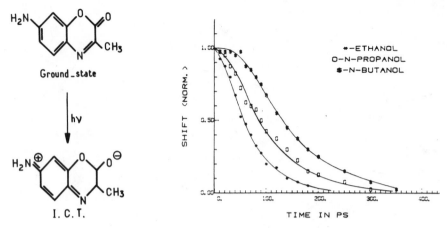

Fig. 4. Charge transfer in 7-amino
3-methyl - 1,4-benzoxazine
2-one (AMBO) upon excitation
hν.

Fig.5. Normalized appeareent shift of
AMBO emission as a function of
time in three alcohols.

can fit experimental curves with characteristic times similar to those corresponding to the breaking of solvent hydrogen bonds, when they are known
(for ethanol : $\tau \sim 200$ ps ; For propanol : $\tau \sim 400$ ps ; For butanol
$\tau \sim 600$ ps). So, in this case, the relaxation processes are essentially governed by the physical properties of the environment of the excited state
such as its degree of aggregation, steric hindrance. But those different
physical parameters are not simply related to the observed kinetics. Actually, work is in progress in our lab in order to precise their relative
contribution to the relaxation processes.

CONCLUSION

The typical cases what we have presented here show that picosecond study of excited states of polyatomic molecules in solution allows a best understanding of their microscopic properties when they are in large interaction with their environment and that through a wide range of important physical problems.

REFERENCES

1. C. Rullière, A. Declémy, Ph. Pée, Revue Phys. Appl. 18 : 347 (1983).
2. C. Rullière, A. Declémy, Ph. Kottis, Laser Chem. 5 : 185 (1985).
3. C. Rullière, A. Declémy, Ph. Kottis, L. Ducasse, Chem. Phys. Letters
   117 : 583 (1985).
4. R.A. Goldbeck, A.J. Twarowski, E.L. Russel, J.K. Rice, R.R. Birge,
   E. Switkes, D.S. Kliger, J. Chem. Phys. 77 : 3319 (1982).
5. A. Declémy, C. Rullière, Ph. Kottis, Chem. Phys. Letters 101 : 401 (1983).
6. S.K. Garg, C.P. Smith, J. Phys. Chem. 69 : 1294 (1965).

SPECTRAL LINEWIDTH OF SEMICONDUCTOR LASERS*

J. Harrison and A. Mooradian

Lincoln Laboratory, Massachusetts Institute of Technology
Lexington, Massachusetts 02173-0073

The semiconductor diode laser (SL) has a number of properties that make it a desirable light source for spectroscopy. Aside from the advantages in size, cost and complexity, SL's cover a broad spectral range and may easily be modulated at rates up to several gigahertz. However, coherence remains an issue in many applications. Here we discuss the issue of the ultimate SL linewidth. Included is a discussion of SL spectral characteristics including the enhanced Lorentzian linewidth, the power-independent linewidth, and the fine structure in the field power spectrum. We also discuss the prospects for line narrowing through active frequency control and present heterodyne measurements between two independently operating external cavity (GaAl)As lasers. Finally, we demonstrate that such stable, narrow-line sources can easily be phase-locked.

Our studies have primarily involved (GaAl)As devices. These were the first to be commercially developed and include a number of different structures that operate in a single mode and achieve high output powers at room temperature. Recent advances in the technologies of both (GaAl)As lasers and nonlinear crystals make it increasingly likely that efficient second harmonic generation will make III-V SL's important spectroscopic sources capable of extended operation in the 330-750 nm range.

We introduce the concept of the laser linewidth by considering the response of the laser field to a spontaneous emission event. The spontaneous photon will couple to the field with a random phase, instantaneously changing both the amplitude and phase of the laser field. The magnitude of both perturbations decreases with increasing field intensity as the ratio of the spontaneous to the stimulated emission rate decreases. The field amplitude sees a strong restoring force due to gain clamping so that the return to steady-state is achieved through a damped relaxation oscillation that may occur over several nanoseconds. The phase, however, sees no such restoring force so that over time the phase of the laser field diffuses uniformly relative to that of the initial field.

In a typical laser, it is the rate of the phase fluctuations that determines the laser linewidth. In this case, the field power spectrum is Lorentzian with a FWHM linewidth given in terms of SL experimental parameters by

---

*This work was sponsored by the Department of the Air Force.

$$\Delta\nu_{FWHM} = (h\nu/8\pi P)\ (c/n\ell)^2\ (G)\ (-\ln R)\ n_{sp}\ (1 + \alpha^2)\ K \qquad , \qquad (1)$$

where $h\nu$ is the photon energy, P is the single-ended output power of a symmetric device with facet reflectivity R and G is the net gain at threshold.

The enhanced spectral broadening observed in SL's is contained in the final three terms in Eq. (1). The first, n_{sp}, is the spontaneous emission factor and gives the ratio of the rate of spontaneous emission into the laser mode to that of stimulated emission per photon in the mode. In many laser systems n_{sp} is close to unity. However, this is not true of SL's due to the finite population of the lower level of the laser transition. In SL's, n_{sp} approaches unity only at very low temperatures where the carriers are distributed according to Fermi-Dirac statistics. At room temperature $n_{sp} \simeq 2.5$ for (GaAl)As lasers.

The linewidth enhancement factor, α, represents the linewidth broadening due to the intrinsic coupling of amplitude and phase noise and is given by the ratio of corresponding changes in the real and imaginary parts of the complex refractive index of the active medium. In most laser systems $\alpha \ll 1$ because emission occurs at the peak of the resonant dispersion where changes in the imaginary part of the index (i.e., changes in the gain due to spontaneous emission) are associated with negligible changes in the real part of the index (i.e., the phase of the field). This is not the case in SL's because of the strong interband dispersion. Emission typically occurs on the low energy side of resonance where the unpumped absorption is lower than that at the peak even though the rate of change of gain with carrier injection is greater at the peak.[1] The linewidth enhancement may be considered in terms of the delayed change in the phase of the laser field that accumulates during the relaxation oscillations following the instantaneous phase perturbation associated with a spontaneous emission event. This delayed phase change adds to the instantaneous perturbation and thereby increases the net rate of phase fluctuations and the Lorentzian linewidth.

Measured values of α in several types of (GaAl)As lasers range from 2.2 to 6.2.[2-7] A number of methods have been employed in the determination of the linewidth enhancement factor. Each of these measures, in some way, corresponding changes in the intensity and frequency of the laser field. The range of values observed in the various lasers reflects the dependence of α on structure-dependent factors including the active region geometry[8,9] and the internal losses (i.e., quasi-Fermi energy levels).[10]

The final term in Eq. (1) is the astigmatism factor, K. This accounts for the enhanced coupling of the spontaneous emission to longitudinal modes of the laser. This can be a significant factor in many structures, especially narrow stripe devices in which the gain profile is primarily responsible for the definition of the optical mode, where K = 11 has been observed.[11] However, in single-mode, index-guided SL's, typically, K is very close to unity.

Another unusual spectral feature that is characteristic of the SL is observed in the field spectrum where resonances appear spaced from the central Lorentzian laser line by harmonics of the relaxation oscillation frequency. The intensities of these spectral sidebands fall off rapidly with increasing harmonic number and their relative integrated intensities decrease linearly as the laser power is increased. In addition, an asymmetry in the intensities of corresponding upper and lower sidebands typically occurs. This is a manifestation of the intrinsic correlation of amplitude and phase fluctuations.[12] In Fig. 1 the intensities of the fundamental upper and lower sidebands are plotted as functions of inverse power for a (GaAl)As transverse junction stripe (TJS) laser at room temperature.

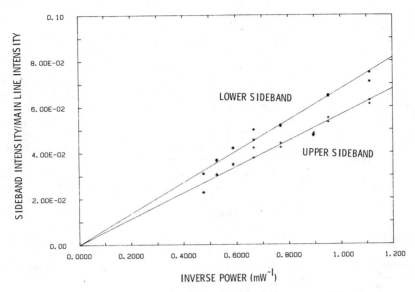

Fig. 1.   Integrated sideband intensities normalized to
the central Lorentzian line for a TJS laser at
room temperature.

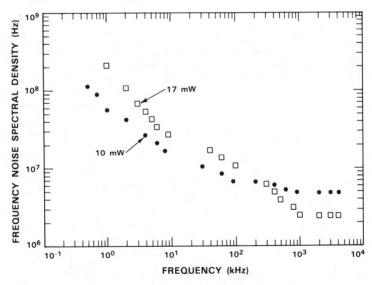

Fig. 2.   Frequency noise spectra of a channeled substrate
planar laser at room temperature operating at
10 mW and 17 mW single-ended output power.

In Fig. 2 we show data taken with similar TJS lasers at three temperatures for $\Delta\nu_{FWHM}$ as a function of inverse power. The characteristic linear dependence of the Lorentzian linewidth on inverse power is clearly evident. A representative value for the linewidth-differential power product for (GaAl)As lasers at room temperature is $\Delta\nu_{FWHM} \Delta P \sim 100$ MHz·mW. As the temperature is reduced, a power-independent linewidth component becomes increasingly evident. This is observed to increase from about 2 MHz at 273 K to over 11 MHz at 77 K.[13] Additional measurements on these TJS devices at 1.7 K indicate that at moderate powers and above, the observed linewidth is dominated by a power-independent component of about 30 MHz.[14] A number of mechanisms have been proposed to explain this effect and are discussed in the following paragraphs.

A phenomenological model for the effects of statistical number fluctuations offers excellent agreement with the observed magnitude of the power-independent linewidth at the four temperatures of measure.[13,14] The model involves a number of parameters that have been measured independently. The pertinent physical mechanism is the dependence of the refractive index, and therefore the cavity frequency, on carrier density. At very low temperatures, there may also be a non-negligible component related to gain peak fluctuations associated with the change in quasi-Fermi energy levels with carrier number. Agreement between experiment and model occurs when the magnitude of the carrier number fluctuations is taken as the square root of the carrier number. Note that such fluctuations must be associated with a noise source that does not saturate at threshold.

Thermally driven fluctuations in the occupation of states around the quasi-Fermi levels have also been proposed as a mechanism by Vahala and Yariv.[15] This theory is unable to account for the large power-dependent linewidth observed at 1.7 K where the driving force for the fluctuations should become negligible. Temperature fluctuations associated with nonradiative recombination and absorption have also been shown theoretically by Lang, Vahala and Yariv[16] to induce a power-independent linewidth component if sufficiently intense.

Another proposed mechanism is the coupling of longitudinal modes.[17,18] Sato and Fujita[18] described this effect in terms of the self-modulation of the refractive index that may result from the interference between the dominant longitudinal mode and one or more submodes that decrease in amplitude as the laser mode power increases. However, this mechanism is inconsistent with the observed spectral behavior of devices similar to that of Fig. 3.

Zeiger[19] has recently published a theoretical examination of the effect of traps on the SL linewidth. Given a sufficient density of trap states within the optical mode volume, a result similar to the observed behavior is obtained if the trap decay rates are at least of the order of the power-independent linewidth. The motivation for the trapping theory came in part from the widely observed ~1/f behavior in the low frequency current, amplitude and frequency noise spectra of SL's. This is discussed in more detail shortly. Similar behavior in the current fluctuation noise in semiconductors has been related to the presence of trap levels.

Another fundamental process in semiconductors that may contribute to the observed power-independent linewidth is light scattering. Among the many scattering mechanisms that have been observed in GaAs, those that are related to a random process (i.e., carrier density variations, temperature fluctuations, etc.) and produce radiation that couples to the laser mode may increase the noise in the mode above that due to spontaneous emission. However, as opposed to the spontaneous light, the on-line scattered component will increase with the power in the mode above threshold and therefore contribute a power-independent phase fluctuation rate. Various Rayleigh mechanisms appear as candidates for scattering noise that can be quite intense given the resonant enhancement that occurs because the laser radiation is near the peak of the interband dispersion. The influence of carrier scattering on SL linewidth is currently under study in our laboratory.

The frequency noise power spectral density of a SL typically exhibits a 1/f dependence below 100 kHz and is flat from 1 MHz to well above 100 MHz. Relaxation oscillations will induce a pronounced peak in the spectrum above 1 GHz. The "white" spectral component represents the phase fluctuations that are responsible for the Lorentzian linewidth and its intensity is equal to π times the Lorentzian FWHM.[20] The 1/f component represents a random walk of the center frequency of the field. This phase noise is responsible for a slight Gaussian rounding at the peak of the laser field spectrum[20] and results in a power independent component in the linewidth. Figure 3 shows typical frequency noise spectra for a TJS laser at two power levels.

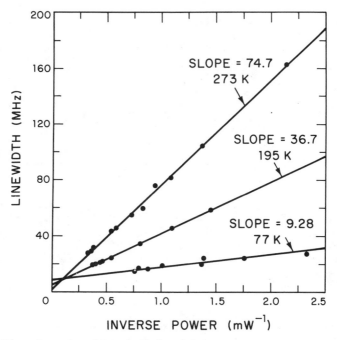

Fig. 3. Linewidth (FWHM) vs inverse power for a TJS
laser at 273 K, 195 K and 77 K.

The nature of the power-independent linewidth suggests that the linewidth at high powers may be reduced by adding to the laser drive current a frequency-dependent component to compensate for the intrinsic center frequency fluctuations.[21,22] This requires an optical discriminator in a feedback loop, the bandwidth of which is at least equal to the linewidth. Results have been obtained by Saito et al.[21] in which the heterodyne signal between a narrow-line master SL and a broader slave SL was reduced from 20 MHz to 1 MHz by applying feedback to the slave laser. In this experiment the discriminator simply consisted of a mixer and a delay line. Ohtsu[22] reported similar line-narrowing with a solitary device employing a Fabry-Perot etalon as an optical discriminator.

Optical feedback can be used to achieve dramatic linewidth reduction. By placing a SL in an external cavity, the linewidth may be reduced by as much as the square of the increase in the photon lifetime (i.e., several orders of magnitude). The linewidth of the external cavity SL, $\Delta\nu_{xcav}$, is related to that of a solitary device at the same pump level, $\Delta\nu_{sol}$, by the following relation

$$\Delta\nu_{xcav} = \Delta\nu_{sol} \cdot [n\ell/n\ell+L]^2 \cdot [G_{xcav}/G_{sol}]^2 \qquad , \qquad (2)$$

where the first term in brackets is the ratio of the active optical path length to the sum of the active and passive optical path lengths of the external cavity. The second bracketed term in Eq. (2) is the ratio of the net gains at threshold of the external cavity and solitary lasers. In our experiments, the square of the length ratio is typically about 10^{-4} while that of the gain term is about three. The external cavity SL is broadly tunable when a dispersive element is included in the cavity. By employing compact, stable cavity structures and carefully isolating the laser from acoustic sources, one can construct a very narrow line, tunable source that can deliver tens of milliwatts. High power, pulsed SL's can also be used in external cavities to produce much higher peak powers in a single longitudinal mode for efficient frequency mixing.

We have constructed two compact external cavity lasers using Hitachi HLP-1400 (GaAl)As lasers. AR coatings were applied to the internal facets with residual modal reflectivities of less than 1%. The lasers were imaged through a thin etalon onto 5% output couplers using commercial microscope objectives that transmitted 70% of the incident light at 833 nm. The cavity structure is shown in Fig. 4. Line coincidence was achieved by a combination of current and etalon tuning, and no effort was made to thermally stabilize the lasers. The heterodyne signal was derived from a Ge avalanche PIN photodiode and observed on a spectrum analyzer. The data obtained on a logarithmic scale (Fig. 5a) is slightly broader than the heterodyne lineshape that is calculated to be about 5 kHz FWHM in this experiment. In order to obtain a measure of the linewidth jitter due to acoustic sources, Figs. 5b and 5c show the heterodyne signal on a linear scale at sweep rates of 17 MHz/sec and 100 kHz/sec, respectively. It is evident that in several milliseconds, the acoustically driven peak-to-peak jitter is within a factor of two of the intrinsic heterodyne linewidth. After a moderate warmup period, the maximum frequency variation of the heterodyne signal was observed to be less than 2 MHz over 10 minutes.

With such stable, narrow-line devices, it is straightforward to phase-lock two external cavity lasers. The experimental schematic is included in Fig. 6a. The heterodyne signal is mixed with a local oscillator set to the heterodyne center frequency. The mixer output is then amplified in order to supply a feedback current that is proportional to the phase shift of the heterodyne signal. Note that because the tuning rate of the external cavity laser was 25 MHz/mA, the peak-to-peak variation in the feedback current over several seconds was less than 5 µA. Figure 6b shows the heterodyne signal under phase-locked conditions. The observed signal indicates that the phase-locked heterodyne linewidth is less than the 15 Hz resolution half-width of the spectrum analyzer.

In conclusion, we have reviewed the spectral characteristics of SL's. A combination of electrical and optical feedback may be employed to obtain ultranarrow, tunable spectroscopic sources.

The views expressed are those of the authors and do not reflect the official policy or position of the U.S. government.

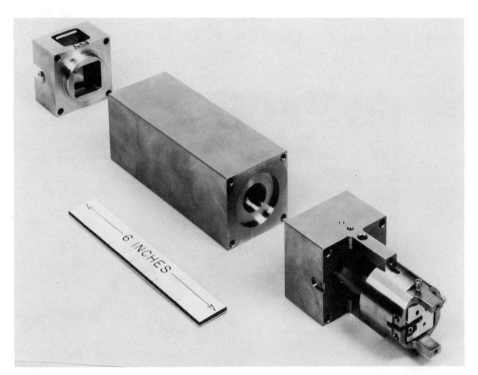

Fig. 4. External cavity laser assembly. Laser mounts on
floating plate at lower right. Objective screws into
threaded central section. Etalon and output coupler
are contained in block at upper left.

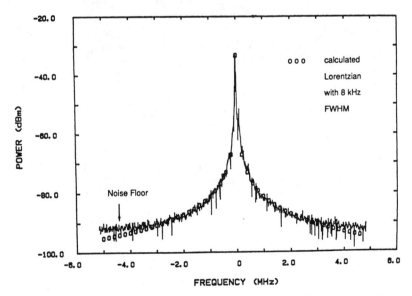

Fig. 5. Heterodyne signal of two free-running external
cavity lasers.
a) log (10 dB/div) vertical scale: 1 s sweep,
30 kHz resolution bandwidth; circles indicate
8 kHz FWHM Lorentzian curve.

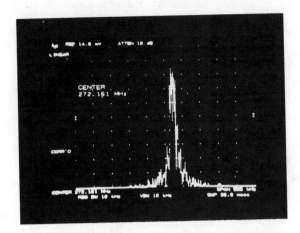

Fig. 5.   b) Linear vertical scale: 50 kHz/div, 30 ms
            sweep, 10 kHz resolution bandwidth.

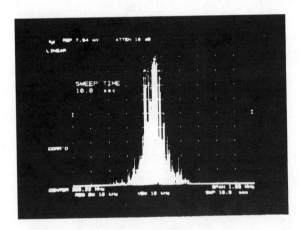

Fig. 5.   c) Linear vertical scale: 100 kHz/div, 10 s
           sweep, 10 kHz resolution bandwidth.

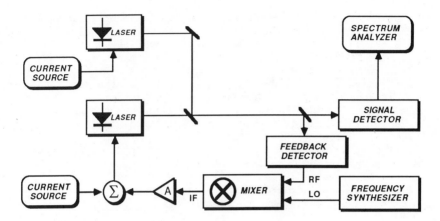

Fig. 6.   a)   Schematic of phase-locking experiment.

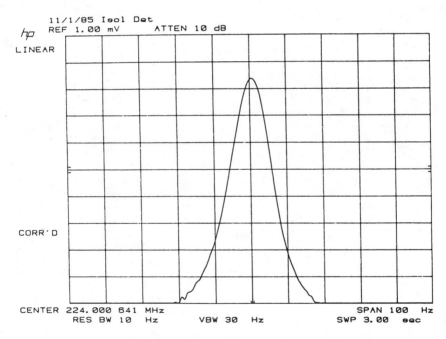

Fig. 6.   b)   Heterodyne signal of two phase-locked
             external cavity lasers.   Linear vertical
             scale: 10 Hz/div, 10 Hz resolution band-
             width (i.e., 15 Hz resolution halfwidth).

REFERENCES

1. C. H. Henry, R. A. Logan and F. R. Merritt, Measurement of gain and absorption in AlGaAs buried heterostructure lasers, J. Appl. Phys. 51:3042 (1980).

2. R. Schimpe, B. Stegmuller and W. Harth, FM noise of index-guided GaAlAs diode lasers, Electron. Lett. 20:206 (1984).

3. K. Kikuchi and T. Okoshi, Estimation of linewidth enhancement factor of AlGaAs lasers by correlation measurement between FM and AM noises, IEEE J. Quantum Electron. QE-21:669 (1985).

4. I. D. Henning and J. V. Collins, Measurements of the semiconductor laser linewidth broadening factor, Electron Lett. 19:927 (1983).

5. D. Welford and A. Mooradian, Output power and temperature dependence of the linewidth of single-frequency cw (GaAl)As diode lasers, Appl. Phys. Lett. 40:865 (1982).

6. C. H. Henry, R. A. Logan and K. A. Bertness, Spectral dependence of the charge in refractive index due to carrier injection in GaAs lasers, J. Appl. Phys. 52:4457 (1981).

7. C. Harder, K. Vahala and A. Yariv, Measurement of the linewidth enhancement factor α of semiconductor lasers, Appl. Phys. Lett. 42:328 (1983).

8. J. Arnaud, Role of Petermann's K-factor in semiconductor laser oscillators, Electron Lett. 21:538 (1985).

9. K. Furuya, Dependence of linewidth enhancement factor α on waveguide structure in semiconductor lasers, Electron Lett. 21:200 (1985).

10. Y. Arakawa and A. Yariv, Fermi energy dependence of linewidth enhancement factor of GaAlAs buried heterostructure lasers, Appl. Phys. Lett. 47:905 (1985).

11. K. Petermann, Calculated spontaneous emission factor for double-heterostructure injection lasers with gain-induced waveguiding, IEEE. J. Quantum Electron. QE-15:566 (1979).

12. K. Vahala, C. Harder and A. Yariv, Observation of relaxation resonance effects in the field spectrum of semiconductor lasers, Appl. Phys. Lett. 42:211 (1983).

13. D. Welford and A. Mooradian, Observation of linewidth broadening in (GaAl)As diode lasers due to electron number fluctuations, Appl. Phys. Lett. 40:560 (1982).

14. J. Harrison and A. Mooradian, Spectral characteristics of (GaAl)As diode lasers at 1.7 K, Appl. Phys. Lett. 45:318 (1984).

15. K. Vahala and A. Yariv, Occupation fluctuation noise: a fundamental source of linewidth broadening in semiconductor lasers, Appl. Phys. Lett. 43:140 (1983).

16. R. J. Lang, K. J. Vahala and A. Yariv, The effect of spatially dependent temperature and carrier fluctuations on noise in semiconductor lasers, IEEE J. Quantum Electron. QE-21:443 (1985).

17. W. Elsasser and E. O. Gobel, Multimode effects in the spectral linewidth of semiconductor lasers, IEEE J. Quantum Electron. QE-21:687 (1985).

18. H. Sato and T. Fujita, Spectral linewidth broadening due to self-modulation of the refractive index of semiconductor lasers, Appl. Phys. Lett. 47:562 (1985).

19. H. J. Zeiger, Theory of the effect of traps on the spectral characteristics of diode lasers, Appl. Phys. Lett. 7:545 (1985).

20. F. G. Walther and J. E. Kaufmann, Characterization of GaAlAs laser diode frequency noise, Topical Meeting on Optical Fiber Communication, New Orleans, LA, Tech. Dig. 70 (1983).

21. S. Saito, O. Nilsson and Y. Yamamoto, Frequency modulation noise and linewidth reduction in a semiconductor laser by means of negative frequency feedback technique, Appl. Phys. Lett. 46:3 (1985).

22. M. Ohtsu, Conference on Lasers and Electro-Optics, Baltimore, MD (1985)

SPECTRAL AND DYNAMIC FEATURES OF

QUANTUM WELL SEMICONDUCTOR LASERS

Amnon Yariv

California Institute of Technology

Pasadena, California 91125

ABSTRACT

Recent experimentation with semiconductor lasers has revealed many new features of the spectral behavior. New theories have been formulated to explain this behavior and are directing attention to the new quantum-well lasers for improved spectral behavior.

INTRODUCTION

The key role played by semiconductor lasers (SCL's) in optical communication has resulted in a theoretical and experimental attention being devoted to them and a resulting increase in the understanding of their basic properties and potential. In this paper we will review two aspects of this development:

(1) Spectral purity

(2) High frequency current modulation response

In what follows I will address these issues and describe some of the recent progress and the prospects for the future.

Spectral Purity

The ultimate linewidth of a laser oscillator is limited by the "contamination" of the coherent stimulated emission by incoherent spontaneous emission. The resulting limiting linewidth was given by Schawlow and Townes as[1]

$$(\Delta \nu)_{laser} = \frac{\pi (\Delta \nu_{1/2})^2 h \nu}{P_e} \left(\frac{N_2}{N_2 - N_1} \right)$$ [1]

where $\Delta \nu_{1/2}$ is the linewidth of the laser resonator at the absence of a gain medium ("passive line"), P_e is the total power emitted by the atomic medium and ν the frequency.

Recent measurements of $(\Delta \nu)_{laser}$ in semiconductor lasers by Welford and Mooradian[2] have shown major departures from the prediction of (1). The main discrepancy is an increase by a factor of 25-40 in $(\Delta \nu)_{laser}$ while following, except at very high powers, the P_e^{-1} dependence. This discrepancy was settled by a new model[3] for the noise in lasers and a new theory[4] which accounts for this effect. The basic explanation is that in addition to direct "contamination" of the laser field by spontaneous emission there exists an additional and indirect effect: spontaneous emission events are accompanied by corresponding change in the electron and hole populations so that they lead to a modulation in the index of refraction of the laser medium and, hence, to a modulation of the resonance frequency of the laser. In the language of nonlinear optics we account for this effect by representing the induced polarization of the medium by

$$P = \epsilon_0 \left[\chi_r^{(1)} + i\chi_i^{(1)} + \chi_r^{(3)}|E|^2 + i\chi_i^{(3)}|E|^2 + \ldots \right] E \quad [2]$$

The term $\chi_r^{(3)}|E|^2$ represents changes in the index of refraction due to changes in optical intensity, such as those due to spontaneous emission, and its inclusion in the traditional laser oscillator model for a laser leads to a new result for the laser linewidth[4]

$$(\Delta \nu)_{laser} = \frac{\pi (\Delta \nu_{1/2})^2 h\nu}{P_e} \left(\frac{N_2}{N_2 - N_1}\right)(1 + \alpha^2)$$

where

$$\alpha = \frac{\partial \chi_r^{(3)}}{\partial n} \Big/ \frac{\partial \chi_i^{(3)}}{\partial n} \quad [3]$$

where n is the carrier density in the active region. A quantum mechanical derivation for α leads to[6]

$$\alpha(\omega_L) = \int_{-\infty}^{\infty} \rho(\omega)[P_c{}'(n,\omega) - P_v{}'(n,\omega)] \frac{\omega_L \omega}{(\omega_L - \omega)^2 + 1/T_2{}^2} d\omega \Big/$$

$$[4]$$

$$\int_{-\infty}^{\infty} \rho(\omega)[P_c{}'(n,\omega)] - P_v{}'(n,\omega)] \frac{1/T_2{}^2}{(\omega_L - \omega)^2 + 1/T_2{}^2} d\omega,$$

where $P_c(n,\omega)$, $P_v(n,\omega)$ are the (quasi) Fermi occupancy functions for electrons and holes at an energy $\hbar\omega$ and carrier density n and $\rho(\omega)$ is the joint density of states at $\hbar\omega$. Figure 1 shows a plot of the calculated value of α based on [4]. The predicted values of $\alpha \sim 3$ are not far from the observed values of $\alpha \approx 3-5$.

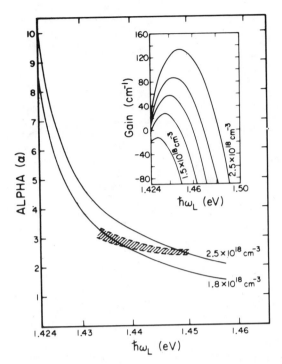

Fig. 1 (Ref. 6). A plot of the theoretical expression
[Eq. 4] for the amplitude to-phase coupling factor
α as a function of the lasing wavelength for two
carrier densities. The shaded region corresponds
to the range of operation of conventional semiconductor lasers.

The range of linewidths predicted by Eq. [3] and which are observed
in practice are around $(\Delta \nu)_{laser} \sim 1\text{-}3 \times 10^8$. Such values are too
high for some extremely demanding spectroscopic or coherent
communication systems. There are two main approaches for
reducing $(\Delta \nu)_{laser}$. The first one involves operating the
semiconductor laser in an external cavity, i.e., use an external
reflector at some length (say 1-15 cm from the laser chip. The
main effect here is decrease $(\Delta \nu_{1/2})$ in Eq. [3] roughly by the ratio
of the new modal volume to the original one. This approach works
and has resulted in SC lasers with $(\Delta \nu)_{laser}$ as low as a few KHz.
Its main disadvantage is that it sacrifices the monolithic pure
electronic nature of the SC laser so that it is more appropriate for
laboratory experiments and specialized applications.

The other approach for reducing $(\Delta \nu)_{laser}$ is by modifying the
density of state distribution $\rho(\omega)$. This distribution has,
according to Eq. [4], a strong effect on the α parameter. The
distribution $\rho(\omega)$ is a characteristic of the semiconductor and in
the parabolic band approximation is given by

$$\rho(\omega) = \frac{1}{2\pi^2} \left(\frac{2m_c m_v}{m_c + m_v}\right)^{3/2} (\omega - E_g/\hbar)^{1/2} \qquad [5]$$

145

where m_c, m_v are the conduction band and valence band affective masses, respectively while E_g is the gap energy. The main prospect for a significant modification of $\rho(\omega)$ at present involves the use of quantum well lasers,[7] In these lasers the active region, in which electrons and holes recombine to give off light, is less than 100Å wide. Such small transverse dimensions are obtained using molecular beam epitaxy and employ most often a GaAs active region sandwiched between higher gap GaAlAs layers. The transverse quantization results in modified Bloch states

$$u(\vec{r}) = \psi_{\vec{k}_\perp}(\vec{r}_\perp) \begin{Bmatrix} \sin \\ \cos \end{Bmatrix} \frac{i\pi}{L_z} z, \quad i=1,2,3\ldots \qquad [6]$$

where $\psi_{\vec{k}_\perp}(\vec{r}_\perp)$ is a two dimensional (x,y) Bloch function while L_z is the width of the quantum well (confining layer) and an infinite barrier height was assumed at the two interfaces. In such a structure the ith sub-band is given by

$$\rho_{well}(\omega) = \frac{m_j^*}{\pi \hbar^2 L_z} \sum_i H \left[\hbar\omega - \frac{\hbar^2}{2m_j^*} \left(\frac{\pi i}{L_z}\right)^2 \right] \qquad [7]$$

where m_j^* (j=l,h) is the reduced mass with respect to heavy holes (h) or light holes (l) and H(E) is the Heaviside function and the summation is over the sub-bands i whose minimum energy $(\hbar\pi i)^2/(2m_i^* L_z^2)$ is smaller than $\hbar\omega$. Instead of the square root dependence of $\rho(\omega)$ in the bulk, the density of states function exhibits a staircase behavior which leads to drastic modification in almost all the laser dynamic, spectral and noise properties.

Frequency Response

The plot for f_r, the maximum frequency response of the laser is based on the relation[9]

$$f_r = \frac{1}{2\pi} \left(\frac{P\omega}{2\tau\mu_0} \frac{d\chi_I(n)}{dn} \right)^{1/2} \qquad [8]$$

$$\alpha = \frac{d\chi_R(n)/dn}{d\chi_I(n)/dn}, \qquad [9]$$

where P,ω,τ are the photon density frequency, and passive cavity lifetime of the lasing mode; μ_0 is the nonresonant value of the refractive index; n is the carrier density. $\chi_R(n)$ and $\chi_I(n)$ are the real and imaginary parts of the complex susceptibilities of the active medium. Their derivatives with respect to the carrier density are given, respectively, by

$$\frac{d\chi_I(n)}{dn} = \int dE \frac{dg(n,E)}{dn} \frac{\hbar/T_2}{(E-E_1)^2 + (\hbar/T_2)^2} \qquad [10]$$

$$\frac{d\chi_R(n)}{dn} = \int dE \frac{dg(n,E)}{dn} \frac{E - E_1}{(E-E_1)^2 + (\hbar/T_2)^2}, \qquad [11]$$

where E_1 and T_2 are the lasing photon energy and the collisional broadening time due to carrier-carrier and carrier-phonon interaction, $g(n,E)$ is the gain envelope function which is given by

$$g(n,E) = CM\rho_{red}[f_c(n,E) - f_v(n,E)] \qquad [12]$$

The calculated[8] dependence of α of the well width which results from the use of Eq. [7] in Eq. [4] is shown in Figure 2. Also shown in the same figure is the improvement in the current modulation frequency response of the laser

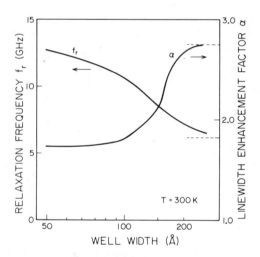

Fig. 2. The calculated value of the α parameter of a quantum well laser as a function of the well width.

Conventional SC lasers have been demonstrated with upper modulation frequency f_r of ~11GHz. The predicted improvement in the value of f_r in the quantum well lasers should give rise to values of f_r ~20-30 GHz.

Another fundamental aspect of semiconductor lasers is that their field spectrum (spectral density function of the laser field) is not Lorentzian but possess side bands.[10-11] The side bands can be understood by considering first the density of electrons in the conduction band of the laser medium when the input current i(t) is modulated at ω

$$i(t) = i_0 + i_1 e^{i\omega t} \qquad\qquad\qquad [13]$$

A solution of the coupled photon density-electron density equation $n(t)$ for a SC laser yields

$$n(t) = n_0 + n_1 e^{i\omega t}$$

$$[14]$$

$$\frac{n_1}{i_1} = \frac{-i\omega}{(eV)[\omega^2 - i\omega/\tau - 4\pi^2 f_r^2]}$$

where V is the volume of the active region, τ the electron-hole recombination time and f_r is defined by [8]. A plot of n_1/i_1 shows a major peak near $\omega=2\pi f_r$. This corresponds to a large modulation of the carrier density. The above argument was meant to illustrate the existence of an intrinsic resonance at f_r in a semiconductor laser, or for that matter, in any laser. Now if we think of the effect of spontaneous emission in a non modulated CW laser as equivalent in some way to an external current, we might expect that the resulting carrier density fluctuation spectrum will be particularly rich near $\omega=2\pi f_r$. Since we have already established that carrier fluctuations lead to index fluctuations and thus to fluctuations in the resonant frequency of the laser, we expect that the spectrum of the laser should be modified by an effective <u>frequency</u> <u>modulation</u> (FM) near f_r. A rigorous Van der Pol type[13] analysis bears out this conclusion. The actual predicted and measured features of such a laser are shown in Figure 3.

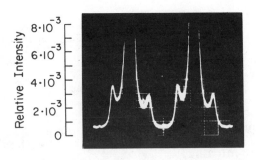

Fig. 3(a). The experimental display of the field spectrum $|E(\omega)|^2$ of a semiconductor laser

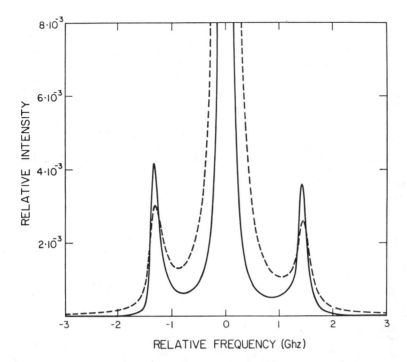

Fig. 3(b). A theoretical prediction of the field
            spectrum (solid convolved with the
            instrumental lineshape (dashed)).

The agreement between theory and experiment bears strong support
to the validity of the detailed physical models used to analyze SC
lasers and to the richness of the consequences of the interplay
between photons and electrons in semiconductors.

        In conclusion: We have reviewed some of the fundamental
issues and physical mechanism involved in the spectral features of
semiconductor lasers and outlined recent progress in their
understanding.

References

1. Schawlow, A., and Townes, C. H., Physical Review 112:1940
    (1958).
2. Fleming, M., and Mooradian, A., "Fundamental line broadening
    of single-mode (GaAl)As diode lasers," Appl. Phys. Lett.
    38:511 (1981).
3. Henry, C. H., "Theory of the linewidth of semiconductor
    lasers," IEEE J. Quantum Electron QE-18:259 (1982).
4. Vahala, K., and Yariv, A., IEEE J. of Quantum Electron. QE-
    19:1096 (1983).
5. See, for example, Yariv, A., "Quantum Electronics," J. Wiley
    & Sons, New York (1975), p. 308.
6. Vahala, K., Chiu, L. C., Margalit, S., and Yariv, A., Appl.
    Phys. Lett. 42:631 (1983).

7. Holonyak, Jr., N., Kolbas, R. M., Dupuis, R. D., and Dapkus, P. D., IEEE J. Quantum Electron QE-116:170 (1980).
8. Arakawa, Y., Vahala, K., and Yariv, A., Appl. Phys. Lett. 45:950 (1984).
9. Lau, K., Bar-Chaim, N., Ury, I., and Yariv, A., Appl. Phys. Lett. 43:11 (1983).
10. Vahala, K., Harder, Ch., and Yariv, A., Appl. Phys. Lett. 42:211 (1983).
11. Diano, B., Spano, P., Tambarini, M., and Piazolla, S., IEEE J. Quantum Electron. QE-19:226 (1983).
12. Yariv, A., "Optical Electronics," Holt, Rinehart and Winston, New York (1985) p. 491.
13. Vahala, K., and Yariv, A., IEEE J. Quantum Electron. QE-19:1102 (1983).

PROGRAMMABLE, SECONDARY FREQUENCY STANDARDS BASED INFRARED

SYNTHESIZERS FOR HIGH RESOLUTION SPECTROSCOPY

Charles Freed

MIT/Lincoln Laboratory

Lexington, MA 02173-0073

ABSTRACT

The intent of this paper is to give a brief overview of salient components and techniques which led to the development of programmable infrared synthesizers; absolute frequency accuracy, reproducibility and resettability with standard deviations of about 5 kHz (1.7×10^{-7} cm^{-1}) relative to the primary Cesium frequency standard at the National Bureau of Standards in Boulder, Colorado, has been achieved.

INTRODUCTION

The inherently great spectral purity, intensity, tunability and small beam divergence of infrared lasers have been utilized and combined with readily available commercial R.F. and microwave instruments to construct the equivalents of programmable synthesizers in the infrared frequency

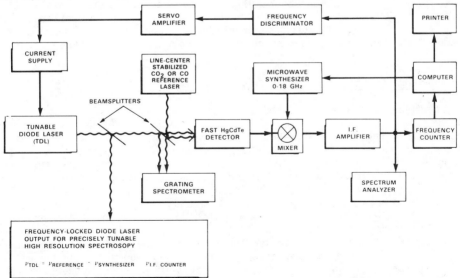

Fig. 1. Block diagram of an accurate, tunable, computer controlled, kHz resolution infrared frequency synthesizer.

151

domain. Figure 1 shows the block diagram of the prototype[1] of such synthesizers in the 9.0 to 12.3 (or 5.0 to 8.0) μm wavelength range. The synthesizer output is derived from a lead-salt tunable diode laser (TDL); a small portion of the TDL output is heterodyned against a line-center stabilized grating controlled CO_2 (or CO) molecular laser. The beatnote of the two lasers is detected by a high speed HgCdTe (varactor) photo-diode.[2] The detected beat frequency, which is generally in the 0 to 18 GHz range is further heterodyned to some convenient intermediate frequency (I.F.) using readily available commercial R.F./microwave frequency synthesizers and wide-band double balanced mixers. The I.F. output is amplified and amplitude limited by means of low-noise, wide-band amplifiers and limiters. The limiter output is in turn used as input to a wide-band, delay-line type frequency discriminator (2 to 600 MHz typical bandwidth). The output of the frequency discriminator is further ampli-fied by means of a servo amplifier/integrator, the output of which is then used to control the current which determines the output frequency of the TDL. By closing the servoloop in this fashion the TDL output is frequency offset-locked to the combination of CO_2 (or CO) laser, RF/microwave synthesizer, and the center-frequency of the wideband I.F. discriminator which is monitored by a frequency counter. The entire infrared synthe-sizer system shown in Fig. 1 is controlled by a computer. If, for instance, the microwave synthesizer is frequency swept under computer control, the I.R. output frequency of the TDL would be swept too in synchronism with the microwave synthesizer, since the frequency offset-locking servoloop forces the TDL output to maintain the frequency relationship

$$\nu_{TDL} = \nu_{CO_2/CO} \pm \nu_{synthesizer} \pm \nu_{I.F.\ counter} \tag{1}$$

In Eq. (1) above the frequency of the RF/microwave synthesizer is pre-determined by the operator or computer program, the I.F. frequency is very accurately measured (and averaged if so desired) even in the presence of appreciable frequency modulation (which may be necessary in order to line-center lock either or both lasers); thus the absolute accuracy of the TDL output frequency, ν_{TDL} will, to a very large degree, depend on the abso-lute accuracy, resettability and long term stability of the reference lasers(s). The most accurate results obtained to date were achieved with the use of CO_2 reference lasers; these will be briefly summarized in the next section. For detailed description of techniques and results the readers are referred to the references cited, which were either recently published or are about to be published.

Line-Center Stabilized, Grating Controlled CO_2 Isotope Lasers

This section summarizes the long-term frequency stabilization of CO_2 lasers using the standing-wave saturation resonances in a low-pressure, room temperature, pure CO_2 absorber. As Fig. 2 graphicaly illustrates, these resonances are observed by subjecting the CO_2 absorber gas to the standing-wave CO_2 laser field and detecting the change induced in the entire collisionally coupled (00°1)→000 band spontaneous emission fluo-rescence at 4.3 μm as the laser frequency is tuned within the Doppler profile of a specific 00°1-[10°0,02°0] regular band lasing transition.[3-6] By using the same molecule (CO_2) both as the emitting (laser) and absorbing medium, the long-term frequency stabilization can be accomplished almost perfectly on the line centers of the lasing tran-sitions. This is a very desirable attribute for frequency reproducibility and resettability. Although first demonstrated with CO_2 lasers,[4] the frequency stabilization technique utilizing the standing wave saturation resonances via the intensity changes observed in the spontaneous

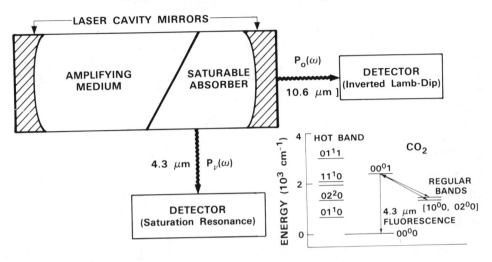

Fig. 2. Graphic illustration of the saturation resonance observed in CO_2 fluorescence at 4.3 μm. The figure shows an internal absorption cell within the laser cavity. External cells may also be used.

fluorescence (side) emission can be (and has been) used with other laser systems as well (e.g., N_2O).[7] This method of frequency stabilization is particularly advantageous whenever the absorbing transition belongs to a hot band with a weak absorption coefficient (such as CO_2 and N_2O). In principle at least, both Lamb-dip and inverted Lamb-dip stabilization of CO_2 lasers is possible and was indeed demonstrated. The former required very low-pressure gas fills and was prone to severe asymmetrical distortions due to competition from adjacent transitions.[8] The inverted Lamb-dip stabilization method on the other hand required very long CO_2 absorption cells heated to hundreds of degrees above room temperature. Of course, saturable absorbers other than CO_2 (e.g., SF_6, OsO_4) can and have been used, but will not be discussed here; the utilization of such absorbers requires the finding of fortuitous near coincidences between each individual lasing transition and a suitable absorption feature in the saturable absorber gas to be used. Indeed, the above considerations prompted the search for an alternate method of frequency stabilization which in turn led to the 4.3 μm fluorescence stabilization technique in 1970.[4]

In CO_2 molecular lasers transitions occur between two vibrational states. Since each vibrational state has a whole set of rotational levels, a very large number of laser lines, each with a different frequency (wavelength) can be generated. Moreover, isotopic substitution of the oxygen and/or carbon atoms make 18 different isotopic combinations possible for the CO_2 molecule. Approximately 80 to 150 regular band lasing transitions may be generated for each of the CO_2 isotopic species. By using optical heterodyne techniques, the beat frequencies between laser transitions of individually line-center stabilized isotopic CO_2 laser pairs were accurately measured. As a result, the absolute frequencies, vacuum wavenumbers, band centers, and ro-vibrational constants for $^{12}C^{16}O_2$, $^{13}C^{16}O_2$, $^{12}C^{18}O_2$, $^{13}C^{18}O_2$, $^{12}C^{17}O_2$, $^{16}O^{12}C^{18}O$, $^{16}O^{13}C^{18}O$, $^{14}C^{16}O_2$ and $^{14}C^{18}O_2$, have been simultaneously calculated from well over 900 beat frequency measurements.[11] The accuracies of these frequency determinations are, for the majority of the transitions, within about 5 kHz relative to the primary Cs frequency standard. Consequently, in the 8.9-12.3

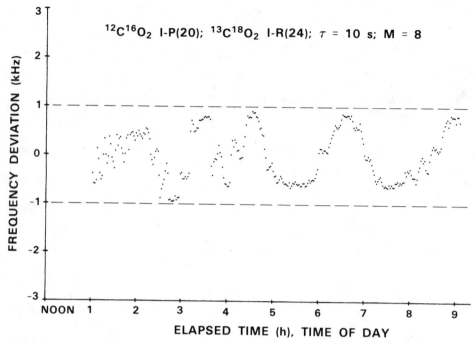

$^{12}C^{16}O_2$ I-P(20); $^{13}C^{18}O_2$ I-R(24); τ = 10 s; M = 8

Fig. 3. Slow drifts in beat frequency due to small deviations from true
zero of the electronics (caused by temperature variations).

μm wavelength region line-center stabilized CO_2 isotope lasers can be
conveniently used as secondary frequency standards. One can also utilize
difference frequencies and harmonics of CO_2 lasing transitions to synthe-
size precisely known reference lines well beyond the 8.9-12.3 μm range.

Reference (3) provides detailed information, guidelines and additional references for the long-term frequency stabilization of CO_2 lasers so
that the reproducibility of the over one-thousand CO_2 isotope reference
lines may be accomplished to within a few kHz accuracy. As an
example[3], the vertical scale in Fig. 3 shows the last digit in kHz of
the beat frequency between two line-center stabilized CO_2 isotope lasers
taken over an 8 1/2 hour time period. Approximately 100 seconds apart a
data point was printed out which represented the deviation from the beat
frequency which was averaged over the 8 1/2 hour observation period. Note
that the peak frequency deviation in Fig. 3 was less than ±1 kHz, (i.e.,
≈±3 part in 10^{11}.

Grating-Controlled CO Isotope Lasers

In principle at least, CO isotope lasers can most conveniently serve
as reference lasers in the 4.9 to 8.0 μm wavelength domain. In terms of
spectral purity, sealed-off operation[12] and abundance of readily
available lasing transitions CO isotope lasers are at least as good as
their CO_2 counterparts. Lamb-dip stabilization of CO lasers was also
accomplished[13] nearly 15 years ago. However, the resettability of the
Lamb-dip stabilization method is at least 100 times less accurate than the
4.3 μm fluorescence stabilization of CO_2 lasers. The absolute accuracy of
presently available CO laser transition frequencies is also only good to
within about one or two MHz. If one MHz or so accuracy is not sufficient,
a direct comparison of a CO reference laser line with an appropriately
selected frequency doubled line-center stabilized CO_2 laser transition is
always possible and was so demonstrated several years ago.[14]

Linewidth Considerations and Lead-Salt Tunable Diode Lasers

The output waveform of a stable, single-frequency laser far above the threshold of oscillation may be approximated by an almost perfect sine wave with nearly constant amplitude and frequency. For a laser operating in an ideal environment, the spectral purity is measured by a linewidth which is determined by frequency fluctuations caused by random walk of the oscillation phase under the influence of spontaneous emission (quantum) noise. In their fundamental 1958 paper, Schawlow and Townes predicted that the quantum phase noise limited line profile will be a Lorentzian with a full width between the half power points (FWHM) that may be approximated by:

$$\Delta\nu_{FWHM} \sim \frac{a\pi h\nu_o}{P_o} \left(\frac{\nu_o}{Q_c}\right)^2 \tag{2}$$

where a, h, ν_o, P_o, and Q_c denote the population inversion parameter, Planck's constant, the center frequency, power output and "cold" cavity Q of the laser, respectively. In a well designed small CO_2 or CO laser the "cold" cavity Q is given by:

$$Q_c \sim \frac{2\pi L \nu_o}{c\, T_r} \tag{3}$$

where L, c, T_r denote the cavity length, velocity of light and mirror transmission, respectively (diffraction losses are usually negligible compared to output coupling loss). In a small CO_2/CO laser with $L = 50$ cm, $T_r = 5\%$, Q_c is of the order of 10^7; thus for a typical power output of 1 to 10 Watts (which is easily obtainable with a small TEM_{ooq} mode CO_2/CO laser) the quantum phase noise limited linewidth is less than 10^{-6} Hz. In actual practice, the so-called "technical noise sources" caused by power supply ripple and noise, acoustic and structure born vibrations, etc., dominate over the quantum phase noise. However, with adequate care, the linewidth broadening of the CO_2/CO reference lasers caused by modulation due to extraneous "technical" noise sources can be kept well below one kHz.

For semiconductor lasers in general, and lead-salt diode lasers in particular, linewidth calculations based even on the first order Schawlow-Townes formulation of Eq. (2) will yield numerical values that are typically several kHz to tens of kHz wide. The reasons for this are as follows.

To begin with, the typical single mode output power of lead-salt diode lasers presently available is generally (much) less than 10 mw. Also, Q_c, the "cold" cavity Q calculated for a "solitary" diode laser (the cavity of which is defined by two cleaved end facets) is generally much less than 10. In terms of the usually measured semiconductor laser parameters the full Lorentzian linewidth between half-maximum power points (FWHM) can be written as

$$\Delta\nu = \left(\frac{h\nu}{8\pi P_o}\right) \left(\frac{v_g}{L}\right)^2 \left(\gamma L + \ln\frac{1}{R}\right) \left(\ln\frac{1}{R}\right) (n_{sp}) (1 + \alpha^2) \,. \tag{4}$$

In Eq. (4) $h\nu$ is the photon energy, P_o the single ended output power, v_g the group velocity, L the length of the cavity defined by the two cleaved end facets with identical reflectivities of R, and γ the distributed loss coefficient. The spontaneous emission factor n_{sp} should approach unity for the cryogenic temperatures at which lead-salt diode lasers are typically operated, and was indeed found to be $n_{sp} \approx 1.3$ at T≃15K.[15]

The linewidth given in Eq. (4) differs from the Schawlow-Townes approximation in Eq. (2) by the factor $n_{sp}(1 + \alpha^2)$. The significance of the linewidth enhancement factor α in Eq. (4) was recently recognized by Henry[16] in order to explain the linewidth broadening measured by Fleming, Mooradian, and Welford[17,18] for AlGaAs lasers. Since there is a detailed discussion of the spectral broadening mechanism in another paper[19] in the Proceedings of this Symposium, this paper will only discuss results relevant to lead-salt diode lasers. For example, it was determined both experimentally[1,15] and theoretically[20] that the linewidth enhancement factor α can be considerably less than 1.0 at least in some of the lead-salt lasers, whereas α was found to be significantly greater than unity in AlGaAs lasers. Thus, in the case of broad-area $PbS_{1-x}Se_x$ homojunction diode lasers the fundamental linewidths measured were only about a factor of 2 wider than predicted by the first order Schawlow-Townes approximation of Eq. (2) and $n_{sp} \cdot (1+\alpha^2)$ was ≃2.17.[15] For example, Fig. 4 shows the beat-note spectrum of a strong mode of a multi-line diode laser, with approximately half of the total power emitted in the particular mode whose spectrum is shown. The dotted line is a computer generated Lorentzian profile with a 22 kHz linewidth. The line-width was calculated from the measured laser parameters using Eq. (4). A flat, white noise spectrum 44 dB below the Lorentzian peak was also added to the theoretical profile in order to correct for the noise floor of the measuring system. Note that the vertical scale is logarithmic with a 10-dB/cm calibration.

Figure 5 shows three data points for the linewidths measured for a single mode $PbS_{1-x}Se_x$ diode laser (indicated by circles) in addition to the linewidth of the laser illustrated in Figure 4 (indicated by a square). Figure 5 shows the linewidths plotted as a function of the inverse of the single-ended laser power output. A linear extrapolation of the three data points of the single mode laser to zero inverse power indicated negligible linewidth intercept. Hence, the power-independent

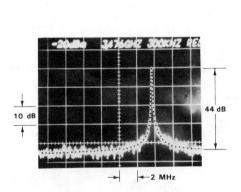

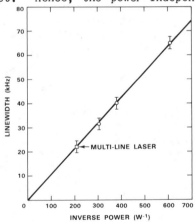

Fig. 4. Beat-note spectrum of a strong mode of a multi-line diode laser; Δν = 22 kHz, P_0 = 8.8 mW integrated over all lines, T = 6.6K, ν_{beat} = 3.5 GHz, ν_{laser} = 1882 cm^{-1}.

Fig. 5. Linewidths of two $PbS_{1-x}Se_x$ diode lasers plotted as a function of the inverse of single-ended output power.

contribution to the linewidth reported by Welford and Mooradian[21] for AlGaAs lasers must be insignificant in the $PbS_{1-x}Se_x$ lasers we measured. This seemingly complete absence of the power independent contribution to the linewidth is even more remarkable considering that the results shown in Figures 4 and 5 were obtained with the $PbS_{1-x}Se_x$ lasers operating between 6.6 and 14.9 K; it is precisely at the lowest cryogenic temperatures where Mooradian et al, found the power independent contribution to the linewidths of AlGaAs lasers to be the greatest, reaching several MHz to tens of MHz in width.

It would be premature, however, to conclude that the extremely narrow (20-30 kHz) linewidths achieved with broad-area, homojunction solitary $PbS_{1-x}Se_x$ tunable diode lasers can be obtained with all other lead-salt compounds and diode structures as well. Figure 6 shows some of the most frequently used lead-salt semiconductors which can provide a class of tunable lasers with emission wavelengths anywhere between about 2 1/2 µm and 30 µm. The figure illustrates the dependence of laser wavelengths on composition of the lead-salt compounds. The lower left portion of the figure shows PbEuSeTe which can be used to cover the 2.6 to 6.6 µm wavelength range. This is a new material recently developed at the General Motors Research Laboratories (GMR). Molecular beam epitaxial (MBE) growth of $Pb_{1-x}Eu_xSe_yTe_{1-y}$ lattice-matched to PbTe substrates has been used to fabricate double heterojunction diode lasers, including quantum well, large optical cavity (LOC) devices, the first such TDL structures reported for a lead-salt compound. These Mesa stripe

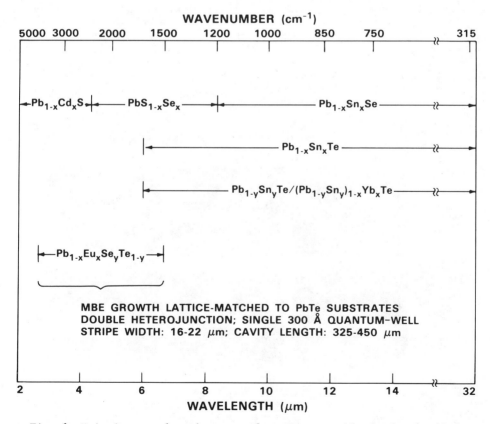

Fig. 6. Emission wavelength range of various tunable lead-salt diode lasers.

geometry diode lasers were fabricated with stripe widths from 16 to 22 μm and the cleaved cavity lengths were typically 325 to 450 μm long. The PbTe quantum well width, L_z, was varied in the sequence 300, 600, 1200 and 2500 Å in a series of otherwise similar growths. Strong quantum effects were observed for $L_z \leq 1200$ Å.[22] Even these first-reported PbTe quantumwell diode lasers achieved major improvements in lasing threshold current (2.1 ma at 13 K), maximum operating temperatures (174 K cw; 284 K pulsed), over 30% (>700 cm^{-1} at 2000 cm^{-1} center frequency) tuning range with a single laser, and demonstrated much better frequency reproducibility and far-field patterns than the broad-area homojunction lead-salt lasers. However, the narrowest linewidths obtained (~500 kHz) during the initial phase of the experiments[23] were approximately an order of magnitude wider than were achieved with broad-area homojunction $PbS_{1-x}Se_x$ and $Pb_{1-x}Sn_xTe$ lasers.[15,24] Obviously there are still many unanswered questions but also numerous interesting and applicable publications.[25]

One of the major difficulties in using and furthering the development of lead-salt diode lasers comes from the fact that they must operate at cryogenic temperatures. Although the recent results with quantum-well structures offer the possibility of laser action within thermo-electric cooler range in the near future, the most commonly used technique to produce the required TDL temperatures is by means of closed-cycle, compressed helium refrigerators which are very far from an optimum design that could be made without too much difficulty. As a consequence, the laser diodes are subjected to excessive vibrations which in turn produce a great deal of frequency jitter and mode instability. As an example, the photographs in the upper row of Fig. 7 clearly demonstrate the frequency jitter in a noisy laboratory which is most frequently encountered in practice. In Fig. 7, the camera shutter times of 1/15s, 1s, and 60s are indicated above each column and correspond to single trace, 13 trace and 1000 trace film exposures, respectively. The horizontal scale is 10 MHz/cm for each of the six photographs. The bottom row clearly indicates that the freqency jitter and drift was reduced many orders of magnitude by closing the frequency offset-locking servo-loop as previously described (see Fig. 1).

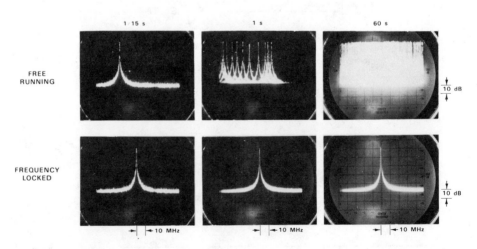

Fig. 7. Comparison of free running and frequency locked TDL output (in a noisy laboratory) for single scan and longer observation times. ν_{TDL} = 1866.7416 cm^{-1} + 2614 MHz; I.F. = 300 kHz and 5 ms/cm.

Conclusions

In conclusion, the results presented in this paper show good agreement between theoretical and measured linewidths of well-behaved lead-salt diode lasers. In some cases at least, the linewidths were only about a factor of two wider than the first order Schawlow-Townes prediction which equates $(n_{sp}) \cdot (1+\alpha^2) = 1$. Furthermore, linewidths as narrow as 22 kHz have been achieved which represents a five-fold improvement in fractional linewidth, $\Delta\nu/\nu$, over the one previously reported[24] for lead-tin-telluride lasers. Frequency offset-locking to a stable reference laser and slaving the tunable diode laser frequency to a programmmable microwave synthesizer were also achieved. These results demonstrate the feasibility of true frequency synthesizers in a very significant portion of the infrared spectrum, because in the 9 to 12.3 µm region the frequencies of well over a thousand CO_2 reference laser lines are known with an accuracy of about 5 kHz with respect to the Cesium frequency standard, and similar data could be generated in a straightforward fashion for thousands of CO reference laser transitions between 5 and 8 µm.

Acknowledgements

This work was supported by the NASA Langley Research Center under contract NAG-1-164. The author would like to express his appreciation to the General Motors Research Laboratories for providing the state of the art diode lasers used in the experiments. The author also wishes to express his gratitude for the wonderful cooperation of J. W. Bielinski at MIT Lincoln Laboratory, D. L. Partin and, above all, W. Lo at the General Motors Research Laboratories. The enthusiastic, cheerful and always fruitful collaboration of Wayne Lo will be very much missed by all of us who were privileged to know him.

References

1. C. Freed, J. W. Bielinski, and W. Lo, "Programmable, Secondary Frequency Standard Based Infrared Synthesizer using Tunable Lead-Salt Diode Lasers," Proceedings of Tunable Diode Laser Development and Spectroscopy Applications," SPIE 438, (1983).

2. D. L. Spears and C. Freed, "HgCdTe Varactor Photodiode Detection of cw CO_2 Laser Beats beyond 60 GHz," Appl. Phys. Lett. 23, 445 (1973).

3. K. L. SooHoo, C. Freed, J. E. Thomas and H. A. Haus, "Line-Center Stabilized CO_2 Lasers as Secondary Frequency Standards; Determination of Pressure Shifts and Other Errors", IEEE. J. Quant. Electron. QE-21, 1159 (1985).

4. C. Freed and A. Javan, "Standing-Wave Saturation Resonances in the CO_2 10.6 µ Transitions Observed in a Low-Pressure Room-Temperature Absorber Gas," App. Phys. Lett. 17, 53 (1970).

5. A. Javan and C. Freed, "Method of Stabilizing a Gas Laser," U.S. Patent No. 3686585 issued Aug. 22, 1972.

6. C. Freed and R. G. O'Donnell, "Advances in CO_2 Laser Stabilization Using the 4.3 µm Fluorescence Technique," Metrologia 13, 151 (1977).

7. J. E. Thomas, "A Study of N_2O Molecular Hyperfine Structure in Excited Vibrational States Utilizing a Stabilized Twin Laser Spectrometer," PH.D. THESIS, Dept. of Physics, M.I.T. (1979).

8. C. Bordé and L. Henry, "Study of the Lamb Dip and of Rotational Competition in a Carbon Dioxide Laser", IEEE. J. Quantum Electron, $\underline{QE-1}$, 874 (1968).

9. H. T. Powell and G. J. Wolga, "Central Tuning Peak and Inhomogenous Saturation with the CO_2 Laser", Bull. AM. Phys. Soc. $\underline{15}$, 423 (1970).

10. L. S. Vasilenko, M. N. Skvortsov, V. P. Chebotaev, G. I. Shershneva, and A. V. Shishaev, "Frequency Stabilization of a CO_2 Laser", Optical Spectroscopy $\underline{32}$, 609 (1972).

11. L. C. Bradley, K. L. SooHoo, and C. Freed, "Absolute Frequencies of Lasing Transitions in Nine CO_2 Isotopic Species," IEEE. J. Quant. Electron., $\underline{QE-22}$, Feb. (1986), (To be Published).

12. C. Freed, "Sealed-Off Operation of Stable CO Lasers," Appl. Phys. Lett. $\underline{18}$, 458 (1971).

13. C. Freed and H. A. Haus, "Lamb Dip in CO Lasers," IEEE J. Quantum Electron. $\underline{QE-9}$, 219 (1973).

14. R. S. Eng, H. Kildal, J. C. Mikkelsen, and D. Spears, "Determination of Absolute Frequencies of $^{12}C^{16}O$ and $^{13}C^{16}O$ Laser Lines," Appl. Phys. Lett. $\underline{24}$, 231 (1974).

15. C. Freed, J. W. Bielinski, and W. Lo, "Fundamental Linewidth in Solitary, Ultranarrow output $PbS_{1-x}Se_x$ Diode Lasers," Appl. Phys. Lett. $\underline{43}$, 629 (1983).

16. C. H. Henry, "Theory of the Linewidth of Semiconductor Lasers," IEEE J. Quantum Electron., $\underline{QE-18}$, 259 (1982).

17. M. W. FLeming and A. Mooradian, "Spectral Characteristics of External-cavity Controlled Semiconductor Lasers", IEEE J. Quantum Electron., $\underline{QE-17}$, 44 (1981).

18. D. Welford and A. Mooradian, "Output Power and Temperature Dependence of the Linewidth of Single-frequency cw (GaAl)As Diode Lasers", Appl. Phys. Lett. $\underline{40}$, 865 (1982).

19. J. Harrison and A. Mooradian "Spectral Linewidth of Semiconductor Lasers" Fritz Haber International Symposium on Methods of Laser Spectroscopy, Dec. 16 - 20, (1985).

20. Y. Shani, R. Rosman, and A. Katzir, "Calculation of the Refractive Indexes of Lead Chalcogenide Salts and its Application for Injection Lasers", IEEE J. Quantum Electron. $\underline{QE-21}$, 51 (1985).

21. D. Welford and A. Mooradian, "Observation of Linewidth Broadening in (GaAl)As Diode Lasers due to Electron Number Fluctuations", Appl. Phys. Lett. $\underline{40}$, 560 (1982).

22. D. L. Partin, "Lead Salt Quantum Well Diode Lasers", Superlattices and Microstructures, $\underline{1}$, 131 (1985).

23. C. Freed, J. W. Bielinski, W. Lo, and D. L. Partin "Output Characteristics of Lead-Telluride Quantum-Well Diode Lasers", paper presented at the IQEC '84 Conference, Anaheim, CA, 21 June 1984.

24. E. D. Hinkley and C. Freed, "Direct Observation of the Lorentzian Line Shape as Limited by Quantum Phase Noise in a Laser above Threshold," Phys. Rev. Lett. $\underline{23}$, 277 (1969).

25. A periodically updated list of papers and reports on lead-salt diode lasers and their applications is available on request from the Laser Analytics Division of Spectra-Physics, Inc., Bedford, MA 01730.

THE HYDROGEN ATOM IN A NEW LIGHT

T. W. Hänsch, R.G. Beausoleil, U. Boesl,[†]
B. Couillaud, C. J. Foot,[††] E. A. Hildum,
and D. H. McIntyre[†††]

Department of Physics
Stanford University
Stanford, CA 94305, USA

INTRODUCTION

It is well known that lasers and coherent light techniques have revolutionized high resolution spectroscopy. Today we have at our disposal a powerful arsenal of techniques which can overcome the Doppler broadening of spectral lines, achieving ever higher spectral resolution.[1] At Stanford, we have long been fascinated by the prospects of applying such tools to atomic hydrogen.[2] As the simplest of the stable atoms, hydrogen permits unique confrontations between experiment and quantum electrodynamic theory. After briefly reviewing past spectroscopic studies of hydrogen, we will report on some recent experimental advances which have opened the door to dramatic future improvements in resolution, creating unprecedented opportunities for precision measurements of fundamental constants and for stringent tests of basic physics laws.

SUMMARY OF PAST WORK

Optical spectroscopy of the simple hydrogen atom has played a central role in the development of atomic physics and quantum mechanics.[3] The visible Balmer spectrum was the Rosetta stone which inspired the pathbreaking discoveries of Bohr, Sommerfeld, De Broglie, Schrödinger, Dirac, and Lamb. More than once, seemingly minute discrepancies between theory and experiment led to important breakthroughs in our understanding of quantum physics.

Nonetheless, the resolution of optical spectral lines remained limited by Doppler broadening to about 1 part in 10^5 until 1971. Fig. 1 illustrates the improvements achieved since then by methods of Doppler-free laser spectroscopy.[4-13] A first "quantum jump" in resolution (A) occurred when the red Balmer-α line was observed by saturation spectroscopy in a gas discharge with a monochromatic tunable dye laser.[4] Selecting slow atoms by spectral hole burning, the method resolved single fine structure components for the first time, and the 2S Lamb shift could be directly observed in the optical spectrum. The technique was applied soon afterwards (B) to measure the absolute wavelength of Balmer-α, yielding an almost tenfold improvement of the Rydberg constant.[5] Later studies of Balmer-α by polarization spectroscopy (D) and by laser-quenching of a metastable atomic beam (F) have brought only minor improvements in resolution to about 1 part in 10^7, as limited by the natural linewidth of Balmer-α, but they have improved the precision of the Rydberg by another order of magnitude to 1 part in 10^9.[7-9]

It was soon recognized[6] that much higher resolution should be achievable by two-photon spectroscopy of the transition from the 1S ground state to the metastable 2S state (Fig. 2). The 2S lifetime of 1/8 sec. implies a natural linewidth of only 1.3 Hz, or less than one part in 10^{15}. Doppler-broadening can be conveniently eliminated by excitation with two counterpropagating photons whose first-order Doppler shifts cancel. Until recently, however, intense monochromatic ultraviolet light at the required wavelength of 243 nm was only available from pulsed laser systems with considerable instrumental linewidth. Nonetheless, even the earliest crude experiments (entry C in Fig. 1) achieved a resolution comparable to that of Balmer-α.[6] Wieman (Fig. 1, E) was able to measure the Lamb shift of the 1S ground state to better than 0.5% or 30 MHz by comparing 1S-2S with the n=2→4 Balmer-β line at 486 nm, even though the accuracy was limited by frequency chirping in the pulsed dye laser amplifiers.[8] The hydrogen deuterium isotope shift could be measured to with 6 MHz, providing first experimental evidence for a predicted relativistic recoil correction.

In a recent beautiful experiment, S. Chu et al.[10] have applied the same laser technique to the 1S-2S transition in positronium (Fig. 1, G). High resolution laser spectroscopy of the purely leptonic positronium atom avoids the complications of nuclear structure effects, but the resolution will ultimately be limited by annihilation to about 2 parts in 10^9.

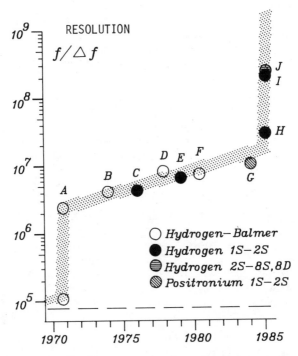

Fig. 1. Resolution achieved by Doppler-free laser spectroscopy of atomic hydrogen and positronium. The letter symbols refer to the following References: A = 4; B = 5; C = 6; D = 7; E = 8; F = 9; G = 10; H = 11; I = 12; J = 13.

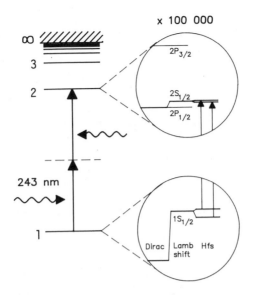

x 100 000

Fig. 2. Energy levels of atomic hydrogen with 1S-2S two-photon transition.

CURRENT EXPERIMENTS

Measurement of the Hydrogen 1S-2S Frequency

After more than a decade of only gradual improvements in resolution, major advances could finally be reported in 1985. Hildum et al.[11] observed the hydrogen 1S-2S transition with a resolution of 3 parts in 10^8 (Fig. 1, H) in a new pulsed experiment designed to reduce the serious systematic errors of earlier studies. A scheme of the apparatus is shown in Fig. 3. The 243 nm radiation is generated by frequency doubling the output of a pulsed dye laser amplifier, fed by a cw dye laser oscillator at 486 nm. The dye amplifier broadens the linewidth to about 180 MHz, and it introduces frequency shifts due to rapid refractive index changes. A narrow spectral "slice" of the output pulse is therefore selected with a confocal filter interferometer which is servo-locked to the frequency of the cw oscillator. The filtered light is intense enough to permit efficient second harmonic generation in an angle-tuned urea crystal. The ultraviolet light is then used to excite the 1S-2S transition in an atomic hydrogen beam. A large fraction of the metastable 2S atoms are photoionized by the same radiation, and the signal is detected by monitoring the ions with a microchannel detector.

Figure 4 shows the 1S-2S two-photon spectrum with its two hyperfine components, recorded in this way. Shown below is a Doppler-free saturation spectrum of $^{130}Te_2$, observed simultaneously with the cw dye laser at the fundamental wavelength. The tellurium spectrum appears shifted by 60 MHz towards lower frequencies, since a 120 MHz acousto-optic Bragg cell is employed as a chopper. The component i_2 is thus found in nearly perfect coincidence with the hydrogen F=1→1 resonance. The absolute frequency of the nearby $^{130}Te_2$ component b_2 has recently been measured interferometrically to within 4 parts in 10^{10}.[14] Using this line as a reference and taking advantage of the auxiliary $^{130}Te_2$ line i_2, the frequency of the centroid of the two hyperfine components was measured to be f(1S-2S) = 2 466 061 395.6(4.8) MHz. The uncertainty is dominated by residual spatial and spectral frequency variations of the filtered light pulses.

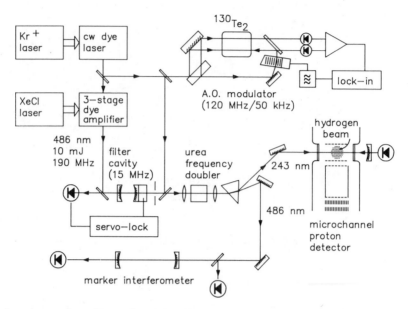

Fig. 3.   Apparatus for pulsed Doppler-free two-photon spectroscopy of
hydrogen 1S-2S at 243 nm. A calibrated frequency reference is
provided by cw saturation spectroscopy of $^{130}Te_2$ with the cw dye
laser oscillator at 486 nm.[11]

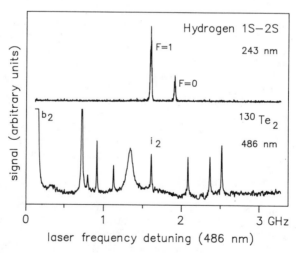

Fig. 4.   Pulsed Doppler-free two-photon spectrum of hydrogen 1S-2S, recorded
with a frequency doubled pulsed dye laser, and saturation spectrum
of $^{130}Te_2$, recorded simultaneously at the fundamental wavelength.

The 1S-2S frequency can be used to determine a new value of the Rydberg constant if we trust the theoretical value of the 1S Lamb shift.[15] The result, R_∞ = 109 737.314 92(21) cm^{-1}, is not in good agreement with the most recent previous value derived from the frequency of Balmer-α.[9] Figure 5 gives a comparison of different laser measurements of the Rydberg constant, adjusted for the redefinition of the meter.[16] The value of Amin et al. has also been corrected for a sign error in the diffraction correction, and its error bars have been extended towards higher frequencies to reflect the results of more recent experiments.[17] Hildum's experiment represents the first laser measurement of the Rydberg constant based on a transition other than Balmer-α. However, its systematic corrections appear to place a practical limit on further improvements of pulsed techniques for such precision measurements.

CW Two-Photon Spectroscopy of Hydrogen 1S-2S

An important breakthrough was achieved very recently (Fig.1, I) when C.J. Foot et al.[12] succeeded in observing the 1S-2S transition by continuous-wave two-photon spectroscopy. The key is a reliable source of intense tunable cw radiation in the difficult wavelength region near 243 nm. No nonlinear crystal is known which would permit efficient critically phase-matched second harmonic generation at this wavelength. However, it is possible to generate 243 nm light as the sum frequency of two different primary lasers. As shown in Fig. 6, the frequency of a 351 nm argon laser and a 789 nm cw dye laser are summed in a KDP crystal, heated to the phase-matching temperature of 62°C. A servo-locked build-up cavity enhances the dye laser radiation at the crystal about 60-fold. With 500 mW from each primary laser, up to 10 mW have been generated near 243 nm. At the final interaction region, another tenfold increase in intensity is provided by a resonant standing wave cavity.

Figure 7 shows a cw two-photon spectrum of hydrogen 1S-2S, as recorded in the first cw experiments. The hydrogen atoms are simply observed in a gas cell at room temperature and the signal is monitored by counting collision-induced Lyman-α photons. The resolution of 5 parts in 10^9 is limited by the

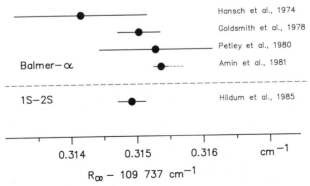

Fig. 5. Laser measurements of the Rydberg constant. Published results have been adjusted for the redefinition of the meter.[16]

finite transit time of the atoms traversing the focused laser field, but collisions and laser bandwidth also contribute.

A comparable resolution (Fig.1, J) was recently achieved by Biraben and Julien[13] who observed 2S-8S and 2S-8D transitions by Doppler-free two-photon spectroscopy of a metastable hydrogen beam.

A New Measurement of the Rydberg Constant

Although large improvements in the resolution of the hydrogen 1S-2S transition appear feasible by straightforward means, a new absolute frequency measurement appears interesting even at the present stage, considering the discrepancy between the two most recent Rydberg measurements. Such an experiment is now in an advanced stage of preparation in our laboratory. The frequency reference is provided by a krypton-laser-pumped 486 nm cw dye laser which is servo-locked to the calibrated Doppler-free i_2 line of $^{130}Te_2$. An external urea crystal produces a few nW of second harmonic radiation in

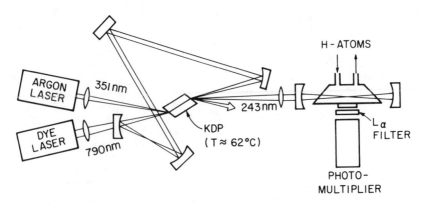

Fig. 6. Ultraviolet sum frequency generator for cw two-photon spectroscopy of hydrogen 1S-2S.[12]

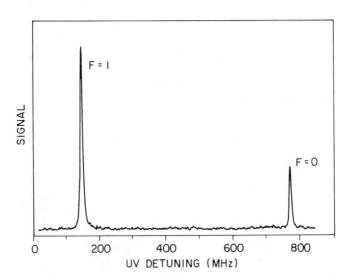

Fig. 7. CW two-photon spectrum of hydrogen 1S-2S, recorded in a gas cell at room temperature.[12]

near coincidence with the 243 nm radiation from the sum frequency generator. An rf beat signal is then readily observed with a photomultiplier, permitting a precise frequency comparison. We expect an accuracy of 4 parts in 10^{10}, limited only by the calibration of the $^{130}Te_2$ reference line.

FUTURE EXPERIMENTS

In a next round of experiments, we will try to advance the resolution of 1S-2S by at least another 3 orders of magnitude. The laser linewidth can be drastically reduced by internal or external frequency stabilizers with fast servo response.[18] Pressure broadening will be eliminated by observing the atoms in a beam rather than in a gas cell. In order to reduce transit broadening, the interaction time must be extended by sending the atoms along the light waves, or by Ramsey spectroscopy with two spatially separated interaction regions.[19] With hydrogen at room temperature and an interaction length of 1 cm, transit broadening would thus be kept below 60 kHz, comparable to the second order Doppler shifts due to the time dilation of special relativity (70 kHz). Considerable improvement is gained if the hydrogen beam is cooled to liquid helium temperature by mounting the escape nozzle on a helium cryostat,[20] as illustrated in Fig. 8. In our example, transit broadening will then be reduced to 7 KHz, and second order Doppler shifts remain smaller than 1 kHz. At the same time, the longer interaction time due to cooling increases the excitation probability per atom almost eightyfold.

For experiments in the more distant future, we are exploring means to slow the hydrogen atoms far below liquid helium temperature. During the past year, we have witnessed a number of spectacular experimental advances in laser cooling and trapping of neutral atoms.[21-23] Unfortunately, laser cooling of hydrogen requires vacuum ultraviolet Lyman-α radiation. On the other hand, a hydrogen atom at 4K can be stopped by scattering only 100 photons.

Once the kinetic energy has been reduced to a few mK, the earth's gravitational field could be used to further slow the atoms in an atomic "fountain," so that the projectiles can be observed for extended periods in free fall near the turning point of their parabolic trajectories. Optical two-photon Ramsey spectroscopy of such freely falling atoms is possible with a single standing wave laser field which the atoms traverse on their way up and again on their way down (Fig. 9).[24,25] As in ordinary Ramsey spectroscopy,[19] the first passage leaves the atoms in a coherent superposition of ground state and excited state, and the second passage will either further excite the atoms or return them to the ground state, dependent

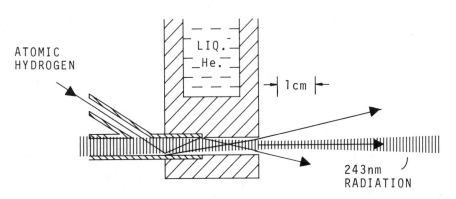

Fig. 8. Apparatus for high resolution spectroscopy of a cold hydrogen beam.

on the relative phase between the light field and the atomic oscillations. For atoms of a given velocity, the combined excitation probability as a function of the laser frequency exhibits the expected periodic Ramsey fringes. In an atomic beam, the slowest atoms spend the longest time between the two interactions and contribute the narrowest fringes. In a fountain, by contrast, the narrowest fringes are due to the fastest atoms, and when averaging over a broad velocity distribution, we predict a simple near-Lorentzian lineshape with just the natural linewidth. First order Doppler shifts cancel even if the counterpropagating beams are slightly misaligned, and a resolution near 1 Hz appears feasible with a fountain of modest dimensions.

COMPARISON WITH OTHER TRANSITIONS

An elaborate frequency chain could be designed to compare the 1S-2S frequency directly with the cesium frequency standard. Such a measurement could yield a much improved Rydberg constant, improving the link between the time standard and the system of fundamental constants. However, it could not by itself serve as a test of basic physics laws, since the cesium hyperfine splitting is not well understood theoretically.

It appears more interesting to compare the 1S-2S transition with other optical frequencies that can also be calculated from first principles, such as other transitions in the hydrogen atom itself. One obvious candidate is the 486 nm $n=2 \rightarrow 4$ Balmer-β line, used in earlier pulsed measurements of the 1S Lamb shift.[8] However the large natural linewidth (≈ 14 MHz) of even the narrowest fine structure components limits the potential accuracy. Two-photon transitions from the metastable 2S state to highly excited nS Rydberg states can provide much sharper resonances, if perturbing external fields are eliminated.[13] The excitation of levels with $n \approx 100$ would require a laser wavelength near 729 nm, slightly more than 3 times larger than the 243 nm of the 1S-2S resonance, and it should be feasible to compare the two frequencies with extreme precision by detecting a microwave beat signal between the ultraviolet light and the third harmonic of the red laser.

WHAT CAN BE LEARNED?

Quantum electrodynamics is believed to provide a very good theory for the hydrogen atom, although the predictions of calculations[15] are limited by the uncertainties of fundamental constants, by unknown nuclear size and structure effects, and by computational approximations. Fortunately

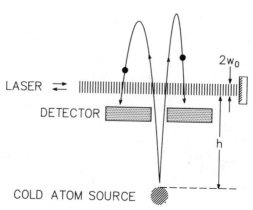

Fig. 9. Scheme for optical Ramsey spectroscopy of an atomic fountain.[24,25]

important cancellations of these uncertainties can occur if different hydrogenic transitions are compared. Precision spectroscopy of atomic hydrogen can hence yield accurate new values for the Rydberg constant, the electron/proton mass ratio, and the rms charge radius of the proton. Possible paths towards this end are illustrated in Fig. 10.

At present, the dominant contribution to the uncertainty of the predicted 1S-2S interval (2.5 MHz) is the uncertainty of the Rydberg constant (10^{-9}), and a measurement of the 1S-2S frequency can improve the Rydberg constant about tenfold, until we approach the uncertainties due to the electron mass (5×10^{-8}: 70 kHz), the charge radius of the proton (10^{-1}: 130 kHz), and approximations in the computation of QED corrections (60 kHz). The fine structure constant (10^{-7}) contributes an uncertainty of only 4 kHz.

Above we have proposed to measure the the small difference frequency df = f(1S-2S) − 3 f(2S-nS). This frequency depends critically on the Lamb shifts of the participating levels, and can provide a stringent test of QED. For n≈100, the theoretical uncertainty of df is dominated by the nuclear size effect (≈70 kHz) and by approximations in the computation of electron structure corrections and uncalculated higher order QED corrections (65 kHz). The contribution of the Rydberg constant (1 kHz), the electron mass (0.05 kHz), and the fine structure constant (4 kHz) are negligible by comparison. The QED computations can be improved, and if theory is correct, a precision measurement of df can provide accurate new values for the charge radii of the proton and deuteron.

Measurements of the 1S-2S hydrogen-deuterium isotope shift have previously been suggested as a means to measure the proton/electron mass ratio.[18] However, after recent improved measurements of this ratio,[26] the uncertainty of the isotope shift is no longer dominated by the electron mass (35 kHz), but instead by nuclear size effects (180 kHz).

We have pointed out[12] that it is easily possible to construct composite frequencies which no longer depend on energy shifts which scale with the inverse cube of the principal quantum number. Such frequencies are thus independent of nuclear size, and they no longer depend on uncalculated higher order QED effects. One interesting example is a composite of the two isotope shifts, $\Delta f(2S-nS) - 1/7 (1 - 8/n^3) \Delta f(1S-2S)$. A measurement of this frequency can serve to determine a much improved value for the electron/proton mass ratio. Once this ratio and the proton size are better known, a much improved value of the Rydberg constant can be determined from the frequency f(1S-2S). By taking advantage of the composite frequency $f(1S-2S) - 7/(4+24/n^3)$ df, which is again independent of nuclear size effects, we can actually determine the Rydberg constant without having to find the proton size explicitly.

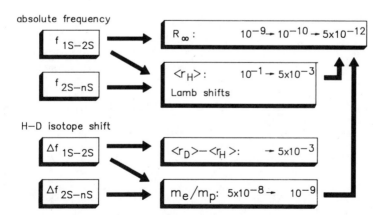

Fig.10.   Proposed paths from precision spectroscopy of atomic hydrogen to accurate new values of fundamental constants.

Stringent tests of basic physics laws would be possible if better precision values for the same constants could be obtained by independent means. Rf spectroscopy of transitions between highly excited hydrogen Rydberg levels may provide a promising approach for such measurements. It will also be most interesting to compare hydrogen 1S-2S with sharp transitions in positronium, muonium, pionium, or other hydrogenlike systems.

ACKNOWLEDGEMENTS

We are grateful to G.W. Erickson, A.I. Ferguson, J.L. Hall, and D. Kleppner for stimulating discussions.

The work reported here has been supported by the National Science Foundation under Grant No. NSF PHY83-08721 and by the U.S. Office of Naval Research under Contract ONR N00014-C-0403.

†Permanent address: Institut für Physikalische Chemie und Elektrochemie, Technische Universität München, 8046 Garching, F.R.G.

††Permanent address: Clarendon Laboratory, University of Oxford, Parks Road, Oxford OX1 3PU, U.K.

†††National Science Foundation Predoctoral Fellow

REFERENCES

1. M.D. Levenson, "Introduction to Nonlinear Laser Spectroscopy," Academic Press, New York, 1982
2. T.W. Hänsch, A.L. Schawlow, and G.W. Series, Scientific American 240, 94 (1979)
3. G.W. Series, "Spectrum of Atomic Hydrogen," Oxford University Press, 1957
4. T.W. Hänsch, I.S. Shahin, and A.L. Schawlow, Nature 235, 63 (1972)
5. T.W. Hänsch, M.H. Nayfeh, S.A. Lee, S.M. Curry, and I.S. Shahin, Phys. Rev. Letters 32, 1336 (1974)
6. S.A. Lee, R. Wallenstein, and T.W. Hänsch, Phys. Rev. Letters 35, 1262 (1975)
7. J.E.M. Goldsmith, E.W. Weber, and T.W. Hänsch, Phys. Rev. Letters 41, 1525 (1978)
8. C. Wieman and T.W. Hänsch, Phys. Rev. A22, 192 (1980)
9. S.R. Amin, C.D. Caldwell, and W. Lichten, Phys. Rev. Lett. 47, 1234 (1981)
10. S. Chu, A.P. Mills, and J.L. Hall, Phys. Rev. Lett. 52, 1689 (1984)
11. E.A. Hildum, U. Boesl, D.H. McIntyre, R.G. Beausoleil, and T.W. Hänsch, Phys. Rev. Letters, submitted for publication
12. C.J. Foot, B. Couillaud, R.G. Beausoleil, and T.W. Hänsch, Phys. Rev. Letters 54, 1913 (1985)
13. F. Biraben and L. Julien, Opt. Commun. 53, 319 (1985)
14. J.R.M. Barr, J.M. Girkin, A.I. Ferguson, G.P. Barwood, W.R.C. Rowley, and R.C. Thompson, Opt. Commun. 54, 217 (1985)
15. G.W. Erickson, J. Phys. Chem. Ref. Data 6, 831 (1977)
16. "Documents concerning the new definition of the metre," Metrologia 19, 163 (1984)
17. W. Lichten, private communication
18. J.L. Hall and T.W. Hänsch, Optics Letters 9, 502 (1984)
19. S.A. Lee, J. Helmcke, and J.L. Hall, in Laser Spectroscopy IV, H. Walther and K. W. Rothe, eds., (Springer Verlag, Heidelberg, 1979), p. 130
20. J.T.M. Walraven and I.F. Silvera, Rev. Sci. Instr. 53, 1167 (1982)

21. A. Migdal, J.V. Prodan, W.D. Phillips, T.H. Bergeman, and
     H.J. Metcalf, Phys. Rev. Lett. $\underline{54}$, 2596 (1985)
22. W. Ertmer, R. Blatt, J.L. Hall, and M. Zhu,
     Phys. Rev. Lett. $\underline{54}$, 996 (1985)
23. S. Chu, L. Hollberg, J.E. Bjorkholm, A. Cable, and A. Ashkin,
     Phys. Rev. Lett. $\underline{55}$, 48 (1985)
24. R.G. Beausoleil and T.W. Hänsch, Optics Letters $\underline{10}$, 547 (1985)
25. R.G. Beausoleil and T.W. Hänsch, Phys. Rev. A,
     submitted for publication
26. R.S. Van Dyck, Jr., F.L. Moore, And P.B. Schwinberg,
     Bull. Am. Phys. Soc. $\underline{28}$, 791 (1983)

RAMAN HETERODYNE SPECTROSCOPY OF HERTZIAN RESONANCES

J. Mlynek, E. Buhr and Chr. Tamm

Institut für Quantenoptik, Universität Hannover
Welfengarten 1, 3000 Hannover 1, Fed.Rep.Germany

INTRODUCTION

In spectroscopy one often wants to measure small energy
splittings in ground states and electronically excited states
in atoms and molecules with high precision and sensitivity.
Such sublevel or Hertzian resonances with typical splitting
frequencies in the radiofrequency range (1 MHZ-1 GHz) yield
valuable information on the atomic structure; the sublevel
splittings may arise, e.g., from hyperfine interactions or
from external fields (Zeeman or Stark splittings). Moreover,
the linewidth of these Hertzian resonances gives us a knowledge on relaxation mechanisms that affect the sublevel populations and coherences; the corresponding decay rates are
thus useful to characterize the dynamics of the atomic system.
    In the past, optical methods have proven to be a power-
ful tool in the study of resonances among closely spaced sublevels[1]. A first example is the radiofrequency-optical double
resonance technique: Here a Hertzian resonance is induced by
means of a rf field; the light field is used to create the
nonequilibrium population of the sublevels ("optical pumping")
that is required for rf excitation, and to detect the rf-
induced changes in the optical properties of the sample.
    Another example is the optical -optical double resonance
method: here two resonant optical fields are used to induce a
Hertzian resonance. When the light fields are coherent with
respect to each other, a coherent superposition of the substates will be created; a convenient way of establishing the
necessary coherence between the light fields is, e.g., amplitude modulation of one light beam[2]. The essential feature of
this modulated pumping technique is the occurence of a narrow
resonance in the Hertzian coherence when the modulation frequency equals the splitting frequency among the substates.
Again the Hertzian coherence can be detected via changes in
the optical properties of the sample.
    It is important to note, that in both double resonance
methods the obtained resolution is neither limited by the
homogeneous or inhomogeneous broadening of the optical transition nor by broadband optical excitation; thus the initial
development of high-resolution double resonance spectroscopy

has been possible with the use of incoherent broadband light sources.

On the other hand, many possible applications of this type of spectroscopy have remained elusive because of a lack of spectral intensity, monochromaticity, tunability and of spatial coherence of the thermal light sources. This situation changed drastically with the advent of the laser that not only brought along a renaissance of classical double resonance spectroscopy, but also the development of new coherent, nonlinear spectroscopic techniques.

In this contribution we present two laser spectroscopic methods that use coherent resonance Raman scattering to detect rf-or laser -induced Hertzian coherence phenomena in the gas phase; these novel coherent double resonance techniques for optical heterodyne detection of sublevel coherence clearly extend the above mentioned previous methods using incoherent light sources. In the case of Doppler broadened optical transitions new signal features appear as a result of velocity-selective optical excitation caused by the narrow-bandwidth laser. We especially analyze the potential and the limitations of the new detection schemes for the study of collision effects in double resonance spectroscopy. In particular, the effect of collisional velocity changes on the Hertzian resonances will be investigated.

In the first part of the paper we discuss the Raman heterodyne detection of rf-generated sublevel coherence in atomic Sm vapor in the presence of rare gas perturbers. In a second part we report on the observation of novel Ramsey-type resonances due to collisional velocity diffusion in Sm vapor; here the experimental technique uses counterpropagating laser fields and relies on coherent Raman processes to optically excite and detect Hertzian coherence.

RAMAN HETERODYNE DETECTION OF RF RESONANCES IN SM VAPOR

Recently, a novel rf-laser double resonance method for optical heterodyne detection of sublevel coherence phenomena was introduced[3]. This so-called Raman heterodyne technique relies on a coherent Raman process being stimulated by a resonant rf field and a laser field (see Fig.1(a)). The method has been applied to impurity ion solids for studying nuclear magnetic resonances at low temperature[3-5] and to rf resonances in an atomic vapor[6,7]. In this section we briefly review our results on Raman heterodyne detection of rf-induced resonances in the gas phase. As a specific example, we report studies on Zeeman resonances in a J=1 - J'=0 transition in atomic samarium vapor in the presence of foreign gas perturbers.

The schematic of our experimental arrangement and the basic coherent Raman process for the studied case of a Zeeman split J=1-J'=0 transition are shown in Fig.1. The laser field $\vec{E}_0$ of frequency ω_E is polarized parallel to a transverse static magnetic field $\vec{B}_0$ that lifts the ground state Zeeman-level degeneracy. The field $\vec{E}_0$ only drives the optical π transition (m-m'=0); resonant optical absorption and isotropic spontaneous reemission thereby induce a transfer of the m=0 population to the m=$\pm$1 levels for atoms of the optically resonant velocity subgroup. In the limit of slow ground-state relaxation, this velocity-selective optical pumping process efficiently produces an alignment in the J=1 manifold. A longitudinal rf field of frequency ω_H can now resonantly excite $|\Delta m|$=1 sublevel coherences and, via a two-photon process, the simultaneous

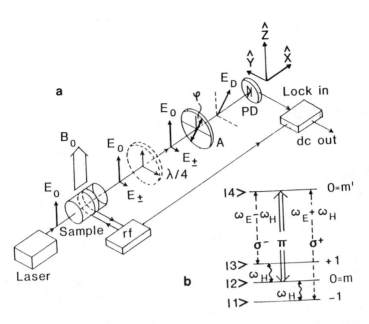

Fig. 1. Raman heterodyne detection of rf-induced sublevel
coherence. (a) Experimental scheme. A, polarization
analyzer; PD, photodetector; $\vec{B}_O$, static magnetic
field. The $\lambda/4$ plate is inserted with its main axes
parallel to the polarization directions of $\vec{E}_O$ and $\vec{E}^{\pm}$
and is used only for the detection of the FM-Raman
heterodyne signal (see text). (b) Energy level diagram
for a Zeeman split J=1-J'=0 transition, showing the
rf-driven coherent Raman process.

presence of the light field $\vec{E}_O$ gives rise to coherent, copropa-
gating Raman sidebands $E\pm$ with frequencies $\omega_E \pm \omega_H$. As a
consequence of the selection rules the Stokes ($\vec{E}_-$) and anti-
-Stokes ($\vec{E}_+$) fields are σ^- and σ^+ polarized, respectively,
i.e., their direction of polarization is perpendicular to the
photodetector is used to project carrier and orthogonally
polarized sidebands of the Raman signal field along a common
direction; at the photodiode, the strong optical carrier
finally serves as a local oscillator for heterodyne detection
of the Raman sidebands. The corresponding rf beat signal of
frequency $\omega_H=|\omega_E-(\omega_E\pm\omega_H)|$ is detected by a phase-sensitive,
narrow-bandwidth rf lock-in amplifier.
    The modulation structure of the resulting field $\vec{E}_D(t)$ in
front of the photodetector deserves a special discussion. In
general, the Raman signal field from the sample cell consists
of two components which can be characterized by a phase relationship between carrier and Raman sidebands corresponding to the
cases of pure amplitude (AM) and frequency (FM) modulation
(see Fig. 2). With the superposition of $\vec{E}_O$ and $\vec{E}^{\pm}$ by means
of an analyzer, these AM and FM components lead to an amplitude
and frequency modulation of the detector field $\vec{E}_D(t)$, respectively. The heterodyne beat signal of a phase-insensitive,
square-law photodetector now is connected only with the AM
component of the Raman signal field and it can therefore be

termed "AM-Raman heterodyne signal" (AM-RHS). The complementary detection mode that isolates the FM-RHS can be realized by introducing a constant additional phase shift of $\theta=\pi/2$ for $\vec{E}_O$ with respect to $\vec{E}_\pm$: Fig. 2 shows that this converts the FM component of the Raman signal field into amplitude modulation, and vice versa. In our experiment, the required phase retardation in the Raman signal field can easily be obtained with a $\lambda/4$ plate in front of the analyzer as indicated in Fig. 1 a.

The separate study and comparison of AM- and FM-Raman heterodyne signals is essential for the analysis of collision effects in our double resonance experiment. To illuminate this point, we see in Fig. 3a an outline of the expected velocity distribution of rf-excited atoms; in the presence of collisions, the shape of this distribution is determined by the collisional redistribution of atomic velocities during the velocity-selective optical pumping and rf excitation processes. It can be seen from Fig. 3(a) and (b) that the AM-RHS is in our case sensitive only to optically resonant atoms whose Doppler shift with respect to the laser field is smaller than the homogeneous optical linewidth: Here, the optical pumping and the detection of the rf-excited sublevel coherence are equally reduced to one small velocity subgroup of the atomic ensemble. In contrast to this, the dispersive shape of the FM-RHS detection sensitivity shown in Fig. 3c implies that the contributions of optically resonant atoms are nearly completely cancelled out in the observation of a velocity average. The detected FM-RHS therefore mainly arises from the asymmetry of the optically off-resonant wings of the distribution shown in Fig. 3 a; for an optical excitation of atoms with nonzero velocities, this asymmetry is due to rf-excited atoms that are partially thermalized by VCC.

It follows that collisions will affect the rf resonance linewidths of AM- and FM-RHS in a different manner: While the AM-RHS linewidth depends on the average time interval between

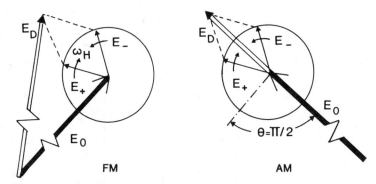

FM   AM

Fig. 2. Phase vectors of laser field $E_O(t)$ and frequency-shifted Raman sidebands $E_+(t)$, $E_-(t)$ in a reference frame rotating with frequency ω_E for the cases of pure frequency (FM) and amplitude modulation (AM) of the superposition field $E_D(t)$. It can be seen that a constant phase shift of $\theta=\pi/2$ of E_O with respect to E_+ and E_- converts FM to AM, and vice versa. If laser and Raman sideband fields have orthogonal polarization directions, this phase shift can be realized with a $\lambda/4$ plate (see text).

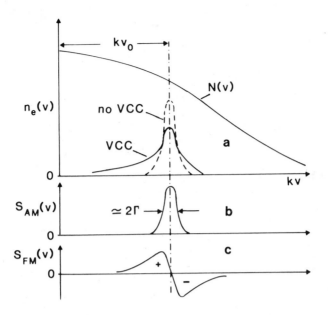

Fig. 3 (a) Outline of the velocity distribution of rf-excited
atoms $n_e(v)$ under conditions of velocity-selective
optical pumping, showing the case of a partial thermalization of $n_e(v)$ due to velocity— changing collisions
(VCC). It is assumed that the laser is tuned to the
high-frequency wing ($kv_o>0$) of the thermal velocity
distribution $N(v)$. (b), (c) Velocity-dependent
detection sensitivities $S_{AM}(v)$ and $S_{FM}(v)$ of AM-
and FM-Raman heterodyne detection. Γ denotes the
homogeneous optical linewidth (HWHM). The measured
Raman heterodyne signals are proportional to the
velocity-integrated products $S_{AM}(v)n_e(v)$ and $S_{FM}(v)n_e$
(v), respectively.

two collisional velocity changes sufficiently strong to take
the active atoms in and out of optical resonance, the linewidth
of the FM-RHS is determined by the time constant of velocity
thermalization and by the time rate of depolarizing collisions.
The comparison of AM- and FM-Raman heterodyne signals thus allows
to study the influence of collisional velocity diffusion of
optically detected rf resonance line shapes. In turn, the
measured AM- and FM-RHS linewidths can be used to obtain quantitative information on both depolarizing and sublevel coherence-
-preserving, velocity-changing collisions.

Experiments were performed on the Sm I line $\lambda=570.6$ nm
($4f^66s^2\ ^7F_1-4f^66s6p\,^7F_o$) with the use of the arrangement shown
in Fig. 1 a. The samarium vapor was contained in a heated
ceramic tube of 1 cm diameter; the temperature of the vapor cell
was $\approx$ 1000 K with the length of the heated zone being 3 cm. In
the experiments, buffer gas (He,Ne,Ar or Xe) was added at a low
pressure. Natural Sm consists of a mixture of seven isotopes with
mass number 144,147,148,149,150,152 and 154. The total Sm number
density was estimated to be $\approx 10^{13}cm^{-3}$; the Doppler broadening
of the optical lines is about 1.2 GHz (full width). For optical
excitation, a free-running single-mode dye ring laser with an

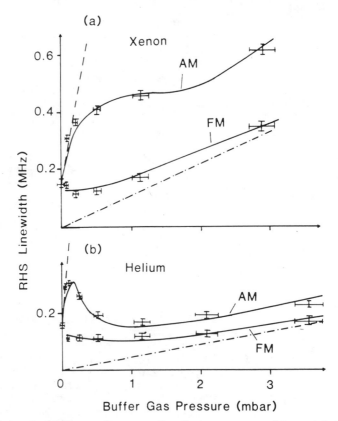

Fig. 4. Pressure dependence of rf resonance linewidths (HWHM)
of AM- and FM-Raman heterodyne signals for xenon and
helium collision partners. Experimental points are
given by crosses; the full curves correspond to fits
that are based on a theory including velocity
diffusion processes. Dashed lines, initial slopes for
the AM-RHS pressure broadening giving γ_{VCC}; dash-dotted
lines, asymptotic slopes for FM-RHS linewidths giving
γ_c.

unfocused beam diameter of 500 μm was used; its beam power was
kept in the range of 0.5 - 1mW. The rf field was produced by a
simple coil surrounding the vapor cell. The polarization
analyzer was set at $\gamma=45^o$ and a p-i-n diode served for optical
detection.

Raman heterodyne signals of the Zeeman resonances were
observed at a fixed radio - frequency $\omega_H = 2\pi \times 20.45$ MHz by varying
the Zeeman splitting via the external magnetic field B_o; in
these experiments the laser was tuned close to the optical
resonance of the ^{154}Sm isotope. The FM-RHS was monitores with
the use of the λ/4 plate as discussed above; corresponding to
the high accuracy in path length difference of the λ/4 plate
used, less than 0.3% of the AM signal in present during FM
detection.

The measured linewidths and lineshapes of the RHS are
found to depend strongly on the buffer gas pressure. Two

examples are given in Fig. 4 which displays the observed rf-
linewidths of the AM- and FM-RHS for xenon and helium in the
pressure range up to 4 mbar. These results can only be explained
if collisional Zeeman-coherence decay and velocity diffusion are
taken into account. Using an appropriate density matrix treatment[7], our calculations clearly show that thermalizing VCC
between Sm and buffer gas atoms do play an important role in
the formation of the RHS. The results of this theory with the
use of a Keilson-Storer collision kernel are displayed as full
curves in Fig. 4. In particular our theory predicts that the
linewidth of the AM-RHS is strongly influenced by VCC and that
it is always larger than the width of the FM-RHS; keeping in
mind that the AM-RHS originates from the optically resonant
velocity subgroup, its linewidth can be interpreted as being
transit-time broadened in velocity space as a result of velocity
diffusion. Interestingly, if there is a sufficiently large
number of VCC during the sublevel coherence lifetime, the
velocity diffusion directed back into the resonant velocity
subgroup becomes important and reduces the AM-RHS broadening.
This effect is most obvious in the case of He where it gives
rise to a decrease in the AM-RHS linewidth in the pressure range
of 0.2 - 0.8 mbar. At higher perturber densities, the pressure
broadening of both AM- and FM-RHS is mainly due to coherence-
-destroying collisions. The calculations also show that in the
presence of VCC an enhanced number of atoms can yield dispersive signal contributions to the coherent Raman process: As a
consequence the FM-RHS can become comparable in amplitude with
the AM-RHS, which in fact is an agreement with our observations.

From the data displayed in Fig. 4 we can derive a rate
constant γ_{VCC} for VCC and a rate constant γ_C for collisions
destroying the Zeeman coherence: Theory predicts that the
asymptotic slope of the FM-RHS yields γ_C whereas the initial
slope of the AM-RHS gives γ_{VCC}. By fitting our experimental
data, the rate constants or equivalently, the corresponding
collisional cross sections σ_C and σ_{VCC} can be determined (see
Table 1); we assume a relative error in σ_C of typically $\pm$ 10%,
whereas the values for σ_{VCC} are only first estimates based on
the small number of data points at low rare gas pressure.

Our studies of the effect of velocity-changing collisions
in an rf-laser double resonance experiment contribute to a new
vista into the role of collisions in laser spectroscopy of sublevel structures: the limitation of the observation time of
the active atoms due to narrow-bandwidth optical excitation and
simultaneous velocity diffusion can be of importance for a
variety of spectroscopic techniques that use a velocity-selective excitation and detection of either sublevel populations
or sublevel coherence. On the other hand, the collisional
velocity diffusion of sublevel coherence within an optical
Doppler distribution can also give rise to new and surprising
phenomena as will discussed in the next section.

Table 1. Experimentally determined values
for the collisional cross sections

Perturber	$\sigma_C (\text{Å}^2)$	$\sigma_{VCC} (\text{Å}^2)$
He	0.98	180
Ne	5.3	240
Ne	12	320
Xe	11	340

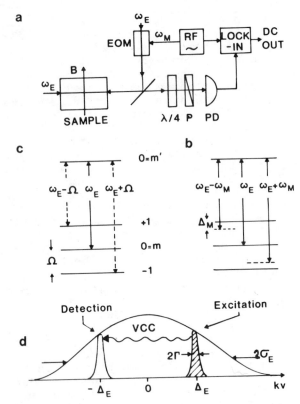

Fig. 5.a) Experimental scheme: EOM, electrooptic modulator; B, static tranverse magnetic field; λ/4, retardation plate inserted with one of its main axes parallel to the polarization of the probe field; P, polarization analyzer; PD, photodetector. b) Modulated excitation process of Zeeman coherence for the Zeeman-split J=1-J'=0 transition of Sm. c) Detection process showing the induced Raman sidebands. d) Schematic of the Doppler distribution indicating the velocity selectivity of the optical excitation and detection of sublevel coherence.

COLLISION-INCUCED RAMSEY RESONANCES IN SM VAPOR

Ramsey's method for the observation of narrow radiofrequency (rf) resonances is well known from atomic and molecular beam experiments[8]. In this contribution, we demonstrate the occurence of similar Ramsey resonances in an atomic vapor due to collisional velocity diffusion of sublevel coherence within an optical Doppler distribution. This new phenomenon is observed using coherent resonance Raman processes to optically induce and detect Zeeman coherence in the Sm λ=570.7 nm J=1-J'=0 transition.

In the experimental configuration shown in Fig. 5a, counterpropagating laser fields are used for the coherent excitation and phase-sensitive detection of oscillating Zeeman coherence. For a nonzero laser detuning with respect to the

Doppler-broadened optical transition ($\Delta_E{\neq}0$), both fields interact with different velocity subgroups whose widths are determined by the homogeneous optical linewidth 2Γ (see Fig. 5d). In a Doppler-free optical double resonance process (Fig. 5b) that uses carrier and sidebands of the modulated excitation field, the sublevel coherence is driven with frequency ω_M for atomic velocities $kv=\Delta_E$. In the presence of VCC, the active atoms are redistributed from $kv=\Delta_E$ to other velocity subgroups where the sublevel coherence further evolves at its eigenfrequency Ω. In the velocity interval centered around $kv=-\Delta_E$, the oscillating coherence is monitored by coherent resonance Raman scattering of the probe field (Fig.5 c). The Raman sidebands of frequency $\omega_E{\pm}\Omega$ and the probe field (ω_E) then yield a heterodyne beat signal of frequency Ω at the photodetector (Fig. 5a); the phase-sensitive detection of this rf beat signal allows to measure magnitude and phase of the sublevel coherence with high sensitivity [9]. As a result of the resonance conditions for the excitation process shown in Fig. 5b, rf resonance signals will be observed if the sublevel detuning $\Delta_M=\Omega-\omega_M$ varies through zero.

In our experiment, the velocity diffusion of sublevel coherence between the excitation and detection processes plays a role similar to the spatial motion of the active atoms in conventional Ramsey experiments; for the case of nonresonant

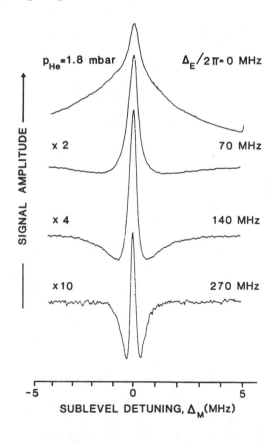

Fig. 6. Measured Raman heterodyne signals of the detection beam as a function of sublevel detuning Δ_M for different laser detunings $\Delta_E/2\pi$, showing narrow interference patterns.

laser detuning ($\Delta_E{\neq}0$), the measured signal is due to Zeeman-
-coherence-preserving VCC that change the atomic velocities from
$kv=\Delta_E$ to $kv'=-\Delta_E$ between the excitation and detection processes
(Fig. 5 d). During the time τ required for velocity diffusion,
the sublevel coherence evolves at its eigenfrequency Ω. This
leads to a phase shift $\Delta_M\tau$ for the rf beat signal at the photodetector, thereby introducing Ramsey-type interference patterns
proportional to $\cos(\Delta_M\tau)$ in the detected signal line shapes. As
a consequence, increasingly narrow resonance structures can be
observed for large velocity diffusion times τ. However, the
values for τ are widely distributed in the ensemble average which
limits the obtainable resolution and washes out higher-order
Ramsey fringes in the resonance signals.

Similar to the measurement of the rf-optical double resonance
signals, the optical-optical double resonance signals were
observed with an optical excitation of the ^{154}Sm transition,
varying the sublevel detuning Δ_M via the external static magnetic field at fixed values for the modulation frequency ω_M. To
obtain a velocity selectivity of the optical excitation and
detection processes as shown in Fig. 5 d, the electrooptic
modulator was used analogously to a Pockels cell, thereby
yielding orthogonal polarization directions for carrier and

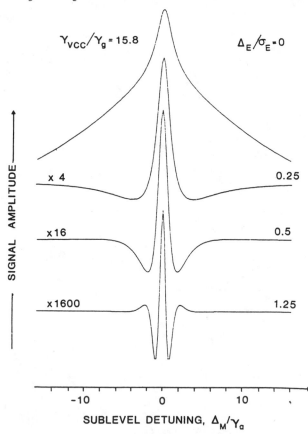

Fig. 7. Calculated Raman heterodyne signals of the detection beam
showing Ramsey-type resonance line shapes. The sublevel
detuning is normalized to the ground state relaxation
rate γ_g and the laser detuning is given in units of the
Doppler width σ_E. γ_{VCC} denotes the rate of velocity
changing collisions.

modulation sidebands with a phase relationship between carrier and sidebands corresponding to the "FM" phase vector diagram displayed in Fig. 2. Under these conditions, the optical detection process is velocity-selective as shown in Fig. 5 d, if a $\lambda/4$ plate is inserted in the probe beam in front of the polarization analyzer (see Fig. 5 a).

Typical experimental results obtained with He collision partners are shown in Fig. 6. It can be seen that for an increasing laser detuning Δ_E, the linewidths of the resonance signals decrease drastically as a result of large velocity diffusion times. Moreover, the measured line shapes also demonstrate the presence of Ramsey interference patterns. These results are in good agreement with density-matrix calculations that include the velocity diffusion and simultaneous collisional decay of the sublevel coherence between the optical excitation and detection processes (Fig. 7). As a particular result, the calculations predict that the linewidths of the Zeeman resonances here can assume values that lie below the "natural" limit γ_g given by the time rate of depolarizing collisions and transit time broadening. It seems clear from this that our experimental technique can find a useful application in high--resolution sublevel spectroscopy and in collision studies.

ACKNOWLEDGEMENTS

We thank W. Lange for helpful discussions. One of us (Chr. Tamm) acknowledges support by a fellowship of the Studienstiftung des Deutschen Volkes and another (E. Buhr) by a fellowship of the Niedersächsische Graduiertenförderung. J.M. is supported by a Heisenberg-fellowship of the Deutsche Forschungsgemeinschaft. This work was funded by the Deutsche Forschungsgemeinschaft.

REFERENCES

1. See,e.g.,C.Cohen-Tannoudji and A. Kastler, in "Progress in Optics" (North-Holland, Amsterdam, 1966), Vol. 5,p.1.
2. W.E. Bell and A.L. Bloom, Phys. Rev. Lett.6 (1961) 280, 623.
3. J. Mlynek, N.C. Wong, R.G. DeVoe, E.S. Kintzer, and R.G. Brewer, Phys. Rev. Lett. 50 (1983) 993.
4. N.C. Wong, E.S. Kintzer, J. Mlynek, R.G. DeVoe, and R.G. Brewer, Phys. Rev. B 28 (1983) 4993.
5. M. Mitsunaga, E.S. Kintzer, and R.G. Brewer, Phys. Rev. Lett. 52 (1984) 1484.
6. J. Mlynek, Chr. Tamm, E. Buhr, and N.C. Wong, Phys. Rev.Lett. 53 (1984)1814.
7. Chr. Tamm, E. Buhr, and J. Mlynek, "Raman heterodyne studies of velocity diffusion effects in radiofrequency-laser double resonance", submitted to Phys. Rev. A.
8. N.F. Ramsey, "Molecular Beams" (Oxford Univ. Press, London, 1963).
9. J. Mlynek, K.H. Drake, G. Kersten, D. Frölich, and W. Lange, Opt. Lett.6 (1981) 87.

LASER FREQUENCY DIVISION AND STABILIZATION

R. G. DeVoe, C. Fabre[*] and R. G. Brewer

IBM Research Laboratory
5600 Cottle Road, K01/281
San Jose, California 95193, USA

The current method for measuring an optical frequency relative to the primary time standard, the cesium beam standard at ~9.2 GHz, utilizes a complex frequency synthesis chain involving harmonics of laser and klystron sources. The method has been extended recently to the visible region,[1] to the 633 nm He-Ne laser locked to a molecular iodine line, with an impressive accuracy of 1.6 parts in 10^{10}. With the new definition of the meter, the distance traversed by light in vacuum during the fraction 1/299 792 458 of a second, the speed of light is now fixed and both time and length measurements can be realized with the same accuracy as an optical frequency measurement. In view of the complexity of optical frequency synthesis, these developments set the stage for originating complementary techniques for stabilizing and measuring laser frequencies which are more convenient.

This paper reports a sensitive optical interferometric technique *dual frequency modulation* (DFM) for measuring and stabilizing a laser frequency by comparison, in a single step, to a radio frequency (rf) standard. Conversely, a low-noise rf source can be stabilized by a laser frequency reference. A prototype[2] has demonstrated a resolution of 2 parts in 10^{10}, but devices currently under development should have a resolution of 10^{-12} and an absolute accuracy of $\sim 10^{-10}$. The method may be competitive with the optical frequency synthesis chain in accuracy and its simplicity suggests its convenient use in metrology, high-precision optical spectroscopy, and gravity wave detection.

The principle of the technique rests on phase-locking the mode spacing c/2L of an optical cavity to a radio frequency standard and simultaneously phase-locking a laser to the n-th order of the same cavity. When these two conditions are satisfied, the optical frequency ω_0 and the radio frequency ω_1 are simply related,

$$\omega_0 = n\omega_1 , \tag{1}$$

neglecting for the moment diffraction and mirror phase-shift corrections. The idea of locking a laser to a cavity is of course a well-established subject,[3,4] but the concept of phase-locking an optical cavity to a radio frequency source is new. Interferometric rf-optical frequency comparisons of lower sensitivity have previously been performed

---

[*]On leave from the Laboratoire de Spectroscopie Hertzienne de l'ENS, Paris

by Bay et al.[5] using a related idea based on amplitude modulation (AM) rather than frequency modulation.

To introduce the DFM technique, first consider a single frequency modulation scheme. An electrooptic phase modulator driven at ω_1 generates a comb of optical frequencies $\omega_0 \pm m\omega_1$ which are compared to cavity modes of frequency $n\sigma$, where σ is the cavity free spectral range, $m=0,1,2,...$ and n is a large integer $\sim 10^6$. The cavity response perturbs the balanced phase relationships between the sidebands and transforms frequency modulation into intensity modulation at ω_1. A photodetector, viewing the cavity either in reflection or transmission, then generates an error signal at the heterodyne beat frequency ω_1. This signal yields a null when the laser frequency ω_0 equals $n\sigma$ and the radio frequency ω_1 matches the mode spacing σ. In this circumstance, the comb of optical frequencies all resonate with their corresponding cavity modes. Although simple in principle, this single FM technique generates a complex error signal which depends not only on the rf detuning $\delta = \omega_1 - \sigma$ but also on the optical detuning $\Delta = \omega_0 - n\sigma$ and is therefore unsuitable for locking.

The DFM technique overcomes this problem by using two phase modulators, driven at frequencies ω_1 and ω_2 respectively, where $\omega_2/2\pi$ is nonresonant with $c/2L$. Dual frequency modulation creates, in lowest order, sidebands at $\omega_0 \pm \omega_1$, $\omega_0 \pm \omega_2$, and $\omega_0 \pm \omega_1 \pm \omega_2$. A photodetector views the cavity in reflection and two error signals are derived, one at ω_2 and the other at the *intermodulation* frequency $\omega_1 \pm \omega_2$. The first signal at ω_2 allows locking the laser to the reference cavity as described elsewhere[3,4] and is independent of ω_1 tuning. The second signal at $\omega_1 \pm \omega_2$ allows locking the cavity to the rf reference. This signal varies directly with the rf detuning $\delta = \omega_1 - \sigma$ and provides the desired null at the rf resonance condition $\omega_1 = \sigma$, while being independent of laser detuning Δ.

As has been shown elsewhere,[2] this intermodulated DFM signal is given by

$$I(\omega_1 \pm \omega_2) = I_0 \cdot C \cdot \sin\omega_2 t \cdot \{Im[g(\Delta + \delta) - g(\Delta - \delta)] \cos\omega_1 t\}$$

where I is the light intensity on the photodiode, g is the complex cavity lineshape function, and $C = 4J_0(\beta_1)J_0(\beta_2) J_1(\beta_1)J_1(\beta_2)$, where β_1 and β_2 are the modulation indices of the phase modulators. Assuming a Lorentzian lineshape function g

$$g(\Delta) = \frac{\Delta(\Delta - i\Gamma)}{(\Delta^2 + \Gamma^2)} ,$$

the error signal becomes, for $\delta < \Gamma$

$$I(\omega_1 \pm \omega_2) = I_0 \cdot C \cdot \frac{\delta}{\Gamma} \cdot [\sin(\omega_1 + \omega_2)t + \sin(\omega_1 - \omega_2)t]$$

By detecting the beat either at $\omega_1 + \omega_2$ or $\omega_1 - \omega_2$ in a double balanced mixer, an error signal proportional to the rf detuning δ can be derived for locking an optical cavity to an rf standard, or conversely an rf source to a cavity. Second, the error signal is independent of laser detuning Δ and thus optical frequency jitter. Third, there is no background signal. Fourth, the DFM signal has excellent signal-to-noise ratio since $C = 4J_0(\beta_1)J_1(\beta_1)J_0(\beta_2)J_1(\beta_2) = .45$ at $\beta_1 = \beta_2 = 1$.

An examination of the above equations shows that the intermodulated DFM technique realizes an *FM Differential Interferometer* which produces a locking signal similar to conventional FM laser locking, but in which the optical tuning parameter $\Delta = \omega_0 - 2\pi \cdot n \cdot c/2L$ is replaced by the radio frequency (differential) tuning parameter $\delta = \omega_1 - 2\pi \cdot c/2L$. The DFM sideband structure creates an optical subtraction in the photodiode of the $(n+1)$-th cavity resonance curve from the $(n-1)$-th curve and thus permits accurate, low noise measurements of the cavity mode spacing.

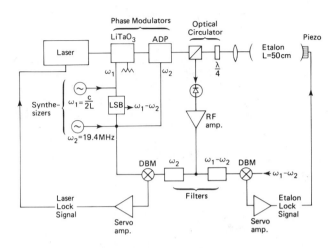

Figure 1. Block diagram of an optical frequency divider showing two servo loops where the laser is locked to a reference cavity and the cavity to a radio frequency standard. The LiTaO$_3$ modulator is driven at ω_1 and the ADP modulator at ω_2. LSB denotes a mixer and filter which generates the difference frequency $\omega_1-\omega_2$ for the double balanced mixer (DBM) in the cavity-rf servo.

A prototype DFM standard (Fig. 1)[2] was constructed to verify the principle and study resolution and systematic errors. A home-made cw dye ring laser contains an intracavity ADP crystal for laser phase-locking to an external cavity at 6000Å. DFM sidebands are applied by two phase modulators driven at $\omega_1/2\pi=c/2L\sim300$ MHz and $\omega_2/2\pi=19.4$ MHz. The laser beam is then mode-matched onto the reference cavity through an optical circulator which both isolates the laser from the cavity and directs the reflected light, the signal, to a high-speed photodiode. The detector photocurrent contains rf beats at ω_2 and $\omega_1\pm\omega_2$ which are amplified and then separately filtered. The error signal at ω_2 is sent to an FM sideband servo which locks the dye laser to the n-th cavity fringe so that $\omega_0=n\sigma$ with a short-term error $\Delta/2\pi\sim300$ Hz RMS or less.[4] The DFM signal at $\omega_1-\omega_2$ is coherently detected in a double-balanced mixer. This signal can control the cavity length L via a piezo so that the resonance condition $\sigma=\omega_1$ is satisfied. In initial tests the piezo control loop was opened and noise levels of 60 millihertz RMS, corresponding to a fractional frequency deviation σ_y (Allan variance) of 2×10^{-10} in 0.1 sec integration time, were measured. The measurement of divider noise of 60 millihertz in the presence of 300 Hz of optical jitter verifies the differential behavior of the DFM standard and indicates that laser jitter is cancelled at the 10^{-3} to 10^{-4} level.

Four improvements have been made to the prototype to increase the resolution to 10^{-12} and to make possible absolute frequency comparisons at the 10^{-10} level. First, the interferometer itself has been placed in an acoustically isolated vacuum chamber and constructed from a 5 cm diameter by 47.5 cm long Zero-Dur spacer. Super-polished ring laser gyro quality mirrors have been used to achieve a finesse of >20,000 or a linewidth $\Gamma/2\pi<10$ kHz. Secondly, $\omega_1/2\pi$ has been raised to 4095 MHz to resonate with two cavity modes separated by 13 times the free spectral range of 315 MHz. Although this does not change the locking error δ, it reduces the *fractional* error and noise δ/ω_1 by a factor of 13. A simple but highly efficient resonant cavity microwave phase modulator has also been developed which will be described elsewhere. Third, two I$_2$-stabilized He-Ne lasers of an NBS design have been constructed to serve as optical frequency standards. Fourth, a low noise microwave source has been built whose frequency is known to $\sim10^{-11}$ by radio comparison to WWVB and Loran C.

These improvements described above have increased the effective Q of the system by a factor of >200, and thus resolution of 10^{-12} is expected. The theoretical shot-noise limited Allan variance of the new system is $\Gamma/(\omega_1\sqrt{N\tau})$ where N is the number of photoelectrons/sec and τ is the integration time. For $N=10^{15}$ sec^{-1} and $\tau=0.1$ sec $\sigma_y=2\times10^{-13}$.

189

New techniques have also been developed to measure systematic errors common to precision interferometry. Equation (1) must be modified by corrections due to phase shifts in the cavity arising from (a) the multilayer dielectric coatings on the mirrors and (b) diffraction effects. Previous workers[6] have corrected for (a) by using two different cavity lengths, which require disassembly and realignment of the interferometer. The DFM standard, on the other hand, is designed to reach 10^{-10} accuracy with mirrors which can be quickly moved in vacuum, without disturbing alignment, by only 4 cm. In addition, the diffraction phase shift (b) can also be determined by measuring the higher order transverse mode spacing.

The DFM technique, therefore, should provide an independent and accurate measurement of an optical frequency that can be compared to the results of the optical frequency synthesis chain.[1]

We are indebted to H. P. Layer for providing us with a detailed design of the NBS He-Ne iodine stabilized laser, to J. L. Hall for informing us of his design of an acoustically isolated interferometer chamber, to K. L. Foster for assistance in all phases of their fabrication, and to F. Walls for help with ultra-stable crystal oscillators. This work is supported in part by the U.S. Office of Naval Research.

REFERENCES

1. D. A. Jennings, C. R. Pollack, F. R. Peterson, R. E. Drullinger, K. M. Evenson, and J. S. Wells, Optics Letters 8:136 (1983).
2. R. G. DeVoe and R. G. Brewer, Phys. Rev. A30:2827 (1984).
3. R. W. P. Drever, J. L. Hall, E. V. Kowalski, H. Hough, G. M. Ford, A. J. Munley, and H. Ward, Appl. Phys. B31:97 (1983).
4. R. G. DeVoe and R. G. Brewer, Phys. Rev. Lett. 50:1269 (1983).
5. Z. Bay, G. G. Luther, and J. A. White, Phys. Rev. Lett. 29:189 (1972).
6. H. P. Layer, R. D. DesLattes, and W. G. Schweitzer, Appl. Optics 15:734 (1976).

HIGH RESOLUTION SPECTROSCOPY OF ALKALINE EARTH MONOHALIDES:

PERTURBATION ANALYSIS, HYPERFINE STRUCTURE AND STARK EFFECT

W.E. Ernst, J.O. Schröder, and J. Kändler

Institut für Molekülphysik, Freie Universität Berlin
Arnimallee 14, D-1000 Berlin 33, W. Germany

I. INTRODUCTION

Optical spectroscopy of molecules gained an immense impulse by the advent of tunable lasers[1]. The most important improvement in resolution was obtained, when the limitations due to the Doppler effect were overcome by several laser spectroscopic techniques created by Schawlow, Hänsch and coworkers[2,3]. Many details on laser spectroscopy and its application to molecules may be found in the book by Demtröder[4].

Among molecules, free radicals have attracted much interest deriving both from the concern about the knowledge of the electronic structure and from the field of studies of chemical reactions. The chemically unstable molecular fragments have usually one or more unpaired electrons and magnetic interactions of the nonzero spin can occur within the molecule or with external fields leading to complex spectra. If they can be successfully analyzed they provide many details of the molecular structure. The alkaline earth monohalides represent a group of radicals which, in some respect, should be the molecular analogue of alkali atoms. To good approximation they consist of a single electron outside two closed-shell ions. The electronic ground and first excited states differ mainly by the orbital of this unpaired electron and the internuclear distances as well as the potential energy curves are very similar. As a consequence vibrational bands with $\Delta v = 0$ are favoured by the Franck-Condon principle and form extremely congested optical spectra with strongly overlapping band sequences. The rotational structure in the bands of heavy species like strontium and barium monohalides can only be resolved by Doppler-free laser spectroscopy techniques. For our investigations we chose Doppler-free polarization spectroscopy[3] which provides the high sensitivity needed for the study of species at low number densities[5]. In general the analysis of optical spectra yields molecular constants the accuracy of which suffers from the high correlation between the upper and lower state parameters[6]. Independent measurements of ground state transitions by microwaves (mw's) can help to break the correlation and are included in our analyses. Moreover, labeling of ground state levels by inducing mw transitions can simplify dense spectra and is described in section II. It will be shown that this method allows a reliable assignment of lines and also helps to reveal local perturbations.

In all alkaline earth monohalides the unpaired electron spin interacts with the nuclear spin of the halogen nucleus. Since the electron is
mainly metal centered, the hyperfine structure (hfs) is very weak and
can barely be resolved with sub-Doppler laser spectroscopy. It is impossible to separate all effects of the magnetic dipole and electric
quadrupole interactions in the ground and excited states, especially if
different coupling schemes of the angular momenta have to be applied to
the electronic states. In this case independent measurements of the
ground state hfs are absolutely essential. The high resolution and sensitivity of laser-mw double resonance techniques allow a precise determination of hfs parameters of the alkaline earth monohalide radicals
(section III).

Whereas the hfs can serve as a local probe for intramolecular
fields and field gradients, information about the total charge-density
distribution is given by the electric multipole moments, especially the
dipole moment of the molecule. Both molecular properties complement each
other in giving a more complete picture of the electronic structure and
chemical bonding. In section IV we show a way to determine electric
dipole moments from high precision Stark effect measurements. The techniques described in sections II-IV will be applicable to many other free
radicals.

II. INVESTIGATION OF EXTREMELY CONGESTED SPECTRA

Optical spectroscopy was performed in a free space reaction cell in
which a large number of radicals can be produced. Alkaline earth monohalides were generated by the reaction of alkaline earth metal vapor
with a halogen donor in a stream of argon buffer gas as shown in Fig. 1
for SrBr. In this way a total number density of 10^{11} cm^{-3} SrBr is produced in a reaction zone of 25 cm diameter at a background pressure of
0.1 Torr. The experimental arrangement shown in Fig. 1 allows the investigation of optical and mw transitions. Mw radiation of vertical polarization is introduced via a horn antenna. Two counterpropagating laser
beams of the same wavelength pass through a slit in the top of the horn
radiator. The best overlap of the three electromagnetic waves is given
in the reaction zone. The probe laser beam has a linear polarization
inclined by 45° with respect to the polarization plane of the counter
propagating pump laser beam and the mw's. In this way the probe laser
beam which is passed through a nearly crossed polarizer behind the cell

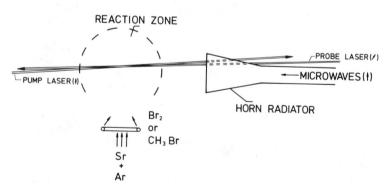

Fig. 1. Experimental arrangement inside the reaction cell: interaction
zone between gas sample and laser and mw radiation (from ref.16).

can be used to detect an anisotropy induced in the gas by the absorption of pump laser or mw radiation or both. The cell windows are mounted with adjustable clamps such as to apply the amount of strain needed to minimize the birefringence. If the pump laser and the mw's are amplitude modulated at frequencies f_p and f_m, respectively, and phase sensitive detection of the probe signal is employed the following experiments can be performed:

1) detection at f_p, laser scanning: optical spectra are recorded at sub-Doppler resolution using Doppler-free polarization spectroscopy[3,5];

2) detection at f_m, laser fixed to an optical transition, mw's scanning: mw transitions are recorded which are connected to a level of the induced optical transition in a double resonance scheme;

3) detection at $f_p + f_m$ or $|f_p - f_m|$, mw's fixed to a rotational transition, laser scanning: optical transitions which are connected to a level pumped by mw's are detected at sub-Doppler resolution.

Mode 2 is a particularly sensitive method to detect mw transitions as will be shown in chapter III and is called microwave-optical polarization spectroscopy (MOPS)[7]. Polarization spectroscopy techniques require less intensity of laser and mw radiation than the corresponding non-linear methods based on fluorescence detection. Power broadening is avoided which is the reason for the largely improved resolution of MOPS compared to conventional microwave optical double resonance (MODR) spec-

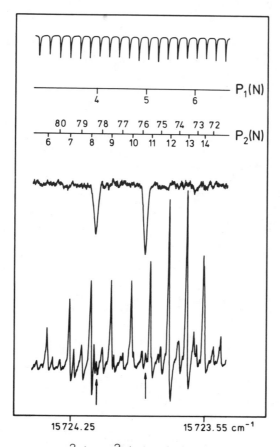

Fig. 2. A section of the $B^2\Sigma^+ - X^2\Sigma^+$ (0,0) band of SrCl obtained with Doppler-free polarization spectroscopy (lower trace), 1.5 GHz frequency markers in the upper trace. The middle trace shows a simplified spectrum observed when $P_1(4)$ and $P_1(5)$ are labeled by mw's inducing a N = 5-4, J = 5.5-4.5 transition (from ref.11).

troscopy[8]. Mode 3 is an important tool for labeling specific lines in sub-Doppler optical spectra in order to simplify the assignment. Probe laser detection of the molecules interacting with both mw's and pump laser is needed since no velocity selection is provided by the mw's. In the mw frequency range pressure broadening is larger than the Doppler effect. Mode 3 is called microwave modulated polarization spectroscopy (MMPS)[9].

MMPS has been used to assign lines in the dense spectra of the $A^2\Pi$ - $X^2\Sigma^+$ [10] and $B^2\Sigma^+$ - $X^2\Sigma^+$ [11] systems of SrCl. In Fig. 2 an example is shown where P_1 lines are nearly covered by a strong P_2 sequence even at sub-Doppler resolution. As the ground state constants were known from previous investigations mw's were used to induce a ground state rotational transition $N = 5 - 4$, $J = 5.5 - 4.5$ in order to label the weak $P_1(4)$ and $P_1(5)$ lines. Due to the identification of line sequences even in dense parts of the spectra we were able to reveal perturbations in the $A^2\Pi$ - $X^2\Sigma^+$ spectra of SrCl which are caused by a $^2\Pi_{1/2} \sim {}^2\Pi_{3/2}, \Delta v = 1$ interaction. The S^+-operator in the rotational Hamiltonian is responsible for this electron spin uncoupling. Another example for the power of mw labeling was the analysis of the SrBr $B^2\Sigma^+$ - $X^2\Sigma^+$ system[12], where the presence of the two nearly equally abundant isotopes ^{79}Br and ^{81}Br leads to even denser spectra. Perturbations were observed in all vibrational bands and could be attributed to a $B^2\Sigma^+ \sim A^2\Pi_{1/2}$, $\Delta v = 3$ interaction [13].

Very recently optical-optical double resonance experiments involving two lasers were performed for the study of highly excited electronic states.

III. INVESTIGATION OF HYPERFINE STRUCTURE BY LASER-MW DOUBLE RESONANCE TECHNIQUES

If hyperfine structure can be resolved in optical spectra the interpretation suffers from the fact that the splittings of hfs levels are not measured directly. Differences of splittings in the order of 10^7 Hz in two different electronic states are measured at $5 \cdot 10^{14}$ Hz. Not only a resolution of $2 \cdot 10^{-8}$ is required but also the different hfs interactions in ground and excited states have to be disentangled. It is certainly much more useful to measure the hfs in one electronic state by observing rf or mw transitions first. For a long time the application of low frequency studies was limited by sensitivity problems which could finally be overcome by the development of laser-mw and laser-rf double resonance techniques. For the investigation of hfs in mw spectra we use MOPS if the species is produced in a reaction cell. If a molecular beam can be generated we apply another laser-mw double resonance method also described in the following. Once the ground state hfs is interpreted we analyze the resolved optical line splittings and separate the effects of excited state hyperfine interaction.

1. Microwave-optical polarization spectroscopy

MOPS corresponds to the experimental mode 2 described in chapter II. The polarization detection scheme allows a sensitivity increase by three to four orders of magnitude compared to sophisticated mw spectroscopy techniques, at the same resolution. Application of saturation modulation which is a nonlinear mw spectroscopy technique developed by Törring[14], permits the detection of about 10^{-6} absorption of the incident mw intensity. This allows the study of rotational transitions of alkaline earth monohalides in the 100 to 300 GHz range, i.e. transitions between levels with high rotational quantum numbers in the case of strontium and barium monohalides. Then the hfs is not resolved because

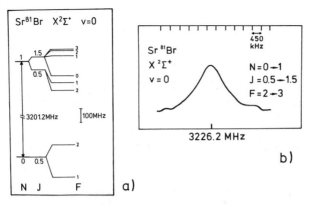

Fig. 3. MOPS study of hfs in the N = 1 - 0 rotational transition in
Sr[81]Br X$^2\Sigma^+$ (v=0); (a) energy level diagram; (b) MOPS signal of
a hfs component (time constant 12s), laser frequency fixed to
the $P_1(1)$ line of the B$^2\Sigma^+$ - X$^2\Sigma^+$ (0,0) band (from ref.16).

the splittings are in the order of 100 kHz. Hyperfine structure can be
well observed in the spectrum of N = 1 - 0 or N = 2 - 1 transitions
which lie around 3 to 6 GHz in the discussed cases. At this frequency
the amount of absorption would be only 10^{-9} under the same conditions
and MOPS allowed the investigation of such 1 - 0 transitions at pressure
limited resolution for the first time[15]. Transitions with $\Delta J = 0, \pm 1$
and $\Delta F = 0, \pm 1$ have been detected because of the intermediate coupling
in low-N states. Figure 3a shows an energy level diagram of the relevant
levels in the X$^2\Sigma^+$, v = 0 ground state of SrBr. For the detection of all
possible N = 1 - 0 mw transitions the probe laser can stay at a frequency on the Doppler envelope of optical lines connected to the levels of
the manifold with N = 1. Then different F-levels are probed for different velocity ensembles by the narrow band laser but at the mw frequency
the Doppler width is negligible compared the pressure broadening of
1 MHz. The MOPS signal for one hfs component is depicted in Fig. 3b. The
small hfs interaction in SrBr X$^2\Sigma^+$ and B$^2\Sigma^+$ has recently been analyzed
and interpreted[16].

The application of MOPS to excited electronic states is also possible and has been demonstrated for BaO[17].

2. Molecular-beam laser-mw double resonance

In a beam the molecules can travel over considerable distances in
the vacuum chamber without experiencing collisions. Depletion of the
population of a specific level by means of optical pumping at one place
can be probed at another place by observing the laser induced fluorescence. If the level is repopulated by some process in the region between
the two places the laser induced fluorescence increases. This Rabi type
experiment with optical state selection has been applied by Childs and
Goodman for the detection of molecular hfs transitions in the radiofre-

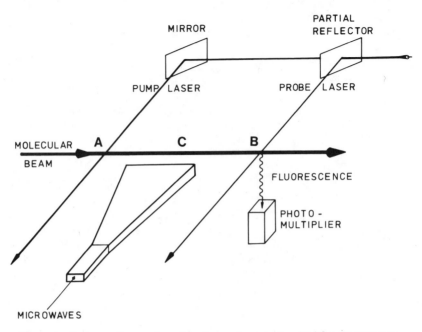

Fig. 4. Set-up for molecular-beam laser-mw double resonance.

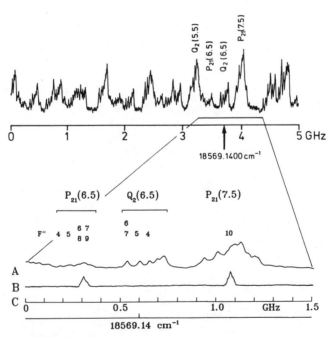

Fig. 5. A section of the $C^2\Pi_{3/2} - X^2\Sigma^+(0,0)$ band of BaI. Upper part: 5 GHz laser scan; lower part, trace A: 1.5 GHz portion of the spectrum; trace B: same portion obtained by mw labeling (mw frequency tuned to the $N = 7 - 6$, $J = 7.5 - 6.5$, $F = 10 - 9$ transition), $\tau = 3s$ for A and B. Trace C: frequency markers.

quency region[18]. As depicted in Fig. 4 we adopted this pump and probe scheme for the study of mw transitions[19]. The three fields of pump, mw, and probe are well separated and the linewidth of mw transitions cannot be broadened by optical saturation, but is mainly determined by the time of flight of the molecules through the interaction region. At this resolution, shielding against the earth's magnetic field has to prevent from unwanted Zeeman broadening.

Effusive beams of alkaline earth monohalides are generated from a high temperature reaction of the stable dihalide with the metal. Mw transitions have been observed with linewidths between 10 and 50 kHz. The most recent example is the investigation of the $X^2\Sigma^+$ ground state of BaI, the heaviest alkaline earth monohalide. For a long time the hfs in sub-Doppler spectra of the $C^2\Pi$ - $X^2\Sigma^+$ system observed by Johnson et al.[20] could not be interpreted. A 5-GHz laser scan near the band origin of $C^2\Pi_{3/2}$ - $X^2\Sigma^+$ (0,0) is shown in Fig. 5. In this part of the spectrum not even the rotational assignment was clear. We performed a detailed study of the 1.5 GHz interval in trace A. By successive application of double resonance experiments we were able to assign the rotational lines and their hfs components. Trace B shows the mw labeling of optical hfs transitions. The density of states in BaI is so large that in the probe-laser - molecular-beam interaction volume only about 5 to 10 molecules are in a specific hfs state of N = 6, despite of the reasonable beam intensity of $2 \cdot 10^{15}$ sr^{-1}s^{-1}. The precise determination of ground state hfs parameters has now allowed the interpretation of the optical hfs[21].

IV. STARK EFFECT STUDIES

In addition to the high resolution, another advantage of the described experimental arrangement for molecular-beam laser-mw double resonance spectroscopy is the convenient application of additional electric or magnetic fields. Whereas deposits in a chemical reaction cell can cause discharges or at least alter field conditions, the additional electrodes stay clean in a beam chamber. We performed Stark effect spectroscopy in the ground states of several alkaline earth monohalides by mounting Stark plates in the molecular-beam - mw interaction region. The electric field vector was parallel to the mw polarization. In this way the electric field dependence of the Stark splitting and shift of mw transitions could be measured with the selection rule $\Delta M_F = 0$ [22]. As the hfs splittings were in the same order of magnitude as the level shifts in the electric field, analyzing the data required the diagonalization of the complete energy matrix[23]. With the knowledge of rotational and hfs constants, the electric dipole moment μ is the only adjustable parameter to fit the measured line positions.

An example of the resolution of this type of Stark spectroscopy is given in Fig. 6 at the fairly low electric field strength of 150 V/cm. Three M_F-components of the N = 3 - 2, J = 2.5 - 1.5, F = 5 - 4 transition in the $X^2\Sigma^+$, v = 0 state of CaI near 12.3 GHz are well resolved in a 800 kHz mw scan[24]. The electric field shifts were followed up to 900 V/cm and are depicted in Fig. 7. The measured data points lie well on the solid lines which represent the line positions calculated with the fitted dipole moment. The high accuracy of this method allowed the determination of the vibrational dependence of the electric dipole moment e.g. in the case of SrF[25].

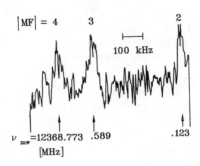

Fig. 6. 800 kHz mw scan of the Stark components of the N = 3 - 2, J = 2.5 - 1.5, F = 5 - 4 transition in the $X^2\Sigma^+$ (v=0) state of CaI at a field strength of 150 V/cm (from ref. 24).

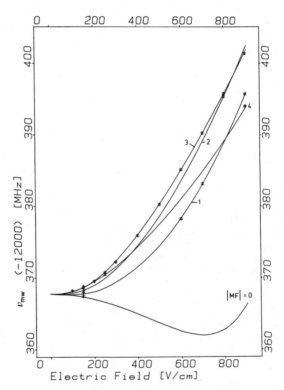

Fig. 7. Mw frequency vs. electric field for the N = 3 - 2, J = 2.5 - 1.5, F = 5 - 4 transitions with $|M_F|$ = 0,1,2,3,4, ΔM_F=0. Frequency measurement errors are given as vertical bars for each data point. The solid lines represent the calculated line positions for μ = 4.5968 D (from ref. 24).

The knowledge of precise dipole moments of alkaline earth monohalides stimulated the development of an ionic bonding model for these radicals[26]. Using the known equilibrium internuclear distances of the species and the polarizabilities of the metal and halogen ions as input parameters, this electrostatic model allows the prediction of the dipole moment. All former and more recent measurements agreed with the predictions well within about 10%. Electric dipole moments and hfs interaction parameters provide different information about the charge density distribution in a molecule. Internal fields are responsible for the polarization of electron orbitals and we have shown for SrBr that the idea of spin polarization causing the magnetic hfs agrees with the electrostatic picture of the bonding model[16]. An even more complete interpretation was possible for BaF, where the dipole moment measurements were complemented by the analysis of the hfs interaction from the two nuclei of ^{137}Ba and ^{19}F [27].

Very recently we applied an electric field in the probe laser - molecular beam interaction region and performed optical Stark spectroscopy. Stark shifts of up to 2.5 GHz at an optical linewidth of 25 MHz were measured and, due to the former ground state experiments, allowed the first determination of the electric dipole moments for the excited $A^2\Pi$ and $B^2\Sigma^+$ states of CaCl [28].

V. SUMMARY

A number of laser spectroscopic methods some of which have been newly developed were applied to the group of alkaline earth monohalide radicals. The effects of weak hyperfine interaction were studied by using laser-mw double resonance techniques. The Stark effect was investigated at high resolution in molecular beam experiments. Doppler-free polarization spectroscopy provided the sensitivity and resolution needed for the analysis of optical spectra. In particularly congested parts of the spectra mw labeling was helpful for the identification of lines. It is to be expected that these methods will be applied to other groups of free radicals in the near future and help to cast some light on the structure of new interesting molecular species.

Acknowledgments

The authors gratefully acknowledge stimulating discussions with Professor T. Törring. The work has been performed within the research program of the "Sonderforschungsbereich Hyperfeinwechselwirkungen" which is supported by the Deutsche Forschungsgemeinschaft.

REFERENCES

1. H. Walther (ed.), "Laser Spectroscopy of Atoms and Molecules", Topics in Applied Physics, Vol. 2, Springer, Berlin, Heidelberg, New York (1976).
2. M.S. Sorem and A.L. Schawlow, Saturation Spectroscopy in molecular iodine by intermodulated fluorescence, Opt. Commun. 5: 148 (1972).
3. C. Wieman and T.W. Hänsch, Doppler-free laser polarization spectroscopy, Phys.Rev.Lett. 36:1170 (1976).
4. W. Demtröder, "Laser Spectroscopy", Springer Series in Chemical Physics, Vol.5, Springer, Berlin, Heidelberg, New York (1982).
5. W.E. Ernst, Doppler-free polarization spectroscopy of diatomic molecules in flame reactions, Opt. Commun. 44:159 (1983).
6. D.L. Albritton, A.L. Schmeltekopf, and R.N. Zare, An introduction to the least-squares fitting of spectroscopic data, in: "Molecular

Spectroscopy: Modern Research", Vol.II, K. Narahari Rao, ed.,
Academic Press, New York (1976).

7. W.E. Ernst and T. Törring, Microwave-optical polarization spectroscopy, Phys.Rev.A 25:1236 (1982).

8. R.W. Field, A.D. English, T. Tanaka, D.O. Harris and D.A. Jennings,
Microwave optical double resonance spectroscopy with a cw dye
laser: BaO $X^1\Sigma$ and $A^1\Sigma$, J.Chem.Phys. 59:2191 (1973).

9. W.E. Ernst, Microwave modulated polarization spectroscopy,
Opt. Commun. 46:18 (1983).

10. J.O. Schröder, B. Zeller, and W.E. Ernst, Doppler-free polarization
spectroscopy of SrCl: II. the $A^2\Pi - X^2\Sigma^+$ system, to be published.

11. W.E. Ernst and J.O. Schröder, Doppler-free polarization spectroscopy of SrCl: The $B^2\Sigma^+ - X^2\Sigma^+$ system, J.Chem.Phys. 81:136 (1984).

12. J.O. Schröder and W.E. Ernst, The $B^2\Sigma^+ - X^2\Sigma^+$ system of $Sr^{79}Br$ and
$Sr^{81}Br$: Rotational and vibrational analysis, J.Mol.Spectrosc.
112:413 (1985).

13. W.E. Ernst and J.O. Schröder, Analysis of perturbations in the
$B^2\Sigma^+ - X^2\Sigma^+$ system of SrBr, submitted to J.Mol.Spectrosc..

14. T. Törring, Saturation effect modulation in microwave spectroscopy,
J.Mol.Spectrosc. 48:148 (1973).

15. W.E. Ernst and T. Törring, Hyperfine structure in the $X^2\Sigma$ state of
CaCl measured with microwave-optical polarization spectroscopy,
Phys.Rev.A 27:875 (1983).

16. W.E. Ernst and J.O. Schröder, Polarization spectroscopy of $Sr^{79}Br$
and $Sr^{81}Br$: Analysis of the $X^2\Sigma^+$ and $B^2\Sigma^+$ hyperfine structure,
Z.Phys.D no.1 (Jan. 1986, in press).

17. W.E. Ernst and J.O. Schröder, Microwave-optical polarization spectroscopy of excited states, Phys.Rev.A 30:665 (1984).

18. W.J. Childs and L.S. Goodman, Rotational dependence of spin-rotation and hyperfine splittings in the $^2\Sigma^+$ ground state of CaF,
Phys.Rev.Lett. 44:316 (1980).

19. W.E. Ernst and S. Kindt, A molecular-beam laser-microwave double
resonance spectrometer for precise measurements of high temperature
molecules, Appl.Phys.B 30:79 (1983).

20. M.A. Johnson, C. Noda, J.S. McKillop, and R.N. Zare, Rotational
analysis of the BaI $C^2\Pi - X^2\Sigma^+(0,0)$ band, Can.J.Phys. 62:1467
(1984).

21. W.E. Ernst, J. Kändler, C. Noda, J.S. McKillop, and R.N. Zare,
Hyperfine Structure of the BaI $X^2\Sigma^+$ and $C^2\Pi$ states, to be
published.

22. W.E. Ernst, S. Kindt, and T. Törring, Precise Stark-effect
measurements in the $^2\Sigma$ ground state of CaCl, Phys.Rev.Lett.
51:979 (1983).

23. W.E. Ernst, S. Kindt, K.P.R. Nair, and T. Törring, Determination
of the ground state dipole moment of CaCl from molecular-beam
laser-microwave double resonance measurements, Phys.Rev.A 29:1158
(1984).

24. W.E. Ernst, J. Kändler, J. Lüdtke, and T. Törring, Precise measurement of the ground state dipole moment of CaI, J.Chem.Phys. 83:2744
(1985).

25. W.E. Ernst, J. Kändler, S. Kindt, and T.Törring, Electric dipole
moment of SrF $X^2\Sigma^+$ from high precision Stark-effect measurements,
Chem.Phys.Lett. 113:351 (1985).

26. T. Törring, W.E. Ernst, and S. Kindt, Dipole moments and potential
energies of the alkaline earth monohalides from an ionic model,
J.Chem.Phys. 81:4614 (1984).

27. W.E. Ernst, J. Kändler, and T. Törring, Hyperfine structure and
electric dipole moment of BaF $X^2\Sigma^+$, submitted to J.Chem.Phys..

28. W.E. Ernst and J. Kändler, Experimental determination of the electric dipole moments in the $A^2\Pi$ and $B^2\Sigma^+$ states of CaCl,
to be published.

PRESSURE BROADENING OF HYPERFINE MULTIPLETS

Nir Katzenellenbogen and Yehiam Prior

Department of Chemical Physics
The Weizmann Institute of Science
Rehovot 76100, Israel

INTRODUCTION

High resolution sub-Doppler spectroscopy enables the observation of the effects of collisions on the individual components of hyperfine (hf) multiplets. Whereas "standard" pressure broadening experiments supply an average broadening coefficient, one may expect that the different quantum numbers of the hf components are reflected in their broadening coefficients. It is the purpose of this study to investigate this phenomena. In general, one observes collisional processes that affect all lines equally and processes that affect them differently. Hyperfine multiplets may supply a convenient system to examine in detail different models for the case of overlapping lines as discussed below.

Pressure broadening of $^{79}Br_2$ was investigated under self broadening and foreign gas broadening conditions. Sub-Doppler saturation spectroscopy was used to resolve the hyperfine spectrum of bromine (having splitting of about 300 MHz). The even rotational lines of bromine(79-79) consist of six hf components split by the interaction of the nuclear quadrupole moment with the orbital electric field gradient at the nucleus (see Figure 1). The two external lines are individual hyperfine components of F=J with pseudo-spin number ε=0,2 and the central peak is an overlap of four nearly degenerate components of F=J+1,J+2 (with the degeneracy removed by the spin-rotation coupling)[1]. In this short paper we report the observation that different hyperfine components undergo different pressure broadening and discuss the interpretation of this phenomenon as due to reorientation of the nuclear spin following a "sudden" change in the molecular angular momentum.

EXPERIMENTAL SETUP

The experiment involved performing saturation spectroscopy in an intermodulated fluorescence scheme following Sorem and Schawllow[2]. We used a Spectra Physics' 380 ring dye laser of about 10 MHz bandwidth. The beam is split in two and each part is chopped at a different frequency, f_1=550Hz and f_2=850Hz. The two beams are focused by lenses of 27 mm focal length, and travel counter-propagating in the cell. The fluorescence at $90°$ to the beams is monitored by a photomultiplier (PM) at a reference frequency of f_1+f_2=1400Hz. The output of the PM tube is fed into a lock-

in amplifier, and to a computer for further processing. A typical
spectrum is Shown in Figure 1. It is the absorption spectrum of the zero
velocity group of molecules.

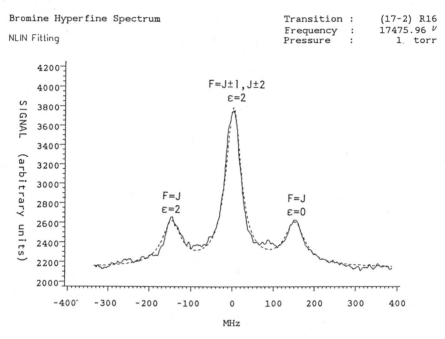

Bromine Hyperfine Spectrum

NLIN Fitting

Transition :   (17-2) R16
Frequency :   17475.96 ν
Pressure :   1. torr

Figure 1. Hyperfine spectrum of the (17-2) R16 line of bromine:
the dotted curve is the fitted curve used in the data extraction.

The laser wavelength was determined to an accuracy of 0.01 cm^{-1} using a
home built He-Ne referenced wavemeter. Frequency markers were obtained by
monitoring the FM sidebands imposed on a fraction of the beam by a LiTiO$_3$
electro-optic modulator operating at 100 MHz.

The identification of the ro-vibrational lines was based on the
spectroscopic parameters of Barrow[3] et al.

EXPERIMENT AND RESULTS

Pressure broadening was measured for the R16, R24 and R50 rotational
transitions of the 17-2 vibrational band in the $X^1\Sigma_g$ - $B^3\Pi(0_u^+)$ electronic
transition of $^{79}Br_2$. Self broadening and foreign gas broadening were
measured in the range of 0.5 to 8 torr.

The results of one such experiment is depicted in Figure 2. The slopes
of the broadening of the three peaks are very different, with the smallest
being of the ε =0 line and the largest of the ε =2 line. A SAS NLIN
procedure was used to analyse the data by fitting the lines in each
spectrum to three Lorentzian. Table I gives the pressure broadening
coefficients of the three transitions under self broadening and helium (as
foreign gas) broadening conditions. It is evident that self broadening is
more pronounced than helium broadening. The differences in broadening of
the three peaks are larger for the lower rotational level R16 than they
are for the higher ones. Also, the lower rotational level R16 is pressure
broadened more than the higher rotational levels.

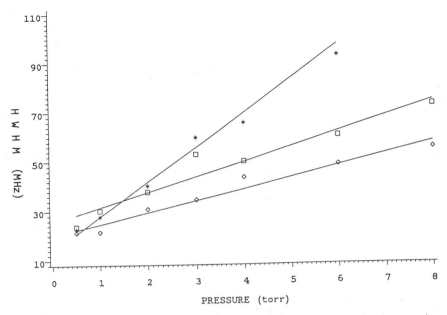

Pressure Broadening Plot   self broadening

Figure 2. Pressure self broadening for the (17-2) R16 line: star is F=J,ε=2, square is F=J,ε=0 and diamond is F=J±1,J±2, ε=2.

Table I. Pressure broadening coefficients: I for self broadening and II for broadening by helium.

	transition	left peak F=J ε=2 MHz/torr	center peak F=J±1,J±2 ε=2 MHz/torr	right peak F=J ε=0 MHz/torr
	R16	12.8 ± 1.0	4.7 ± 0.3	6.1 ± 0.7
I	R24	10.5 ± 0.9	5.7 ± 0.4	7.1 ± 1.0
	R50	6.8 ± 0.7	4.8 ± 0.4	2.6 ± 1.3
	R16	6.6 ± 0.8	4.5 ± 0.3	5.2 ± 0.7
II	R24	7.1 ± 1.3	4.7 ± 0.2	5.9 ± 0.7
	R50	7.0 ± 1.0	4.0 ± 0.3	4.2 ± 0.7

DISCUSSION

The analysis presented here is qualitative in nature, and a full discussion will be given elsewhere[4].

Phase interrupting collisions and J changing collisions affect all hyperfine components alike and thus can account for the basic broadening seen for all components. The coupling between the nuclear spin and the molecular (rotational) angular moment is taken as very weak, and thus the following simplified picture may be envisaged: The collision causes a sudden change in the orientation of the rotational angular momentum vector J, but does not change the orientation of the nuclear spin in space. After the collision the nuclear spin has a new orientation with respect to the molecular rotational angular momentum and the broadening occurs due to spin reorientation and transitions between hyperfine levels of the same total nuclear spin.

The external peaks in Fig. 1, labeled by the pseudo spin quantum numbers $\varepsilon = 0,2$, are symmetric and anti-symmetric combinations of states with spin quantum numbers $I = 0,2$. The reorientation process does not affect states with zero spin, whereas states with $I = 2$ may make a transition to one of the five different hf components causing a broadening.

In the spectra of even rotational bromine lines one may distinguish between two types of processes. The first is the broadening of an individual, well resolved line giving maximal reorientation broadening as can be seen in the left peak ($F = J, \varepsilon = 2$). The other is of reorientation within an overlapping set of lines, giving rise to smaller broadening as can be seen in the central peak ($F = J \pm 1, J \pm 2$). The overlapping lines can even show pressure narrowing for relatively low pressures. The case of overlapping lines has been discussed in great detail for high pressure rotational spectra and several models have been utilized both theoretically and experimentally[5-7]. A full discussion of this case will be presented elsewhere[4].

Helium gave lower values for the broadening coefficients with smaller differences between the peak as compared to self broadening. This is not surprising as helium is less efficient in the orientation of the molecular angular momentum and less efficient in all the other processes as well.

CONCLUSION

The spectrum of a single ro-vibrational line allows us to see both the low pressure (individual line) limit and the high pressure (overlapping lines) limit under the same pressure condition. With this information we can test various models for the collisional process under extreme conditions without having to actually go to high pressures, while working with a small number of lines.

This work is supported in part by the Israel Academy for Sciences. Y.P. is the incumbent of the Jacob and Alphonse Laniado career development chair.

REFERENCES

1. G.W. Robinson and C.D. Cornwell, J.Chem.Phys. 21, 1436 (1953).
2. M.S. Sorem and A.L. Schawlow, Opt.Comm. 5, 148 (1972).
3. R.F. Barrow, T.C. Clark, J.A. Coxon and K.K. Yee, J.Mol.Spect. 51, 428 (1974).
4. A. Kofman, N. Katzenellenbogen, Y. Prior, to be published.
5. R.G. Gordon, J.Chem.Phys. 46, 448 (1967).
6. R.J. Hall and A.C. Eckbreth in Laser Applications, eds. J.F. Ready and R.K. Erf, Academic Press, (1984).
7. R.E. Tench and S. Ezekiel, Chem.Phys.Lett. 96, 253 (1983).

MODIFIED OPTICAL BLOCH EQUATIONS FOR SOLIDS

P. R. Berman

Physics Department
New York University
4 Washington Place
New York, New York 10003

R. G. Brewer

IBM Research Laboratory
5600 Cottle Road, K01/281
San Jose, California 95193, USA

In a recent experiment of DeVoe and Brewer,[1] it was concluded that the optical Bloch equations are incapable of describing the saturation phenomena observed. Optical free induction decay (FID) measurements of the impurity ion crystal Pr^{3+}:LaF_3 were conducted where the Pr^{3+} ions are coherently prepared by a laser field under steady-state conditions and then freely precess when the driving field is removed. At low optical fields, the observed Pr^{3+} optical linewidth is dominated by magnetic fluctuations arising from pairs of fluorine nuclear flip-flops. At high optical fields, this nuclear broadening mechanism is quenched and the Bloch equations are seriously violated. On physical grounds, this failure is due to a time-averaging of the magnetic interaction as the optical nutation frequency increases.[2] The phenomenological dipole dephasing time T_2 of the Bloch equations is therefore not a true constant but lengthens with increasing field strength.

Several theories[2-8] have been proposed to explain the DeVoe-Brewer results. Most of these theories are an extension of a method developed by Redfield[9] for modifying the Bloch equations to include the effects of local, magnetically-induced fluctuations of the transition frequency of the radiating spins.

In this report, we present theoretical results for the optical FID decay rate obtained by assuming specific models for the local field fluctuations imposed on the ions' optical resonance frequency, and compare the results with "diffusion-type" theories[2-5] of frequency fluctuations. Whereas all the theories explain the overall qualitative structure of the experimental results, especially in strong fields, the predictions in the low field regime differ sufficiently to allow for the possibility of an experimental test of one theory over another. The low field data of DeVoe and Brewer are not yet precise enough to provide a critical test of any of the theories, but future experiments may be more conclusive.

Without going into the details of the calculation, a good qualitative understanding of the FID decay rate can still be obtained using simple physical considerations. The FID signal results when an ensemble of ions such as Pr^{3+} is initially prepared by an external optical field and then freely radiates when the field is suddenly removed. A large-scale inhomogeneous width Δ^* characterizes the sample, leading to a first-order FID signal that decays rapidly in a time $(\Delta^*)^{-1}$ — an effect that we do not consider here. The contribution to the FID signal which *does* concern us arises from a nonlinear interaction with the external field. As in the case of an inhomogeneously broadened vapor, the external field "burns" a hole in the ground state population and creates a bump in the excited state population, corresponding to that subgroup of ions with transition frequencies that resonantly interact with the external field. The FID decay rate is determined by the power broadened linewidth of the hole resulting from the preparative phase *plus* the free-precession decay when the field is turned off.

In the absence of frequency fluctuations for a two-state system, the FID decay rate given by the Bloch equations is

$$\gamma_F = \left[(1/T_2^o)^2 + \chi^2 (T_1^o/T_2^o) \right]^{1/2} + 1/T_2^o , \qquad (1)$$

where T_1^o and T_2^o are relaxation times in the absence of any frequency fluctuations and χ is the Rabi frequency of the external field. The bracketed term in (1) is the power-broadened hole width and the second term is the contribution of the free-decay period.

Frequency fluctuations modify the FID signal by affecting both the width of the hole and the free-decay process. We adopt a simple model, often applicable to solids, in which local field fluctuations produce frequency shifts $\delta\varepsilon(t)$ in the ion resonance frequency. Only Markoffian processes are considered; *i.e.*, the frequency following a fluctuation depends at most only on the frequency before the fluctuation. As such, the frequency jump processes can be characterized by the three parameters:

Γ = rate at which fluctuations occur ("jump rate"),
$\delta\varepsilon$ = rms frequency shift per fluctuation,
$\varepsilon_o = \sqrt{2}$ frequency associated with the frequency fluctuations distribution.

The quantity ε_o, in effect, is a measure of the maximum frequency displacement produced by the local field fluctuations. It is assumed that Γ, $\delta\varepsilon$ and ε_o are much smaller than the large inhomogeneous width Δ^*.

In this work, we restrict the discussion to three types of Markoffian processes: (1) a Gaussian-Markoffian or diffusion-type model; (2) a "Difference" model in which the probability of finding a frequency ε following a fluctuation when the frequency was ε' before the fluctuation is a function of $(\varepsilon-\varepsilon')$ only; and (3) a "Strong" model in which the frequency distribution following a fluctuation is the equilibrium distribution characterized by width ε_o. In the diffusion model, Γ, $\delta\varepsilon$ and ε_o appear only in the combinations:[10]

$\beta = \Gamma\delta\varepsilon/\varepsilon_o$ = effective jump rate,
$q = \beta\varepsilon_o^2/2$ = diffusion coefficient.

We now proceed to give the FID decay rates calculated for each model. Only the weak field regime is considered since, for strong fields, all models predict the same decay rate $\gamma_F \sim \chi(T_1^o/T_2^o)^{1/2}$ owing to the fact that, for large χ, frequency fluctuations become unimportant on the coherence time scale associated with the Rabi oscillation.[†]

As has been noted,[1] this result disagrees with the conventional Bloch prediction $\gamma_F \sim \chi(T_1/T_2)^{1/2}$, where the low field values T_1 and T_2 include the effects of frequency fluctuations.

DIFFUSION MODEL

Two limits, $\varepsilon_o \ll \beta$ and $\varepsilon_o \gg \beta$, can be distinguished. If $\varepsilon_o \ll \beta$, one finds[2-5] an FID decay rate

$$\gamma_F \sim 2/T_2; \quad (1/T_2) = (1/T_2^o) + \varepsilon_o^2/2\beta , \tag{2}$$

which implies that the T_2 associated with the weak-field regime can be much smaller than T_2^o if $\varepsilon_o^2/2\beta \gg 1/T_2^o$. When $\varepsilon_o \gg \beta$, the problem is best solved by the "Difference" model (discussion below) and, for $\varepsilon_o(\beta T_1^o)^{1/2} \gg 1/T_2^o$, we find a decay rate of order

$$\gamma_F \sim 2/T_2; \quad (1/T_2) = \frac{1}{2} (\beta T_1^o/2)^{1/2} \varepsilon_o . \tag{3}$$

DIFFERENCE MODEL

The difference model[10] does not obey detailed balancing and cannot be used in the limit $\varepsilon_o \ll \Gamma$ since frequency fluctuations *do* restore equilibrium in this limit. Thus, use of the difference model, which is characterized by a "kernel" $W(\varepsilon' \to \varepsilon)$ [$W(\varepsilon' \to \varepsilon)$ is the probability density per unit time for fluctuations to change the frequency from ε' to ε] given by

$$W(\varepsilon' \to \varepsilon) = \Gamma(2\pi\delta\varepsilon^2)^{-1/2} \exp[-(\varepsilon - \varepsilon')^2/2\delta\varepsilon^2] , \tag{4}$$

is limited to the case $\varepsilon_o \gg \Gamma$. Within this restriction, one can distinguish two subcases, $\delta\varepsilon < (1/T_2^o) + \Gamma$, $\delta\varepsilon > (1/T_2^o) + \Gamma$.

If $\delta\varepsilon < (1/T_2^o) + \Gamma$, one recovers the diffusion limit. The broadening of the hole is of order $\sqrt{n}\,\delta\varepsilon$, where $n = \Gamma T_1^o$ is approximately the number of frequency jumps occurring in time T_1^o. Considering only the interesting case when this width is considerably larger than the natural width $1/T_2^o$, i.e., when

$$(\Gamma T_1^o)^{1/2} \delta\varepsilon T_2^o = (\beta T_1^o)^{1/2} \varepsilon_o T_2^o \gg 1 , \tag{5}$$

we find an FID signal which decays as

$$\text{FID signal} \propto \frac{\pi}{2} - \tan^{-1}\left[(\beta T_1^o/2)^{1/2} \varepsilon_o t\right] . \tag{6}$$

The decay rate (3) agrees qualitatively with the numerical work of Javanainen[5] using a diffusion model.

If $\delta\varepsilon > (1/T_2^o) + \Gamma$, fluctuations take the ion frequency outside the hole burned by the field and simply increase the decay rate, one finds

$$\gamma_F \sim 2/T_2; \quad 1/T_2 = 1/T_2^o + \Gamma , \tag{7}$$

---

†The zero field intercept extrapolated from the strong field regime differs for the various models and could be used in a consistency check of the theories. Moreover, values of γ_F in the intermediate field regime can also be used to distinguish the theories.

STRONG MODEL

In the strong model, the kernel is given by

$$W(\varepsilon' \to \varepsilon) = \Gamma \left(\pi \varepsilon_0^2 \right)^{-1/2} \exp\left(-\varepsilon^2 / \varepsilon_0^2 \right) , \tag{8}$$

and is independent of the initial frequency ε'. If $\varepsilon_0 \ll \Gamma$, we find[8] a decay rate

$$\gamma_F \sim 2/T_2; \quad 1/T_2 = 1/T_2^o + \varepsilon_0^2 / 2\gamma , \tag{9}$$

which is similar to (2). For large fluctuation rates, it is unimportant whether a few strong or many weak fluctuations redistribute the frequency. If $\varepsilon_0 \gg \Gamma$, $1/T_2^o$, χ, fluctuations shift the frequency out of the hole burned by the field and simply increase the decay rate as

$$\gamma_f \sim 2/T_2; \quad 1/T_2 = 1/T_2^o + \Gamma . \tag{10}$$

The physical mechanism is the same as the leading to (7).

All the weak field results for the FID decay rate give values of $\gamma_F \equiv 2/T_2 \neq 2/T_2^o$. Since the form of T_2 is model dependent, the different models lead to different predictions for the weak-field regime. However, with the exception of (6) which indicates a nonexponential decay, one cannot distinguish the various theories in the weak-field regime unless one has an independent means for experimentally varying the parameters $\delta\varepsilon$, Γ and ε_0. This appears feasible in molecular vapors by varying the perturber-to-active-atom mass ratio or the perturber pressure, but may prove difficult in solids. If an independent variation of the parameters is not possible, one must compare the various theories over the range from weak to strong external fields to obtain a best fit to the data, a procedure that may prove useful when new experimental data become available.

This work is partially support by the U.S. Office of Naval Research.

REFERENCES

1. R. G. DeVoe and R. G. Brewer, Phys. Rev. Lett. 50:1269 (1983); 52:1354 (1984).
2. A. Schenzle, M. Mitsunaga, R. G. DeVoe and R. G. Brewer, Phys. Rev. A30:325 (1984); R. G. Brewer and R. G. DeVoe, Phys. Rev. Lett. 52:1354 (1984).
3. M. Yamanoi and J. H. Eberly, Phys. Rev. Lett. 52:1353 (1984); J. Opt. Soc. Am. B1:751 (1984).
4. E. Hanamura, J. Phys. Soc. Jpn. 52:2258 (1983); 52:3265 (1983); 52:3678 (1983).
5. J. Javanainen, Opt. Comm. 50:26 (1984).
6. K. Wodkiewicz, B. W. Shore and J. H. Eberly, Phys. Rev. A30:2390 (1984); K. Wodkiewicz and J. H. Eberly, Phys. Rev. A32 (in press).
7. P. A. Apanasevich, S. Ya. Kilin, A. P. Nizovtsev and N. S. Onishchenko, Opt. Comm. 52:279 (1984).
8. P. R. Berman and R. G. Brewer, Phys. Rev. A (in press).
9. A. G. Redfield, Phys. Rev. 98:1787 (1955).
10. P. R. Berman, Phys. Rev. A9:2170 (1974); P. R. Berman, J. M. Levy and R. G. Brewer, Phys. Rev. A11:1668 (1975).

CARS SPECTROSCOPY OF TRANSIENT SPECIES

James J. Valentini

Department of Chemistry
University of California
Irvine, California 92717

INTRODUCTION

Coherent anti-Stokes Raman scattering (CARS) spectroscopy is finding numerous applications in the physical and biological sciences and engineering.[1] The sensitivity, spectral generality, and insensitivity to background luminescence of CARS, as well as its high intrinsic temporal, spatial, and spectral resolution, have made it a very useful tool for high-resolution molecular spectroscopy, time-resolved spectroscopy, combustion diagnostics, and other applications. Our work with CARS takes advantage of several of these attributes, in using CARS for the observation of short-lived or unstable chemical species.

We use pulsed-laser CARS spectroscopy with nanosecond temporal resolution, to measure the nascent rotational and vibrational state distributions of molecular fragments produced by laser photodissociation under collisionless conditions, and to determine the quantum state distributions of chemical reaction products under single-collision reaction conditions. We also have been using CARS spectroscopy to detect electronically excited molecules in microwave discharges. In this paper I will discuss some results from our recent CARS studies of the UV and visible photodissociation of O_3 to produce $O_2(^1\Delta_g)$ and $O_2(^3\Sigma_g^-)$, the reaction of H atoms (produced by HI photolysis) with D_2 to form HD, and the observation of electronically excited $O_2(^1\Delta_g)$ in microwave-excited electric discharges in O_2

The very high temporal resolution achievable with CARS spectroscopy is particularly advantageous for these experiments. As a light scattering technique, CARS affords complete control of the temporal resolution of the experiments; the only limit being the temporal width of the laser pulses used to obtain the spectra. By coupling pulsed-laser CARS with pulsed-laser photolysis, using commercially available pulsed Nd:YAG and Nd:YAG-pumped dye lasers, it is possible to effect photodissociation of a molecule <u>and</u> spectroscopically interrogate its photofragments on a time scale short compared to the mean time between collisions, even at pressures of 1 to 10 torr. Similarly, using pulsed-laser photolysis to generate the reactive species, we can use CARS to probe the energy states of the products of simple gas-phase chemical reactions, after a single collision, if the CARS laser pulses are delayed with respect to the photolysis pulse by a time comparable to or less than the mean time between collisions.

Also advantageous to us is the generality of CARS spectroscopy. Since it is based on the Raman effect, CARS can be used for spectroscopic detection of any molecule. Also, as a Raman technique, CARS spectroscopy can be carried out in the visible region of the optical spectrum, for which reliable, powerful, continuously tunable laser sources are commercially available. Most of the species involved in our experiments, like $O_2(^3\Sigma_g^-)$

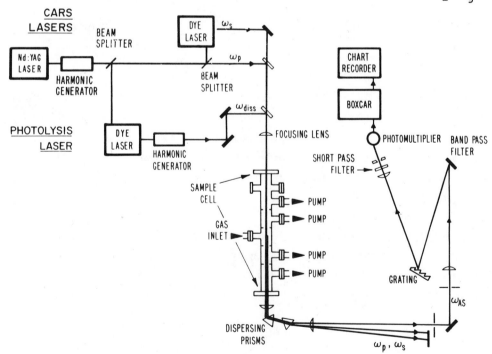

Fig. 1. CARS spectroscopy apparatus used for the experiments described here.

and $O_2 (^1\Delta_g)$, N_2, and H_2, do not absorb in the visible or near UV, and so can be detected by laser spectroscopies which in some way involve photon absorption only in the far or vacuum UV.

EXPERIMENTAL APPARATUS

The experimental apparatus used for many of our experiments is shown in Figure 1. A Nd:YAG laser (Quanta-Ray DCR-2A) is used to pump two dye lasers (Quanta-Ray PDL-2). One of these lasers produces a beam which is frequency doubled and/or mixed (Quanta-Ray WEX-1) with the 1.06 μm fundamental of the YAG to produce a photolysis beam which is continuously tunable over the range 216 to 432 nm. The other dye laser produces the tunable Stokes beam for CARS spectroscopy. The pump beam for the CARS process is a part of the second-harmonic output of the YAG.

Optical delays allow adjustment of the timing of the arrival of the different laser pulses at the sample cell. The CARS pulses typically are delayed with respect to the photolysis pulse by 0-3 ns for the photodissociation experiments (collision-free), and 3-20 ns for the chemical reaction experiments (single-collision). Typical pulse energies are 1-15 mJ in the UV photolysis beam, 5 mJ in the CARS Stokes beam, and 20 mJ in the CARS pump beam. All our photodissociation and chemical reaction experiments are carried out in low-density gas samples, at 1-10 torr. Thus, the CARS phase-matching requirement is met for collinear pump and Stokes beams. The UV photolysis beam also is made collinear with the CARS beams.

Separation of the anti-Stokes signal beam from the input beams is effected by the use of spectrally dispersive optics (prisms and/or gratings) and associated spatial filters, as well as by interference filters. The signal is detected by a photomultiplier (RCA 1P28) and processed by a boxcar (Stanford Research Systems SR 280/SR 250). The boxcar output is fed to a strip-chart recorder, as well as to a laboratory computer for subsequent analysis.

For our photodissociation and chemical reaction studies the gas sample flows through the cell, at a rate sufficient to completely replenish the sample in the CARS-probed volume between successive laser pulses, at the 10 Hz repetition rate of the experiments. For the photodissociation experiments (e.g. $O_3 + h\nu \rightarrow O_2 + O$) the gas contains only the species to be dissociated, while for our chemical reaction experiments the sample is a mixture of a photolytic precursor and a second reactant gas (e.g. HI/D_2 for $H + D_2 \rightarrow HD + D$).

Our studies of O_2 ($^1\Delta_g$) in an electric discharge are similar, except that no photolysis laser is required. For these studies, our sample cell is a quartz tube containing static or slowly flowing oxygen at a few torr pressure. The sample is excited by 20 to 100 W of microwave power.

PHOTODISSOCIATION OF OZONE

Figure 2 shows a vibrational Q-branch CARS spectrum of O_2($^3\Sigma_g^-$) formed in the photolysis of O_3 at 638 nm.[2] The spectrum shows transitions from O_2 in vibrational states v = 0, 1, 2, and 3, with rotational states J = 15 to 53. The spectrum is completely rotationally resolved, and thus can be analyzed very easily to reveal the nascent rotational and vibrational distribution of the O_2. This spectrum and a dozen others that we have taken, for photolysis between 532 nm and 638 nm, reveal that the photodissociation dynamics are fairly simple. Franck-Condon-like effects dominate

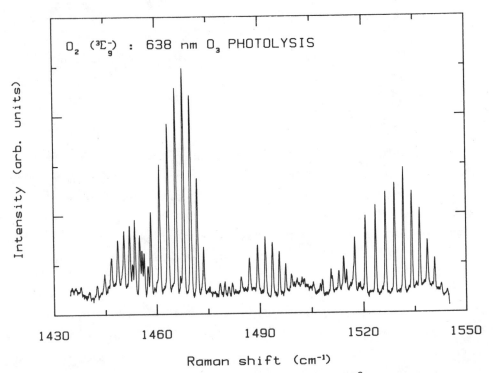

Fig. 2. Vibrational Q-branch CARS spectrum of O_2($^3\Sigma_g^-$) formed by the photolysis of O_3 at 638 nm. The bands at 1545 cm^{-1}, 1525 cm^{-1}, 1505 cm^{-1}, and 1485 cm^{-1} are due to v = 0, 1, 2, and 3, respectively.

in determining the vibrational state distribution, while kinematic considerations and impulsive energy release govern the rotational/translational
energy disposal.

While simple to analyze, the spectrum of Figure 2 illustrates a small
complication involved in CARS spectra of molecules with non-equilibrium
state distributions. In this spectrum, the CARS signal is dominated by
$(\chi'')^2$, the square of the magnitude of the imaginary part of the third-
order optical hyperpolarizability, and thus the peak in the CARS signal is
proportional to $\Delta N^2 = [N(v'',J) - N(v',J)]^2$, the <u>square</u> of the population
<u>difference</u> between the molecular energy levels (v'J and v'',J) connected by
the CARS Q-branch transition. Thus, the signal magnitude at resonance
cannot reveal the sign of the population difference, which is needed to
extract the quantum state populations. Since the sample is not at equilibrium, $N(v'',J)$ is not necessarily greater than $N(v',J)$ for $v'' < v'$.

Fortunately, the sign of ΔN can be determined from the lineshape of
the CARS transition. An interference between the real part of the hyperpolarizability, χ', and the non-resonant hyperpolarizability, χ^{NR}, gives a
contribution to the CARS signal which is directly proportional to ΔN.
This contribution has a dispersive lineshape, which is zero at resonance,
but which gives the CARS signal an asymmetric lineshape, whose "phase"
yields the sign of ΔN. Note that the spectrum of Figure 2 reveals a
population inversion, $N(3,J) > N(2,J)$ for all J.

It may not seen appropriate to refer to the results of Figure 2 as
CARS spectroscopy of "transients", since $O_2(^3\Sigma_g^-)$ is the ground electronic
state of a stable species. However, the $O_2(^3\Sigma_g^-)$ photofragments are very
transient, i.e., short-lived, in the quantum states in which they are
produced. At 300 K, the temperature of this gas sample, the most probable
rotational state is J = 9, and the relative populations of v = 0, 1, 2,
and 3, are $1:5.6x10^{-4}:3.6x10^{-7}:2.5x10^{-10}$. The photofragments will begin
to relax toward the equilibrium distribution after but a single collision.
Hence, the photofragments should be viewed as "transients" with a lifetime
of only about 100 ns, the mean time between collisions in the sample.

The photodissociation of O_3 in the UV produces electronically excited
photofragments, $O_2(^1\Delta_g)$ and $O(^1D)$. A spectrum of $O_2(^1\Delta_g)$ formed by photolysis of O_3 at 248 mm is shown in Figure 3.[3] The diatom is produced in
its six lowest-lying vibrational states, in high J states for low v and
low J states for high v. As with visible photolysis, the spectra of the

$O_2({}^1\Delta_g)$ photofragment, produced upon dissociation at more than 20 wavelengths between 230 nm and 311 nm, can be understood in terms of fairly simple impulsive dynamics. However, there is one feature of the spectrum which is far from simple and not so easily understood at first. This is the greater intensity for transitions from even (J = 2,4,6,...) compared to the intensity for transitions from odd (J = 3,5,7,...) rotational levels.

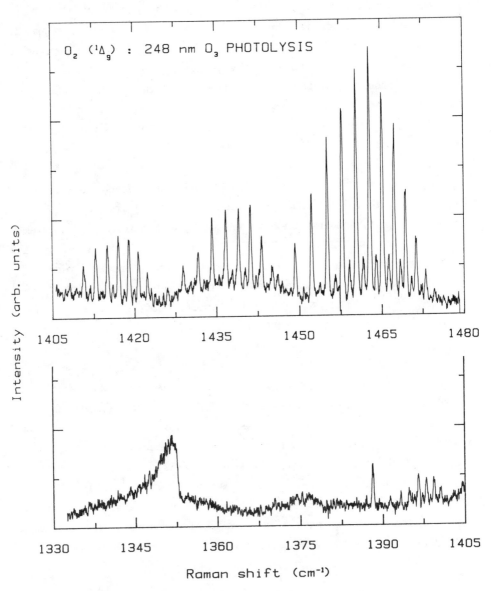

Fig. 3. Vibrational Q-branch CARS spectrum of $O_2({}^1\Delta_g)$ formed by the photolysis of O_3 at 248 nm. Bands due to v = 0, 1, 2, 3, 4, and 5 can be seen at 1460 cm^{-1}, 1440 cm^{-1}, 1415 cm^{-1}, 1395 cm^{-1}, 1375 cm^{-1}, and 1350 cm^{-1}.

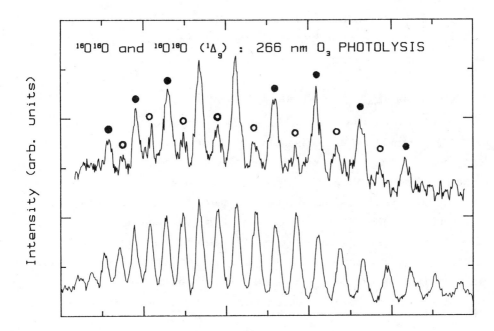

Fig. 4. The v = 0 band of the vibrational Q-branch CARS spectrum of
O_2 ($^1\Delta_g$) formed by the photolysis of O_3 at 266 nm. Bottom panel:
$^{16}O^{18}O$ from photolysis of $^{16}O^{18}O^{16}O$, $^{16}O^{16}O^{18}O$, $^{18}O^{18}O^{16}O$, and
$^{18}O^{16}O^{18}O$. Top panel: $^{16}O^{16}O$ from photolysis of $^{16}O^{16}O^{16}O$ and
$^{16}O^{16}O^{18}O$. Transitions from even J states are marked by solid
circles, odd J states by open circles.

We believe this intensity alternation reflects a rotational-state-
dependent surface crossing in the photolysis exit channel, between the
initially photo-excited surface which yields O_2($^1\Delta_g$), and a second surface
which yields O_2($^3\Sigma_g^-$). A consideration of the surface crossing in a diabatic representation within the Born-Oppenheimer approximation indicates
that such a rotational-state dependence could be associated with it.[5] A
rotational wavefunction parity constraint and the existence of only odd J
states for O_2 ($^3\Sigma_g^-$) could restrict the possibility of surface crossing to
those O_2($^1\Delta_g$) rotational states which have odd J. Thus, the surface
crossing could selectively deplete odd J $^1\Delta_g$ states, leaving even J $^1\Delta_g$
states uneffected.

Photodissociation of ^{18}O enriched O_3 supports this explanation. CARS
spectra of $^{16}O^{18}O$($^1\Delta_g$) from photodissociation of $^{16}O^{18}O^{16}O$, $^{16}O^{16}O^{18}O$,
$^{18}O^{18}O^{16}$, and $^{18}O^{16}O^{18}O$, shown in Figure 4, reveal no even J – odd J
alternation, while CARS spectra of $^{16}O^{16}O$($^1\Delta_g$) formed at the same time by
photodissociation of $^{16}O^{16}O^{16}O$ and $^{16}O^{16}O^{18}O$ in the same sample, also
shown in Figure 4, do display this alternation.[4] The surface crossing

explanation for the alternation predicts its disappearance when a "hetero-
nuclear" O_2 species like $^{16}O^{18}O$ is formed, since such species exist in
both even and odd J states in $^3\Sigma_g^-$.

SPECTROSCOPY OF $O_2(^1\Delta_g)$ IN MICROWAVE DISCHARGES

The unusual even J – odd J intensity alternation in the CARS spectra
of $O_2(^1\Delta_g)$ formed by photodissociation of O_3 led us to investigate the
CARS spectra of $O_2(^1\Delta_g)$ formed by microwave-excited electric discharge in
oxygen. Since no CARS or spontaneous Raman spectra of $O_2(^1\Delta_g)$ had ever
been obtained before, we had to be sure that the even J – odd J intensity
alternation in the $O_2(^1\Delta_g)$ photofragment spectra was not the result of
some unanticipated and highly unusual spectroscopic effect. To demonstrate that it was not a spectroscopic artifact, we needed to be able to
get CARS spectra of $O_2(^1\Delta_g)$ produced in some other way. The microwave
discharge seemed to be an ideal source.[5]

Figure 5 shows part of a CARS spectrum of the O_2 species in a microwave-excited discharge in pure O_2. The J = 33, 35, and 37 lines of the v
= 0, $^3\Sigma_g^-$ Q-branch are shown, together with the v = 1, $^3\Sigma_g^-$ Q-branch, and
the v = 0 $^3\Sigma_g^-$ O-branch lines. The region around the v = 0, J = 13 O-
branch transition has been expanded to show the $^1\Delta_g$ v = 0 Q-branch. Note
that in this $^1\Delta_g$ CARS spectrum there is no preference for even or odd J
states, indicating that the even J – odd J alternation in the photofragment spectra discussed above is a consequence of the dissociation dynamics, not a spectroscopic artifact.

Analysis of the spectrum of Figure 5 indicates that O_2 $^1\Delta_g$ is present
at a relative concentration of several percent, as is vibrationally excited $^3\Sigma_g^-$. The $^3\Sigma_g^-$ vibrational temperature is ~1,500 K, while the rotational temperature is ~1,100 K.

ANALYSIS OF INCOMPLETELY RESOLVED CARS SPECTRA

The v = 5 band around 1350 cm^{-1} in the $O_2(^1\Delta_g)$ spectrum in Figure 3 is
incompletely resolved. In this vibrational state the photofragments
appear principally in low J states, for which the spacing between adjacent
rotational lines is comparable to or less than our effective resolution, ~
0.3 cm^{-1}. Analysis of such spectra to extract rotational state populations entails extensive deconvolution, due to the interferences between
closely spaced CARS transitions.

216

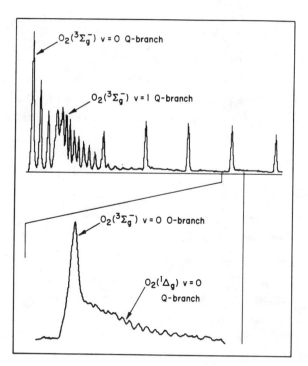

Fig. 5.   CARS spectrum of the species present in a microwave discharge in
pure O_2.

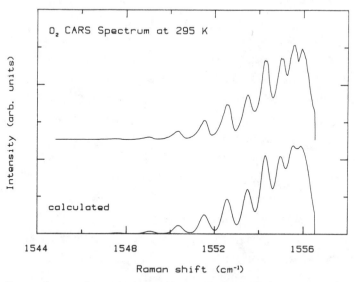

Fig. 6.   Comparison of a vibrational Q-branch CARS spectrum of O_2 at 295
K (top) with a spectrum calculated (bottom) by a computer code
which deconvolutes the experimental spectrum to extract individual rotational state populations, which are assumed to be independent.   The extracted populations give a Boltzmann distribution at 299 K.

For rotational distributions which follow a Boltzmann distribution, this deconvolution is not too difficult, for then the CARS rotational bandshape depends on only a single parameter, the temperature. However, photofragment rotational distributions generally are not Boltzmann, so the deconvolution has to treat each rotational state population as an independent variable. For a band composed of n rotational transitions, there will be n coupled non-linear equations which must be solved by the deconvolution procedure.

To allow us to extract rotational populations from vibrational bands like the v = 5 band of Figure 3, we have written a computer code which solves these n coupled non-linear equations.[6] Figure 6 illustrates the results obtainable with this code. The calculated and observed CARS spectra are nearly identical. The rotational populations derived from the experimental spectrum by the deconvolution program give a linear Boltzmann plot with a temperature of 299 K and a correlation coefficient of 0.997. The experimental spectrum was taken at an ambient temperature of 295 K.

H + D$_2$ → HD + D REACTION

Figure 7 shows a vibrational Q-branch CARS spectrum of HD formed under single-collision conditions in the H+D$_2$ reaction at a collision energy of 1.3 ev.[7] The translationally hot H atoms were generated by laser photolysis of HI at 266 nm. Due to the small moment of inertia of HD, the rotational lines in the spectrum are well separated. The CARS transitions show a large contribution from the $\chi'\chi^{NR}$ term in $|\chi^{(3)}|^2$, due to the small amount of HD formed under single-collision conditions, and the appreciable non-resonant hyperpolarizability of HI.

The quantum state population distribution extracted from this CARS spectrum is shown in Figure 8, where it is compared with the distribution predicted by an ab initio theoretical calculation using quasi-classical trajectories. The agreement is extremely good, indicating the validity of describing the state-to-state dynamics in terms of classical mechanics, even though the system is highly quantized.

As with our CARS experiments which probe O$_2$ ($^3\Sigma_g^-$) formed in the visible photodissociation of ozone, the species we detect in this experiment, HD, is a stable chemical entity, and not a "transient" in the common sense. However, as for the O$_2$ photofragment, the HD is a transient, short-lived species in the rotational and vibrational states in which the

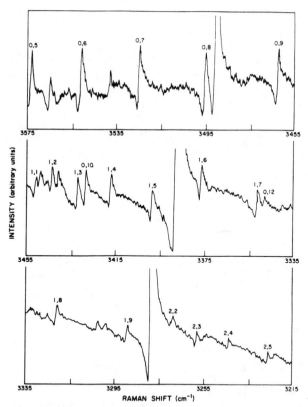

Fig. 7.   Vibrational Q-branch CARS spectrum of HD formed in the $H + D_2 \rightarrow$ HD + D reaction at 1.30 eV.   The peaks are labeled in v,J format.   Unlabeled peaks are due to D_2.

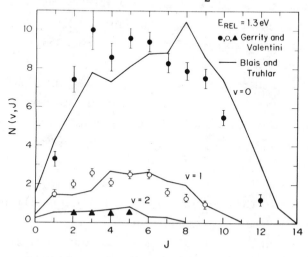

Fig. 8.   Comparison of the observed rotational-vibrational distribution (symbols) of HD formed in the $H + D_2 \rightarrow$ HD + D reaction at 1.30 eV, extracted from the CARS spectrum of Fig. 7, and the theoretically predicted distribution (lines) from quasi-classical trajectory calculations, Ref. 8.

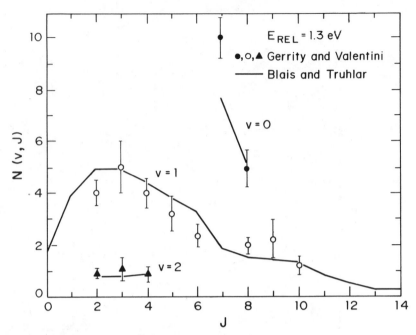

Fig. 9. Comparison of the observed rotational-vibrational distribution (symbols) of excited D_2 formed in inelastic $H + D_2$ collisions at 1.3 eV, extracted from a CARS spectrum, and the theoretically predicted distribution (lines) from quasi-classical trajectory calculations, Ref. 10.

reaction produces it. At 300 K, the temperature of our sample, the most probable rotational state of HD is J = 1. The 300 K thermal population of J = 10, which is appreciably populated by the reaction, is only 10^{-9} of that in J = 1. The ratio of the v = 0, 1, and 2, populations at 300 K is $1:3 \times 10^{-8}:2 \times 10^{-15}$. Therefore, the nascent rotational and vibrational population distribution of the HD reaction product is very far from equilibrium, and this distribution will begin to relax toward the equilibrium distribution after but a single collision. This makes the detected HD products very short-lived.

We also have used CARS to probe inelastic $H + D_2$ collisions, that is those which do not lead to reaction, but do produce rotationally and/or vibrationally excited D_2. The quantum state population distribution of rotationally and/or vibrationally excited D_2 formed in inelastic $H + D_2$ collisions at 1.3 eV collision energy is shown in Figure 9. This experimental quantum state distribution, derived from a CARS spectrum of D_2,[9] is again compared with the state distribution predicted by an ab initio theoretical calculation using quasi-classical trajectories.[10] As for the reactive collisions, the agreement between experiment and theory is very good.

CONCLUSIONS

The examples discussed above illustrate the power and utility of CARS spectroscopy for the detection of transient species produced by photodissociation or chemical reaction. Increases in the effective sensitivity of CARS through electronic resonance enhancement or non-resonant background suppression[1] should expand even more the capabilities of CARS for detection of transient, short-lived species. We are pursuing the use of these "advanced" CARS techniques, to allow us to detect photofragments and reaction products at lower density, and in an even wider range of experiments.

ACKNOWLEDGMENTS

I wish to acknowledge the graduate students and postdoctoral fellows who contributed to the research described here: D.P. Gerrity, H.B. Levene, J.-C. Nieh, D.L. Phillips, and K.D. Tabor. This work is supported by the National Science Foundation, Grant No. CHE-8506010; the U.S. Department of Energy, Grant No. DE-FG03-85ER13453; the Donors of the Petroleum Research Fund, administered by the American Chemical Society; the Research Corporation; and the Camille and Henry Dreyfus Foundation.

REFERENCES

1. For a recent review of CARS see J. J. Valentini, Coherent Anti-Stokes Raman Spectroscopy, in: "Spectrometric Techniques", Vol. IV, G. A. Vanasse, ed., Academic Press, New York (1985).
2. H. B. Levene, J.-C. Nieh, and J. J. Valentini, "CARS Spectroscopy of O_2 ($^3\Sigma_g^-$) Formed in the Photolysis of O_3 Between 532 and 638 nm: Dynamics of the Visible Photodissociation of O_3," to be published.
3. D. P. Gerrity, J.-C. Nieh, D. L. Phillips, K. D. Tabor, and J. J. Valentini, "CARS Spectroscopy of O_2($^1\Delta_g$) Formed in the Photolysis of O_3 Between 230 nm and 311 nm: Dynamics of the UV Photodissociation of O_3", to be published.
4. J.-C. Nieh, D. P. Gerrity, D. L. Phillips, K. D. Tabor, and J. J. Valentini, "UV Photodissociation of ^{18}O enriched O_3: Isotope Effects on the Photofragment State Distributions", to be published.

5. I. Rosenthal, Chemical and Physical Sources of Singlet Oxygen, in: "Singlet O_2", Vol. I, A. A. Frimer, ed.; CRC Press, Boca Raton, Florida (1985).

6. A. A. Anda, D. L. Phillips, and J. J. Valentini, "A General Method for the Deconvolution of Incompletely Resolved CARS Spectra", to be published.

7. D. P. Gerrity and J. J. Valentini, "Experimental Study of the Dynamics of the $H + D_2 \rightarrow HD + D$ Reaction at Collision Eenrgies of 0.55 and 1.30 eV", J. Chem. Phys. 81: 1298 (1984).

8. N. C. Blais and D. G. Truhlar, "Calculated Product-State Distributions for the Reaction $H + D_2 \rightarrow HD + D$ at Relative Translational Energies 0.55 and 1.30 eV", Chem. Phys. Lett. 102: 120 (1983).

9. D. P. Gerrity and J. J. Valentini, "Dynamics of Inelastic $H + D_2$ Collisions: Product Quantum State Distributions at 1.1 and 1.3 eV Collision Energy", J. Chem. Phys. 83: 2207 (1985).

10. N. C. Blais and D. G. Truhlar, "Product State Distributions for Inelastic and Reactive $H + D_2$ Collisions as Functions of Collision Energy", J. Chem. Phys. 83: 2201 (1985).

ENGINEERING CARS FOR

COMBUSTION DIAGNOSTIC APPLICATIONS

Alan C. Eckbreth

United Technologies Research Center
East Hartford, Connecticut 06108
U.S.A.

INTRODUCTION

Although the controlled use of combustion has been employed for a
long time in the propulsion, automotive and energy conversion fields,
much remains to be learned about the fundamental chemical and fluidynamic
processes and their interaction. This is particularly important in this
era of energy scarcity and environmental concern if combustion processes
are to be made more efficient and cleaner. Despite the instrumental hos-
tility of practical combustion devices, the combustion process itself is
delicately stabilized and easily perturbed. Not surprisingly, the application of nonintrusive laser spectroscopy to this field is anticipated to
afford considerable advances in combustion technology in the coming years.
In addition to being nonperturbing, laser diagnostic techniques can possess concurrently high spatial and temporal resolution, traits not normally displayed by traditionally-employed physical probing.

Of various laser approaches, coherent anti-Stokes Raman spectroscopy
(CARS) has emerged as the most promising approach for the remote, spatially
and temporally precise probing of combustion processes.[1,2] Indeed over
the last several years, the development of CARS has progressed through
laboratory flames of increasing complexity to practical application in gas
turbine combustors, internal combustion engines, furnaces and a myriad of
other devices.[3] Spontaneous Raman scattering possesses many advantages
relative to coherent Raman techniques, e.g. simplicity, single ported,
multiple species capability. Unfortunately it does not possess the req-
uisite signal levels to overcome the serious interferences, both naturally
occurring and laser induced, common to highly luminous, particle laden
flames.[4] Among the many nonlinear Raman approaches,[5] CARS possesses a
number of diagnostic advantages, some of which, from a spectroscopic perspective, are typically viewed as disadvantages. CARS of course is a
signal generative as opposed to a signal modulation process, i.e. gain,
loss or polarization rotation. One can generally record a nascent signal
far more easily and over a greater dynamic range than can be achieved
with beam modulation techniques where one is attempting to accurately
measure a small change against a large background probing beam. In the

turbulent environment of practical combustion devices, flow induced
refractive effects can cause laser beam steering and defocussing and
subsequent loss of signal, either generated or modulated. For quantitative measurements, one must compensate for such signal loss through
localized referencing at the measurement location. In CARS this can be
easily accomplished either explicitly or implicitly through the nonresonant susceptibility. In stimulated Raman gain (SRGS) or inverse Raman
spectroscopy (IRS), there exists no analogous referencing approach.
External referencing approaches, for either generation or modulation
schemes, typically do not compensate well for these effects. Furthermore, a signal generative process such as CARS is easily multiplexed with
intensified optical multichannel detectors. Although IRS was originally
demonstrated in broadband fashion on photographic film, such approaches
are difficult to implement accurately on multichannel detectors where the
"signal" is weak and buried in the noise of the large background. In
addition, CARS is not all that difficult to perform experimentally which
accounts for its practical utilization in numerous hostile applications.

In the next section of this paper, we will focus in more detail on
the implementation of CARS for combustion diagnostics and highlight some
of the laser physics problem areas where further improvements would be
desirable. Subsequent to that, the status of current spectroscopic
research areas in high pressure modelling and electronic resonance CARS
will be described. The paper will conclude with some illustrative field
applications including a description of a compact, mobile CARS instrument
designed for practical combustion measurements.

DIAGNOSTIC PERSPECTIVE AND PROBLEMS

Gas phase CARS spectroscopic investigations are typically carried
out using collinear phase matching and very narrowband pump and probe
lasers of linewidth 0.1 cm^{-1} or less to achieve high spectral resolution. As illustrated in Fig. 1, CARS is implemented quite differently
when applied to combustion diagnostics.

Phase Matching

Due to the large temperature and, thus, density gradients characteristic of combustion processes, collinear phase matching generally results
in very poor and often ambiguous spatial resolution. In many instances,
the CARS signal emanating from the focal region is not observed at all
compared with the integrated signal generated in the pre and post focal
regions. Thus crossed-beam phase matching approaches, i.e. BOXCARS,
either planar[6] or folded[7,8] as depicted in Fig. 1 are employed to achieve
high and concise spatial resolution. Although diagnostics are typically
performed from the well spectrally-separated and distinct vibrational Q
branches, folded BOXCARS approaches are well suited for performing pure
rotational CARS since the signal is spatially separated even when the
mixing is fully degenerate. In many practical applications, two beam,
three-dimensional phase matching schemes, illustrated in Fig. 2, are particularly simple and convenient to employ.[9] These also tend to be less
susceptible to refractive flow effects. The constrained Stokes aperture
does not focus as tightly as the larger diameter, annular pump. Thus, a

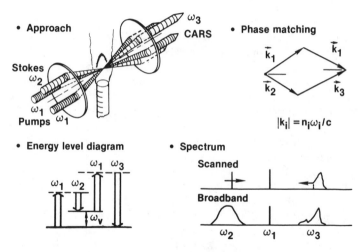

• Approach

Stokes
ω_2

ω_1
Pumps ω_1

ω_3
CARS

• Phase matching

$\vec{k}_1$ $\vec{k}_1$

$\vec{k}_2$ $\vec{k}_3$

$|k_i| = n_i\omega_i/c$

• Energy level diagram

ω_1 ω_3

ω_1 ω_2

ω_v

• Spectrum

Scanned

Broadband

ω_2 ω_1 ω_3

Fig. 1. Coherent anti-Stokes Raman spectroscopy (CARS) from a diagnostic perspective. Crossed-beam phase matching is employed
for high spatial precision, broadband Stokes sources for high
temporal resolution.

certain amount of dithering between the two can be tolerated without major
signal detriment since the CARS signal is linear in Stokes intensity.

Broadband or Multiplex CARS

   Turbulent combustion environments can fluctuate widely in temperature and chemical composition on a rapid time scale; the characteristic
frequency spectrum extends generally beyond a kilohertz and occasionally
beyond ten kilohertz. To avoid averaging over these temporal fluctuations, which would not yield true parameter averages due to the nonlinearities of the signal generation process, single pulse measurements need
to be performed. Thus, in most instances, one sacrifices the very high
spectral resolution inherent in nonlinear Raman spectroscopy, and employs
broadband Stokes sources (150–200 cm^{-1} FWHH) to access simultaneously
all of the ro-vibrational Raman resonances within a given band region.[10]
The broadband or multiplex CARS signature is dispersed in a spectrograph
and captured on an optical multichannel detector at resolutions typical-

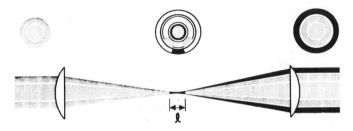

ℓ

Fig. 2. Laser beam arrangement for USED CARS phase-matching approach.
This is a two beam, three-dimensional phase matching technique
particularly convenient for practical application. The acronym
USED stands for unstable-resonator spatially enhanced detection.

ly between 1 and 4 cm^{-1}. This is generally more than adequate for measurement purposes. Since the population distribution over the various energy states is quite temperature sensitive, thermometry derives from signature analysis. An example of the temperature variation of the N$_2$ CARS signature is shown in Fig. 3. N$_2$ is commonly used for airfed combustion thermometry since it is chemically unreacted for the most part and thus present in large concentrations throughout the combustion region. In the absence of appreciable nonresonant background, i.e. a species in high concentration or at low concentration with the background suppressed, typically through polarization orientation approaches,[11,12] concentration measurements are based upon absolute signal strengths. CARS is a rather unique spectroscopy in that for species less than ∼ 30%, the signatures are also concentration sensitive due to interference with the nonresonant susceptibility permitting measurements without first order regard to the absolute signal level. This aspect is illustrated in Fig. 4 where the concentration sensitivity of the CARS signature of O$_2$ is displayed. This is an important advantage in practical media where refractive and extinctive effects can markedly affect the signal level adversely.

The laser intensities required to exploit gaseous nonlinearities in a short time period dictate employment of pulsed lasers. These are typically frequency-doubled neodymium:YAG lasers at 532 nm which are ideally spectrally situated for CARS work from both a dye laser pumping and optical detection standpoint. These lasers operate at repetition rates in the 20-50 Hz range. The combustion medium cannot be followed in real time, but is statistically sampled by an ensemble of single shot measurements which form a probability distribution function (pdf). From the pdf, the parameter time average can be ascertained as well as the magnitude of the turbulent fluctuations.

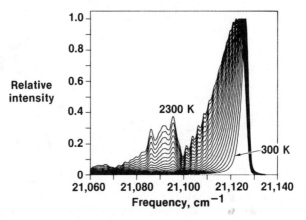

Fig. 3. Temperature dependence of N$_2$ CARS spectrum from 300 to 2300 K in 100-K increments for 70% N$_2$ concentration and 2 cm^{-1} spectral resolution.

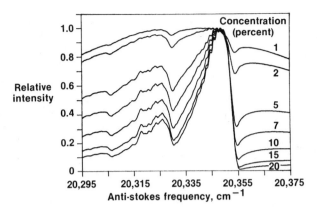

Fig. 4. Variation of the O_2 CARS spectrum with concentration at 1500 K and 2 cm^{-1} resolution.

Unlike the depiction in Fig. 1, broadband dye lasers are not perfectly smooth. There is a fine scale amplitude ripple which at ~ 1 cm^{-1} resolution on an optical multichannel detector results in about a $\pm 5\%$ rms variation. The Stokes laser consists of a Brewster angled cell with rapidly flowing dye, pumped slightly off axis, and installed in a high Fresnel number, planar Fabry–Perot optical cavity. The smoothness of the profile appears to scale with the total mode density packed into the profile. The oscillator output is typically amplified through a similar Brewster-angled cell. Wave mixing with a multimode pump exascerbates the spectral amplitude ripple into the ± 9-10% range due to the different temporal variations in the field in various spectral slices of the broadband dye laser. Wave mixing with a single mode pump laser results in amplitude noise very close to that inherent in the dye laser.[13] The spectral noise obviously distorts the single pulse resonant CARS spectrum introducing measurement error or an instrumental spread into the pdf even from an isothermal source.[3,14] Simultaneous spectral referencing provided by generating CARS in a nonresonant gas does not markedly improve the width of the isothermal histogram.[14] From an engineering viewpoint it is a complication to be avoided. With a single multichannel detector, such approaches are difficult to implement with fiber optic CARS links particularly in conjunction with signal beamsplitting schemes[15] necessary to overcome the limited dynamic range of optical multichannel detectors. Diagnosticians are thus quite interested in approaches that might yield very spectrally smooth broadband Stokes sources.

Multiple Species Measurements

One of the disadvantages of CARS, in addition to increased complexity, when compared with spontaneous Raman scattering, is the inability to address more than one molecular species at a time as usually implemented. Different constituents are generally interrogated sequentially by switching dye cells in the Stokes oscillator cavity[3] or by changing the dye solution. There are a few fortuitous instances of closely spaced Raman resonances, e.g. N_2 and N_2O, N_2 and CO, O_2 and CO_2, where a single broadband dye laser can be employed to monitor more than one species at a time. In general, however, for simultaneous measurement, a separate

Stokes source has to be introduced for each constituent to be monitored.[16]
Folded BOXCARS phase-matching schemes expedite such multiple species
approaches and have been employed with two Stokes sources to monitor O_2
and N_2 simultaneously. Or, one can introduce an additional, but
spectrally-shifted, pump to wave mix with a single Stokes source and
address a second constituent. This approach has been used to generate
CARS simultaneously in N_2 and propane.[17] Beyond two Stokes or pump
sources, such approaches become cumbersome and complex. Recently we have
developed a dual broadband Stokes approach to CARS which permits several,
i.e. more than just two, constituents to be monitored at the same time.[18]
The basic approach is illustrated in Fig. 5 and employs a combination of
two-color and three-color wave mixing processes. The two separate, two-
color processes are straightforward broadband CARS processes with a narrow pump. In the three-color process, the two broadband sources excite
all Raman resonances which coincide with their frequency differences.
Since the frequency difference range between two broadband sources is
quite extensive, many species can be excited. The narrowband pump scat-
ters off the resultant coherent Raman excitation and the CARS so generated exhibits the spectral resolution of the pump laser or spectrograph
employed. This is due to the fact that the coherent Raman excitation is
defined by the molecular resonances and not by the manner in which the
coherence was established, i.e. by broadband sources. This was first
pointed out by Yuratich.[19] The three-color CARS signatures reside at
the same frequencies they would if generated in a two color process.
What is serendipitious in combustion is that if the Stokes sources are
positioned to observe the major products of hydrocarbon-air reactions,
namely CO_2 and H_2O, the dual broadband frequency differences cover
all the important diatomic constituents, i.e. N_2, CO and NO if sufficiently abundant. If the CO_2 Stokes laser is centered near 1320 cm^{-1}
midway between the ν_1 and $2\nu_2$ modes, then O_2 can be observed as well
in the low frequency tail of the dye laser. Most broadband dyes possess unnarrowed FWHH on the order of 150 to 200 cm^{-1} and base widths at
least double this value. We have found that the dye DCM dissolved in
DMSO, well suited to the H_2O Raman frequencies, possesses a FWHH of
350 cm^{-1}. Using this dye, a very broad frequency difference range can
be achieved.

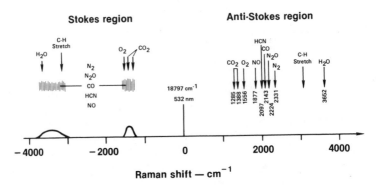

Fig. 5. Dual broadband Stokes approach to simultaneous, multiple
species CARS measurements.

Due to the spectral energy partitioning inherent in the use of two broadband sources, the three-color process is weaker than what would normally pertain in a two-color process. However, it is not as weak as one might initially suspect. The signal will depend quite strongly on the dual Stokes overlap integral with the Raman resonances excited. If the two broadband sources have the same bandwidth and center frequencies to coincide with a particular Raman resonance, all of the energy in each broadband laser is available to excite those particular Raman resonances. This is unlike normal broadband two-color CARS with a narrowband pump in which only a fraction of the broadband dye energy is used to drive each Raman resonance in the wave mixing process. It is also interesting to speculate on the single pulse quality of dual broadband CARS. For two statistically independent dye sources, many frequency combinations drive each Raman resonance unlike the normal two color situation. Thus one might expect better single shot quality due to the averaging in effect over the random amplitude profiles of the two broadband source.

Phase matching all of these processes is readily achieved by using all planar, all folded, or a combination of planar and folded BOXCARS approaches.[18] A particularly simple implementation, recently demonstrated, is dual broadband USED CARS,[20] illustrated in Fig. 6. In this scheme the broadband dye lasers are coaxially aligned inside an annular pump beam. Such annular outputs are naturally obtained from the diffractively-coupled unstable resonantors often employed (or easily installed) on the high intensity pump laser. The frequency separation between the the dye lasers when set to CO_2 and H_2O respectively is sufficiently large that the two beams are easily combined on a dichroic. Figure 7 displays simultaneously-generated CARS spectra of CO_2, N_2 and H_2O from a C_2H_4/air flame using the beam arrangement sketched in Fig. 6. Instead of the complexity of pumping two broadband dye lasers, it would be desirable if just a single broadband source could be employed and the second Stokes source generated by a nonlinear process at the appropriate frequency separation, e.g. broadband stimulated Raman scattering. One advantage of such an approach is the resultant coalignment of the original and Stokes-shifted beams.

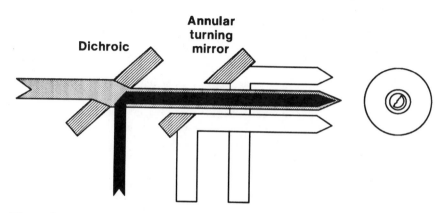

Fig. 6. Laser beam arrangement for dual broadband USED CARS.

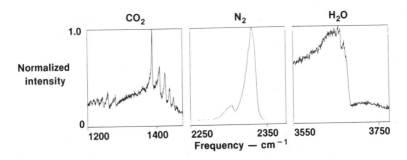

Fig. 7. Simultaneously generated dual broadband USED CARS signatures from CO_2, N_2 and H_2O in the postflame zone of a premixed C_2H_4-air flame.

Pure Rotational CARS

The dual broadband technique also suggests a new approach to pure rotational CARS. Pure rotational CARS is of interest for accurate thermometry at low and modest temperatures because of the richness of the rotational spectrum. This is in contrast with the ro-vibrational spectrum where the individual transitions are closely overlapped (with the exception of H_2) and temperature sensitivity is not particularly acute at low temperatures. There is also interest in pure rotational CARS for multi species measurements, although, in combustion processes, the spectra are broad, overlapped and dominated by N_2 and, thus, difficult to interpret. The principal difficulty with pure rotational CARS at small Raman shifts, < 200 cm^{-1}, is obtaining the Stokes dye laser in close proximity to the pump laser. This is usually achieved, at a loss of laser power available to the wave mixing, by using a higher harmonic, simultaneously generated, of the pump laser to pump the Stokes source or by pumping two dye laser sources in close spectral proximity. In the latter case, the dye lasers can be wavemixed directly or the pump can wave mix and scatter off the coherent Raman excitation created by the two. Based upon the spectral resolution and physics of dual broadband CARS, pure rotational coherences can probably be excited by a single broadband source with the rotational CARS or CSRS created by wave mixing with the pump. Figure 8 shows how this can be phase matched so that the pure rotational signal is spatially well separated from the spectrally adjacent pump laser. In this approach, note that there is no specific requirement for the frequency positioning of the broadband dye source. Large frequency shifts are desirable to create larger phase-matching angles and better spatial discrimination against the pump.

This approach should also lead to better single pulse quality of pure rotational CARS spectra since many frequency combinations drive each rotational transition as discussed earlier. If the quality were sufficiently high, it might eliminate the necessity of resonant referencing currently necessary in pure rotational CARS (or CARS in any constituent with well separated resonances) to obtain accurate temperature measurements.[21]

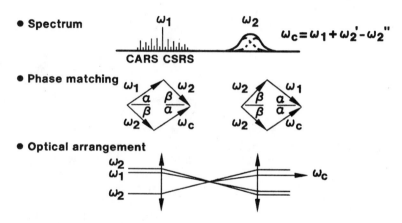

Fig. 8. New approach to pure rotational CARS. The broadband Stokes laser can be spectrally positioned arbitrarily.

SPECTROSCOPIC RESEARCH AREAS

High Pressure Modelling

Most practical combustion devices operate at pressures considerably well above atmospheric. The CARS signatures, used to extract temperature and species concentration information, Figs. 3 and 4, are quite pressure sensitive for most of the heavy diatomic and triatomic molecules due to the phenomenon of collisional narrowing. Even in the absence of such band narrowing, the signatures would and do exhibit pressure sensitivity due to constructive interferences among neighboring Raman transitions. The interferences depend on the degree of spectral overlap which is a function of the individual line spacing and Raman linewidths; the latter tend typically to be mostly pressure broadened at atmospheric pressure except at very high temperatures where Doppler contributions can become a significant fraction of the total linewidth. At pressures sufficiently high that many state changing collisions occur during the radiation-matter interaction, the band exhibits a collapse about the most populous rotational state in a given band. Such a phenomenon is well known in many spectroscopies, e.g. nmr, ir, Raman and many of the concepts developed earlier are now being successfully extended to CARS.

The third order susceptibility is expressed in terms of the "G" matrix appearing as a denominator whose diagonal elements correspond to the Raman linewidths of the individual transitions and whose off-diagonal elements are the state to state inelastic transfer rates. Fortunately for the molecules of interest in combustion, vibrational energy transfer is slow and the state to state rates are dominated by rotationally inelastic events, which, in turn, do not exhibit a strong vibrational state dependence. Thus the problem of specifying the G matrix "simplifies" to modelling the Raman linewidths and the state to state rotational energy transfer.[22]

This is generally performed through the use of various phenomenological models which have been proposed to model such transfer, e.g. the

exponential gap of Polanyi—Woodall, inverse power law, etc. . Fitting
such models to measured Raman linewidths, when available, or infrared
linewidth data, permits determination of the constants in the rotational
transfer models. Although inversion of the G matrix is computationally
time-consuming, this procedure can be greatly expedited by applying the
approach of Gordon and McGinnis which involves diagonalization of the
matrix and a single inversion step.[23] This approach is more rigorous
and is yielding closer predictions of experimental spectra in the one
case examined, N_2, than obtained by the more approximate but less computationally involved rotational diffusion model.[24] The latter is
quite simple to program and provides fairly good fits in the few cases in
which its been examnined. Figure 9 displays a CARS spectrum of N_2 obtained in a high pressure, internally-heated cell at 100 atm and 1730 K.
Also shown are theoretical calculations in which collisional narrowing is
ignored (isolated lines) and when it is accounted for by the rotational
diffusion model and G-matrix inversion using an inverse power law fit to
the Raman linewidths. Although much work has been done in this area,
further investigations are needed particularly in the area of acquiring
Raman linewidth data as a function of temperature and rotational quantum
number and closer examination of which approach best models the state-to-
state rotational transfer. It is hoped that sufficient fundamental
understanding can be achieved to formulate fairly general models of collisional narrowing without resort to molecule specific approaches in a
general sense, e.g. exponential gap for N_2, inverse power law for CO,
etc. .

Electronic Resonance CARS

In its normal diagnostic implementation in combustion, where broadband and crossed-beam approaches are employed to obtain high temporal and
spatial resolution, CARS is limited in sensitivity to the major flow constituents at atmospheric pressure. Sensitivities should be improved at
elevated pressures due to the nominal quadratic scaling of the signal
with pressure. However, at elevated pressures, laser intensities may have

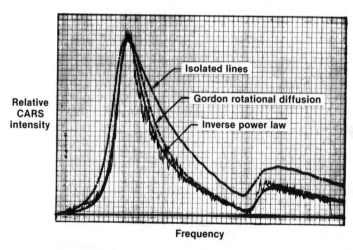

Fig. 9. CARS spectrum of N_2 from a heated, high pressure cell at 103
atm and 1700 K. The experimental spectrum at 0.4 cm^{-1} resolution is compared with various theoretical model predictions.

to be decreased to avoid gas breakdown or stimulated Raman effects, so
that detection sensitivities actually may not scale as pressure squared.
This limited sensitivity precludes CARS from observing the pool of radical
species so important in combustion chemistry. Radicals are typically
present in the ppm or tens of ppm range. Laser-induced fluorescence
spectroscopy (LIFS) is ideally suited to these species at these levels,
but there may be situations, e.g. high pressure, where a coherent
approach to radical detection is preferable.

CARS can be resonantly-enhanced electronically when either the pump,
Stokes or the CARS frequency itself coincides with an electronic transition in the probed species. Stokes resonances are weighted by the
excited vibrational state involved in the Raman resonance and this
enhancement is generally weak even at flame temperatures. More typically
one tries to achieve primary resonance with the pump laser. In so doing,
Stokes resonances are automatically satisfied. The strength of the resonance scales as the product of the four dipole matrix elements involved
with each field in the wave mixing process. Thus only certain transitions tend to be enhanced leading in most cases to a simplification of
the CARS spectrum. In the case of the combustion relevant OH molecule
under study in our laboratory, a simple triplet spectrum is predicted
since each Raman-resonant, downward Stokes transition must satisfy the
appropriate dipole selection rules for strong electronic enhancement as
shown in Fig. 10.[25]

Experimentally, pump resonance is achieved by tuning one of two
frequency-doubled dye lasers into the $A^2\Sigma$ to $X^2\Pi$ (1,0) band of OH.
This band was selected rather than the stronger (0,0) band since the
Stokes laser is then positioned in the spectral region covered by the
dye DCM which has a very broad tuning range necessary to cover the simple, but quite spectrally-separated electronically-resonant CARS spectrum. Electronic resonance enhancement has been observed in OH in flames
as seen in Fig. 11. Without enhancement, OH, even at the percent concentrations possible in very high temperature flames, would be submerged
within the H_2O CARS spectrum whose modulated bandhead is seen near 3652
cm^{-1}. Rather than observing a single spike in each region, a more com-

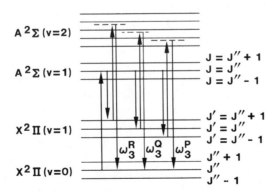

Fig. 10.   Energy level diagram for electronic resonance enhancement in
           the CARS spectrum of hydroxyl, OH, in which the pump laser is
           tuned into resonance. Strong enhancement occurs only for
           allowed downward Stokes transitions leading to a triplet
           spectrum.

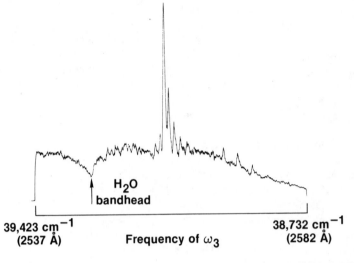

$$39{,}423 \text{ cm}^{-1} \qquad\qquad\qquad 38{,}732 \text{ cm}^{-1}$$
(2537 Å)   Frequency of ω_3 (2582 Å)

Fig. 11. Portion of electronically-resonant CARS spectrum of OH in a
flame for pump laser tuned to the (1-0) $Q_1(2)$ transition.

plex structure is seen whose origin we are currently trying to understand.
In our experiments, the pump transition is highly saturated and, potentially, more than one pump transition is being excited. With diminution
in power, fewer adjacent peaks are observed but the spacing does not
appear intensity dependent. The possibility of rotational transfer to
adjacent states during the coherent interaction was investigated but not
observed. Currently, a single mode dye laser is being fabricated to investigate the satellite structure further. One of the diagnostic issues
to be addressed is whether one can perform quantitative, saturated resonance CARS. Usually in resonance CARS, the pump laser intensity is decreased to avoid saturation, thus mitigating to some extent the detection
advantages achieved through resonance enhancement. Detectivity comparisons with LIFS are also of interest; at atmospheric pressure, our preliminary experiments would indicate LIFS to be considerably more sensitive.
If, and at what pressure, electronic resonance CARS becomes more sensitive also remains to be addressed.

FIELD APPLICATIONS

Despite its sophistication, CARS has been engineered to operate in
some very hostile environments, both inside and outside the combustion
device. Measurement demonstrations have been performed in a myriad of
practical devices ranging from simulations of gas turbine combustors,
internal combustion engines, furnaces and coal gasifiers as detailed
elsewhere.[2,3] Here we will briefly mention two, our work in actual
afterburning jet engine exhausts and a mobile CARS instrument for combustion tunnel applications.

Jet Engine Demonstrations[3]

A CARS instrument, consisting of a transmitter and receiver, was
fabricated and installed on a traversing framework which permitted the

Fig. 12. CARS instrument and measurement point traversing system at the
          exhaust of a jet engine mounted in an outdoor test stand.

CARS measurement point to be moved throughout the exhaust of afterburning jet engines. A photograph of the instrument installed about an
engine is shown in Fig. 12. Because of the very high acoustic noise
levels, ~ 150 dB, the instrument needed to be capable of completely
remote operation. The transmitter houses the frequency-doubled neodym-
ium:YAG pump laser, the Stokes dye laser and all of the ancillary optics
necessary to form and align the wave mixing beams. One of the features
of the transmitter is a chevron of five dye cells which can be translated
into the Stokes optical cavity to address sequentially the various exhaust
constituents of interest, N_2, O_2, CO, H_2O and unburned hydrocarbon
species. Each cell is optically pumped in two separate spatial locations,
one for the oscillator gain region, the other to amplify the folded-back
oscillator output. The receiver separates the CARS signature from the
incident wave mixing beams and focusses it into a fiber optic link for
transmission to the remotely-located spectrographic detection equipment.
The residual wave-mixing beams in the receiver can be employed to generate a reference signal from a gas-filled cell for signal normalization.
USED CARS (Fig. 2) phase matching is employed. The instrument is capable
of single pulse thermometry from N_2 at a 20 Hz rate. All spectra are
recorded on a single shot basis to a streaming tape drive. At low concentrations, the spectra are averaged to enhance S/N assuming the fluctuations at the measurement location are small enough for averaging to be
valid. Figure 13 shows a temperature histogram and time series at one
location in the engine exhaust with the afterburner at full power.

Mobile CARS Instrument

     The aforementioned CARS instrument is not easily transportable and
represents a dedicated test stand instrument. We have recently completed
construction of a relatively compact, mobile CARS instrument which can be
easily moved from one application to another. It is conceptually very

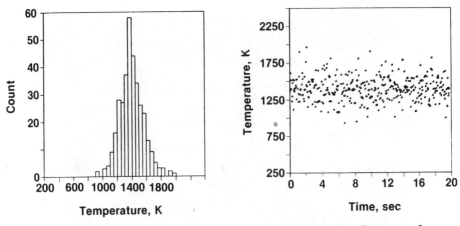

Fig. 13. CARS N$_2$ temperature histogram and time series from an afterburning jet engine exhaust.

similar to the instrument developed for jet engine measurements. It consists of a CARS transmitter on wheels approximately 2 m long x 1.5 m high
x 1.5 m wide with self-contained hydraulic jacks for height adjustment
and leveling. The receiver is carryable, 0.6 m square and 0.3 m high and
is typically installed on an inexpensive but rigid stand custom built for
each application. The transmitter and receiver contain identical peri-
scope towers for beam exit and entry. Spanning the towers is an optical
traversing framework containing the focussing and recollimating field
lenses to move the measurement volume through the combustion field under
study. The traversing framework is also envisioned to be specially
designed for each application since each is likely to vary markedly.
Optical fibers are again employed to pipe the CARS signatures to a
remotely located, three cabinet instrumentation and control rack from
which the instrument is operated. Stepper motors permit optimization of
the CARS signal generation and positioning of the optical fiber for maximum signal capture. A photograph of the mobile instrument installed
about a combustion tunnel is shown in Fig. 14. In this case as can be
seen, the traversing optics were mounted on a beam bridging the test
section. The instrument has been employed to demonstrate single pulse
CARS thermometry in supersonic combusting flows in its initial demonstration.

SUMMARY

Despite its sophistication, CARS can be engineered out of the
research laboratory and into the hostile realm of practical, real world
combustion devices. In such applications, it is, conservatively speaking,
capable of spatially-resolved, single pulse, instantaneous thermometry
and time-averaged fractional concentration measurements. There are,
nevertheless, several areas in which progress is still required and to
which laser science, nonlinear optics and spectroscopy can fundamentally
contribute. These include improved understanding of collisional narrowing
physics and electronic resonance enhancement, particularly in saturation.
Instrumentally, spectrally-smooth broadband dye lasers are of great interest as well as pump lasers with much higher repetition rates, preferably
single mode and diffraction limited, to greatly expedite data collection
rates.

Fig. 14. Mobile CARS instrument installed about a combustion test facility.

REFERENCES

1. S. A. J. Druet and J. P. E. Taran, CARS Spectroscopy, Prog. Quant. Elect. 7:1 (1981).
2. R. J. Hall and A. C. Eckbreth, Coherent Anti-Stokes Raman Spectroscopy. (CARS): Application to Combustion Diagnostics, in: "Laser Applications, Vol. 5," J. F. Ready and R. K. Erf, Eds., Academic Press, New York (1984).
3. A. C. Eckbreth, G. M. Dobbs, J. H. Stufflebeam and P. A. Tellex, CARS Temperature and Species Measurements in Augmented Jet Engine Exhausts, Appl. Opt. 23:1328 (1984).
4. A. C. Eckbreth, P. A. Bonczyk and J. F. Verdieck, Combustion Diagnostics by Laser Raman and Fluorescence Techniques, Prog. Energy Combust. Sci. 5:253 (1979).
5. A. B. Harvey, Ed., "Chemical Applications of Nonlinear Raman Spectroscopy" Academic Press, New York (1981).
6. A. C. Eckbreth, BOXCARS: Crossed-Beam Phase Matched CARS Generation in Gases, Appl. Phys. Lett. 32:421 (1978).
7. J. A. Shirley, R. J. Hall and A. C. Eckbreth, Folded BOXCARS for Rotational Raman Studies, Opt. Lett. 5:380 (1980).
8. Y. Prior, Three-Dimensional Phase Matching in Four Wave Mixing, Appl. Opt. 19:1741 (1980).
9. D. Klick, K. A. Marko and L. Rimai, Broadband Single-Pulse CARS Spectra in a Fired Internal Combustion Engine, Appl. Opt. 20:1178 (1981).
10. W. B. Roh, P. W. Schreiber and J. P. E. Taran, Single-Pulse Coherent Anti-Stokes Raman Scattering, Appl. Phys. Lett. 29:174 (1976).

11. L. A. Rahn, L. J. Zych and P. L. Mattern, Background-Free CARS Studies of Carbon Monoxide in a Flame, Opt. Comm. 39:249 (1979).

12. A. C. Eckbreth and R. J. Hall, CARS Concentration Sensitivity With and Without Nonresonant Background Suppression, Combust. Sci. Tech. 25:175 (1981).

13. D. R. Snelling, R. A. Sawchuk and R. E. Mueller, Single Pulse CARS Noise: A Comparison Between Single-Mode and Multimode Pump Lasers, Appl. Opt. 24:2771 (1985).

14. M. Pealat, P. Bouchardy, M. Lefebvre and J. P. Taran, Precision of Multiplex CARS Temperature Measurements, Appl. Opt. 24:1012 (1985).

15. A. C. Eckbreth, Optical Splitter for Dynamic Range Enhancement in Optical Multichannel Detectors, Appl. Opt. 22:2118 (1983).

16. G. L. Switzer, et al., Simultaneous CARS and Luminosity Measurements in a Bluff-Body Combustor, AIAA Paper 83-1481 (1983).

17. R. E. Teets, Laser Mode Effects on CARS Spectroscopy, presented at First International Laser Science Conference, Dallas, TX (1985).

18. A. C. Eckbreth and T. J. Anderson, Dual Broadband CARS for Simultaneous, Multiple Species Measurements, Appl. Opt. 24:2731 (1985).

19. M. A. Yuratich, Effects of Laser Linewidth on Coherent Anti-Stokes Raman Spectroscopy, Mol. Phys. 38:625 (1979).

20. A. C. Eckbreth and T. J. Anderson, Dual Broadband USED CARS, submitted to Applied Optics (1986).

21. D. V. Murphy and R. K. Chang, Single-Pulse Broadband Rotational Coherent Anti-Stokes Raman-Scattering Thermometry of Cold N_2 Gas, Opt. Lett. 6:233 (1981).

22. J. H. Stufflebeam, R. J. Hall and J. F. Verdieck, CARS Diagnostics of High Pressure and Temperature Gases, in "Combustion Diagnostics by Nonintrusive Methods," J. A. Roux and T. D. McCay, Eds., Vol. 92 Progress in Astronautics and Aeronautics, AIAA, New York (1984).

23. M. L. Koszykowski, R. L. Farrow and R. E. Palmer, Calculation of Collisionally Narrowed Coherent Anti-Stokes Raman Spectroscopy Spectra, Opt. Lett. 10:478 (1985).

24. R. J. Hall and D. A. Greenhalgh, Application of the Rotational Diffusion Model to Gaseous N_2 CARS Spectra, Opt. Comm. 40:417 (1982).

25. J. F. Verdieck, R. J. Hall and A. C. Eckbreth, Electronically-Resonant CARS Detection of OH, in "Combustion Diagnostics by Nonintrusive Methods," J. A. Roux and T. D. McCay, Eds., Vol. 92 Progress in Astronautics and Aeronautics, AIAA, New York (1984).

COHERENT CONTROL OF DIRECT AND RESONANT UNIMOLECULAR REACTIONS

Moshe Shapiro[a] and Paul Brumer[b]

[a]Department of Chemical Physics, The Weizmann Institute
Rehovot, 76100 Israel. and [b]Chemical Physics Theory Group,
Department of Chemistry, University of Toronto, 80 St. George St.
Toronto, M5S1A1 Canada

Abstract

A General method of controlling branching (direct and resonant) unimolecular reactions, using the coherence of lasers is presented. It is shown that pre-selected products may be obtained, in preference to others, as a result of interference between decay channels of a <u>superposition of degenerate continuum states.</u> It is shown that such states can be prepared by exciting a superposition of bound states using two coherent light sources with a well defined relative phase. Theoretical limits of coherent control, in direct and compound (statistical like) processes are derived. Enhanced (often 100%) yields of the (chemical) products of interest are obtained. Computational results on the I^*/I branching ratios in the photodissociation of FI and CH_3I, including the effects of initial conditions are presented. In addition, complete blocking-off of a <u>single pre-selected</u> quantum state of a given reaction product, resulting in population inversion with respect to all higher energetic levels, is shown to be attainable.

I. Introduction

Since the advent of tunable and high power laser technology, reaction dynamicists have sought a means of altering reaction pathways via laser irradiation (1). In particular, the possibility of controlling dissociation (and ionization) via laser "pulse shaping" has excited the imagination of many investigators. Recently(2a,2b), we have demonstrated, however, that regardless of the pulse shape a <u>single</u> pulse is insufficient for achieving active control over yields of reactions. Nevertheless, we have shown that this goal <u>can</u> be attained using <u>multiple</u> pulse configurations.

In this paper we generalize our previous results and show that direct as well as resonant processes can be controlled via the coherent generation of a degenerate continuum superposition state. Moreover we show that decay via a resonance (predissociation) enhances the probability of full control. The result is a practical way of altering product ratios in chemical (e.g. dissociation and ionization) decay processes using (weak) coherent radiation.

The basis of our approach is qualitatively straightforward. Consider a molecule at total energy E which can decay (fragment, ionize or emit a photon) into several different products (channels). The fact that a number of channels are accessible at a fixed energy means that the system possesses a set of degenerate continuum eigenstates, each of which correlates with a particular asymptotic channel(3,4). The key question is how to gain control over the relative branching amongst product channels. Our main theoretical results are a) that achieving this goal requires control over the composition of a coherent linear superposition of energetically degenerate continuum states and b) that such control can be attained by initial preparation of the molecule in a superposition of bound states, followed by photoexcitation. Experimentally, such a process requires coherent radiation at at least two frequencies and is essentially a two step, double resonance experiment. The first step prepares a bound superposition state and the second excites this state to the continuum. This approach is consistent with our previous results(2a). in which we noted that, barring very specific cases, product yield control can not be accomplished solely through the use of a single frequency.

Section II of this paper contains the theory of the preparation and decay of a continuum superposition state and a theoretical discussion on the limits of control attained. In section III we present a number of computational results in which we demonstrate how to achieve control over the yield of I vs. I* in the photodissociation of a diatomic (FI) and a polyatomic (CH$_3$ I) molecule. We also show how a single vibronic state of a photo-fragment can be enhanced or completely turned off.

II. Preparation and Decay of A Continuum Superposition State

A. General Formulation

Consider a molecule with Hamiltonian H_m subject to the following radiation field, which begins to affect the molecule at t=0:

$$\epsilon(t) = \int d\omega \ \epsilon(\omega) \cos(\omega t + \phi) \qquad (1)$$

Here $\epsilon(\omega)$ denotes the electric field at frequency ω . Assuming a dipole interaction, the total Hamiltonian is:

$$H = H_m - \mu \int d\omega \ \epsilon(\omega) \cos(\omega t + \phi) \qquad (2)$$

where μ is the component of the dipole in the direction of the electric field. If the molecule is initially (t=0) in a single bound molecular eigenstate $|E_i>$ at energy E_i, then traditional photodissociation ensues. Consider, however, the dynamics which results when the molecule is previously prepared in a superposition of nondegenerate bound states given by

$$|\chi(t=0)> \ = \ \Sigma_j \ c_j \ |E_j>.$$

Modifying the standard photodissociation formalism(3,4) to account for this case proceeds as follows. First, the total system wavefunction is expanded in the bound and continuum molecular eigenstates as:

$$|\psi(t)> \ = \ |\chi(t)> \ + \ \int dE \ B(E,\mathbf{n},q|t) \ |E,\mathbf{n},q^-> \ \exp(-iE \ t/\hbar). \qquad (3)$$

where

$$|\chi(t)> \ = \ \Sigma_j \ c_j|E_j> \ \exp(-iE_j t/\hbar).$$

Here $|E,\mathbf{n},q^->$ is the continuum state which correlates with the asymptotic product state $|E,\mathbf{n},q^\circ >$ consisting of products in arrangement channel q (q=1, 2, ...) and internal quantum states labeled by $\mathbf{n}$. The principal quantity of interest is usually the probability of decay into final arrangement channel q, given, to within an overall normalization factor, by:

$$P(q;E) \ = \ \Sigma_{\mathbf{n}} \ |B(E,\mathbf{n},q| \ \infty)|^2 \ . \qquad (4)$$

Substituting Eq. (3) into the time dependent Schroedinger equation and adopting the standard use of first order perturbation theory and the rotating wave approximation(3,4) gives:

$$B(E,n,q|t) = (i/2\hbar) \int d\omega \; \bar{\epsilon}(\omega) \int_o^t dt' <E,n,q^-|\mu \; |\chi \; (t')> \; \exp\{it'(E/\hbar -\omega \;)\} \quad (5)$$

with

$$\bar{\epsilon}(\omega \;) = \epsilon(\omega) \exp(-i\phi).$$

Introducing the explicit form of $\chi(t)$ we obtain:

$$B(E,n,q|t) = (i/2\hbar) \; \Sigma_j \; c_j \; <E,n,q^-|\mu|E_j> \int d\omega \; \epsilon(\omega) \int_o^t dt' \; \exp\{-i(\omega_{Ej}- \omega)t'\} \quad (6)$$

where $\omega_{Ej} = (E-E_j)/\hbar$. For t sufficiently large the time integral reduces to a δ function giving:

$$B(E,n,q| \; \infty) = (\pi \; i/\hbar \;) \; \Sigma_j \; c_j < E,n,q^-|\mu \; |E_j> \bar{\epsilon}(\omega_{Ej}) \quad (7)$$

with

$$P(q;E) = (\pi/\hbar)^2 \; \Sigma_n \; |\Sigma_j \; c_j\bar{\epsilon}(\omega_{Ej}) \; <E,n,q^-|\mu|E_j>|^2. \quad (8)$$

Expanding the square allows the more convenient form:

$$P(q;E) = (\pi/\hbar)^2 \; \Sigma_{i,j} \; F_{i,j} \; \mu^{(q)}_{i,j} \quad (9)$$

with

$$F_{i,j} = c_i\bar{\epsilon}(\omega_{Ei}) \; c_j^* \; \bar{\epsilon}^* \; (\omega_{Ej})$$

$$\mu^{(q)}_{i,j} = \Sigma_n \; <E_j|\mu \; |E,n,q^-> \; <E,n,q^-|\mu \; |E_i> \quad (10)$$

Here $\mu^{(q)}_{i,j}$ contains only molecular attributes whereas $F_{i,j}$ contains all aspects of the preparation including the magnitudes and phases of the electric field and initially prepared coherent state $|\chi \; (0)>$. Experimental control over these parameters allows manipulation of the magnitude of $P(q;E)$. In particular, variation of the magnitudes and phases of $F_{i,j}$, through the c_j and the fields, allows one to maximize or minimize a selected final product channel. The case of two product arrangement channels (q=1,2) is discussed below.

B. Two Product Arrangement Channels

As an explicit example we consider a case, depicted in Fig. 1, in which two product channels are accessible at energy E and the prepared initial superposition state is a combination of only two states, denoted $|E_a >$ and $|E_b >$. In addition, $\epsilon(\omega)$ is now comprised of two sharp lines at frequencies

$$\omega_1 = (E-E_a)/\hbar$$

and

$$\omega_2 = (E-E_b)/\hbar \; .$$

In this arrangement one first uses, for example, a laser (or more likely a microwave) pulse at

$$\omega_{12} = \omega_1 - \omega_2$$

to excite a ground level $|E_a>$ to $|E_b>$, thus forming a superposition state of $|E_a>$ and $|E_b>$. One then follows by irradiating the sample with two coherent beams at ω_1 and ω_2. Equation (9) then consists of four terms and the ratio of product yields in the two channels is conveniently written as:

$$R(1{:}2;E) = P(q{=}1;E) \; / \; P(q{=}2;E) =$$

$$\frac{\{|\mu^{(1)}_{1,1}| + x^2|\mu^{(1)}_{2,2}| + 2x\,\cos(\theta_1 - \theta_2 + \alpha^{(1)}_{1,2})|\mu^{(1)}_{1,2}|\}}{\{|\mu^{(2)}_{1,1}| + x^2|\mu^{(2)}_{2,2}| + 2x\,\cos(\theta_1 - \theta_2 + \alpha^{(2)}_{1,2})|\mu^{(1)}_{2,2}|\}} \tag{11}$$

where the following notation has been adopted:

$$f_j \exp(i\theta_j) = \bar{\epsilon}(\omega_{Ej})\,c_j \; ;$$

$$\mu^{(q)}_{i,j} = |\mu^{(q)}_{i,j}|\exp(i\alpha^{(q)}_{i,j}) \; ;$$

$$x = |f_2/f_1|. \tag{12}$$

The external control parameters are now $\theta_1 - \theta_2$ and x, which are the relative phase and amplitude ratio due to both the superposition state preparation and subsequent applied fields. The functional form of Eq. (11) prevents analytic determination of the most general conditions under which the yield in channel q=2 is an extremum. However, if the molecule is such that

$$\{\mu^{(1)}_{1,2}\}^2 = |\mu^{(1)}_{1,1}||\mu^{(1)}_{2,2}| \tag{13a}$$

then tuning to the experimental parameter amplitude ratio

$$x = |\mu^{(1)}_{1,2}|/|\mu^{(1)}_{2,2}| \tag{14a}$$

and phase

$$\cos(\theta_1 - \theta_2 + \alpha^{(1)}_{1,2}) = -1 \tag{14b}$$

yields all products in channel 2. Since $\mu^{(q)}_{i,j}$ is a scalar product of two vectors

in a linear space, it follows from the Schwarz inequality that in general,

$$\{\mu^{(1)}_{1,2}\}^2 \leq |\mu^{(1)}_{1,1}|\;|\mu^{(1)}_{2,2}| \tag{13b}$$

with the equality (Eq.(13a)), being the exception rather than the rule. Nevertheless, as shown below, even if the strict inequality holds, and the numerator (denominator) of Eq. (11) can never vanish, it is still possible to substantially enhance (deplete) the branching fraction R(1:2;E).

There are a number of interesting cases in which full control is still attainable. For example, when only one **n** state at energy E for a given q exists. In particular when the number of product channels is only two, (as in diatomic molecules), the Schwarz equality (Eq. 13a) holds and full control can be attained. In the polyatomic case, the Schwarz equality holds whenever the bound-free transition-dipole matrix elements, $<E,\mathbf{n},q^-|\mu|E_i>$ are facorizeable,

$$<E,\mathbf{n},q^-|\mu|E_i> = \mu_i^{(q)}X^q(\mathbf{n}). \tag{15}$$

Factorization implies that the branching process to all final **n** channels is independent of the excitation step. Such is the case when the process is "statistical" or

"Markoffian" like, and is rigorously the case at an isolated resonance. It follows from Eq. (15) that the $\mu^{(q)}_{i,j}$ matrix element can be written as,

$$\mu^{(q)}_{i,j} = \mu^{(q)}_i \mu^{(q)}_j \Sigma_n \{|X^q(n)|^2\} = \mu^{(q)}_i \mu^{(q)}_j M^q \qquad (16)$$

where,

$$M^q = \Sigma_n \{|X^q(n)|^2\}. \qquad (17)$$

Thus, $|\mu^{(q)}_{i,j}|^2 = |\mu^{(q)}_{i,i}| \, |\mu^{(q)}_{j,j}|$, and the Schwarz <u>equality</u> holds. Full optimization is therefore attainable.

The basic reason why such separable processes are amenable to full phase control is that, unlike direct reactions, only one phase, that associated with the preparation stage, is expressed. This follows, since the M^q coefficients are real positive numbers, (see Eq. (17)). The above is a basic property of pure quantum states, namely whether "regular" or "chaotic", the phase of $\mu^{(q)}_i$ is not lost.

In the next section we demonstrate the validity of our theory by working out a number of realistic examples.

III. Computational Examples.

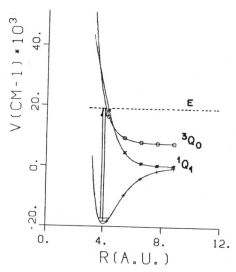

Figure 1: Cuts through the potential energy surfaces for the ground and excited states of CH_3 I. Also shown are the locations of two bound states at E_1 and E_3 and the particular excitation energy $E=37593.9$ cm^{-1} associated with Figures 2 - 4.

Dissociation of Methyl Iodide via direct photoexcitation through the $^3Q_0$ and $^1Q_1$ electronic states to form $I(^2P_{3/2})$, denoted I, or $I(^2P_{1/2})$, denoted I^*, has been the subject of many experimental[6] and theoretical[7] studies. A cut through the CH_3 I potential surfaces[8] is shown in Fig. 1. Also shown are two of the five bound states studied, $|E_1 >$ being the ground vibrational state, $|E_2 >$ and $|E_3 >$ corresponding to one and two quanta of excitation in the C-I stretch, $|E_4 >$ having one quantum of excitation in the umbrella mode, and $|E_5 >$ having one quantum of excitation in the C-I mode and one quantum of excitation in the umbrella mode.

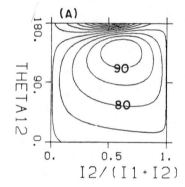
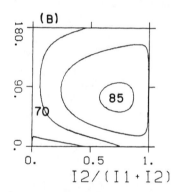

Figure 2: Contour plot of the yield of I^* (i.e. percent of I^* as product) in the photodissociation of CH_3 I from a linear superposition of a) $|E_2 >$ and $|E_5 >$, b) $|E_3 >$ and $|E_5 >$. The abscissa is labelled by the relative amplitude parameter $S = |f_2|^2/(|f_1|^2+|f_2|^2)$ and the ordinate by the relative phase parameter $\theta_1 - \theta_2$

Converged multichannel calculations previously performed[8] for CH_3 I were utilized, in conjunction with Eq. (10), to compute the yield $I^*/(I + I^*)$, at $E = \hbar \omega_1 = 37593.9$ cm^{-1} ($\lambda = 266$nm) for excitation from several different pairwise combinations of the five lowest bound states. In this case the I^* channel is labeled q=1. A sample of the results are shown in Fig. 2 as a function of the amplitude parameter, $S = |f_2|^2/(|f_1|^2+|f_2|^2)$, and phase parameter $\theta_1 - \theta_2$ and clearly demonstrates the broad range of control afforded over the I to I^* product ratio.

244

As shown in Fig. 2a, excitation from a superposition of the $|E_2 >$ and $|E_5 >$ states allows for a change in the yield of I^* production from 70%, the "natural" value, to $> 90\%$, (at S=0.6 and $\theta_1 - \theta_2 = 140^o$) or to 30% (at S=0.5 and $\theta_1 - \theta_2 = 180^o$). Thus, a mere change of 40° in phase difference, while maintaining S at 0.5 can increase the I^*/I branching ratio by as much as 20. Less drastic yield changes ,(see Fig. 2b) are observed with some of the other linear combinations studied. Preliminary studies show that the extent of variation attainable correlates with the quantity

$$|\mu^{(1)}_{1,1} \mu^{(1)}_{2,2} - \{\mu^{(1)}_{1,2}\}^2|^{-1} .$$

To demonstrate the maximum control afforded in the diatomic case we redid the $CH_3 I$ computations including only the third vibrational state in the sum {Eq. (10)} defining $\mu^{(q)}_{i,j}$. In this way the CH_3 radical is essentially replaced by a single particle, equivalent in mass to F. This provides a diatomic model akin to FI. The results are shown in Fig. 3 for a different choice of initial states as compared to Fig. 2. The ability to reduce the I yield to zero, or increase it fully to one, is clearly evident.

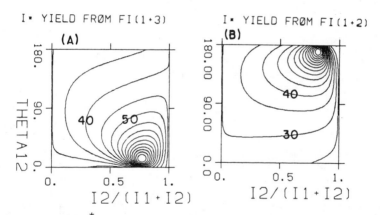

Figure 3: Yield of I^* in the photodissociation of a model diatomic "FI" (see text) from a linear superposition of a) $|E_1 >$ and $|E_2 >$, b) $|E_1 >$ and $|E_3 >$. Axis labels as in Fig. 2.

Control over the probability of observing a single (q,n) quantum state can in to a large extent be attained too. For example, since

$$|\mu_{1,2}^{(q,n)}|^2 = |\mu_{1,1}^{(q,n)}||\mu_{2,2}^{(q,n)}|$$

then tuning to

$$x = |\mu_{1,2}^{(q,n)}|/|\mu_{2,2}^{(q,n)}|$$
and
$$\cos\{\theta_1 - \theta_2 + \alpha_{1,2}^{(q,n)}\} = -1$$

results in the elimination of all product in channel (q,n).

Figure 4 shows the yield of the $(I^*, v=3)$ and $(I^*, v=4)$ quantum states, where v=3(4) denotes the second (third) excited vibrational state of the CH_3 radical, obtained under coherent dissociation of the 1+3 superposition state.

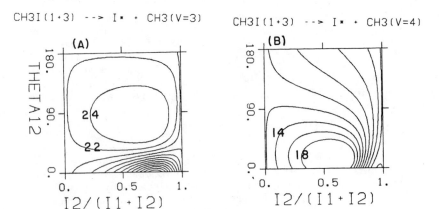

Figure 4: Contour plot of the yield of a quantum state of the products in the photodissociation of CH_3 I from a linear superposition of $|E_1 >$ and $|E_3 >$, to yield $I^* +$ a) v=3, or b) v=4. The labelling of the abscissa and ordinate is as in Fig. 3

The range of quantum state yield, for v=3 (case a), is seen to be controllable over a range of 0 to 26% whereas for v=4 (case b) a 0 to 20% range is possible. The maximum yield, in both cases, is substantially larger than the single laser case (S=0 or S=1).

IV. Summary

In considering the use of lasers to control molecular pathways one may identify two alternative approaches, which we shall term passive and active. In the former case one utilizes, for example, the high frequency resolution of the laser to probe different frequency regimes and allows the frequency dependence of the natural branching ratio of the molecule to determine the product yields. Obtaining enhancement of the desired product is thus heavily determined by molecular properties. The same holds true for methods which use strong laser fields at a single frequency; they serve to complicate the dynamics but the effect of the laser cannot be foreseen. In the active approach, which we introduce in this letter, lasers are used to override the natural tendency to branch in a fixed manner. In our case we do this by linearly combining different states, with different branching ratios, to enhance the desired product. The essential result of this paper is that this new approach requires a coherent linear combination of the energetically degenerate continuum eigenstates and that such a state may be created via excitation from a superposition state. The approach we propose is not limited solely to control over product yields in photodissociation, as discussed in future publications(9) .

References

1. a) A. Ben Shaul, Y. Haas, K.l. Kompa and R.D. Levine, "Lasers and Chemical Change" (Springer-Verlag, Berlin, 1981).

b) N. Bloembergen and E. Yablonovitch, Physics Today, 31, 23 (1978).

c) R.L. Woodin and A. Kaldor, Adv. Chem. Phys. 47(b), 3 (1981).

d) R.W. Falcone, W.R. Green, J.C. White, J.F. Young and S.E. Harris, Phys.Rev. A15, 586 (1977); W.R. Freen, J. Lukasik, J.R. Wilson, M.D. Wright, J.F. Young, S.E. Harris, Phys. Rev. Lett. 42, 970 (1979).

e) T.F. George, I.H. Zimmermann, J.R. Laing, P.L. De Vries, Acc. Chem. Res. 10, 449 (1977); T.F. George J. Phys. Chem. 86, 10 (1982).

f) A.M.F. Lau and C.K. Rhodes, Phys. Rev. A 15, 1570 (1977); ibid 16, 2392 (1977).

g) K.C. Kulander and A.E. Orel, J. Chem. Phys. 74, 6529 (1981).

2. a) M. Shapiro and P. Brumer, J. Chem. Phys. (submitted);

b) P. Brumer and M. Shapiro, Phys. Rev. Lett. (submitted);

3. M. Shapiro and R. Bersohn, Ann. Rev. Phys. Chem. 33, 409 (1982).

4. G.G. Balint-Kurti and M. Shapiro, Adv. Chem. Phys. 60, 403 (1985); P. Brumer and M. Shapiro, Adv. Chem. Phys. 60, 371 (1985).

5. M. Shapiro and P. Brumer, J. Chem. Phys. (in press)

6. a) R.K. Sparks K. Shobatake, L.R. Carlson and Y.T. Lee, J. Chem. Phys. 75, 3838 (1981).

b) H.W. Hermann and S.R. Leone J. Chem. Phys. 76,4759 (1982).

c) G.N.A. van Veen, T. Baller, A.E. deVries and N.J.A. van Veen Chem. Phys. 87, 405 (1984).

7. a) M. Shapiro and R. Bersohn J. Chem. Phys. 73, 3810 (1980).

b) S.Y.Lee and E.J. Heller, J. Chem. Phys. 76, 3035 (1982).

c) V. Engel and R. Schinke, Mol. Phys. 51, 189 (1984).

8. M. Shapiro, "Photophysics of Dissociating CH_3I: Resonance-Raman and Vibronic Photofragmentation Maps" J. Phys. Chem. (submitted for publication).

9. M. Shapiro and P. Brumer, to be published.

STIMULATED RAMAN SCATTERING, PHASE MODULATION, AND COHERENT ANTI-STOKES

RAMAN SCATTERING FROM SINGLE MICROMETER-SIZE LIQUID DROPLETS

Richard K. Chang, Shi-Xiong Qian,[*] and Johannes Eickmans

Yale University
Section of Applied Physics and Center for Laser Diagnostics
New Haven, Connecticut 06520, USA

ABSTRACT

Nonlinear optical interactions in micrometer-size liquid droplets are readily observable. New data on the stimulated Raman scattering of benzene droplets (up to the 6th-order Stokes), and water droplets containing KNO_3 (up to the 2nd-order Stokes) will be presented, as well as data on the phase modulation broadened line shape of the 1st and 2nd Stokes of CS_2 droplets, and the phase-matching angular dependence of the coherent anti-Stokes Raman scattering (CARS) signal of ethanol droplets. The phenomenological description of these nonlinear interactions involving $\chi^{(3)}$ will be presented in terms of the internal field distributions of the incident and Raman fields which are modified by the droplet interface and the feedback provided by the morphology-dependent resonances.

INTRODUCTION

In contrast to meter-long optical fibers, nonlinear optical interactions from a single micrometer-size liquid droplet are usually considered to be weak because of the short length for possible nonlinear interactions and because of the inability to maintain a high intensity in the form of a guided wave along the propagation direction. We will review several experimentally observed nonlinear optical processes associated with the $\chi^{(3)}$ of the liquid, which is in the form of spherical droplets flowing in a linear stream. In particular, we will present the stimulated Raman spectra from benzene

---

[*] On leave from the Department of Physics, Fudan University, Shanghai, People's Republic of China.

249

droplets and water droplets containing KNO_3 salt, the phase modulation broadened spectra from CS_2 droplets, and the phase-matching curve for coherent anti-Stokes emission from ethanol droplets.

Central to all nonlinear optical interactions is the morphology of the droplets, specified most readily for spherical droplets by the size parameter $x = 2\pi a/\lambda$ (where a is the radius and λ is the incident or wavelength-shifted wavelength) and by the refractive index ratio m ($m = n/n_o$ where n and n_o are the refractive indices of the liquid and the surrounding air, respectively). For $x \gg 1$ and $m > 1$, geometric optics results suggest that the spherical interface focuses the incident plane wave near the exit face of the droplet (see Fig. 1) and a Lorenz-Mie calculational result for specific ranges of x values predicts that the incident wave is also localized near the entrance face.[1] Input resonance[2] is reached when the incident wavelength λ_L or the radius is tuned to one of the morphology-dependent resonances (MDR's), which are specified by mode number n and mode order ℓ. At an input resonance, the internal fields at λ_L are even more enhanced throughout the region near the spherical interface. Intense internal fields can be achieved in a droplet whether the incident radiation is at an input resonance or not.

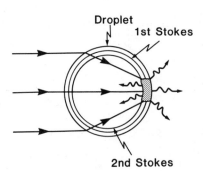

Fig. 1. Geometric optics representation of the focusing of the incident plane wave at one spot near the exit face of the droplet. This spot provides the amplification of the spontaneous Raman waves. Some of the spontaneous and amplified spontaneous Raman waves leak out of the droplet, and some of the amplified spontaneous Raman waves travel around the circumference. When the round-trip gain exceeds the round-trip loss, stimulated Raman oscillations at the 1st Stokes result. The 1st Stokes wave can serve as the pump field for the subsequent stimulated Raman oscillations at the 2nd Stokes.

Output resonance is reached (for a droplet with a fixed radius) when specific wavelengths λ_s within the inelastic emission profile correspond to MDR's with different n,ℓ values. For those wavelengths, the droplet can be envisioned as an optical cavity with a large Fresnel number and Q-factors which are dependent on the specific n and ℓ values. The portion of the inelastic radiation detected is that allowed to "leak" out of the droplet cavity. However, it is the internal field distributions of the electromagnetic waves at λ_L and λ_s, which are best described by spherical harmonic functions and not by plane waves as in the case of an extended medium, that affect the nonlinear optical interactions. Such interactions in droplets can be illustrated by several well known examples in nonlinear spectroscopy of liquids in an optical cell.

STIMULATED RAMAN SCATTERING (SRS)

The generation of SRS within a single liquid droplet, e.g., benzene, is illustrated in Fig. 1. Spontaneous Raman scattering is generated throughout the droplet and Raman gain is dominant at the focal volume suggested by geometric optics. Part of the spontaneous Raman and amplified Raman emission exits the droplet. However, part of this emission (at MDR's within the 1st Stokes linewidth) propagate around the circumference. Raman oscillations at specific wavelengths corresponding to MDR's are possible when the round-trip loss at some MDR's (due to leakage from the droplet cavity, assuming no optical absorption) is less than the round-trip gain which occurs mainly at the focal volume. The MDR's correspond to standing waves of the sphere and thus the SRS waves can be considered as a superposition of counterpropagating waves traveling around circumferences which contain the focal volume shown in Fig. 1.

SRS spectra from D_2O, H_2O, H_2O containing 0.5 M KNO_3 and from ethanol droplets have previously been reported.[3,4] Up to 14th-order Stokes emission has been detected in the SRS spectra of CCl_4 droplets.[5] Microscope photographs of the droplets revealed that the SRS radiation (red) is confined around the interface when pumped by the second-harmonic emission (green) of a Nd:YAG laser.[6]

The SRS spectrum of a benzene droplet (~35 μm in radius) is shown in Fig. 2. In addition to the 1st-order Stokes SRS emission at $\nu_2 = 992$ cm^{-1} for benzene droplets, up to the 6th-order Stokes (with $6\nu_2$ shift) were simultaneously detected with an optical multichannel analyzer (OMA). Anti-Stokes emission from benzene droplets was not detectable. The Raman shifts of the

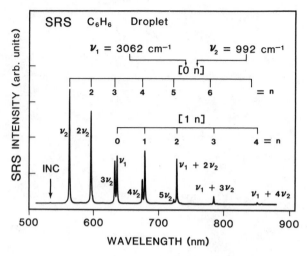

Fig. 2.   Multi-order stimulated Raman scattering (SRS) from a
          benzene droplet. The entire spectrum was detected by an
          OMA with one laser pulse. The SRS spectrum consists of
          multi-order Stokes involving the ν_2 mode ([0n] peaks, up
          to n = 6) and the 1st-order SRS Stokes involving the ν_1
          mode which can be pumped by the incident wave (INC, to
          give the [10] peak) and by the nth-order ν_2 SRS waves
          (to give the [1n] peaks).

nth-order Stokes waves are at $\omega_{ns} = \omega_L - n\nu_2$. The intensity ratio of the
2nd and 1st Stokes from the droplet is much larger than that from the cell,
which is about 1/50. The weak signal strength of the 5th and 6th Stokes from
the droplet is partially due to the decrease of the photocathode sensitivity
of the vidicon detector in the red wavelength region.

   Figure 1 illustrates the sequential Stokes interaction within the droplet. Unlike the incident pump at λ_L which is localized mainly at the focal
spot, the field of the 1st Stokes with ν_2 shift is distributed around the
interface. Thus the pumping length of this 1st Stokes wave is ~2πa while
that of the incident wave is only a fraction of a. When several ℓ's are
included, the density of MDR's is such that at least one MDR exists within
the Raman gain profile at λ_{2s} to provide the necessary feedback for the
Stokes wave at λ_{2s}. The SRS processes then repeat for the nth-order Stokes
with the (n-1)th-order Stokes as the pump field.

Another example of the sequential SRS process involves water droplets containing 1 M KNO_3. Figure 3 shows the SRS spectra which consist of the multi-order Stokes of the ν_1 mode of NO_3^-, analogous to the multi-order ν_2 modes of benzene droplets. Figure 3 also shows the 1st-order SRS of the O-H stretching mode of H_2O and the 1st-order Stokes of the NO_3^- ν_1 mode sequentially pumped by the 1st-order SRS fields of H_2O. Since the spontaneous Raman profile of H_2O in the O-H stretching region is broad (nearly 400 cm^{-1} wide), MDR's with different n,ℓ values can simultaneously provide feedback for a large number of SRS waves within the 1st Stokes profile of H_2O. The MDR's for fixed ℓ and different n are separated by ~25 cm^{-1} for droplets ~50 µm in radius. NO_3^- was specifically chosen because the linewidth of its ν_1 mode is less than such wavenumber separation between MDR's of different n value. Each of the more intense SRS peaks within the broad O-H Raman gain profile serves as a pump source for the sequential 1st-order SRS process of the NO_3^- ν_1 mode. It is therefore possible to trace each of the NO_3^- SRS peaks back to a specific O-H SRS peak which serves as the pump, i.e., each of the

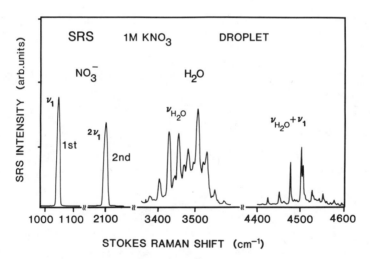

Fig. 3. Multi-order SRS from water droplets containing 1 M KNO_3. The SRS spectra from four spectral regions are separately detected by the OMA. The SRS of the 1st and 2nd Stokes of the NO_3^- anions at ν_1 and $2\nu_1$, respectively, are observed. The 1st Stokes within the O-H stretching mode of water (ν_{H_2O}) consists of two sets of MDR's. Some of the more intense ν_{H_2O} SRS peaks sequentially pump the ν_1 NO_3^- mode and give rise to SRS at $\nu_{H_2O} + \nu_1$.

former set of NO_3^- related peaks is displaced exactly one ν_1 from the latter set of H_2O related SRS peaks.

PHASE MODULATION

Phase-modulation broadening of the incident and Stokes waves has been reported from liquid and solid optical fibers.[7] Phase modulation is associated with the real part of $\chi^{(3)}$ which is known to be large for CS_2. Self phase modulation of the incident wave at ω_L takes place within the focal volume as shown in Fig. 1. The elastic scattering spectrum from CS_2 droplets was noted to be asymmetrically broadened on the longer wavelength side of λ_L, consistent with the fact that the molecular orientational time of CS_2 (~2.7 psec) leads to such asymmetry.

The 1st Stokes spectrum from CS_2 droplets was also noted to be asymmetrically broadened. For the Stokes wave, phase modulation can be induced by the incident wave in the focal volume, i.e., $Re\chi^{(3)}_{(-\omega_S,\omega_L,-\omega_L,\omega_S)}$, as well as by the Stokes waves themselves as they traverse around the circumference, i.e., self phase modulation through $Re\chi^{(3)}_{(-\omega_S,\omega_S,-\omega_S,\omega_S)}$. Within the droplet, the intensity at ω_L is larger than that at ω_S. However, the path length over which the Stokes wave at ω_S is self phase modulated is longer than that by the ω_L wave, i.e., for the Stokes wave, self phase modulation has a longer path length than phase modulation by the incident wave at ω_L.

Figure 4 shows the Stokes spectrum for the 1st- and 2nd-order SRS waves from CS_2 droplets. The broadening is only on the longer wavelength side (~400 cm^{-1}) of the ν_1 (658 cm^{-1}) and $2\nu_1$ (1322 cm^{-1}) SRS peaks. The structures superimposed on the broadening have been tentatively assigned as MDR's at specific wavelengths of the asymmetrically broadened emission. Such broadening and structures are not present in the SRS spectra from benzene droplets (see Fig. 2) or from CCl_4 and ethanol droplets, liquids having lower $Re\chi^{(3)}$ than CS_2. Consequently, for CS_2 droplets, the internal fields at ω_L and at ω_{ns} can be sufficiently intense to cause phase modulation broadening as well as to give rise to sequentially pumped multi-order Stokes.

COHERENT ANTI-STOKES RAMAN SCATTERING (CARS)

Neither the SRS or the phase modulation process requires a phase-matching condition. In order to observe four-wave mixing processes which require phase matching, CARS studies were conducted in ethanol and H_2O

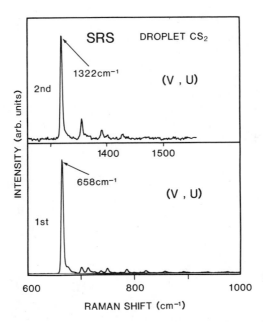

Fig. 4.   Phase modulation broadened SRS spectra at the 1st and
the 2nd Stokes of CS_2 droplets. Spectral broadening
is observed only on the Stokes side of each SRS peak.
The structure on the broadened continuum is due to
MDR's. The incident polarization is vertical
(labeled V) to the scattering plane. The polarization of the SRS signal was unanalyzed (labeled U).

droplets.[7] While the SRS intensity from micrometer-size droplets was
exceptionally large, the CARS intensity from similar size droplets was weak
as expected from such a small sample of liquid. The liquid droplet did not
exhibit any marked features in the CARS intensity or spectrum, regardless
of whether the Stokes wave was on one input resonance and/or the pump wave
was on another input resonance. Even though the field strengths within a
droplet can be large, the importance of the phase-matching condition in
four-wave mixing processes cannot be overlooked.

Phase matching cannot be satisfied for the three waves at ω_L, ω_S, and
ω_{AS} (AS denotes anti-Stokes) propagating around the circumference. However,
phase matching can still be satisfied to some extent in the direction of
the pump and Stokes waves. The phase-matching angular dependences of the
CARS output intensity are shown in Fig. 5 for ethanol droplets and ethanol
in a cell. The angular curve for the cell case is symmetrical and has a
halfwidth of ~±0.3°. The angular curve for the droplet case is asymmetrical
and broad and does not peak at the phase-matching angle of the cell.

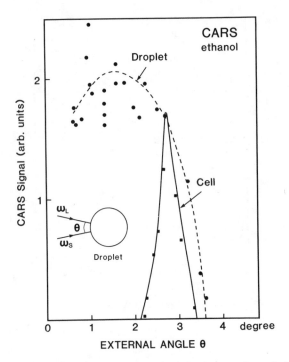

Fig. 5. Dependences of the CARS signal from ethanol droplets and ethanol in a cell as a function of the external angle between the two input beams at ω_L and ω_S. The phase-matching curve for ethanol in a cell is $\theta = 2.8°$ with a HWHM of $\pm0.3°$. The phase-matching curve for an ethanol droplet is asymmetrical and broad and does not peak at the same angle for ethanol in a cell.

The phase-matching condition is dictated by the dispersion of ethanol and thus should be the same for ethanol in a cell or in droplet form, i.e., $\theta = 2.8°$. However, we believe that the observed angular curve from ethanol droplets can be explained by the following properties of the internal field distributions of the ω_L and ω_S waves: (1) the overlap between the focal volumes of the ω_L and ω_S waves favors a collinear geometry, i.e., $\theta = 0°$; (2) the small focal volumes of the ω_L and ω_S waves lead to broad phase-matching angular curves; and (3) the spherical interface introduces additional propagation directions, where $\vec{k} \rightarrow \vec{k} + \Delta\vec{k}$, and thus the angular tuning curve is broadened analogous to phase matching with highly focused beams in an extended medium, i.e., θ is broadened symmetrically about the phase-matching angle involving plane waves. The phase-matching curve shown in Fig. 5 is the net result of all these three phenomena.

The existence of such a phase-matching curve for the CARS output indicates that the four-wave mixing process is possible within a droplet and takes place mainly in the focal volume overlap region of the two input waves at ω_L and ω_s. The ω_{AS} wave is poorly coupled to the MDR's at the anti-Stokes wavelengths and does not give rise to noticeable MDR structures, i.e., no detectable output resonances are exhibited even though, in principle, some output resonances should exist in the CARS spectra. Since no threshold is involved in the CARS process, in contrast to the SRS process, the output resonances in the CARS spectra should be much less pronounced than the resonance structures in the SRS spectra.

Another example of four-wave mixing is coherent Raman mixing among the following four waves: (1) the pump field at ω_{IR}; (2) the Stokes field at $\omega_{IR} - \omega_{vib}$; (3) the pump field at ω_L; and (4) the Stokes fields at $\omega_L - \omega_{vib} = \omega_s$. Coherent Raman mixing can provide additional Raman gain to the 1st Stokes at ω_s in addition to the normal SRS gain provided by the input beam at ω_L. An intense pump field at ω_{IR} noticeably lowered the SRS threshold at ω_L via four-wave mixing,[7] i.e., $\chi^{(3)}_{(-\omega_s,-\omega_{IR},\omega_{IR}-\omega_{vib},\omega_L)}$. Again, we noted that the additional gain provided by the coherent Raman mixing process was sensitive to the angle between the ω_{IR} and ω_L beams.[7] In principle, such a coherent Raman mixing process can provide additional Raman gain to the sequentially pumped multi-order Stokes process through $\chi^{(3)}_{(-\omega_{ns},\omega_{[n-1]s},-\omega_{[n-2]s},\omega_{[n-1]s})}$. Evidence exists[5] that such coherent Raman mixing gain can be comparable to the usual SRS gain via $\chi^{(3)}_{(-\omega_s,-\omega_L,\omega_L,\omega_s)}$ or $\chi^{(3)}_{(-\omega_{ns},-\omega_{[n-1]s},\omega_{[n-1]s},\omega_{ns})}$.

CONCLUSION

Nonlinear laser spectroscopy in micrometer-size liquid droplets is indeed possible. Experimental studies on spherical droplets clearly indicate that nonlinear optical processes involving $\chi^{(3)}$ are readily observable. MDR's are especially noticeable for those nonlinear optical phenomena that have thresholds. Four-wave mixing processes which have a phase-matching condition are also observable in droplets. Nonlinear optical interactions must be taken into account when high intensity beams with submicron wavelengths irradiate micrometer-size droplets. The theoretical treatment of nonlinear optical interactions, especially the growth of the Stokes waves and the phase-matching condition, needs to be reconsidered in the context of MDR's. It is hoped that our experimental results will stimulate such theoretical considerations which take the droplet morphology into account in solving nonlinear wave equations.

We gratefully acknowledge the partial support of this work by the Army Research Office (Contract No. DAAG29-85-K-0063) and the Air Force Office of Scientific Research (Contract No. F49620-85-K-0002).

REFERENCES

1. P. Chýlek, J.D. Pendleton, and R.G. Pinnick, "Internal and Near Surface Field of a Spherical Particle at Resonance Conditions," Appl. Opt., in press.

2. C.F. Bohren and D.R. Huffman, "The Scattering of Light by Small Particles," Wiley, New York (1983).

3. J.B. Snow, S.-X. Qian, and R.K. Chang, Opt. Lett. 10:37 (1985).

4. S.-X. Qian, J.B. Snow, and R.K. Chang, in "Laser Spectroscopy VII," T.W. Hänsch and Y.R. Shen, eds., Springer-Verlag, Berlin (1985).

5. S.-X. Qian and R.K. Chang, "Multi-Order Stokes Emission from Micrometer-Size Droplets," submitted for publication.

6. S.-X. Qian, J.B. Snow, and R.K. Chang, "Lasing Droplets: Highlighting the Liquid-Air Interface by Laser Emission," Science, in press.

7. S.-X. Qian, J.B. Snow, and R.K. Chang, Opt. Lett. 10:499 (1985).

A FEW RECENT EXPERIMENTS ON SURFACE STUDIES BY

SECOND HARMONIC GENERATION

Y. R. Shen

Department of Physics, University of California, Berkeley
and Lawrence Berkeley Laboratory
Berkeley, California 94720 USA

Surface second harmonic generation (SHG) has recently been shown to be a viable tool for surface studies.[1] It is sensitive to submonolayers of molecules on surfaces and has many obvious advantages in comparison with the conventional surface tools. Being optical, it is nondestructive, capable of in-situ remote sensing, and applicable to any interface assessible by light. The time response of SHG is nearly instantaneous. Therefore, the technique also has the potential of being useful for surface dynamic studies with extremely high time resolution. Thus, surface SHG can provide many unique opportunities in surface science research. In the past several years, we have already succeeded in demonstrating that the technique can be used to monitor molecular adsorption and desorption at a variety of interfaces,[2] to probe the spectrum of submonolayers of molecules adsorbed on surfaces,[3] to measure the orientation and distribution of adsorbed molecules,[3,4] to study surface reconstruction and phase transformation of semiconductors,[5] and so on. In this paper, we describe a few additional experiments that we have recently carried out in our laboratory to further explore the applicability of surface SHG.

Let us first give a sketch of the underlying theory. We consider an interface formed by two bulk media with centrosymmetry. Because of the broken symmetry at the interface, the second-order nonlinearity of the interface layer becomes nonvanishing (under the electric-dipole approximation). We can use a surface nonlinear susceptibility tensor $\overleftrightarrow{\chi}_s^{(2)}$ to describe the interface nonlinearity which includes both local and nonlocal responses of the interface layer to the incoming field.[6] Since the layer thickness is generally much smaller than an optical wavelength, the dipoles nonlinearly induced in the layer can be considered as forming a surface polarization sheet at the interface boundary. This surface polarization acts as the source for generating the second harmonics. It can be shown that the SH signal generated from a surface polarization at 2ω induced by a laser pulse incident at an angle θ with intensity $I(\omega)$, cross-section A, and pulsewidth T, is given approximately by

$$S(2\omega) \sim [32\pi^3 \omega \sec^2\theta / \hbar c^3 \epsilon(\omega)\epsilon^{1/2}(2\omega)] |\chi_s^{(2)}|^2 I^2(\omega) AT \quad \text{photons/pulse.} \quad (1)$$

However, the second-order nonlinearity of the centrosymmetric bulk is not strictly vanishing because of the higher-order multipole contributions, and its effect on the surface SHG may not be negligible.[7] Fortunately, in most cases of practical interest, the bulk contribution to SHG is either much smaller or can be subtracted from the surface contribution.

259

Measurement of $\overset{\leftrightarrow}{\chi}_S^{(2)}$ then allows us to deduce useful information about the interface layer.

Equation (1) shows that the SH signal should increase with increase of I, A, and T, but is eventually limited by surface optical breakdown. If we assume that the threshold energy for breakdown for a short-pulse excitation is $(IT)_{th} \sim 1$ J/cm^2, then the maximum SH signal is given by[8]

$$S_{max} \sim 10^{27} |\chi_S^{(2)}|^2 A/T \quad \text{photons/pulse.} \tag{2}$$

With A $\sim$ 0.1 cm^2, T $\sim$ 10 nsec, and a minimum detectability of 1 photon/pulse, we should be able to measure a $\chi_S^{(2)}$ as small as $\sim 3 \times 10^{-18}$ esu. Since $\chi_S^{(2)} = N_S \alpha^{(2)}$, where N_S is the number of surface atoms or molecules per cm^2, and $\alpha^{(2)}$ is the nonlinear polarizability which has a value $\gtrsim 10^{-29}$ esu for the more highly nonlinear molecules, the above estimate suggests that surface SHG has the sensitivity to detect $N_S \lesssim 3 \times 10^{11}$ molecules/cm^2. Note that the surface density of a monolayer is of the order of $10^{14} - 10^{15}$ molecules/cm^2.

One may ask whether a CW laser will also be intense enough as a pump source for surface SHG. In this case, the pump intensity is limited by laser-induced surface melting, which as a threshold power P_{th} roughly proportional to $A^{1/2}$ for sufficiently small A. If we assume $P_{th}/A^{1/2} \sim 10^4$ W/cm (as for Si), the maximum signal is

$$S_{max} \sim 10^{36} |\chi_S^{(2)}|^2 \quad \text{photons/sec,} \tag{3}$$

independent of A.[8] The minimum detectable $|\chi_S^{(2)}|$ is again around 3×10^{-18} esu. The above estimate suggests that in some cases where $|\chi_S^{(2)}|$ is large ($\gtrsim 10^{-15}$ esu), even a CW diode laser with a power of ~ 10 mW should be powerful enough to detect a monolayer of molecules by SHG. In a recent experiment,[8] using a 20-mW diode laser at 780 nm focused to a spot of $\sim 10^{-6}$ cm^2 on a silver electrode in an electrolytical cell, we were able to monitor the adsorption and desorption of monolayers of AgCl and pyridine. As shown in Fig. 1, the SH signal (~ 4000 photons/sec)

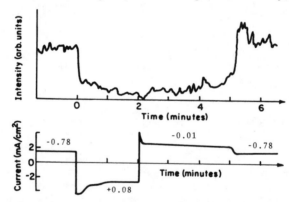

Fig. 1. Top trace: SH intensity vs. time from an Ag electrode in an electroltyic cell with 0.1 M KCl and 0.1 M pyridine. Bottom trace: cell current vs. time. This bias potentials (-0.78, +0.08, -0.01 v) are relative to a standard calomel electrode. (After Ref. 8.)

drops abruptly at t = 0 when pyridine is desorbed from Ag. The residual signal at t > 0 arises from AgCl layers formed on Ag. Since SHG is only sensitive to the first one or two surface AgCl layers, the signal does not vary appreciably with the multiplayer formation or reduction of AgCl on Ag. The reduction of AgCl is complete at t = 5 min, and with the bias potential changed to -0.78 v, the monolayer of pyridine is again adsorbed on Ag, leading to a sudden rise in the SH signal.

Both Eq. (2) and Eq. (3) also suggest that one can detect a monolayer of molecules in an area of $\lesssim 10^{-7}$ cm^2 by SHG. Thus, surface SHG is potentially useful as a surface microscopic technique which is capable of probing the distribution of a surface monolayer of molecules.[8] Figure 2 depicts the result of our first attempt in this respect. The SHG was used to map out a laser-ablated hole in a monolayer of dye molecules adsorbed on a fused quartz plate. With the probe laser focused to a 8-μm spot, the surface scan could indeed display the hole with a resolution of ~ 10 μm, as seen in Fig. 2.

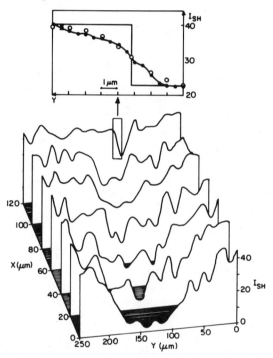

Fig. 2. SH image of an ablated hole in a rhodamine 6G dye monolayer. Inset: a high resolution scan of the region bracketed by the rectangle. Circles are a theoretical fit assuming a step change in the layer. (After Ref. 8.)

Surface SHG is a viable method to monitor adsorption and desorption of molecules on well-defined metal and semiconductor surfaces in ultra-high vacuum (UHV).[2,5,9] Figure 3 is an example, showing how the SHG responds to the adsorption of CO on Cu(100).[10] The sample properly cleaned and kept at 140°K in UHV was allowed to be exposed to CO, and SHG with a Nd:YAG laser beam was used to monitor the adsorption of CO. It is known that at T ~ 140°K, CO adsorbs on Cu(100) only at the top sites. The adsorption kinetics is then likely to obey the simple Langmuir model.[11] This was actually the case. As shown in Fig. 3, the experimental data can indeed be fit very well by a theoretical calculation following the Langmuir model. From the fit, the sticking coefficient of CO on Cu(100) can be deduced.

As a monitoring tool, surface SHG has the advantage of being nearly instantaneous in response, and is therefore ideal for surface dynamic studies. In the case of studying surface adsorbates, one may need to

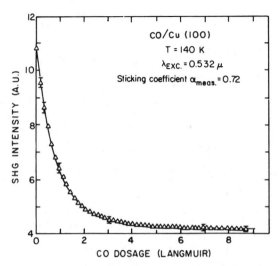

Fig. 3. SHG as a function of CO dosage on Cu(100) at 140 K. The solid
theoretical curve, derived from the Langmuir kinetic model with
the surface coverage given by $\theta = \theta_s[1 - \exp(-\alpha D)]$ and the SH
signal by $S = |A + B\theta/\theta_s|^2$, is used to fit the experimental
data. Here, D denotes the CO dosage, and α, A, and B are constants. (After Ref. 10.)

know the absolute surface coverage of the adsorbates. This requires
calibration of the surface SH signal. In a UHV chamber, the surface SHG
can be calibrated against the thermal desorption spectroscopy (TDS).[10]
The latter measures $\partial\theta/\partial t = (\partial\theta/\partial T)(\partial T/\partial t)$ when the surface temperature
is increased at a given rate. Here, θ denotes the surface coverage. A
normalized integration of the thermal desorption spectrum gives

$$\int_{T_i}^{T} \frac{\partial\theta}{\partial T}\,dt / \int_{T_i}^{T} \frac{\partial\theta}{\partial T}\,dT = 1 - \theta(T)/\theta_s, \qquad (4)$$

where θ_s is the saturated surface coverage. If, during the thermal desorption, the surface is simultaneously monitored by SHG, then $\theta(T)/\theta_s$
derived from TDS can be used to calibrate the SH signal. We again take
CO on Cu(100) as an example. In this case, the one-to-one correspondence
between θ and the SH signal was already established by the good agreement
between the experimental result and the Langmuir kinetic model shown in
Fig. 3. Therefore, we could convert $\theta(T)/\theta_s$ obtained from TDS into a
curve of SHG versus T, which could then be compared with the direct measurement of the SHG. As shown in Fig. 4, the agreement is excellent, in-
cating that TDS can indeed be used as a calibration for SHG from adsorbates. We note that while both TDS and SHG can monitor the surface coverage, the former technique has a much longer response time compared with
the latter.

Surface SHG has also been used to study monolayers of molecules at a
liquid/air interface.[2,4] In a recent experiment, we have used it to
probe the so-called liquid expanded (LE) – liquid compressed (LC) transition of a Langmuir film floating on a water surface.[12] The system is
usually prepared by first spreading a monolayer of molecules on a water
surface. The molecules can then be swept together by a moving barrier.
A measurement of the surface tension (π) as a function of the surface
area per molecules (A) at a constant temperature yields the π–A isotherm
for the two-dimensional system. Analogous to the P–V curve for a three-
dimensional system, the π–A curve for such a two-dimensional system may
also undergo the gas-liquid and liquid-solid transitions.[12] It has been

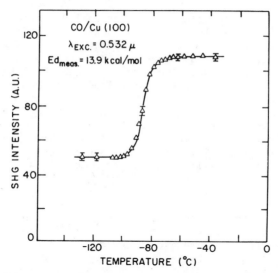

Fig. 4. SHG from CO on Cu(100) as a function of desorbing temperature.
The solid curve is calculated from thermal desorption spectra
data, and the triangular points are obtained from direct measurements. (After Ref. 10.)

found by many researchers that in some cases, the π-A curve may even exhibit an additional phase transition in the liquid phase, designated as
the LE-LC transition.[12] Fattic acid on water is a well-known example.
This is illustrated in Fig. 5, where the onset of the LE-LC transition is
marked by a kink in the π-A curve.[13] Although many careful π-A measurements and theoretical calculations on the system have been reported, the

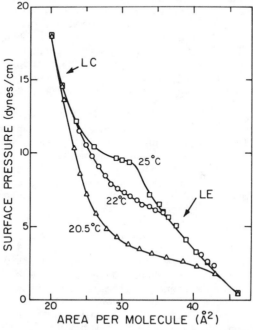

Fig. 5. Surface pressure of PDA as a function of area per molecule on a
water surface of pH = 2 for various temperatures. (After Ref.
13.)

263

nature of this LE-LC transition is not yet well understood. Some authors
suggested that the transition should be marked by a sudden change in the
molecular orientation, but no direct measurement of the molecular orientation has even been reported. Whether the transition is of first or second order is also not clear, since most researchers failed to observe a
clear LE-LC coexistence region (characterized by a flat section) on the
π-A curve.

The surface SHG allows us to measure directly an averaged molecular
orientation. The basic principle is as follows. The second-order sur-
face nonlinear susceptibility $\overset{\leftrightarrow}{\chi}^{(2)}$ of a layer is related to the second-
order nonlinear polarizability $\overset{\leftrightarrow}{\alpha}^{(2)}$ of the molecules by

$$\overset{\leftrightarrow}{\chi}^{(2)} = N_s \langle\overset{\leftrightarrow}{G}\rangle : \overset{\leftrightarrow}{\alpha}^{(2)}, \tag{5}$$

where N_s is the number of molecules per cm^2, $\overset{\leftrightarrow}{G}$ is a tensor describing the
transformation from the molecular coordinates to the lab coordinates, and
the angular brackets denote an average over the molecules. If $\overset{\leftrightarrow}{\alpha}^{(2)}$ is
dominated by a single element $\alpha^{(2)}_{\hat{z}'\hat{z}'\hat{z}'}$ with $\hat{z}'$ chosen to be along a certain axis attached to the molecule, then we can easily show

$$\frac{\chi^{(2)}_{zzz}}{\chi^{(2)}_{zxx}} = \frac{2\langle\cos^3\theta\rangle}{\langle\sin^2\theta\cos\theta\rangle} . \tag{6}$$

Here, $\hat{z}$ is parallel to the surface normal, and θ is the angle between $\hat{z}'$
and $\hat{z}$. Thus, a measurement of $\chi^{(2)}_{zzz}/\chi^{(2)}_{zxx}$ gives a weighted average of the
molecular orientation.

For the case of pentadecanoic acid (PDA) $[CH_3(CH_2)_{13}-COOH]$ in Fig. 5,
we have found that the second-order polarizability is dominated by the
C-OH bond in the molecule. By measuring $\chi^{(2)}_{zzz}/\chi^{(2)}_{zxx}$ for PDA monolayers on
water at different surface molecular densities N_s and assuming a δ-func-
tion for the molecular orientational distribution, we obtained the results in Fig. 6.[13] It is seen that θ approaches zero at low N_s and 60°
at high N_s. This is understandable because the C-OH bond, being polar,
likes to dip normally into the water. As the surface density of molecules increases, the steric interaction between molecules tends to align
the main body of the molecules along the surface normal. Consequently,
at high surface densities, the C-OH bond is expected to tilt at an angle
θ ~ 60° from the surface normal.

Figure 6 also shows that at the surface density where the LE-LC transition first sets in, a kink in the θ versus N_s curve appears. The value
of θ at this point is around 45°, and does not seem to vary with temperature even though the transition point (A_c, π_c) does change appreciably
with temperature. The results indicate that associated with the LE-LC
transition, there must be a change in the molecular orientation, in addition to a change in the surface density.

Analysis of the SH data also enable us to answer the question of whether the LE-LC transition is first or second order. It is known that the
local-field effect arising from induced dipole-induced dipole interaction
can affect SHG from a PDA monolayer. Depending on whether the layer is
heterogeneous (in the coexistence region of the first-order transition)
or homogeneous (for the second-order transition), the local-field effect
should be different. Our calculation assuming a heterogeneous layer in
the transition region showed a much better agreement with the observed SH
data. This indicates that the LE-LC transition must be of the first order.

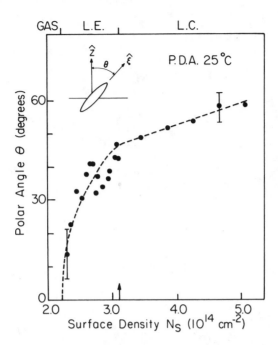

Fig. 6. Tilt angle θ between the C-OH bond in PDA and the surface normal as a function of surface density for PDA on water at 25°C. The dashed line is an extrapolation through the data points. (After Ref. 13.)

The experiments described above are just a few examples illustrating how surface SHG can be used to probe adsorbates at various interfaces. There are undoubtedly many other possible applications one can think of. Among them, the potential use of the technique to study surface dynamics is very exciting. The possibility of using surface SHG to study coating, epitaxial growth, corrosion, and catalysis is equally fascinating.

The author acknowledges major contributions to this work from his collaborators G. T. Boyd, R. Carr, S. Grubb, T. W. Hansch, M. W. Kim, Th. Rasing, and X. D. Zhu. This work was supported by the Director, Office of Energy Research, Office of Basic Energy Sciences, Materials Sciences Division of the U.S. Department of Energy under Contract No. DE-ACO3-76SF00098.

REFERENCES

1. See, for example, Y. R. Shen, Annual Rev. Material Sciences 16 (to be published).
2. C. K. Chen, T. F. Heinz, D. Ricard, and Y. R. Shen, Phys. Rev. Lett. 46, 1010 (1983), T. F. Heinz, H. W. K. Tom, and Y. R. Shen, Laser Focus 19, 101 (1983); H. W. K. Tom, C. M. Mate, X. D. Zhu, J. E. Crowell, T. F. Heinz, G. Somorjai, and Y. R. Shen, Phys. Rev. Lett. 52, 348 (1984).
3. T. F. Heinz, C. K. Chen, D. Ricard, and Y. R. Shen, Phys. Rev. Lett. 48, 478 (1982).
4. H. W. K. Tom, T. F. Heinz, and Y. R. Shen, Phys. Rev. A 28, 1883 (1983); Th. Rasing, Y. R. Shen, M. W. Kim, P. Valint, and J. Bock, Phys. Rev. A 31, 537 (1985).
5. T. F. Heinz, M. M. T. Loy, and W. A. Thompson, Phys. Rev. Lett. 54, 63 (1985).
6. P. Guyot-Sionnest, W. Chen, and Y. R. Shen (to be published).

7. See, for example, Y. R. Shen, "The Principles of Nonlinear Optics," (J. Wiley, New York, 1984), Chapter 25.
8. G. T. Boyd, Y. R. Shen, and T. W. Hansch, Optics Lett. (to be published).
9. H. W. K. Tom, X. D. Zhu, Y. R. Shen, and G. A. Somorjai, in "Proc. XVII Internatl. Conf. on Physics of Semiconductors" (Springer-Verlag, Berlin, 1984), p.99.
10. X. D. Zhu, Y. R. Shen, and R. Carr, Surf. Sci. 163, 114 (1985).
11. See, for example, I. Langmuir, J. Am. Chem. Soc. 28, 28 (1918).
12. See, for example, G. M. Bell, L. L. Coombs, and L. J. Dunne, Chem. Rev. 81, 15 (1981).
13. Th. Rasing, Y. R. Shen, M. W. Kim, and S. Grubb (to be published).

SPIN-SPIN CROSS-RELAXATION OF OPTICALLY-EXCITED

RARE-EARTH IONS IN CRYSTALS

F. W. Otto, F. X. D'Amato, M. Lukac and E. L. Hahn

Physics Department
University of California, Berkeley
Berkeley, CA 94720

ABSTRACT

A laser saturation grating experiment is applied for the measurement of electron hyperfine state spin orientation diffusion among Tm^{+2} impurity ion hyperfine ground states in SrF_2. A strong laser pulse at λ_1 produces a spatial grating of excited spin states followed by a probe at λ_2. The probe transmission intensity is to assess diffusion of non-equilibrium spin population into regions not excited by the pulse at λ_1. In a second experiment, a field sweep laser hole burning method enables measurement of Pr^{+3} optical ion hyperfine coupling of optical ground states to the reservoir of F nuclear moments in LaF_3 by level crossing. A related procedure with external rf resonance sweep excitation maps out the nuclear Zeeman-electric quadrupole coupled spectrum of Pr^{+3} over a wide range by monitoring laser beam transmission absorption.

INTRODUCTION

Two ongoing experiments are described. In both experiments optical pumping is employed as a means of creating non-equilibrium spin distributions. In both cases the spins couple with one another via either nuclear or electron spin-spin interaction. The first experiment is to measure spatial transport of spin orientation in order to determine the mean spin-spin cross-relaxation rate between Tm^{+2} ions in SrF_2. In the second experiment transient changes in spin population are observed as Pr^{+3} spin levels come into resonance with neighboring F spin levels. This monitors the cross-relaxation rate between Pr^{+3} and F and provides a means for verifying the spectroscopy of the Pr^{+3} ions.

A) SPIN ORIENTATION DIFFUSION MEASUREMENT

Introduction

In a conventional optical pumping experiment in solids, an intense pumping beam perturbs the net populations of various atomic or molecular energy levels in a sample, which are then monitored to provide information about the relaxation processes that cause the system to return to its equilibrium state. While this technique is sensitive to the net level populations of the atoms, it is relatively insensitive to processes which preserve net populations while redistributing them spatially through the sample volume, for example by mutual spin flip interactions or reabsorption of resonant fluorescence radiation. In this first experiment we attempt to measure the rate of cross-relaxation of Tm^{+2} impurity ion hyperfine levels with neighboring ions in a SrF_2 sample by measuring the rate of spatial diffusion of hyperfine spin orientation in the sample.

The diffusion constant due to mutual spin flip interactions of the ions is of order

$$D \propto W_{ij}(r_{ij})^2$$

with

$$W_{ij} = (1/12 \ S(S + 1) T_2^* \gamma^4 \hbar^2 (3\cos\theta_{ij}-1)^2/(r_{ij})^6$$

averaged over ion pairs in the sample. Note that care must be taken to neglect contributions to this average due to closely coupled neighboring ion pairs. Following Goldman and Jacquinot[1] and correcting for the 60MHz $(1/T_2^*)$ inhomogeneous linewidth of the Tm^{+2} state and taking into account the Tm nuclear spin I=1/2 we calculate an approximate diffusion constant of 5 x $10^{-12} cm^2$/sec for diffusion parallel to the applied magnetic field. This corresponds to a diffusion length of 0.14 microns in the T_1 lifetime of the hyperfine state (about 1 sec.).

Experiment

The spin orientation order in the sample is created and monitored using standing wave optical pumping of the ground state hyperfine levels of the Tm^{+2} ion (see Fig. 1). These levels have a magneto-optic susceptibility which depends on the electron spin orientation. The absorption constant for right or left circularly polarized light can be written as

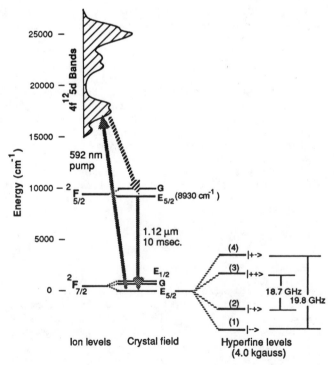

Fig. 1. Energy levels and optical pumping cycle for Tm^{+2} in SrF_2.

$$\alpha_{rcp} = \sigma_+(n_1+n_2) + \sigma_-(n_3+n_4)$$

$$\alpha_{lcp} = \sigma_-(n_1+n_2) + \sigma_+(n_3+n_4)$$

where n_1+n_2 is the total population of the electron spin-up state, and n_3+n_4 is the total population of the spin-down state (the nuclear spin may be ignored for magnetic fields above 1 kgauss). Through the Kramers-Kroenig relations, a similar expression exists for the indicies of refraction for right and left circular polarized light (n_{rcp}, n_{lcp}). An intense circularly polarized optical pulse is applied to the sample and together with its reflection off of a mirror mounted to the back surface of the sample creates a standing wave pump field, which induces a periodic magneto-optic suscepti-bility in the sample volume

$$\alpha_{rcp} = \alpha_0 + \alpha_1 \cos(K_g z) + \alpha_2 \cos(2K_g z) + \ldots$$

$$n_{rcp} = n_0 + n_1 \cos(K_g z) + n_2 \cos(2K_g z) + \ldots$$

$$K_g = 4\pi n_0/\lambda_1 \qquad (n_1, n_2 << n_0) .$$

A weak cw circularly polarized probe beam is applied to the sample and together with its reflection off the mirror creates a standing wave probe field which monitors α_1 and n_1 when the probe wavelength is matched to the pump wavelength. Solving Maxwell's Equations for the probe field inside the sample volume with a modulated adsorption constant and index of refrac-tion, for the case where $\lambda_1 \approx \lambda_2$ yields a coupled wave equation for the inci-dent and reflected wave envelopes;

$$\partial E_{inc}/\partial z = (\alpha_0/2)E_{inc} + i\kappa e^{i\delta z}E_{ref}$$

$$\partial E_{ref}/\alpha z = -(\alpha_0/2)E_{ref} - i\kappa e^{-i\delta z}E_{inc}$$

with

$$\kappa = \pi n_1/\lambda_2 - i\alpha_1/4$$

and

$$\delta = 4\pi n_0(1/\lambda_1 - 1/\lambda_2) .$$

The solution of these equations show that when the pump and probe wave-lengths are matched ($\delta L << 1$ where L is the sample length) the reflected intensity of the sample increases, thus measuring the first spatial fourier coefficient of the magnetic susceptibility (see Fig. 2).

The experimental apparatus is diagrammed in Figure 3. The sample is a .02% Tm^{+2}:SrF_2 crystal (2x5x5 mm) mounted against a mirror and kept at 2K in a 4 kilogauss field. The optical pump pulse is provided by a flash-lamp pumped dye laser which produces a 1 μsec, .5 mj pulse at 592 nm. The probe beam is provided by a Coherent 590 cw dye laser with a .5 mm cavity etalon, which produces 70 mw of output near 592 nm with a linewidth of about 0.3 cm^{-1}. This beam is chopped at 500 Hz and attenuated, resulting in 30 μW incident on the sample over a spot area of about .07 mm^2. The wavelengths of the cw and pulsed beams are monitored using a 1/4 meter monochrometer, a 6 cm^{-1} etalon, and a 1 cm^{-1} etalon. The incident and reflected beams from the sample are misaligned by about 80 mrad so as to allow access to the individual beams. The incident and reflected beams are detected, and sent to lock-in amplifiers. The ratio of the reflected to the incident intensity is stored and averaged. The wavelength of the cw probe beam is swept

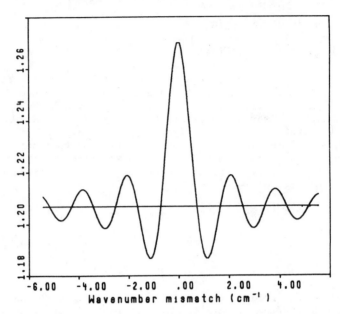

Fig. 2. Calculated ratio of reflected intensity immediately after pump pulse to its equilibrium value, as a function of wavenumber difference between the probe and pump beams ($1/\lambda_1 - 1/\lambda_2$). Straight line indicates ratio if no modulation of optical properties exists ($\alpha_1=0$, $n_1=0$). Calculated for .02% Tm^{+2} sample 2 mm thick, with $\alpha_0=5.9$ cm^{-1}, $\alpha_1=.21$ cm^{-1}, $n_0=1.439$, $n_1=2\times10^{-7}$, equilibrium $\alpha_0=6.4$ cm^{-1}.

through the wavelength matching condition in order to see the predicted increase in reflected intensity when $\lambda_1 = \lambda_2$.

Results (current status)

Currently the experiment is limited by sample heating effects on the hyperfine relaxation time of the spins, as evidenced by an apparent T_1 time which varies by up to 20% on timescales of order 5 min. This seems to indicate that too much excess heat is being generated inside the sample volume by the pump pulse, causing temperature fluctuations. In addition, there are various mechanical and electronic problems that limit our ability to take reproducible data over the timescales necessary to do the measurement.

In addition to the magneto-optic susceptibility associated with the ground state hyperfine levels in this sample, there is an electric susceptibility associated with the metastable $E_{5/2}$ state of the ion, which lies 8930 cm^{-1} above the ground state and has a lifetime of 10 msec. at 2K. When the pump pulse is applied to the sample, this state is excited in the sample volume and gives rise to a spatial modulation of the electric susceptibility of the sample, which obscures the magneto-optic effect. The lifetime of this state sets an upper limit on the size of the hyperfine diffusion constant (2×10^{-9}cm^2/sec) that may be measured using this technique. In a previous experiment, we attempted to measure spatial diffusion of this metastable excitation (due to trapping of resonant fluorescence), using backscattering from a laser induced grating to measure the first spatial fourier component

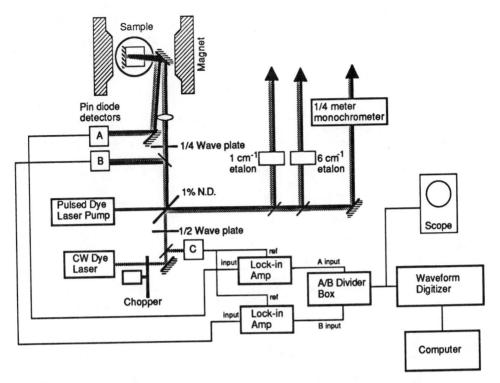

<u>Fig. 3.</u> Experimental apparatus for spin diffusion experiment.

of the electric susceptibility. We observed no evidence for diffusion of
this metastable excitation, a result which is in agreement with measurements
done by workers on similar metastable transitions in other samples.[2,3] The
failure of this previous technique to observe modulation of the magneto-optic
susceptibility due to the ground state hyperfine levels led to the current
experiment.

B) LEVEL-CROSSING CROSS-RELAXATION MEASUREMENT

<u>Introduction</u>

     In a separate experiment changes of resonant laser transmission in a
solid which depend upon the effects of nuclear spin-spin reservoir level-
crossings were observed in $Pr^{+3}:LaF_3$ for the $^3H_4 \leftrightarrow {}^1D_2$ inhomogeneously
broadened (10 GHz fwhm) optical transition[4,5]. These changes in transmission are due to the disturbance of a laser hole-burning cycle by resonant
cross-relaxation with neighboring spins at certain values of externally-
applied magnetic field.

     Optical pumping in solids is a process which is influenced by competi-
tion between the perturbing laser pump, fluorescence processes, and the
restoring resonant (cross-relaxation) and non-resonant (relaxation) population redistribution processes. This competition results in optical hole-
burning if the pumping processes are strong enough. The optical transmission is quite sensitive to sudden non-equilibrium changes in coupling

between the nuclear levels, in contrast to steady-state transmission, which changes only slightly and is obscured by 1/f (and other) noise. It is for these reasons (and the desirability of doing signal averaging) that a transient method must be used.

Theory

The Pr ground state nuclear spin levels consist of three levels split by the second-order hyperfine interaction and the relatively small nuclear electric quadrupole interaction. In non-zero magnetic fields each of these levels is additionally split due to the enhanced Zeeman interaction (see Fig. 4). The Hamiltonian of this system has been given by Teplov[6].

$$H_{Pr} = -\hbar(\gamma_x^{Pr}B_xI_x + \gamma_y^{Pr}B_yI_y + \gamma_z^{Pr}B_zI_z)$$
$$+ D[I_z^2 + I(I+1)/3] + E[I_x^2 - I_y^2] \, ,$$

where $D = 4.1795 \pm 0.0013$ MHz and $E = 0.154 \pm 0.004$ MHz[4] and γ^{Pr} is the enhanced nuclear Zeeman tensor,

$$\gamma_i/2\pi (kHz/Gauss) = (g_N\beta_N + 2g\beta\Lambda_{ii})/h,$$

$$\Lambda_{ii} = \sum_{n\neq 0} A_j \frac{|<0|J_i|n>|^2}{E_n - E_0}$$

where E_n is the energy of level n, i = x,y,z.

A study of enhanced nuclear Zeeman tensor γ^{Pr} has been made in low magnetic fields (B = 100 gauss)[7]. In a previous paper[8], we extended the study to higher values of B where the enhanced Zeeman interaction becomes comparable to the pseudo-quadrupole interaction. While our fit was insensitive to the value of $\gamma_y^{Pr}(\gamma_y^{Pr}/2\pi = 4.98$ kHz/g)[7], we obtained

$$\gamma_z^{Pr}/2 = 10.16 \pm 0.05 \text{ kHz/g, and}$$

$$\gamma_x^{Pr}/2\pi = 3.45 \pm 0.05 \text{ kHz/g.}$$

The Hamiltonian for the fluorine spins in a magnetic field is

$$H_F = -\hbar\gamma^F \sum_i B_i S_i, \text{ where } \gamma^F = 4.0055 \text{ kHz/G}$$

Method

LaF$_3$ has three crystal C$_2$ symmetry axes which pass through pairs of sites where Pr has been substituted for La. Each of these sites is surrounded by a group of 9 fluorines. The C$_2$ axes are oriented at 120° to each other in the plane perpendicular to the C$_3$ axis. Because the optical electric-dipole transition can be excited only by a laser electric field E_L which has a component along a drystal C$_2$ axis, it is possible to selectively excite different crystal sites.

An externally applied Dc magnetic field is swept at an optimum rate dB/dt (500 gauss/sec) through values where the optical ground states of Pr^{+3} cross with fluorine nuclear spin levels. During the magnetic field sweep, the Hamiltonian changes monotonically, except near the values of magnetic field where the energy conservation condition is satisfied and single spin-flips of the rarer (0.5% abundant) Pr spins occur with opposite single (or less probably double) spin-flips of the nearby flourine nuclei.

Prior to the level-crossing imposed by the field-sweep, dB/dt, the laser-excited levels are bleached (Fig. 5a) for sufficient laser intensity.

272

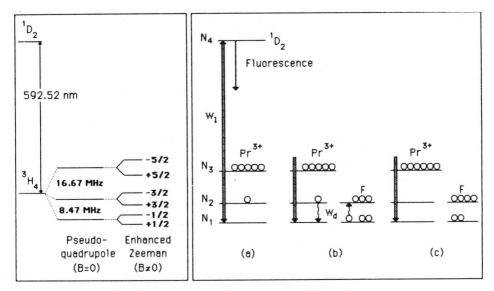

Figure 4   Figure 5

Fig. 4. The optical ground state splittings of Pr^{+3}.

Fig. 5. The cross-relaxation process. n_3 represents the four levels not participating in cross-relaxation with neighboring fluorines. Because of inhomogeneous broadening two other cases (not shown here) occur simultaneously: 1) laser resonant with n_2 (producing reverse F polarization); and 2) laser resonant with any of the n_3 levels (not affecting the fluorines).

Because of the fluorescence mechanism (lifetime τ_F = 0.5 msec), particles which occupied the empty levels in the past have transferred to the remaining five (out of six) nuclear levels (Fig. 5b), all of which have spin-lattice relaxation times, T_1, of at least one minute. The sudden onset of Pr $\leftrightarrow$ F spin-spin coupling during field sweep, or the sudden application of NMR, allows the transient transfer of population back into the empty laser-bleached levels, thus momentarily restoring some degree of laser absorption. After these processes two out of the six levels are empty (Fig. 5c) and the steady-state transmission is the same as before.

If B is held fixed at any level-crossing, no changes in laser transmission are observed; and if the sweep rate dB/dt is too slow or too rapid the effect is reduced. The effect is also reduced if the laser intensity is too low or too high. A numerical model based on rate equations was used to confirm this critical dependence on the sweep rate and laser intensity[9].

Experiment

A 3x3x7 mm Pr^{+3}:LaF_3 crystal (0.5% at.) was positioned with one of the crystal C_2 axes aligned parallel to the external magnetic field and cooled to 1.8 K. A frequency-stabilized (to 1 MHz rms) linearly-polarized cw dye laser (Coherent 599/21) with an intensity $2mW/mm^2$ was used to excite the sample along the C_3 axis. Light transmission was monitored by a PIN photodiode and digitized by a transient waveform recorder. Signal averaging of up to 75 traces was performed by a microcomputer. When the external magnetic field was swept at approximately 500 G/sec, sharp absorption peaks were observed. The C_2 axis parallel to $\overline{B}$ defines the $\alpha=0°$ site; and two remaining equivalent sites are defined with axes at $\alpha=60°$ with respect to $\overline{B}$.

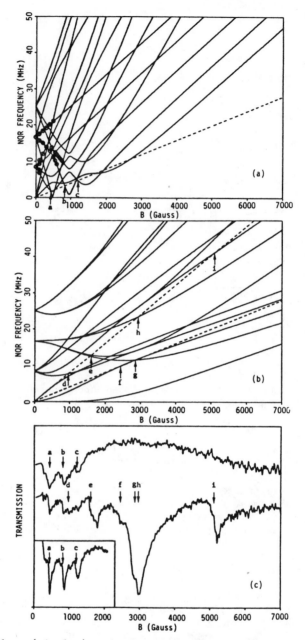

<u>Fig. 6a.</u> Rf data (circles) and numerical calculation (solid line) for
$\alpha=60°$, $\gamma_x^{Pr}/2\pi$ = 3.45 kHz/G, $\gamma_y^{Pr}/2\pi$ = 4.98 kHz/G, and $\gamma_z^{Pr}/2\pi$ = 10.16 kHz/G.
The dashed line is the fluorine transition. Arrows show where energy-
conserving cross-relaxation occurs. Note that only the first crossing
point of a given Pr^{+3} line is indicated (see text). <u>6b.</u> Same as a), but
with $\alpha=0°$. Double fluorine spin-flip transitions are also shown (upper
dashed line). <u>6c.</u> Optical transmission during magnetic field-sweep at a
rate of 500 G/sec. The upper curve represents the average of twenty traces,
where E_L was perpendicular to B and one of the C_2 axes, exciting only sites
with $\alpha=60°$. The lower curve represents the average of 75 traces under the
same conditions. The middle curve shows optical transmission for the case
where E_L was parallel to B and one of the C_2 axes. In this case, sites
with $\alpha=60°$ and $\alpha=0°$ were excited simultaneously. Double spin-flip transitions were observed only for $\alpha=0°$.

Figure (6c) displays the observed level-crossing responses during the magnetic field sweep where the magnetic field B was oriented along one of the crystal C_2 axes. For a laser field E_L polarized perpendicular to B one obtains a response only from the sites with a local quadrupole x-axis at an angle of $\alpha=60°$ with respect to B. When the laser field E_L is polarized parallel to B, an additional set of sites with quadrupole x-axes parallel to B is excited. The eigenlevel diagram for each set of sites is displayed (Figures 6a and 6b) with corresponding level-crossing and enhanced absorption noted. Since the first crossing of a given Pr transition with the fluorine line depopulates the Pr levels involved, subsequent crossings of the same Pr line are less effective and are not observed. Both single and double spin-flip transitions of fluorine nuclei are observed. No crossings with the more complex lanthanum (I = 7/2) quadrupole levels were identified.

The transient field sweep noted above proved to be a very good means by which we have confirmed the presence of level-crossing with the F spin system.

Discussion

From a model based on a set of detailed-balance rate equations which expresses the two-reservoir coupling rate between F and Pr systems, we have confirmed that it is possible to establish an equilibrium net population of F spins in times much shorter than T_1, either at negative or positive temperature. This takes place when four simultaneous NMR saturation transitions couple into one of the two Pr^{+3} transitions (Fig. 7).

A rate-equation model predicts that a net polarization of 1% can be achieved in about 1000 seconds, if T_1 is sufficiently long. A Boltzmann population difference can be achieved in a time of the order of 10 seconds. This technique may be of advantage in cases where quick recovery of spin population is required for NMR studies where ordinarily T_1 would be impractically long.

Cross-Relaxation Polarization

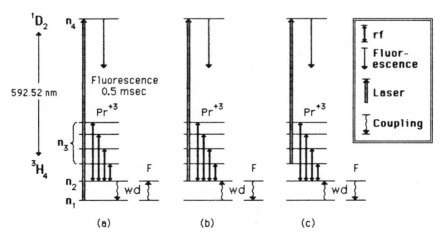

Fig. 7. The proposed scheme for achieving fluorine spin-reservoir polarization. Inhomogeneous broadening causes the pumping cycles shown in a), b), and c) to occur simultaneously.

This work was supported by the National Science Foundation. One of us (M.L.) acknowledges the support of the Center for Electro-optics, Ljubljana, Yugoslavia.

References

1. M. Goldman and J. F. Jacquinot, Journal de Physique 43, 1049 (1982).
2. K. O. Hill, Applied Optics 10, 1694 (1971).
3. Hamilton, et. al, Optics Letters 4, 124 (1979).
4. L. E. Erickson, Op. Comm. 21, 147 (1977).
5. J. Mylnek, N. C. Wong, R. G. DeVoe, E. S. Kintzer, and R. G. Brewer, Phys. Rev. Lett. 50, 993 (1983).
6. M. A. Teplov, Sov. Phys. JETP 26, 872 (1968).
7. B. R. Reddy and L. E. Erickson, Phys. Rev. B 27, 5217 (1983).
8. F. W. Otto, M. Lukac and E. L. Hahn, Proceedings of the Seventh International Conference on Laser Spectroscopy, Hawaii, June 24-28 (1985).
9. F. W. Otto, M. Lukac and E. L. Hahn, to be published.

NONLINEAR OPTICS IN RESONANT GAS MEDIA :

FOUR-WAVE MIXING AND HIGHER-ORDER PROCESSES

Martial Ducloy

Laboratoire de Physique des Lasers
Université Paris-Nord
Vlletaneuse 93430, France

ABSTRACT

This paper reviews a number of recent developments in the field of nonlinear optics in resonant gas media, with a particular emphasis on phase-conjugate (PC) processes. The properties of PC emission via degenerate four-wave mixing (DFWM) are shortly reviewed : Doppler-free emission, dispersive character and directional anisotropy of the emission lineshape for saturating pumps, PC reflectivity in the saturation regime.

Recent extensions to optically thick gas media (e.g. Na vapor, near the D_2 transition) have allowed one to reach PC reflectivities of the order of 200 %, and get PC self-oscillation. Higher-order contributions to the nonlinear susceptibility have been directly monitored via angle-resolved, phase-matched degenerate multiwave mixing. This has been demonstrated in neon, where it has been possible to observe up to eleventh-order contribution.

On the other hand, non-degenerate wave mixing processes in resonant cascade three-level systems have allowed the observation of two-photon induced FWM (in Calcium), as well as dynamic Stark splitting of the emission lineshape in the saturation regime. Finally, novel ways for performing phase-conjugation with frequency up-conversion via non-degenerate, high-order multiwave mixing in a phase-matched backward geometry, are described.

I. PRINCIPLE OF OPTICAL PHASE CONJUGATION BY DEGENERATE FOUR-WAVE MIXING

Among the various techniques to perform phase-conjugation (PC) of an optical wave, backward degenerate four-wave mixing (DFWM) has undergone rapid development in the recent years [1]. Let us recall its principle (Fig. 1) : a nonlinear medium is irradiated by a standing pump wave (frequency ω, wavevectors k_F and $k_B=-k_F$) and a traveling probe wave (ω, k_P). By absorption of two oppositely propagating pump photons, and emission of a probe photon, the nonlinear medium is able to re-emit a fourth photon of frequency $\omega_R=\omega$ and wavevector $k_R=-k_P$. In this process, both energy and linear momentum are conserved. For this reason, in isotropic media, phase-matching is satisfied independently of the probe direction, k_P.

Fig. 1 - Principle of backward degenerate four-wave mixing. F is the forward pump, B, the backward pump and P, the probe.

The corresponding macroscopic dipole polarization induced at third order can be written as

$$P^{(3)} \propto \chi^{(3)} E_F E_B E_P^* = \chi^{(3)} \mathcal{E}_F \mathcal{E}_B \mathcal{E}_P^* e^{i(\omega t + k_P \cdot r + \varphi_F + \varphi_B - \varphi_P)}$$

(Notations are obvious - See Fig. 1 and Ref. [2-3]). This shows the phase reserval of φ_P produced in the DFWM process.

Among the various media exhibiting large values of $\chi^{(3)}$ atomic vapors or gases, close to <u>optical resonance</u>, have been largely used. Due to the resonant enhancement, the nonlinear susceptibility of the gas medium saturates for low incident intensities (about 0.1 mW/mm^2 for an atomic resonance line), which allows one to operate PC mirrors at very low light levels and atomic densities.

In near-resonant media, the PC process can be viewed as the creation of a population (or index) grating by two of the incident fields, and the simultaneous diffraction of the third field by this optically-induced grating. For instance, the e.m. field associated to pump F and probe presents an interference pattern with fringes perpendicular to $q=k_F-k_P$ and separated by the spatial period $\Lambda = 2\pi/q = \lambda/[2 \sin \theta/2]$ (See Fig. 2). This intensity pattern produces in the interaction volume a spatial modulation of atomic populations, <u>i.e.</u> a volumic grating. Pump B, incident along a Bragg grating direction, is diffracted in direction $-k_P$, yielding the PC beam.

One should note the importance of the atomic motion which, by mixing population maxima and minima, tends to wash out the grating contrast, and reduce the diffraction efficiency: the PC intensity increases for larger spatial periods Λ, i.e. smaller angular separation Θ between pump and probe. Most of the experimental works in gases have been done for $\Theta \lesssim 1°$. In that case, pumps and probe are nearly collinear, and define a common axis ($k_p \approx k_F$), along which there is velocity selection by the counter-propagating beams. This is responsible for the Doppler-free character of the PC emission. This Doppler-free character has been observed in a number of vapor or gases [2]. However, with increasing incident powers, the emission lineshape broadens and splits into two components [3,4]. This double-peak structure must be ascribed to the saturation of the third-order susceptibility.

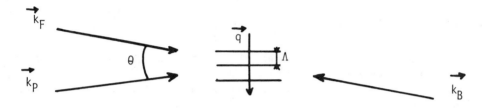

Fig. 2 - Grating interpretation of resonant degenerate four-wave mixing.

II. SATURATION BEHAVIOR OF DEGENERATE FOUR-WAVE MIXING IN RESONANT GAS MEDIA

Several properties of the PC emission lineshape in the saturation regime should be underlined.

II. 1 - In most case, <u>the double-peak lineshape is a signature of the dispersive character of the saturated susceptibility</u>. This character has been demonstrated by means of a heterodyne detection of the PC field, which allows one to monitor its amplitude and phase. With increasing intensities, the dispersion component takes over the absorption component, and becomes predominant at full saturation (Experiments performed in excited Ne, transitions λ = 640 nm, 1 s_5 - 2p_9 and λ = 607 nm, 1s_4 - 2p_3 [4]). The PC intensity lineshape is thus the square of a dispersion curve.

II. 2 - There is a <u>directional anisotropy of the saturation</u>: in the case when only one of the pumps is saturating, the PC lineshape differs if the saturating pump is either the <u>forward</u> or the <u>backward</u> pump [4]. This anisotropy is a direct consequence of the atomic velocity distribution. Its physical interpretation appears clearly when one realizes that resonant PC may be viewed as a <u>saturated absorption</u> (SA) process produced, on the counter-propagating (B) field, by a <u>spatially modulated</u> saturating field (<u>i.e.</u>, F + P field) [5]. The transverse intensity modulation (wavevector q : See Fig. 2) of the forward (F + P) field is transferred, via <u>SA processes</u>, to the B field. The spatial modulation which is impressed on the B field, can be decomposed into its Fourier components, exp(inq.r). The first harmonics (n=1) yields the PC field. For saturating probes, higher-order harmonics may appear, yielding high-order degenerate <u>multiwave</u> mixing signals (see Section IV). On the other hand, for weak probes, the first harmonics, <u>i.e.</u> the PC field, is readily obtained as the <u>derivative</u> of the SA field with respect to the saturating field (F) intensity. This derivation operation directly provides the PC lineshape in the case of <u>forward</u> saturation [5]. In the case of <u>backward</u> saturation, PC and SA signals are similar, and both exhibit a pronounced dispersive character [3-4].

II. 3 - When <u>both pumps are saturating</u>, the PC emission presents a number of distinct features [6,7]. The highest PC reflectivities correspond to backward pump intensities larger than the forward one. In that case ($I_B > I_F$), the PC lineshape is double-peaked with a frequency splitting increasing linearly with the Rabi frequencies of beams F and B (Ω_F, Ω_B). For three-level systems, in the range $\Omega_F < \Omega_B < 5\Omega_F$, the splitting is equal to $1.3 \Omega_B + 3.5 \Omega_F$ (experimentally checked on the Ne 607 nm line [7]). It has also been observed that there is an absolute maximum reflectivity for a <u>finite</u> value of the F pump intensity, but for <u>very large</u> B intensities ($I_B \gg I_F$).

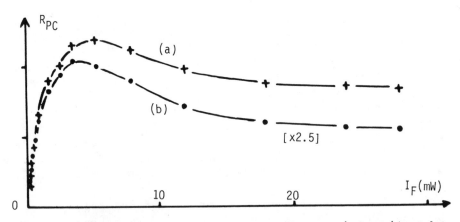

Fig. 3 - Peak PC reflectivity versus F pump intensity (I_F), in the asymptotic regime (B pump focussed, with $I_B = 90$ mW). Neon transition $1s_4(J=1) \rightarrow 2p_3(J=0)$, $\lambda = 607$ nm. (a) Parallel pump polarizations (Two-level system). (b) Cross-polarized pumps (Three-level system).

For optically thin media, this maximum reflectivity is simply related to the linear (weak field) absorption coefficient, $\alpha_0 L$ [7] :

- for three-level systems, $R_{max} = 2.75 \times 10^{-3} (\alpha_0 L)^2$
- for two-level systems, $R_{max} = 1.11 \times 10^{-2} (\alpha_0 L)^2$

These predictions, which have been corroborated by experiments [7] on the Ne 607 nm line (Fig. 3), should be compared with the ones for stationary atoms in similar conditions ($R_{max} = 3.7 \times 10^{-2} (\alpha_0 L)^2$ [8]). To get larger reflectivities, one needs to consider optically thick media, $\alpha_0 L > 1$.

III. PC SELF-OSCILLATION IN SODIUM VAPOR [9]

In the case of optically thick media, e.m. field propagation effects and beam absorption should be taken into account. In particular, for inhomogeneously-broadened gas media, pump absorption gets very large-even complete-inside the Doppler linewidth, and one needs to tune the incident frequencies outside this absorption zone. In this way, one gradually goes from the Doppler-broadened regime (on line center) to the homogeneously-broadened regime (far off-line-center) which is the one generally adopted with intense pulsed lasers and dense vapors [10]. By using dense Na vapors, near the D_2 line ($\lambda = 589$ nm), one has demonstrated the possibility of getting very large c.w. PC reflectivities, up to 200 %. Typical operating conditions were : pump intensities ≈ 150 mW, detuning from D_2 line-center ~ 2 GHz, pump absorption at the operating frequency : 90 - 98 % (on line center, $\alpha_0 L > 100$) [9]. Although efficiencies higher than one have been reported previously in sodium [10], these very large reflectivities have led to the first observation of a c.w. oscillation in an empty PC cavity, bounded by an ordinary mirror and the PC Na mirror : the gain medium is actually the PC mirror itself. Pump depletion correlated to the PC oscillation has been monitored, and the polarization characteristics of the self-oscillating beam have been explored for various pump polarizations (linear, circular, etc ...) [9]. In particular, polarization conjugation on the oscillation has been observed for circular counter-rotating polarizations of the pumps. Also noteworthy is the fast response time (~ 10 ns) of the PC oscillation, as compared with slow photorefractive crystals.

IV. DEGENERATE, HIGH-ORDER MULTIWAVE MIXING IN RESONANT GAS MEDIA. [6, 7, 12]

As noted in Section II, for saturating probes, high-order spatial Fourier harmonics can be transferred, via saturated absorption (SA) processes, from the incident, trans-

versely-modulated, (F + P) field to the B field. This corresponds to higher-order degenerate susceptibilities of the
medium, $\chi^{(2n+1)}$, which can be monitored via angle-resolved
phase-matched multiwave mixing [12]. For a given order, n,
wavevector phase-matching is obtained through an adequate
rotation of the B pump incidence direction. In the grating
viewpoint (Fig. 2), as the response of the resonant medium
becomes saturated, the optically-induced population grating
gets anharmonic, and higher-order Bragg diffraction of beam B
can be explored for an adequately phase-matched incidence. In
this way, high-order optical susceptibilities have been monitored in excited neon : up to $\chi^{(11)}$ for the 640 nm line, and
$\chi^{(7)}$ for the 607 nm line [7]. This yields a direct insight in
the higher-order processes involved in the optical saturation
of resonant media.

    Like in resonant PC, the emission lineshape is mainly
governed by the combined effects of atomic velocity selection
and Doppler shift. In a counter-propagating geometry (B nearly
opposite to F and P), the nonlinear re-emission is Doppler-
free. But one can also consider co-propagating geometries, in
which all three beams (F, P, B) are incident along nearly
co-parallel directions ($k_B \simeq k_F \simeq k_P$). In the latter case,
they interact with the same velocity group, independently of
their common frequency, and the emission lineshape is Doppler-
broadened [12].

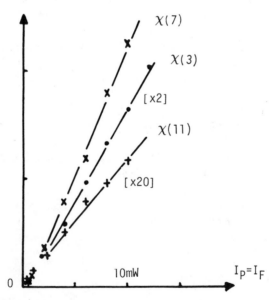

Fig. 4 - Peak emission intensities versus grating beams
intensities ($I_P = I_F$), in degenerate multiwave mixing with
an intense backward diffracting beam ($I_B = 110$ mW).
Neon line $\lambda = 640$ nm.

As in phase conjugation, emission lineshapes in a counter-propagating geometry may be easily interpreted thanks to the close analogy between higher-order resonant multiwave mixing, and SA by a spatially-modulated beam (see section II. 2) [6, 7]. Two main regimes can be considered according that the backward (B) diffracting beam is weak or intense.

(i) For <u>weak diffracting beams</u>, the analogy with SA is complete, and gives the various diffraction efficiencies as the Fourier coefficients of the spatially-modulated SA field [12]. These efficiencies saturate for large intensities I_F, I_p.

(ii) In the regime of a <u>fully-saturating</u> diffracting (B) beam, emission lineshapes become predominantly dispersive (double-peak structure), and independent of intensities I_F, I_p, and of diffraction order, n. The diffraction efficiency approximately varies like n^{-4}, and does not saturate with increasing I_F and I_p values, as long as I_F, $I_p \ll I_B$ (See Fig. 4) [6, 9].

V. TWO-PHOTON RESONANT FOUR-WAVE MIXING

All the work presented above on degenerate four-wave and multiwave mixing may be interpreted in a grating viewpoint : either population grating, for parallel F/P polarizations, or Zeeman coherence grating for cross-polarized F-P fields [9, 13]. However, there is a large class of nonlinear mixing processes for which there is no grating analogy. This is the case when a pure coherent process is involved.

It is well-known that, in three-level saturation spectroscopy, the signal is generally related both to a population change process, induced by the saturating beam (frequency ω_1), and to a coherent process (two-photon or Raman coherence) induced by both saturating and probe (ω_2) fields (See <u>e. g.</u> [14]). If the saturating beam couples two unpopulated states - "transparent" transition - a signal is still observable, which originates solely from the coherent process. This is, for instance, the case for cascade three-level systems, when $\omega_1 > \omega_2$ (figure 5a). Such a situation is generally quite attractive, as the signal does not saturate with increasing saturating beam intensities [14].

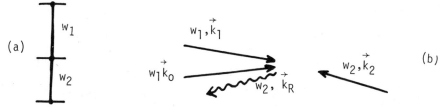

Fig. 5(a) - Cascade three-level system. (b) Non-degenerate wave mixing geometry.

This situation can be generalized to a <u>non-degenerate four-wave mixing</u> geometry [15], in which two, nearly co-parallel $(k_1 \approx k_0)$. ω_1 beams intersect at a small angle in the sample cell, and a third (ω_2) beam is approximately counter-propagating to the ω_1 beams, with a wavevector k_2 imposed by phase-matching : $|k_R| = |k_2|$, with $k_R = k_1 + k_2 - k_0$ (see Fig. 5b).
The analogy developed in the previous sections between PC and spatially-modulated SA, may be extended to interpret non-degenerate FWM as a three-level saturation spectroscopy process produced by a spatially-modulated saturating field. A two-photon coherent emission is thus predicted, with wavevector k_R, when $\omega_1 > \omega_2$.
Experiments [16] have been performed on the cascade system $^1S_0 - {}^3P_1 - {}^3S_1$ of calcium [ω_2 transition, $^1S_0 - {}^3P_1$: λ_2 = 657 nm ; ω_1 transition, $^3P_1 - {}^3S_1$: λ_1 = 612 nm]. On resonance, an emission signal at 657 nm has been observed, whose lineshape exhibits a.c. Stark splitting for large intensities of the k_1 beam (Fig. 6). This (nearly phase-conjugate) signal essentially originates in a two-photon coherent process, and has no diffraction analog. With this technique, it should be possible to

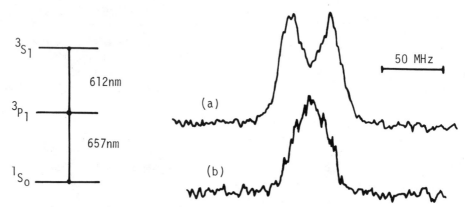

Fig. 6 - Coherent emission lineshape at 657 nm in calcium, observed in the geometry of Fig. 5, with cross-polarized $\omega_1 - \omega_2$ beams. (a) Incident intensities : $I_1 = 50$ mW, $I_0 = 4$ mW (612 nm), $I_2 = 10$ mW (657 nm). In (b), beam k_1 intensity is lowered to $I_1 \approx 5$ mW, and the Stark splitting disappears.

observe emission signals whose intensity could be increased without saturation : thanks to a partial compensation between Doppler effect and velocity-dependent a.c. Stark shift, an increasing number of atomic velocity groups contributes to the signal. The analogy between FWM and saturation spectroscopy shows that it should be possible to consider situations where the above compensation should be total over a large range of atomic velocities [17]. The resulting enhancement of the emission should allow one to monitor very high order non-degenerate susceptibilities.

VI. OPTICAL PHASE CONJUGATION WITH FREQUENCY UP-CONVERSION VIA HIGH-ORDER, NON-DEGENERATE MULTIWAVE MIXING [18]

Non-degenerate four-wave mixing considered in the previous section can be extended to higher-order processes. These higher-order multiwave mixing processes open the way to novel methods for performing PC with frequency multiplication [18]. Consider the geometry of Fig. 5b, in which the nonlinear medium is assumed to present a non-zero $(2p+1)^{th}$ - order response, $\chi^{(2p+1)}$. In a non-degenerate process in which p photons k_1 are absorbed, p photons k_0 emitted and one photon k_2 absorbed, a photon can be radiated at frequency ω_2, with wave-vector $k_R = p(k_1 - k_0) + k_2$ if phase-matching is fulfilled. The expression of the emitted field is proportional to

$$E_R = \chi^{(2p+1)} \varepsilon_2 (\varepsilon_0^*)^P (\varepsilon_1)^P \exp[i(\omega_2 t - k_R.r + \varphi_R)]$$

with a phase given by $\varphi_R = p(\varphi_1 - \varphi_0) + \varphi_2$.
Phase-matching is automatically obtained, <u>independently of k_0</u>, when $k_2 = -p k_1$, i.e. $k_R = -p k_0$. Wavefront reversal is thus obtained at frequency $\omega_2 = p\omega_1 n(\omega_1)/n(\omega_2)$ [n is the refractive index of the medium]. In non-dipersive media, $\omega_2 = p\omega_1$. PC originates in the relationship $\varphi_R = Cst - p \varphi_0$. It should allow the restoration (at frequency ω_2) of optical wavefronts (of frequency ω_1) distorted by <u>non-dispersive</u> aberrating media, in the approximation of <u>geometrical optics</u> (i.e. negligible wavelength-dependent diffraction effects) [18].

Various physical phenomena, similar to the ones discussed in the previous sections, can be at the origin of $\chi^{(2p+1)}$. Several of these rely on the generation by waves k_1 and k_0 of an <u>anharmonic</u> grating (absorption grating, index grating, etc ...), and the simultaneous p-order Bragg diffraction of wave k_2 on this grating. This is the case in resonantly coupled three-level $(\omega_1 - \omega_2)$ systems, in which anharmonic population gratings may be produced in the common level. However, in the same way as in section V, another process relies on the creation of high-order atomic coherence in cascade two-photon transitions [19]. Consider the case of PC with frequency doubling, i.e. the situation of Fig. 5, in which $\omega_2 \approx 2\omega_1$, and only the ground state is populated. The $\chi^{(5)}$ process is thus connected to the creation of atomic coherences precessing at frequencies ω_2 and $\omega_1 + \omega_2$. For very intense ω_2 beams, like in section V, the dynamic Stark splitting plays an important role and should help to increase the conversion efficiency to large values, with reduced saturation [19].

Potential applications of wavefront reversal with frequency doubling or multiplication include [18] (i) wavelength conversion in optical imaging (image transfer from IR to visible. or visible to UV) (ii) UV phase conjugation (iii) photolithography, with the advantage of an image contrast enhancement in the higher-order process.

VII. CONCLUSION

We have presented in this review a number of recent advances in the field of nonlinear optical (NLO) processes in resonant gas media. This field is based on two main ideas :

(i) From the NLO point of view, one needs to satisfy the wavevector phase matching condition necessary to monitor these processes in thick media.

(ii) There is a spectroscopic view point : when working in resonant media, the frequency-dependence of optical suscepti-bilities allows one to carry out nonlinear spectroscopy, in particular to analyze the processes responsible for the line-shapes, and the physics of the e.m. radiation-matter interac-tion (e.g. multiresonant irradiation in the case of non-dege-nerate susceptibilities). Both points of view react on each other : the NLO viewpoint shines a new light on Doppler-free spectroscopic methods in resonant gas media. On the other hand, the use of gas media near resonance provides a simple case study for nonlinear optics, which is entirely tractable via closed-form calculations.

Acknowledgments

The author wishes to thank D. Bloch, S. Le Boiteux, P. Simoneau, F. A. M. de Oliveira and J. R. R. Leite for their contributions to several parts of this work.

[1]   For reviews, see, e.g. Optical Phase Conjugation, edit d 'y R. A. Fisher (Academic, New York, 1983) ; Proceedings of the CNRS colloquium on Optical Phase Conjugation and Instabi-lities, Cargèse, France, 1982, edited by C. Flytzanis, R. Frey and M. Ducloy [J. Phys. (Paris) Colloq. 44, C2 (1983)].

[2]   M. Ducloy, "Nonlinear Optical Phase Conjugation", in Festkörperprobleme-Advances in Solid State Physics (Vieweg, Braunschweig, 1982), Vol. XXII, p. 35, and references therein.

[3]   D. Bloch and M. Ducloy, J. Opt. Soc. Am. 73, 635 and 1844 (1983).

[4]   D. Bloch, R. K. Raj, K. S. Peng and M. Ducloy, Phys. Rev. Lett. 49, 719 (1982).

[5]   M. Ducloy and D. Bloch, Opt. Commun. 47, 351 (1983).

[6]   M. Ducloy, F. A. M. de Oliveira and D. Bloch, Phys. Rev. A32, 1614 (1985).

[7] S. Le Boiteux, P. Simoneau, D. Bloch, F. A. M. de Oliveira
     and M. Ducloy, IEEE J. Quant. Electron, to be published.

[8] R. L. Abrams and R. C. Lind, Opt. Lett. $\underline{2}$, 94 ; $\underline{3}$, 205
     (1978).

[9] P. Simoneau, S. Le Boiteux, D. Bloch, J. R. R. Leite and
     M. Ducloy, to be published.

[10] D. M. Bloom, P. F. Liao and N. P. Economou, Opt. Lett. $\underline{2}$,
     58 (1978) ; B. Kleinmann, F. Trehin, M. Pinard and
     G. Grynberg, J. Opt. Soc. Amer. $\underline{B2}$, 704 (1985).

[11] R. C. Lind and D. G. Steel, Opt. Lett. $\underline{6}$, 554 (1981).

[12] R. K. Raj, Q. F. Gao, D. Bloch and M. Ducloy, Opt. Commun.
     $\underline{51}$, 117 (1984) ; S. Le Boiteux, R. K. Raj, Q. F. Gao,
     D. Bloch and M. Ducloy, J. Opt. Soc. Am. $\underline{B1}$, 501 (1984).

[13] M. Ducloy and D. Bloch, Phys. Rev. A$\underline{30}$, 3107 (1984).

[14] M. Ducloy, J. R. R. Leite and M. S. Feld, Phys. Rev. A$\underline{17}$,
     623 (1978).

[15] D. G. Steel, J. F. Lam and R. A. Mc Farlane, Phys. Rev.
     A$\underline{26}$, 1146 (1982).

[16] P. Simoneau, S. Le Boiteux, D. Bloch and M. Ducloy, to
     be published.

[17] C. Cohen-Tannoudji, Metrologia, $\underline{13}$, 161 (1977) ;
     S. Reynaud, Ann. Physique (Paris) $\underline{8}$, 371 (1983) and refe-
     rences therein.

[18] M. Ducloy, Appl. Phys. Lett. $\underline{46}$, 1020 (1985).

[19] F. A. M. de Oliveira and M. Ducloy, to be published.

EXACT SEMICLASSICAL SOLUTION FOR FOUR-WAVE MIXING

W.M. Schreiber, N. Chencinski, A.M. Levine, and A.N. Weiszmann

The College of Staten Island of The City University of New York
Staten Island, New York 10301

I. INTRODUCTION

Four-wave mixing is of importance both fundamentally and for spectro-scopic diagnostics. Extensive perturbational investigations of the process have been done both directly[1] and diagramatically.[2,3] Exact solutions for various particular physical processes have been previously obtained.[4-10] In this paper, we develop a technique for the semiclassical exact solution of a four-wave mixing process. For the sake of brevity, we will specialize to the dominant path, but the technique is applicable to other paths as well.

II. METHOD OF SOLUTION

Three waves of respective frequencies ω_a, ω_b, and ω_c are incident on a sample and generate a nonlinear polarization and corresponding field at the frequency

$$\omega_p = \omega_a + \omega_b - \omega_c \tag{1}$$

The dominant path for this process is shown in Fig. 1. The atomic states have been labeled 1-4 with the state 1 being the highest energy state. We have included in parenthesis the respective letters j, k, t, g in order to make the connection with the existing literature clear.[1,3] Setting E = 0 halfway between the second and third atomic levels, the respective atomic energy levels are

$$E_1 = \hbar(\omega_{jk} + \tfrac{1}{2}\omega_{kt}) \tag{2}$$

$$E_2 = \tfrac{1}{2}\hbar \, \omega_{kt} \tag{3}$$

$$E_3 = -\tfrac{1}{2}\hbar\omega_{kt} \tag{4}$$

$$E_4 = -\hbar(\tfrac{1}{2}\omega_{kt} + \omega_{tg}) \tag{5}$$

The relevant detuning frequencies are

$$\Delta \omega_a = \omega_a - \omega_{kg} \tag{6}$$

$$\Delta\omega_b = \omega_b - \omega_{jt} \tag{7}$$
$$\Delta\omega_c = \omega_c - \omega_{kt} \tag{8}$$

In matrix representation, the Hamiltonian is expressed as

$$H = \hbar \begin{bmatrix} \omega_{jk} + \tfrac{1}{2}\omega_{kt} & 0 & V_{13}^{(b)} e^{-i\omega_b t} & V_{14}^{(p)} e^{-i\omega_p t} \\ 0 & \tfrac{1}{2}\omega_{kt} & V_{23}^{(c)} e^{-i\omega_c t} & V_{24}^{(a)} e^{-i\omega_a t} \\ \bar{V}_{13}^{(b)} e^{i\omega_b t} & \bar{V}_{23}^{(c)} e^{i\omega_c t} & -\tfrac{1}{2}\omega_{kt} & 0 \\ \bar{V}_{14}^{(p)} e^{i\omega_p t} & \bar{V}_{24}^{(a)} e^{i\omega_a t} & 0 & -(\tfrac{1}{2}\omega_{kt} + \omega_{tg}) \end{bmatrix} \tag{9}$$

We have included a term $V_{14}^{(p)} e^{-i\omega_p t}$ for completeness to denote the presence of a nonlinearly generated effective field of frequency ω_p, albeit that to obtain equivalence with previously obtained lowest order calculations contributions of this term would be neglected to lowest order. For a unitary transformation $T(t)$ on the states $|\psi(t)\rangle$,

$$|\psi'(t)\rangle = T(t)|\psi(t)\rangle \tag{10}$$

the Schrodinger equation implies

$$i\hbar \frac{\partial}{\partial t}|\psi'(t)\rangle = \tilde{H}|\psi'(t)\rangle \tag{11}$$

$$\tilde{H} = H' + i\hbar T \dot{T}^\dagger \tag{12}$$

$$H' = THT^\dagger \tag{13}$$

We seek a transformation $T(t)$ such that $\tilde{H}$ is time independent.[7-9] Then correspondingly

$$|\psi'(t)\rangle = e^{-\frac{i\tilde{H}t}{\hbar}}|\psi'(0)\rangle \tag{14}$$

A transformation which yields a time independent $\tilde{H}$ for the four-wave mixing Hamiltonian of Eq. (9) is

$$T(t) = \begin{bmatrix} e^{i(\omega_b - \tfrac{1}{2}\omega_c)t} & 0 & 0 & 0 \\ 0 & e^{i\tfrac{1}{2}\omega_c t} & 0 & 0 \\ 0 & 0 & e^{-i\tfrac{1}{2}\omega_c t} & 0 \\ 0 & 0 & 0 & e^{i(\omega_a - \tfrac{1}{2}\omega_c)t} \end{bmatrix} \tag{15}$$

Note that $T(0) = I$ and thus $|\psi'(0)\rangle = |\psi(0)\rangle \equiv |\psi\rangle$. Correspondingly, one obtains

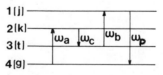

Fig. 1. Dominant four-wave mixing path.

$$\tilde{H} = \hbar \begin{bmatrix} -\Delta\omega_b + \frac{1}{2}\Delta\omega_c & 0 & V_{13}^{(b)} & V_{14}^{(p)} \\ 0 & -\frac{1}{2}\Delta\omega_c & V_{23}^{(c)} & V_{24}^{(a)} \\ \overline{V}_{13}^{(b)} & \overline{V}_{23}^{(c)} & \frac{1}{2}\Delta\omega_c & 0 \\ \overline{V}_{14}^{(p)} & \overline{V}_{24}^{(a)} & 0 & \Delta\omega_a - \frac{1}{2}\Delta\omega_c \end{bmatrix} \tag{16}$$

Denoting the eigenvalues of $\tilde{H}$ as λ_j $j = 1,2,3,4$ and the corresponding eigenvectors as

$$|\phi_j> = \sum_{i=1}^{4} a_{ij}|\psi_i> \quad j = 1,2,3,4 \tag{17}$$

the expectation value $<0(t)>$ of an operator is[9]

$$<0(t)> = \sum_{i,k,l,m = 1}^{4} a_{ik}\overline{a}_{jk}\overline{a}_{ml}a_{jl}0_{mi}f_{iklm}(t) \tag{18}$$

In Eq. (18), denoting the respective diagonal element of $T(t)$ as $T_{ii}(t)$ $i = 1,2,3,4$

$$f_{iklm}(t) = e^{-\frac{i}{\hbar}\lambda_{kl}t}T_{mm}(t)\overline{T}_{ii}(t) \quad i,k,l,m = 1,2,3,4 \tag{19}$$

$$\lambda_{kl} = \lambda_k - \lambda_l \tag{20}$$

For the polarization operator P, assuming that the system is initially in the ground state, Eq. (15) and Eqs. (18)-(20) imply that an exact expression for the component oscillating at the frequency ω_p is

$$P(\omega_p) = P_{41}\sum_{k=1}^{4} a_{1k}|a_{4k}|^2\overline{a}_{4k} \tag{21}$$

where P_{41} denotes the corresponding dipole moment. The corresponding lowest order term is

$$P(\omega_p) = P_{41}\frac{V_{13}V_{24}\overline{V}_{23}}{(\omega_a-\omega_{kg})(\omega_c-\omega_a+\omega_{tg})(\omega_{jg}-\omega_p)} \tag{22}$$

in agreement with the result found in the usual perturbational approach.[1,3]

Plots of the polarization at the frequency ω_p calculated from both the exact expression given in Eq.(21) and the perturbational expression given in Eq. (22) are shown in Fig. 2. The values used were $\omega_a = \omega_b = 30,000$ cm^{-1}, $\omega_{kg} = 33,000$ cm^{-1}, $\omega_{tg} = \omega_{jk} = 17$ cm^{-1}, $P_{14} = 1$, $V_{13} = V_{24} = \overline{V}_{23} = 100$, $V_{14} = 0$, and ω_c was varied. The exact expression remains finite at all values and exhibits a resonance shifted by 3.25 cm^{-1}. The tail of the resonance is flatter than that obtained from the perturbational result.

III. CONCLUSION

We have obtained the semiclassical solution for the expectation value of an operator and in particular $P(\omega_p)$ for the dominant path. The technique can be used to obtain corresponding results for other paths as well. It is also possible that the exact semiclassical four-wave mixing solution will assist in the development and understanding of an exact fully quantized solution.[11]

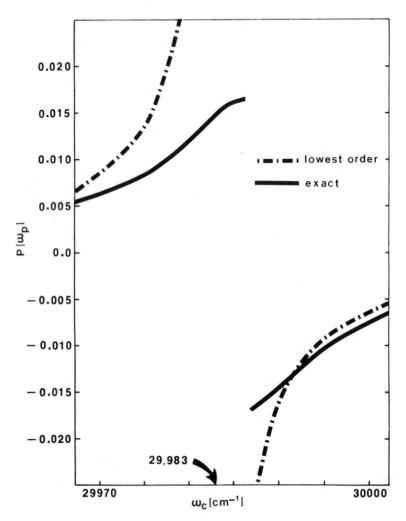

Fig. 2. Plot of polarization $P(\omega_p)$ over the incident frequency range ω_c.

ACKNOWLEDGMENT

This research was supported by grant number 6-62078 from the PSC-CUNY Research Award Program of The City University of New York.

REFERENCES

1. N. Bloembergen, H. Lotem, and R.T. Lynch, Jr., Ind. J. Pure Appl. Phys. 16, 151 (1978).
2. T.K. Yee and T.K. Gustafson, Phys. Rev. A 18, 1597 (1978).
3. Y. Prior, IEEE J. Quantum Electron., QE-20, 37 (1984).
4. E.T. Jaynes and F.W. Cummings, Proc. IEEE 51, 89 (1963).
5. C. Cohen-Tannoudji and S. Haroche, J. Phys. (Paris) 30, 153 (1969).
6. L. Allen and J.H. Eberly, Optical Resonance and Two-Level Atoms (Wiley, New York, 1975), and references therein.
7. G.W. Series, Phys. Rep. 43, 1 (1978), and references therein.
8. D.T. Pegg, in Proceedings of the Second New Zealand Summer School in Laser Physics, edited by D.F. Walls and J.D. Harvey (Academic, New York, 1980).
9. A.M. Levine, W.M. Schreiber, and A.N. Weiszmann, Phys. Rev. A 25, 625 (1982).
10. A.M. Levine, W.M. Schreiber, and A.N. Weiszmann, J. Chem. Phys. 78, 1377 (1983).
11. Y. Prior and A.N. Weiszmann, Phys. Rev. A 29, 2700 (1984).

STOCHASTIC FLUCTUATIONS INDUCED EXTRA RESONANCES

Yehiam Prior and Peter S. Stern

Department of Chemical Physics
Weizmann Institute of Science
Rehovot 76100, Israel

INTRODUCTION

The theory of nonlinear optics has been developed mostly for monochromatic laser light, and in most situations this approximation is adequate. Several authors[1], however, have dealt with the influence of laser phase fluctuations on various processes such as resonance fluorescence[2] or multiphoton transitions[3]. More recently, a new family of Pressure Induced Extra Resonances in Four Wave Mixing (PIER4) has been observed[4]. These resonances are a manifestation of the crucial role played by collisions (and other dephasing processes) in nonlinear (NL) interactions. In these experiments, four wave mixing (FWM) signals were observed from Na vapor in the presence of high buffer gas pressure. Several different resonances were observed, but in general, the dephasing process (collision) removes the destructive interference between two (or more) terms in the explicit expression for the NL susceptibility, enabling the observation of these resonances. The "conventional" theory of nonlinear susceptibilities accounts for the observed phenomena, subject to the condition that time ordering is treated properly (i.e. by double sided diagrams).

The effect of laser phase fluctuations on PIER4 has been considered by Agarwal[5], and detailed line shapes were calculated. In a recent publication[6] we have derived (within the phase diffusion model) equations of motion in the limit of short correlation times. The effect of the stochastic phase fluctuations was shown to be similar to T_2 dephasing processes, and a procedure was given for the inclusion of this similarity in many nonlinear processes. In particular, two predictions were made:
  a) If a given radiation mode contributes n photons to a multiphoton absorption process, its contribution to the width of the multiphoton transition is n^2 times its own stochastic width (observed recently by Elliot[7] et al.).

  b) When <u>correlated</u> lasers are used as the excitation sources in FWM experiments, a stochastic fluctuation induced extra resonance in four wave mixing (SFIER4) will be observed, in analogy to the pressure induced extra resonance (PIER4), even in the absence of collisions or other molecular dephasing processes.

It is the purpose of this paper to demonstrate the existence of these resonances by means of computer simulations of phase fluctuations.

FRAMEWORK OF THE CALCULATION

For a two level system (TLS) interacting with a laser field described by

$$\varepsilon_x = \varepsilon_\ell (e^{i(\omega_\ell t+\phi)} + e^{-i(\omega_\ell t+\phi)}) \tag{1}$$

the equation of motion of the system is

$$\dot{u} = -(\Delta\omega+\dot\phi)v + \kappa\varepsilon w\sin\phi$$

$$\dot{v} = (\Delta\omega+\dot\phi)u + \kappa\varepsilon w\cos\phi \tag{2}$$

$$\dot{w} = -\kappa\varepsilon(v\cos\phi+u\sin\phi)$$

where ϕ is the (fluctuating) phase of the laser. For $\phi = 0$ this reduces to the standard result, where all quantities are defined as usual. In Eq. (2) relaxations are not included, and they are treated explicitly for each case. In this paper, we are interested only in transverse relaxation where energy is not exchanged and only dephasing occurs. Three kinds are discussed: a) phenomenological introduction of T_2 in the equation of motion, b) simulation of collisions, c) simulation of phase fluctuations. Longitudinal relaxation is treated either phenomenologically or by explicit simulation of spontaneous emission.

For any relaxation model studied, the computational procedure is as follows:

1. Initial conditions are chosen for a TLS in the ground state $u = v = 0$, $w = -1$. At $t = 0$ the exciting laser is turned on, and it remains on.
2. The equations of motion (2) are integrated from $t = 0$ to t_{final} which is always long compared to all relevant time scales in the problem.
3. The result of this numerical integration is $u(t)$, $v(t)$ and $w(t)$. The induced polarization $P(t) = u(t) + iv(t)$ is numerically Fourier transformed for positive and negative frequencies , to give $P(\omega)$. The computed $P(\omega)$ is the polarizability induced in the two level system by the pump field as seen by a weak probe of frequency ω in the rotating frame.

When collisions or phase fluctuations are simulated, the following steps are added to the computational procedure:

4. A mean time between collisions T_{2A} is chosen corresponding to a given pressure.
5. A random number t_{coll} with a probability distribution function $\exp(-t_{coll}/T_{2A})$ is generated.
6. Equation (2) is integrated from $t = 0$ to t_{coll}.
7. At $t = t_{coll}$, a collision (or a random phase jump) is simulated
8. The next t_{coll} is generated, and the integration continues.
9. The procedure is repeated and averaged over many atoms.

If the dominant NL interaction is of third order, the present result bears close relationship to the third order susceptibility. The plots presented here are of $\sqrt{|P(\omega)|^2}$.

PHENOMENOLOGICAL DESCRIPTION OF RELAXATION - BLOCH EQUATIONS

With the inclusion of phenomenological T_1, T_2 in equations (2), and by setting $\phi = 0$, one is left with the standard Bloch Equations. In this notation $T_2 = T_1$ means no proper dephasing and the population relaxation rate T_1 is twice the spontaneous emission rate.

Figure 1 displays the susceptibility for off resonance pumping as a function of increasing pressure (decreasing T_2). The two side peaks correspond to the Rabi sidebands induced by the pump field, as seen in the experiments of Wu[8] et al. As the pressure is increased from zero, the lines are broadened (standard pressure broadening) but in addition a new feature appears at $\omega = 0$. This feature at $\omega_{probe} = \omega_{pump}$ is analogous to the degenerate frequency pressure induced extra resonance predicted and observed in FWM experiments.

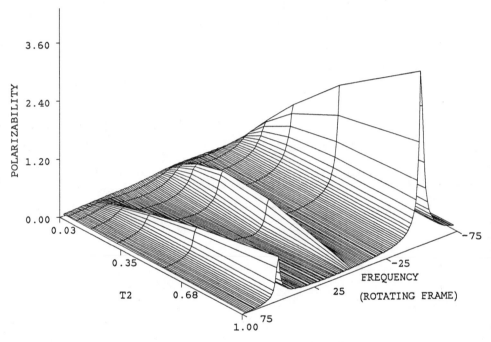

Figure 1: a three dimensional plot of the induced polarizability as a function of probe frequency and dephasing rate T_2.

COLLISION SIMULATION

An elastic collision may be described, in the Bloch vector picture, by a "jump" of the polarization vector P on its precession cone, such that its projection along the w axis remains unchanged. One may describe such a collision by transformation of rotation by an angle θ. The evolution of the angle of rotation θ depends on the particular model assumed for the collision. In the impact approximation where the duration of the collision is short compared to all other time scales of the problem one may assume that each collision completely randomizes the orientation of the atom. For each collision is assigned a random value (between $0-2\pi$) with equal probability, and without any memory of its present value. Simulations of hard collisions were performed, and compared to the analytical result of the numerical solution with phenomenological relaxation times. The results obtained analytically and numerically from averaging over 2000 atoms are indistinguishable.

Figure 2 depicts the induced polarization for a given set of conditions
The two side peaks correspond to the Rabi sidebands, and the central
feature at ω = 0 is the pressure induced extra resonance at the
degenerate frequency.

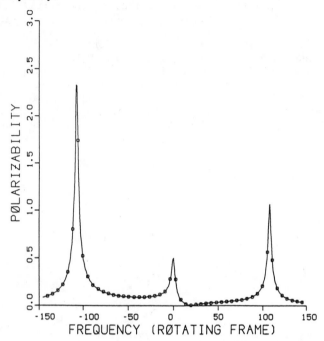

Figure 2: The induced polarizability for $\Delta\omega$ =40, $\kappa\epsilon$ =100 and
T_{2A}=1 (frequencies are measured in units of $1/T_1$).

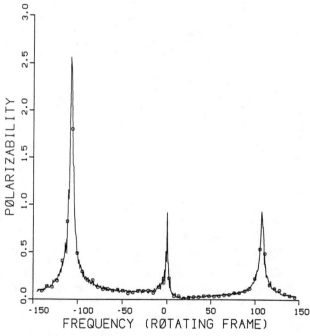

Figure 3: The induced polarizability for $\Delta\omega$ =40, $\kappa\epsilon$ =100 and
T_{2F}=0.5 . (frequencies are measured in units of $1/T_1$).

PHASE FLUCTUATIONS

For a laser with a fluctuating phase, the simulation procedure is similar to that of the collision. Several models may be assumed for the stochastic phase fluctuations. In this calculation we have assumed a random jump between 0 and 2π with no memory of the previous phase. This is a model of Markovian "hard" phase jumps, which is different than the phase diffusion model assumed in our analytic work. This model may be applicable for actual lasers suffering from acoustic mirror jitter or other mundane laboratory noises.

The full details of the simulations will be given elsewhere[9], but in Figure 3 the results of one such simulation are depicted for conditions similar to those of Figure 2. The similarity between the two cases is obvious, and the figure shows clearly that phase fluctuations can cause extra resonances in the induced polarization.

DISCUSSION

This short description of the simulation of phase fluctuations demonstrates the similarity between them and collisions. The existence of extra resonances, as predicted in our previous work, is shown, but this similarity should be considered with caution. While the collision interrupts the vector p and causes discontinuity in its value for a single atom (molecule), the phase fluctuation is a "weaker" effect, causing a change only in the derivative of u and v as they appear in the Bloch equations. Thus, one expects, and indeed this expectation is verified by further simulations, that the effect of the phase fluctuations will be dependent on the pump laser field. For weak fields, this will be a small effect. For strong fields, however, this will be a major effect, where weak and strong are defined by the angle of the effective field with the w axis in the Bloch picture.

The stochastic fluctuation induced extra resonances allow careful investigation of different noise models. It would be interesting to compare the predictions of various noise models and find out if these extra resonances are useful for distinguishing between different noise processes. Work on this question is in progress.

Y.P. is the incumbent of the Jacob and Alphonse Laniado career development chair.

REFERENCES

1. S. Stenholm, Foundations of Laser Spectroscopy, John Wiley and Sons, New York, 1983, pp.194-225 and references therein.
2. P.Avan and C. Cohen Tannoudji, J. Phys B 10,155 (1977); 10,171 (1977).
3. A.T. Georges, P. Lambropoulos and J.H. Marburger Opt. Comm. 18,509, (1976); Phys. Rev. A 15,300 (1977).
4. Y. Prior, A.R. Bogdan, M. Dagenais and N. Bloembergen Phys. Rev. Lett. 46,111 (1981).
5. G.S. Agarwal and C.V. Kunasz, Phys. Rev. A 27,996 (1983).
6. Y. Prior, I. Schek and J. Jortner Phys. Rev. A 31,3775 (1985).
7. D.S. Elliot, M.W. Hamilton, K. Arnett and S.J. Smith Phys. Rev. Lett. 53,439 (1984).
8. F.Y. Wu, S. Ezekiel, M. Ducloy and B.R. Mollow, Phys. Rev. Lett. 38, 1077 (1977).
9. Y. Prior and P.S. Stern, to be published.

EXTRARESONANCES IN DEGENERATE FOUR-WAVE

MIXING INDUCED BY SEQUENTIAL DECAY

A.D. Wilson-Gordon[*], H. Friedmann[*] and M. Rosenbluh[†]

[*]Department of Chemistry, [†]Department of Physics
Bar-Ilan University
Ramat-Gan 52100, Israel

INTRODUCTION

Bogdan et al[1,2] have observed that four-wave mixing (FWM) exhibits a sharp resonance at $\omega_2 = \omega_1$ in the presence of dephasing collisions despite the fact that the incident laser frequencies, ω_1 and ω_2, were detuned with respect to the atomic transition frequency ω_{ba}. The width of the pressure-induced extraresonance in degenerate FWM (PIER D4) signal was found to be determined by the decay rate Γ_b of the upper state $|b>$ of the $|a> \rightarrow |b>$ transition. We have shown[3] that when the lower state $|a>$ of the transition is also allowed to relax at a rate Γ_a, via collision-induced transitions to states that do not interact directly with the incident radiation, an additional Rayleigh-type extraresonant (ER) feature at $\omega_2 = \omega_1$, with width Γ_a, is introduced into PIER D4. Experimental confirmation of PIER D4 due to ground-state relaxation has been obtained recently.[4]

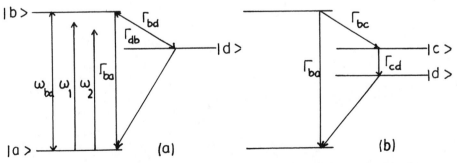

Fig. 1   Energy levels and decay rates for (a) three-level and (b) four-level system.

In the present paper, we show that two extraresonant features can be obtained in the degenerate FWM (DFWM) signal even in the absence of ground-state relaxation.[5] Here, again, we assume that the lasers with frequencies ω_1 and ω_2 are tuned near the transition frequency ω_{ba}. However, in addition to its direct decay to the ground state $|a\rangle$, the upper state $|b\rangle$ also decays to $|a\rangle$ via either one intermediate state $|d\rangle$ (see Fig. 1a):

$$|b\rangle \; \text{---} \; |d\rangle \longrightarrow |a\rangle \tag{1}$$

or via two intermediate states $|c\rangle$ and $|d\rangle$ (see Fig. 1b)

$$|b\rangle \longrightarrow |c\rangle \longrightarrow |d\rangle \longrightarrow |a\rangle \tag{2}$$

Examples of the three-level decay scheme of Eq. (1) are the green absorption band of ruby[6] and alexandrite[7] (no return from level $|d\rangle$ to $|b\rangle$ and fluorescein-doped boric acid[8] (return from level $|d\rangle$ to level $|b\rangle$ also occurs). These systems have been investigated experimentally[6-8] and theoretically[9] in connection with the related problem of population oscillations of state $|a\rangle$ at the difference frequency $\omega_2 - \omega_1$ between the probe frequency ω_2 and pump frequency ω_1. They are characterized by rapid decay from $|b\rangle$ to $|d\rangle$ followed by slow recovery to the ground state so that

$$\Gamma_{bd} \gg \Gamma_{ba} , \Gamma_{da} \tag{3}$$

where Γ_{ij} is the rate of decay from state $|i\rangle$ to state $|j\rangle$. For the particular case of fluorescein-doped boric acid,[8] the following inequalities hold:

$$\Gamma_{bd} \gg \Gamma_{ba} \gg \Gamma_{db} \gg \Gamma_{da} . \tag{4}$$

For simplicity, we shall treat only two extreme cases of the four-level decay scheme of Eq. (2) in which the Bloch equations for the four-level system $|a\rangle$, $|b\rangle$, $|c\rangle$ and $|d\rangle$ reduce mathematically but not physically to those of a three-level system. In the first case,

$$\Gamma_{cd} \gg \Gamma_{bc} \tag{5}$$

so that the population in $|c\rangle$ rapidly reaches its steady-state value and in the second case,

$$\Gamma_{da} \gg \Gamma_{cd} \tag{6}$$

so that the population in level $|d\rangle$ rapidly reaches its steady-state value. An example of the second case is atomic Na in which $|a\rangle = 3^2S_{1/2}$, $|b\rangle = 4^2P_{1/2}$, $|c\rangle = 4^2S_{1/2}$ and $|d\rangle = 3^2P_{1/2}$.

302

THE MODEL

We consider the level schemes depicted in Fig. 1 and assume that the electric field intensity is given by

$$\vec{E} = \tfrac{1}{2}[\vec{\varepsilon}\,\mathcal{E}_1 \exp(-i\omega_1 t) + \vec{\varepsilon}\,\mathcal{E}_2 \exp(-i\omega_2 t) + c.c.] \qquad (7)$$

where

$$\mathcal{E}_1 = |\mathcal{E}_1|[\exp(i\vec{k}_1 \cdot \vec{r}) + \exp(i\vec{k}_1' \cdot \vec{r})], \quad \mathcal{E}_2 = |\mathcal{E}_2|\exp(i\vec{k}_2 \cdot \vec{r}), \qquad (8)$$

$\vec{\varepsilon}$ is the complex polarization unit vector, $\vec{k}_1$ and $\vec{k}_1'$ are the wave-vectors of the incident fields at frequency ω_1 and $\vec{k}_2$ is the wave-vector of the incident field at frequency ω_2. Then further assuming that $\rho_{aa} + \rho_{bb} + \rho_{cc} + \rho_{dd} = \rho_{aa}^{\,eq}$, we obtain the following equations of motion for the density-matrix elements:

$$i\hbar\dot{\rho}_{aa} = -V_{ba}\rho_{ab} + \rho_{ba}V_{ab} - i\hbar\Gamma'(\rho_{aa} - \rho_{aa}^{\,eq}) + i\hbar\Gamma''\rho_{bb} \,,$$

$$i\hbar\dot{\rho}_{bb} = V_{ba}\rho_{ab} - \rho_{ba}V_{ab} - i\hbar(\Gamma_b + \Gamma_{db})\rho_{bb} - i\hbar\Gamma_{db}(\rho_{aa} - \rho_{aa}^{\,eq}) \,, \qquad (9)$$

$$i\hbar\dot{\rho}_{ab} = -\hbar(\omega_{ba} + i\gamma)\rho_{ab} + V_{ab}(\rho_{bb} - \rho_{aa}) \,,$$

where $\Gamma_b = \Gamma_{ba} + \Gamma_{bd}$ and $\Gamma_b = \Gamma_{ba} + \Gamma_{bc}$ are the total rates of decay from level $|b\rangle$ for the three- and four level systems respectively, $\gamma = \tfrac{1}{2}\Gamma_b + \gamma^*$ is the rate of decay of ρ_{ab}, $\Gamma' = \Gamma_{da}$ and $\Gamma'' = \Gamma_{ba} - \Gamma'$ for the three-level system, $\Gamma' = \Gamma_{da}$ and $\Gamma'' = \Gamma_{ba} - \Gamma_{da}(1 + \Gamma_{bc}/\Gamma_{cd}) \approx \Gamma_{ba} - \Gamma'$ for the four-level system when Eq. (5) holds, and $\Gamma' = \Gamma_{da}\Gamma_{cd}/(\Gamma_{da} + \Gamma_{cd}) \approx \Gamma_{cd}$ and $\Gamma'' = \Gamma_{ba} - \Gamma''$ when Eq. (6) holds. The matrix elements of the interaction Hamiltonian are written in the RWA as

$$V_{ba} = -\hbar[\mathcal{U}_1 \exp(-i\omega_1 t) + \mathcal{U}_2 \exp(-i\omega_2 t)] = V_{ab}^* \qquad (10)$$

with $\mathcal{U}_1 = \tfrac{1}{2}\mu_{ba}\mathcal{E}_1/\hbar$, $\mathcal{U}_2 = \tfrac{1}{2}\mu_{ba}\mathcal{E}_2/\hbar$.

In order to determine $\rho_{ab}(\omega_2 - 2\omega_1)$, the Fourier component of ρ_{ab} oscillating at frequency $\omega_2 - 2\omega_1$, we solve the equations for the Fourier components of the density matrix elements. For simplicity, we consider the case where the $|a\rangle \rightarrow |b\rangle$ transition is not saturated. The solution is given by

$$\rho_{ab}(\omega_2 - 2\omega_1) = C(1 + A + B) \qquad (11)$$

where

$$A = \tfrac{1}{2}\frac{2i\gamma - \tfrac{1}{2}i(\Gamma' + \Gamma_b + \Gamma_{db}) + \tfrac{1}{2}X}{\omega_2 - \omega_1 + \tfrac{1}{2}i(\Gamma' + \Gamma_b + \Gamma_{db}) - \tfrac{1}{2}X}\left[1 - \frac{i(\Gamma'' - \Gamma_{db})}{X}\right]$$

$$B = \frac{1}{2} \frac{2i\gamma - \frac{1}{2}i(\Gamma' + \Gamma_b + \Gamma_{db}) - \frac{1}{2}X}{\omega_2 - \omega_1 + \frac{1}{2}i(\Gamma' + \Gamma_b + \Gamma_{db}) + \frac{1}{2}X} \left[1 + \frac{i(\Gamma'' + \Gamma_{db})}{X} \right] \qquad (12)$$

$$C = \frac{2\upsilon_1^{*2}\upsilon_2 \rho_{aa}^{eq}}{(\omega_{ba} - \omega_1 + i\gamma)(\omega_2 - \omega_{ba} + i\gamma)(\omega_2 - 2\omega_1 + \omega_{ba} + i\gamma)}$$

with $X = [-(\Gamma_{db} + \Gamma_b - \Gamma')^2 + 4\Gamma_{db}\Gamma'']^{\frac{1}{2}}$

Let us consider two special cases: $\Gamma_{db} = 0$ and $\Gamma_{db} \neq 0$ when Eq. (4) holds.

SPECIAL CASES

I. $\Gamma_{db} = 0$

When $\Gamma_{db} = 0$, A and B reduce to

$$A = \frac{\frac{1}{2}i(2\gamma - \Gamma')}{\omega_2 - \omega_1 + i\Gamma'} \left(1 - \frac{\Gamma''}{\Gamma_b - \Gamma'}\right)$$

$$B = \frac{i\gamma^*}{\omega_2 - \omega_1 + i\Gamma_b} \left(1 + \frac{\Gamma''}{\Gamma_b - \Gamma'}\right) . \qquad (13)$$

In the absence of collisions, B = 0, so that a single ER feature with width Γ' appears at $\omega_2 = \omega_1$. In the presence of collisions, two ER features appear at $\omega_2 = \omega_1$, one with width Γ' and one with width Γ_b. The amplitudes of the features are pressure-dependent but their widths are independent of pressure provided collision-induced depopulation of the states $|b\rangle$, $|c\rangle$ and $|d\rangle$ is negligible. When the indirect decay schemes are absent, only the peak with width Γ_b survives and we return to the situation discussed by Bloembergen and coworkers.[1,2]

II. $\Gamma_{db} \neq 0$

When the rate of return from level $|d\rangle$ to $|b\rangle$ is significant and the inequalities of Eq. (4) hold, A and B of Eq. (13) reduce to

$$A = \frac{1}{2} \frac{i(2\gamma - \Gamma') - ix}{\omega_2 - \omega_1 + ix + i\Gamma'} \left(1 - \frac{\Gamma'' - \Gamma_{db}}{\Gamma_b + \Gamma_{db} - \Gamma' - 2x}\right)$$

$$B = \frac{1}{2} \frac{i(2\gamma^* - \Gamma_{db}) + ix}{\omega_2 - \omega_1 - ix + i(\Gamma_b + \Gamma_{db})} \left(1 + \frac{\Gamma'' - \Gamma_{db}}{\Gamma_b + \Gamma_{db} - \Gamma' - 2x}\right) \qquad (14)$$

with $x = \Gamma_{db}\Gamma''/(\Gamma_{db} + \Gamma_b - \Gamma')$. We thus obtain two ER peaks at $\omega_2 = \omega_1$ with widths determined by $\Gamma_{da} + x$ and $\Gamma_b + \Gamma_{db} + x \approx \Gamma_b + \Gamma_{db}$ respectively. The peak whose width is determined by $\Gamma_{da} + x$ may be very narrow and become very intense if $\Gamma_{da} + x \ll \gamma$ as in the case of fluorescein. Because of the very small value of $\Gamma_{da} + x$ for fluorescein, DFWM is easily saturated. This has recently been observed.[8]

CONCLUSION

We have shown that collisionless ER features in DFWM are predicted when the upper state |b> of the |a> → |b> transition, excited by near-resonant incident laser beams, decays to the lower state |a> both directly and via one or two intermediate states. The width of the new ER, which depends on the relative decay rates in the cascade process, may be very narrow leading to strong DFWM signals. Dephasing collisions affect the intensity of the new ER and also produce the usual pressure-induced ER.

REFERENCES

1. A. R. Bogdan, Y. Prior, and N. Bloembergen, Opt. Lett. 6, 82 (1981).
2. A. R. Bogdan, M. W. Downer and N. Bloembergen, Opt. Lett. 6, 348 (1981).
3. A. D. Wilson-Gordon and H. Friedmann, Opt. Lett. 8, 617 (1983).
4. L. J. Rothberg and N. Bloembergen, Phys. Rev. A 30, 2327 (1984).
5. H. Friedmann, A. D. Wilson-Gordon and M. Rosenbluh, Phys. Rev. A (in press).
6. L. W. Hillman, R. W. Boyd, J. Krasinski and C. R. Stroud Jr., Opt. Commun. 45, 416 (1983).
7. M. S. Malcuit, R. W. Boyd, L. W. Hillman, J. Krasinski and C. R. Stroud Jr., J. Opt. Soc. B 1, 73 (1984).
8. M. A. Kramer, W. A. Tompkin, J. Krasinski and R. W. Boyd, J. Luminescence 31 & 32, 789 (1984).
9. R. W. Boyd and S. Mukamel, Phys. Rev. A 29, 1973 (1984).

THE EXTRARESONANT ORIGIN OF GAIN DUE
TO STIMULATED RAYLEIGH SCATTERING

H. Friedmann and A.D. Wilson-Gordon

Department of Chemistry, Bar-Ilan University
Ramat Gan 52100, Israel

I INTRODUCTION

It is well-known[1] that gain can be obtained by stimulated Raman scattering without population inversion between the lower, initial state $|a>$ and the higher, final state $|c>$ of the two-photon transition. Near resonance, that is when the laser photon frequency ω_1 almost equals the transition frequency ω_{ba}, the overall transition takes place mainly according to the following scheme: (1) absorption of a laser photon of frequency ω_1, (2) stimulated emission of a Stokes laser photon of frequency ω_2 as depicted in Fig. 1a. This process prevails over the reverse process shown in Fig. 1b, provided $\rho_{aa} > \rho_{cc}$.

However, if $\rho_{aa} = \rho_{cc}$, both processes become equally likely and no net stimulated Raman scattering is expected. Similarly, there is

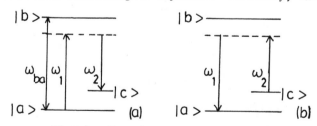

Fig. 1. Two-photon Raman transition.

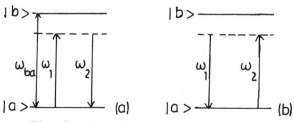

Fig. 2. Two-photon Rayleigh transition.

necessarily population equality between the initial and final states in the two-photon Rayleigh transition since these states are one and the same state. Thus the processes depicted in Figs. 2a and 2b are equally likely and no stimulated Rayleigh scattering (SRS) is expected. However we shall see in the following sections that the solution of the optical Bloch equations for a two-level system leads to an extraresonant (ER) SRS peak in one of the Rayleigh wings. Our formulae indicate that there is a direct connection between ER behaviour in four-wave mixing (FWM) and the appearance of SRS peaks. Near and at line-center where $\omega_1 \simeq \omega_{ba}$, SRS appears as a dip in the absorption spectrum without gain. This dip has been discussed extensively in the literature (see, for example, Ref. 2 and references therein) and will not be considered here. The appearance of SRS peaks is also evident from the theoretical results of Boyd et al[3] in the presence or absence of collisions and from the experimental results of Wu et al[4] in the absence of collisions. However the possibility of obtaining gain by SRS beyond a certain threshold intensity of the pump laser and the ER nature of this stimulated emission was not mentioned explicitly by these authors.

II BLOCH EQUATIONS

Consider the two-level system of Fig. 2 interacting with the bichromatic field

$$\vec{E}(t) = \tfrac{1}{2}\vec{\varepsilon}[\mathcal{E}_1 + \mathcal{E}_2 \exp(i\delta t)]\exp(i\omega_1 t) + c.c. \tag{1}$$

where ω_1 and $\omega_2 = \omega_1 + \delta$ are the frequencies of the bichromatic field and $\mathcal{E}_{1,2}$ its field strengths. Denoting the Rabi frequencies by

$$\Omega_1 = \tfrac{1}{2}\mu_{ab}\mathcal{E}_1/\hbar \quad , \quad \Omega_2 = \tfrac{1}{2}\mu_{ab}\mathcal{E}_2/\hbar \ , \tag{2}$$

we may write the Bloch equation for the system in the rotating-wave approximation as

$$i\,\dot{\rho}_{ab} = -(\omega_{ba} + i/T_2)\rho_{ab} - (\Omega_1 + \Omega_2 e^{i\delta t})e^{i\omega_1 t}(\rho_{bb} - \rho_{aa}) \tag{3}$$

$$i(\dot{\rho}_{bb} - \dot{\rho}_{aa}) = 2(\Omega_1 + \Omega_2 e^{i\delta t})e^{i\omega_1 t}\rho_{ba} - 2(\Omega_1^* + \Omega_2^* e^{-i\delta t})e^{-i\omega_1 t}\rho_{ab} \tag{4}$$

$$-(i/T_1)(\rho_{bb} - \rho_{aa}) - i/T_1 \ .$$

The steady-state solution for these equations can be written in the form (see Ref. 5 and references therein for similar methods)

$$\rho_{ab} = \sum_n \rho_{ab}(n\delta - \omega_1)e^{-i(n\delta - \omega_1)t} \equiv \sum_n a_{-n} e^{-i(n\delta - \omega_1)t} \tag{5}$$

$$\rho_{ba} = \sum_n \rho_{ba}(n\delta+\omega_1)e^{-i(n\delta+\omega_1)t} \equiv \sum_n a_n^* e^{-i(n\delta+\omega_1)t} \tag{6}$$

$$\rho_{bb}-\rho_{aa} = \sum_n (\rho_{bb}-\rho_{aa})^{n\delta} e^{-in\delta t} \equiv \sum_n d_{-n} e^{-in\delta t} \quad . \tag{7}$$

Inserting these expressions into the Bloch equations leads to the following recurrence relations:

$$\rho_{ab}(-n\delta-\omega_1) \equiv a_n = -(\Delta_1-n\delta+i/T_2)^{-1}(\Omega_1 d_n+\Omega_2 d_{n-1}) \tag{8}$$

$$(\rho_{bb}-\rho_{aa})^{-n\delta} \equiv d_n = P_n^{-1}[(-i/T_1)\delta_{no}+Q_n \Omega_1^*\Omega_2 d_{n-1}+R_n\Omega_1\Omega_2^* d_{n+1})] \tag{9}$$

where

$$P_n = (-n\delta+i/T_1)-4(n\delta-i/T_2)\{|\Omega_1|^2[\Delta_1^2-(n\delta-i/T_2)^2]^{-1}+|\Omega_2|^2[\Delta_2^2$$
$$-(n\delta-i/T_2)^2]^{-1}\} \quad , \tag{10}$$

$$Q_n = 2[(2n-1)\delta-2i/T_2][(\Delta_1-n\delta+i/T_2)(\Delta_2+n\delta-i/T_2)]^{-1} \quad , \tag{11}$$

$$R_n = 2[(2n+1)\delta-2i/T_2][(\Delta_1+n\delta-i/T_2)(\Delta_2-n\delta+i/T_2)]^{-1} \quad , \tag{12}$$

$$\Delta_1 = \omega_{ba}-\omega_1 \quad , \quad \Delta_2 = \omega_{ba}-\omega_2 \quad . \tag{13}$$

Moreover, because of the reality of $\rho_{bb}-\rho_{aa}$

$$d_{-n} = d_n^* \quad . \tag{14}$$

Exact expressions for the Fourier coefficients a_n and d_n can be given in terms of continued fractions.[5] We shall only be interested in the expressions for

$$a_{-1} \equiv \rho_{ab}(\omega_2-2\omega_1); \ a_1^* \equiv \rho_{ba}(\omega_2); \ d_{-1} \equiv (\rho_{bb}-\rho_{aa})^\delta; \ d_o \equiv (\rho_{bb}-\rho_{aa})^{dc} \tag{15}$$

for the case of a weak probe and an arbitrarily strong pump laser, that is, to lowest order in Ω_2 but to all orders in Ω_1 .

III EXTRARESONANT STIMULATED EMISSION PRODUCED BY A WEAK PROBE

The absorption or stimulated emission of the probe radiation is determined by $\mathrm{Im}\rho_{ba}(\omega_2)$. From Eqs. (8), (14) and (15), we find that

$$\rho_{ba}(\omega_2) = \Omega_2^*(-\Delta_2+i/T_2)^{-1}(\rho_{bb}-\rho_{aa})^{dc}+\Omega_1^*(-\Delta_2+i/T_2)^{-1}(\rho_{bb}-\rho_{aa})^\delta \quad . \tag{16}$$

The Fourier coefficient $(\rho_{bb}-\rho_{aa})^\delta$ represents the so called "population oscillation" (see ref. 3 and references therein). In the case of a weak probe, application of Eqs. (8) and (15) leads to

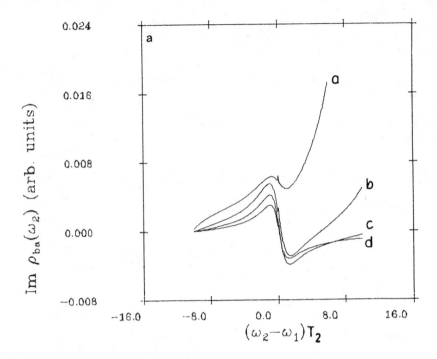

$$(\rho_{bb}-\rho_{aa})^{\delta} = -\Omega_1^{-1}(\delta+\Delta_1+i/T_2)\rho_{ab}(\omega_2-2\omega_1) \tag{17}$$

which shows the relation between population oscillation and FWM determined by $\rho_{ab}(\omega_2-2\omega_1)$.[3] Inserting Eq. (17) into Eq. (16) gives

$$\rho_{ba}(\omega_2) = \Omega_2^*(-\Delta_2+i/T_2)^{-1}(\rho_{bb}-\rho_{aa})^{dc}-(\delta+\Delta_1+i/T_2)(-\Delta_2+i/T_2)^{-1}(\Omega_1^*/\Omega_1)$$
$$\times \rho_{ab}(\omega_2-2\omega_1) . \tag{18}$$

The first term in Eq. (18) describes the usual absorption of a probe photon of frequency ω_2 except that the population inversion is determined by the pump laser. The second term is proportional to $\rho_{ab}(\omega_2-2\omega_1)$ which determines the FWM signal intensity. Therefore, Bloembergen's Rayleigh-type ER's in FWM[6] lead to Rayleigh-type ER probe absorption or stimulated emission peaks.

This can be verified by application of Eqs. (9)-(16) for the case of a weak probe. We obtain, to all orders in the pump intensity,

$$\rho_{ab}(\omega_2-2\omega_1) = -D(\omega_2)^{-1}(\Delta_1+i/T_2)^{-1}\Omega_1^2\Omega_2^*(\delta+2i/T_2)(\rho_{bb}-\rho_{aa})^{dc} \tag{19}$$

with

$$D(\omega_2) = -2|\Omega_1|^2(\delta+i/T_2) -\tfrac{1}{2}(\Delta_2+i/T_2)(\Delta_1+\delta+i/T_2)(\delta+i/T_1) . \tag{20}$$

To order $\Omega_1^2\Omega_2^*$ this reduces to

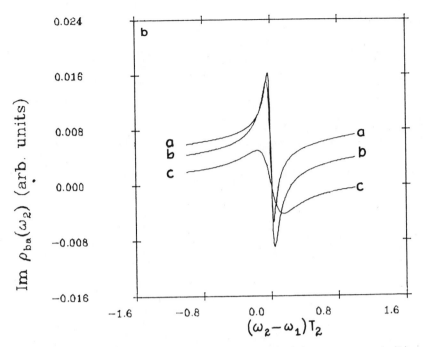

Fig. 3. Probe spectra for two-level system in (a) absence and (b) presence of collisions at detuning $\Delta_1 T_2 = 10$. In (a) $T_1/T_2 = 2$ and in (b) $T_1/T_2 = 0.02$. The Rabi frequencies $\Omega_1^2 T_2$ used were a: 2, b: 4, c: 8 and d: 16 in (a) and a: 0.5, b: 0.8 and c: 2 in (b).

$$\rho_{ab}(\omega_2 - 2\omega_1) = -\frac{2\Omega_1^2 \Omega_2^* \rho_{aa}^{eq}}{(\Delta_1 + i/T_2)(\Delta_2 + i/T_2)(\Delta_1 + \delta + i/T_2)}\left[1 + \tfrac{1}{2}\frac{2i/T_2 - i/T_1}{\delta + i/T_1}\right] \quad (21)$$

Eq. (21) shows that to this order, the Rayleigh-type ER at $\delta = \omega_2 - \omega_1 = 0$ only occurs in the presence of dephasing collisions when $2/T_2 \neq 1/T_1$. However, it has been recognized that when higher-order terms in the pump field are introduced, ER behaviour occurs even in the absence of collisions.[7]

In Fig. 3(a) and 3(b) we observe that beyond a threshold value of Ω_1 $\text{Im}\rho_{ba}(\omega_2)$ becomes negative so that the absorptive response of the system is transformed into stimulated emission. The effect is stronger and arises at a lower intensity threshold in the presence of collisions than in their absence, as expected. The peak of the stimulated emission is slightly shifted from the Rayleigh frequency $\omega_2 = \omega_1$ to that wing of the Rayleigh line which is closer to the resonance frequency ω_{ba}. At exactly the Rayleigh frequency we obtain in some of our probe spectra, a narrow, small absorption peak. This feature is situated in the middle of the dispersion shaped spectra shown in Fig. 3(a) and has not been reported so far.

REFERENCES

1. M.G. Raymer and J. Mostowski, Phys. Rev., A 24, 1980 (1981).
2. R.W. Boyd and S. Mukamel, Phys. Rev. A 29, 1973 (1984).
3. R.W. Boyd, M.G. Raymer, P. Narum, and D.J. Harter, Phys. Rev. A 24, 411 (1981).
4. F.Y. Wu, S. Ezekiel, M. Ducloy, and B.R. Mollow, Phys. Rev. Lett. 38, 1077 (1977).
5. N. Tsukuda, J. Phys. Soc. Japan 46, 1280 (1979).
6. A.R. Bogdan, Y. Prior, and N. Bloembergen, Opt. Lett. 6, 348 (1981).
7. H. Friedmann and A.D. Wilson-Gordon, Phys. Rev. A 26, 2768 (1982).

SHAPES AND WIDTHS OF STARK RESONANCES IN SODIUM

J.-Y. Liu, P. McNicholl, J. Ivri [*],
T. Bergeman, and H. Metcalf

Physics Department
State University of New York
Stony Brook, NY 11794   USA

We have studied the energies and shapes of resonances that appear in the laser induced Stark photoionization (PI) spectrum of sodium in a field-energy domain (FED) above the saddle point in the potential at $F = 1/16n^4$. Our experimental study began as an exploration of these structures in the continuum [1,2], evolved into a systematic comparison of measurements with theory [3], and is currently directed at a study of recently discovered interference effects that result in extraordinary sharpening of some of these resonances [4]. Our theoretical study began with numerical calculations of the hydrogen Stark spectra [5], moved to the formalism developed by Harmin [6] including direct numerical calculation of the widths of Stark resonances, and now employs a matrix version of Fano's theory of discrete-continuum couplings (matrices for multiple continua) [7]. We choose to study Na for reasons of experimental convenience only.

In our experiments sodium atoms in a thermal beam that is mechanically collimated to about 0.01 radian are stepwise ionized by two laser pulses (one tuned for D-line excitation) in a homogeneous dc electric field. The first laser prepares a 3P state and the second, which may be frequency swept, excites the atom to a level that eventually ionizes to produce the charged particles that constitute our signal. We use various combinations of polarizations of these lasers (with respect to the dc field) to determine the angular momentum state of the excited atoms. The laser pulses were 6 - 10 ns duration, essentially sequential, repeated at 12 Hz, directed perpendicular to the atomic beam and opposite to each other for convenience.

The yellow laser was tuned to either the D1 or D2 line and intense enough to power-broaden the transition considerably. Since its purpose was only for population of the 3P state, the details of the excitation are not of concern. The blue laser, tuned near 409 nm, was carefully redesigned in our later experiments to be frequency swept while operating in a single longitudinal mode of width about 300 MHz (shot to shot variations broaden the measured spectrum to about 700 MHz) [8]. Its intensity was not enough to saturate the PI transition. In our early experiments [1,2], its wavelength was calibrated with an etalon and a spectrometer against the 410 nm line of atomic hydrogen. We found that the quantum defect calculations of Stark energies provide such good values for the various resonances that the Na atoms can be used as their own frequency reference for the purpose of identifying the quantum numbers of a particular resonance [3].

Both lasers were focussed to less than 1 mm diameter where they intersected the 2 mm diameter atomic beam between oppositely charged copper plates. The applied voltage and approximately 7 mm plate spacing were known well enough to calculate the field to about 1%, and we estimate the field inhomogeneity to be less than 0.1%. One of the plates was drilled with about 40 holes of 1 mm diameter over a region of about 12 mm diameter. Behind this plate was a few mm drift region and a pair of multichannel plates biased for gain of about 10^6 for detection of ions or electrons, depending on the direction of the fields. The output pulse was amplified, temporally isolated by a boxcar integrator, and recorded.

In our early experiments [1,2] we recorded PI spectra at various electric fields and with various combinations of polarization. We confirmed calculations [5] of the hydrogenic undulations above the zero field ionization threshold, but could not make a quantitative comparison between theory and experiment for the Fano-shaped resonances at lower energies. The primary precision result from these measurements was the confirmation that the energies of the resonances agreed well with values calculated using hydrogenic wave functions.

With the advent of the WKB-quantum defect (WKB-QD) formalism [6] we began more detailed and demanding comparisons with theory [3]. We made several improvements to the apparatus and recorded very many spectra in different FED's. Spectral scanning of the laser was adequate to measure the (large) widths of some of the more rapidly ionizing states. These were predominantly $m_L = 0$ levels that were most strongly downward going with field (as expected), and were typically a few cm^{-1} wide. Measurement of the widths of the narrower $m_L = 1$ levels was difficult because the laser would not scan while retaining the single mode operation characteristic of its fixed wavelength operation. We dodged this problem by sweeping the field instead of the laser, and using the calculated field dependence of the energy to convert the width from field units to energy units. This method failed for states that changed energy only slightly with field, and depended on the (sometimes inappropriate) assumption that the width of a particular resonance was field independent. In FED's where this method worked, we found excellent agreement between the measured widths and those calculated with the WKB-QD theory [3].

In the course of this survey of resonance widths we observed a striking line narrowing effect [4]. In the FED of interest there are broad $m_L = 0$ spectral features that merge into a bumpy continuum, and for laser light polarized parallel to the field, most of the narrow features have $m_L = 1$. However, we have observed a number of narrow features where there was no $m_L = 1$ level present. Careful inspection revealed that these occured near the anticrossing of two otherwise broad $m_L = 0$ levels, and led us to describe them as the result of an interference in the coupling between the ionized continuum and the mixture of states deriving from the anticrossing. That is, there is a partial cancellation of two ionization matrix elements of opposite sign in the antisymmetric mixture of two states near their anticrossing. To within the accuracy of measurement, these interference narrowed features are described completely by the WKB-QD theory.

In these experiments [4] we observed narrowing of broad $m_L = 0$ spectral features by a factor of more than 100 (limited only by the 700 MHz resolution of the laser). Fig. 1 shows theoretical and experimental spectra of a particular Stark resonance at various fields in

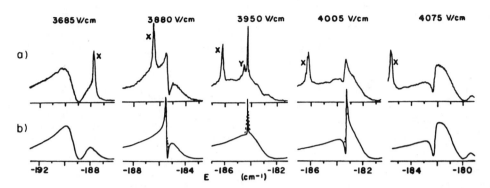

FIG. 1. (a) Experimental and (b) theoretical spectra of Na Stark levels $(n,n_1,m_L)=(20,19,0)$ and $(21,17,0)$ near their anticrossing. X is an $m_L = 1$ peak not included in the calculations. The theoretical curves were computed with th WKB-QD theory of Ref. 6.

the vicinity of an anticrossing. In order to observe further narrowing, we have begun time domain experiments on narrower $m_L = 1$ levels. Here we measure the lifetime of the resonances at different field values by recording the time dependence of the photocurrent. At fields away from the vicinity of the anticrossing, the ionization of the state we have chosen for study is so fast that we cannot resolve this time dependence because of the few ns duration of the laser pulses. However, near the anticrossing the lifetime is extended by the interference and we observe exponential decay of the photocurrent with time constant of more than 100 ns, corresponding to a spectral width of less than 2 MHz and a narrowing of a factor of 500. We expect to improve these observations with a more homogeneous electric field.

Our initial calculations were based on numerical solutions of the Schroedinger equation in parabolic coordinates for the hydrogenic Stark effect [5]. Application of these numerical methods to the FED below the zero field ionization limit was unsuccessful in describing our data, but new WKB-QD [6] method appeared at just the right time for us. It's application to the Na problem was straightforward and the agreement between our measurements and the calculations was very impressive [3]. A principal shortcoming of the method is that it provides only spectra, and the characteristics of particular resonances (strength, width, energy) have to be determined from these spectra. We set out to calculate these numbers directly.

For the case of ionization of discrete states into a single continuum, the method developed by Fano [9] is complete. Application of this method to the case of orthogonal continua has been studied by Mies [10]. We have developed an extension of Fano's formalism to this multichannel case by generalizing scalar quantities to matrices, and have incorporated the WKB-QD parameters into the model [7].

The results are summarized as follows. For atoms with at least one non-zero quantum defect, there are many states whose ionization is caused mainly by mixing of bound and free parabolic states by the non-hydrogenic core electrons. Above the classical saddle point, the ionization rate of a particular atomic resonance rises rapidly with field to a fixed 'plateau' value and then levels off. The plateau ends abruptly by beginning a sharp rise at a field value where the ionization process is dominated by purely hydrogenic tunneling to the continuum.

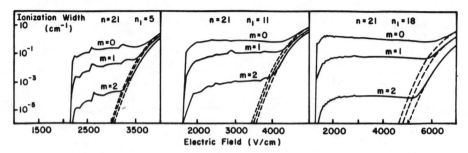

FIG. 2. Calculated ionization widths for various Stark resonances. The peaks and valleys from interferences have been smoothed out for clarity; only their residuals remain as bumps in the plateau areas.

For a fixed value of n, Fig. 2 shows that this plateau rate depends strongly on m_L (states with smallest m_L ionize fastest as expected), and also on the slope of the level with field ($n_1 - n_2$). The plateau is reached at a field value that also depends these parameters. The nearly field independent ionization rate on the plateau is punctuated by very many sharp peaks and dips as the field passes through the region of several level crossings and anticrossings (not shown in Fig. 2 for clarity). The width of the plateau region (in field) is progressively smaller for more downward going states for two reasons. First, they reach the saddle point at progressively lower fields, and second, the hydrogenic ionization process begins to dominate at progressively lower fields. The most strongly downward going state shows no plateau. The interference narrowing signals we have observed come from the sharp dips in various plateau regions.

We plan to extend both theoretical and experimental studies in the direction of the most dramatic changes in ionization rate. It may even be that this technique will have application to precision measurements or electric field calibration methods.

ACKNOWLEDGEMENT

This work supported by the National Science Foundation (USA).

* Permanent address: NRCN, Beer Sheva, Israel

REFERENCES

1. T. S. Luk et al., Phys. Rev. Lett. 47, 83 (1981).
2. L. DiMauro et al., J. Phys. (Paris), Colloq. 43, C2-167 (1982).
3. J. Y. Liu, Ph.D. Thesis, S.U.N.Y. Stony Brook (1984) - unpublished.
4. J. Y. Liu et al., Phys. Rev. Lett. 55, 189 (1985).
5. E. Luc-Koenig and A. Bachelier, Phys. Rev. Lett. 43, 921 (1979), and J. Phys. B, 13, 1743 (1980), and J. Phys. B, 13, 1769 (1980).
6. D. A. Harmin, Phys. Rev. A24, 2491 (1981), and 26, 2656 (1982), and 30, 2413 (1984).
7. P. McNicholl et al., in preparation.
8. P. McNicholl and H. Metcalf, Appl. Opt. 24, 2757 (1985).
9. U. Fano, Phys. Rev. 124 1866 (1961).
10. F. H. Mies, Phys. Rev. 175, 164 (1968).

OFF RESONANT LASER INDUCED RING EMISSION

I. Golub, G. Erez and R. Shuker

Department of Physics
Ben-Gurion University of the Negev
84105 Beer-Sheva, Israel

INTRODUCTION

There has been considerable interest in the conical emission resulting from the passage of a laser light detuned to the blue of a resonant
atomic transition.[1] This conical emission shell has half angle of few
degrees around the laser axis, is red detuned from the transition and is
spectrally broad (~ 10 cm^{-1}). It is obtained under conditions of laser
detuning and atomic vapor density for which self-trapped filaments are
generated. The main model that has been suggested for the ring emission
is that of four-wave parametric amplification of Rabi red sideband
generated in the self-trapped filaments leading to emission in the form of
a cone surrounding the laser beam.[2] The cone angle calculated according
to this model is smaller than the measured one by almost a factor of 2.
Moreover, the absence of a blue counterpart to the red-shifted ring
emission in most of the experiments disagrees with this model.

We propose a Cherenkov-type process for the production of the ring
emission.[3] All the features of this emission are accounted for by this
model. The laser light passing through the metal vapor induces moving
polarization of the medium, which propagates at a velocity in excess of
that of the red-shifted light, resulting in this Cherenkov-type emission.

THE MODEL

We attribute the ring emission to a Cherenkov type process. It is a
result of a medium polarization induced by the laser beam travelling at a
velocity that exceeds the speed of light in medium. This polarization is
essentially the medium excitation moving at the laser radiation group
velocity v_{gr}. The medium emits at frequencies ω within the radiative
transition linewidth that fulfill the Cherenkov condition, namely
$v_{gr}(\omega_L) > v_{ph}(\omega)$. Here $v_{ph}(\omega) = c/n(\omega)$ is the phase velocity of the
emitted light. The Cherenkov condition is fulfilled for the light emitted
to the red of the transition. The maximum spectral intensity of the
Cherenkov emission occurs where $n(\omega)$, the index of refraction, is maximal.
The maximum value of the saturated index of refraction which is correlated
with the cone peak emission frequency ω is at $\omega-\omega_0 = \omega_0-\omega_L$ making the
laser and cone frequency symmetric about the transition frequency ω_0, as
measured experimentally.[1]

The angle of the ring emission is given by the Cherenkov relation, $\cos\theta = v_{ph}(\omega)/v_{gr}(\omega_L)$. We have calculated the dispersive response of the index of refraction in terms of the off-diagonal elements of the density matrix $\rho_{ba}(\omega)$ and have found that $dn/d\omega\big|_{\omega=\omega_L} = 0$. Thus, $\cos\theta = n_L/n_C$. n_L and n_C are the refractive indices at the laser and emission frequencies, respectively. For small angles, $\theta \simeq [2(\Delta n_L - \Delta n_C)]^{\frac{1}{2}}$, where $\Delta n = 1 - n$.

One of the properties of Cherenkov radiation is its characteristic of a surface phenomenon which does not conserve the transverse component of the linear momentum. Thus, the ring emission should originate mainly at the surfaces of the self-trapped filaments, and not in their bulk.

EXPERIMENTAL

Our experimental apparatus consists of a sodium containing heat-pipe cell with an active length of 20 cm; the buffer gas is argon at a few torr pressure. The sodium density is in the range of $10^{14} - 10^{16}$ cm^{-3}. The dye laser is Hansch type and is pumped by a 30 kW peak power copper-halide laser with a 25 nsec duration pulse. The dye laser bandwidth is ~ 0.5 cm^{-1}. The laser beam is spatially filtered and focused into the sodium cell. The laser intensity at the focus is about 10 Mw/cm^2 and is sufficient to form self-trapped filaments. The spectrum of the emission is monitored by 1.2 cm^{-1} resolution McPearson 0.3m monochromator. The forward emission is photographed by an Alphax B216 camera placed after the sodium cell without any imaging optics. The laser beam is blocked with an on-axis disk.

RESULTS AND DISCUSSION

The measured forward light for a laser detuned to the blue of D_2 transition contains a central beam and the ring emission. The central beam is composed of the laser radiation and a coherent peak to the blue of D_1 transition (to be discussed elsewhere). The conical emission has a wide spectrum spanning the frequency region where the index of refraction, n_C, is larger than 1. For sodium densities of $\sim 10^{16}$ cm^{-3} the emission has a component even to the red of D_1 line, where the index of refraction is also larger than 1. The cone angle is 1-3° and increases as the laser frequency approaches the atomic transition and with increasing sodium density.

We have verified experimentally, that the ring emission is a surface radiation and occurs at the boundary of self-trapped filaments, confirming that it is a Cherenkov type radiation. To this end, we examined the spacial coherence of the ring emission. The laser radiation was focused into the sodium cell by a cylindrical lens and the produced pattern is shown in Fig. 1. The off-axis ring emission preserves its symmetry around the laser axis, while the laser beam spot has an ellipse form. Should the ring emission be produced in the interior of the saturated filaments, where n=1, the interference between different filament emissions would result in a pattern representing the spatial distribution of the filaments, i.e. ellipse. We thus conclude, that the ring emission is produced mainly at the interior of the filaments.

We find good agreement between measured cone angles and values calculated according to the Cherenkov emission model, as illustrated in Fig. 2. Here the vertical bars represent the measured range of conical shell angles. The Cherenkov emission angles θ_C are calculated for two saturation degrees Ω, namely 0.5Δ and 2Δ, which occur near the filaments boundary. Here $\Delta = \omega_L - \omega_0$. The ring angle calculated from the phase matching condition of the four-wave mixing model[2], $\cos\theta = 1/n_C$, does not fit the experimental results (Fig. 2).

318

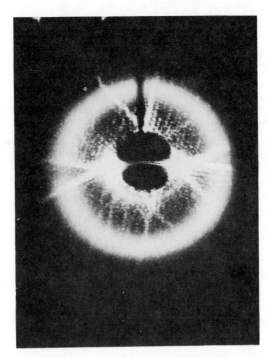

Fig. 1. The pattern of the ring emission produced by focusing with a cylindrical lens. The focal line of the lens is horizontal.

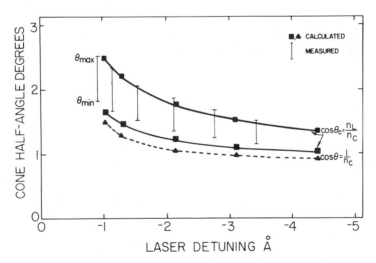

Fig. 2. Cone half-angle as a function of laser detuning. [Na] = $1.8 \cdot 10^{15}$ cm^{-3}. ■- Calculated according to Cherenkov model, for two saturation degrees $\Omega = 0.5\Delta$ and $\Omega = 2\Delta$; ▲-according to four-wave mixing model. Vertical bars represent the measured range of angles.

319

REFERENCES

1. C.H. Skinner and P. D. Kleiber, Observation of Anomalous Conical Emission from Laser-Excited Barium Vapor, Phys. Rev. A 21:151 (1979).

2. D. J. Harter and R. W. Boyd, Four-Wave Mixing Resonantly Enhanced by ac-Stark-Split Levels in Self-Trapped Filaments of Light, Phys. Rev. A 229:739 (1984).

3. I. Golub, G. Erez and R. Shuker, Cherenkov Emission due to Laser-Induced Moving Polarisation in Sodium, Journ. of Physics B, to be published; On the Optical Characteristics of the Conical Emission, Optics Comm., to be published.

LASER INDUCED COHERENT BLUE-SHIFTED EMISSION NEAR D_1 TRANSITION OF SODIUM

R. Shuker, I. Golub and G. Erez

Department of Physics
Ben-Gurion University of the Negev
84105 Beer-Sheva, Israel

INTRODUCTION

When an intense laser beam is blue detuned with respect to the D_2 transition of sodium, at sodium densities $> 10^{13}$ cm^{-3} self-focusing occurs. At still higher densities and for laser detuning in the range 6-15 cm^{-1} to the blue side of the D_2 line a coherent emission in the forward direction is observed.[1] This emission is shifted 1-3 cm^{-1} to the blue of the D_1 transition, depending on the sodium density, and, within the instrumental resolution of our spectrometer of 1.2 cm^{-1}, its frequency is independent on the laser detuning. This emission was observed earlier both in potasium[2] and sodium[3] vapors, but was not studied thoroughly, neither was a satisfactory explanation presented. We present experimental data of this emission and show that non of the ordinary mechanisms, such as collision induced population transfer, three-photon effect etc., is consistent with the experimental results. We propose a laser photon splitting by the doublet system of sodium as a possible explanation of the "blue peak" emission.

EXPERIMENTAL ARRANGEMENT

The experimental apparatus is discussed elsewhere.[4] Briefly, a 10kW peak power Hansch-type dye laser with a 0.5 cm^{-1} bandwidth is focused into a sodium heat-pipe cell. The laser intensity at the focus is 10 Mw/cm^2. The sodium cell is 20 cm in length. Argon at a few torrs pressure is used as a buffer gas.

RESULTS AND DISCUSSION

The forward spectrum of sodium at density of $1.8 \cdot 10^{15}$ cm^{-3} irradiated by laser with frequency ω_1 to the blue of the D_2 transition is shown in figure 1. This spectrum contains, in addition to the laser radiation, a spectrally broad off-axis conical emission attributed by us to a Cherenkov-type emission,[4] and a narrow coherent peak to the blue of the D_1 transition - the "blue peak". This spectrally narrow emission is on-axis and has a threshold on the laser intensity. Thus, this emission is a stimulated one. Within the resolution of our spectrometer, its frequency does not depend on the laser detuning from the D_2 transition.

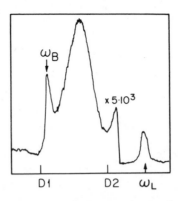

Fig. 1. Spectrum of forward emission of
sodium irradiated by laser at ω_L.
$[Na] = 1.8 \cdot 10^{15}$ cm^{-3}. The jump
to the left of ω_L is due to a
change in the sensitivity.

With increasing sodium density in the range 10^{15}-10^{16} cm^{-3}, the "blue
peak" frequency is shifted from 1.2 cm^{-1} to 3 cm^{-1} away from the D_1
transition.

The properties of the "blue peak" do not depend on the pressure of
the buffer gas (Ar) in the range of 2-20 torr.

On the basis of the experimental data presented above, one can
exclude on the onset several models for the "blue peak" emission. Firstly,
spin-flip collision induced population inversion on the D_1 transition is
not involved - due to both the off-resonant character of the emission and
its independence on the buffer gas pressure. Similarly, pressure induced
extra resonances are rejected. Stimulated electronic Raman and three
photon scattering effects, both by a two or three level system, are
dependent on the laser detuning and neither their frequencies are to the
blue in the vicinity of the D_1 line (figure 2); thus, these processes
are also excluded.

An attempt was made to explain the emission to the blue of the D_1
line as a result of a four-wave mixing process by a three level system;[2]
however, the absence of a prerequisite counterpart fourth photon makes
this explanation highly improbable.

We propose to explain the "blue peak" emission in terms of laser
photon splitting into two photons, as shown schematically in Fig. 3. The
laser creates a virtual level in the vicinity of the $3^2P_{3/2}$ level. Sub-
sequently two photons, quasiresonant with the $^2P_{3/2}$-$^2P_{1/2}$ and $^2P_{1/2}$ -
$^2S_{1/2}$ transitions, are emitted.In this process, the blue-shifted frequency is preferred by the self-focusing. The probability that such
photon splitting would involve a red detuned photon or a photon at the D_1
line line-center is small due to self-defocusing and absorption, respectively. Preliminary estimate of the gain for such a process yields a
value of 10^3 cm^{-1}. Detailed experiments and calculations are under
way.

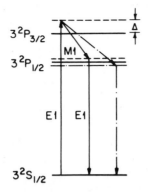

Fig. 2. Possible nonlinear processes for
dressed three-level atom:
a) electronic Stokes Raman;
b) four-wave mixing on two of the
levels, $3^2S_{1/2}$ and $3^2P_{3/2}$;
c) four-wave mixing when the third
level $3^2P_{1/2}$, is involved. In
each case the emitted frequencies
$(\omega_R, \omega_3, \omega_4)$ depend on the laser
detuning.

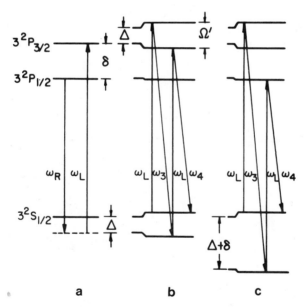

Fig. 3. Laser photon splitting by M1
and E1 transitions. The
observed "blue peak" forward
emission is represented by the
solid downward arrow. The
broken line represents
defocused emission.

REFERENCES

1. I. Golub, G. Erez and R. Shuker, Coherent Blue-Shifted Emission by
   the Sodium Doublet, Proc. Conf. Lasers (San Francisco) 1984, Post-
   deadline Paper; I. Golub, R. Shuker and G. Erez, to be published.

2. V. M. Arutyunyan, T. A. Papazyan, Yu. S. Chilingaryan, A. V. Karmenyan
   and S. M. Sarkisyan, Investigation of Resonance Polarization
   Phenomena During the Passage of Laser Radiation Through Potassium
   Vapor, Sov. Phys. JETP 39:243 (1974).

3. Y. H. Meyer, Multiple Conical Emissions from Near Resonant Laser
   Propagation in Dense Sodium Vapor, Optics Comm. 34:439 (1980).

4. I. Golub, G. Erez and R. Shuker, Cherenkov Emission Due to Laser
   Induced Moving Polarisation in Sodium, Journ. of Physics B (1986),
   in press; and this volume.

LASER INDUCED STIMULATED EMISSION FROM SODIUM VAPOR

Y. Shevy, M. Rosenbluh, and H. Friedmann

Bar-Ilan University
Physics Department
Ramat Gan, Israel

INTRODUCTION

The propogation of an intense, nearly resonant, laser pulse, through sodium vapor results in a number of nonlinear scattering processes. These processes have been previously observed in the spontaneous fluorescence spectra emitted by Na vapor interacting with a laser tuned near the D lines.[1] Here we report the observation of stimulated emission due to the same processes which we obtain whenever the appropriate population differences exist between the states involved in the scattering.[2]

Our results indicate that many of the "conical emission" features previously observed in a number of systems[3-8] have their physical origin in the scattering mechanisms identified in our experiments. The identification is based on the measurement of the frequency spectrum of the stimulated radiation. We do not as yet have a quantitative explanation for the angular dependence of the emission although we believe it to be caused by propagation effects in the highly nonlinear Na vapor.

EXPERIMENTAL RESULTS

The experiments were performed with a Nd:YAG-pumped pulsed dye laser exciting Na atoms contained in a ~75-cm-long, cold-window heated pipe with a vapor-region length of 20 cm operated at 350°C (Na density ~10^{15} cm^{-3}) and containing He buffer gas at a pressure of ~50 Torr. The dye laser was filtered by a combination of a Glan Thompson polarizer, a 2 mm aperture, and a spatial filter in order to minimize the amplified spontaneous emission (ASE) originating in the dye laser. An ASE-free beam is of extreme importance in the experiment, since a number of spectral features become apparent when it is present. The spatial filter slightly focused the laser beam, so in the interaction region the beam diameter was ~1 mm with a ~0.3-mJ pulse energy, a 5-nsec pulse duration, and a 0.2-cm^{-1} linewidth. The forward-stimulated radiation from the heat pipe was sampled with a ~1-mm-diameter optical fiber mounted on a translation stage. This allowed us to sample the stimulated emission as a function of angle relative to the laser beam. The light from the fiber was coupled with a high resolution spectrometer photomultiplier under computer control. The photomultiplier signal was boxcar averaged (200 pulses per sample) and a very large (8 orders of magnitude) dynamic range was obtained with a combination of computer controlled photomultiplier gain and neutral density filters at the spectrometer input slit.

A typical example of stimulated-emission spectra is shown in Fig. 1. Shown in the figure are three scans: one, labeled ω_L, is a spectrometer scan of the input laser with the cell removed. The other two scans are stimulated-emission spectra with the solid lines showing the on-axis spectrum (0 mrad), and the dashed lines showing the spectrum at an angle of 14 mrad with respect to the laser-defined axis. In many instances, such as the peak labeled ω_c, the distinct conical nature of the emission is manifested by the increase in intensity for a particular angle. This is especially true when one considers that the emission at large angles is spread out into a cone at that angle, and integration of the spectrum over this cone results in an even more pronounced peaking of the intensity.

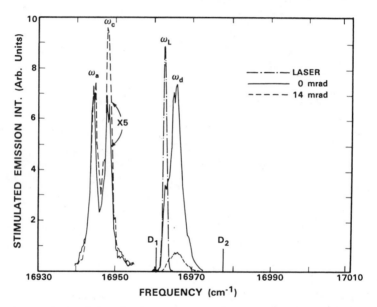

Fig. 1. Stimulated emission spectra for excitation near the D_1 line. The laser intensity is ~10^4 greater than the peak at ω_d.

For an understanding of the spectral features observed in the experiments we refer to Fig. 2 which shows the relevant energy levels of the Na three level system interacting with a strong laser at frequency ω_L. Solid lines indicate real states of the atom while broken lines are used to represent the virtual states induced by the laser. Figs. 2(a) and 2(b) depict excited state Raman scattering (ESR) while 2(c) shows resonance enhanced three photon scattering (RETPS) and 2(d) three photon scattering (TPS). A necessary condition for the observation of stimulated emission is the existance of population differences between the states involved in the interaction. As shown in Figs. 2(a) and 2(b), the stimulated ESR process takes the atom from one excited state to the other. This results in the emission of a photon at either ω_a or ω_b (but not both simultaneously), provided that there exists a population difference between the two excited states. Stimulated TPS and RETPS, shown in Figs. 2(c) and 2(d), require a higher population in the ground state. For these processes it is clearly possible to observe stimulated emission at both ω_c and ω_d simultaneously. The RETPS process shown in Fig. 2(c) is possible only when one considers the true three-level nature of the Na D lines. Not only does the third level provide a new path for TPS but the process can also be resonantly enhanced (for the case shown by the $P_{\frac{1}{2}}$ state). The TPS processes or a combination of the two ESR processes could also result in four-wave mixing, yielding stimulated-emission spectra that would be in-

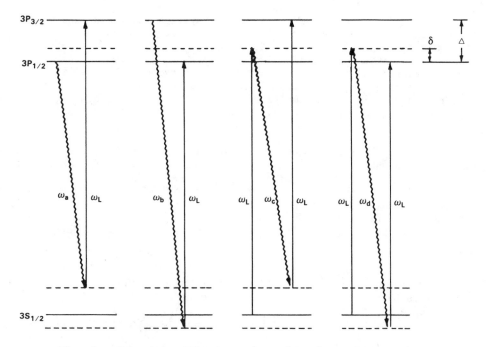

Fig. 2. Schematic representation of Na-laser interaction.
Dashed lines represent virtual levels.

dependent of population differences. The four-wave mixing could be seeded by stimulated TPS or ESR or grow at the two new frequencies simultaneously. Such a mechanism has the attraction of providing a possible explanation of the angular dependence of the stimulated emission.[6] Unfortunately, propagation effects, such as self-focusing, make a quantitative analysis of the problem very difficult and so far a good quantitative fit to the data has not been obtained. An additional experimental problem with this model is that it requires emission at the fourth wave frequency, which under our experimental conditions we have not observed.

The peak labeled ω_a in Fig. 1 corresponds to the ESR process shown schematically in Fig. 2(a). The presence of a signal at ω_a indicates that for this laser frequency the $3P_{1/2}$ state is more populated than the $3P_{3/2}$ state (as expected from the near resonance of the laser to the D_1 transition). The resonance is actually $1.1 cm^{-1}$ red-shifted from the zero-field (photon-free) value of $\omega_L - \Delta$, where Δ is the unperturbed separation of the $P_{1/2}$ and $P_{3/2}$ states. This is a result of the laser-induced light shifts, which for the indicated laser frequency increase Δ ($3P_{1/2}$ is shifted downward, and the $3P_{3/2}$ level is shifted upward).

The peak at ω_c is RETPS to the $3P_{3/2}$ state [see Fig. 2(c)]. The zero-field frequency of this three-photon process is given by $\omega_c = \omega_L - (\Delta - \delta)$, where $\delta \equiv \omega_L - \omega_{D1}$. For this scattering peak the observed frequency corresponds closely to the calculated ω_c since the $1.1 cm^{-1}$ increase in Δ is offset by an almost identical increase in δ. This can be qualitatively understood by noting that, while the $3P_{1/2}$ light shift is downward, the ground-state and $3P_{3/2}$ light shifts are upward and equal, resulting in a net increase in δ.

The peak at ω_d [see Fig. 2(d)] is attributed to TPS to the $3P_{1/2}$ state with a frequency given by $\omega_d = \omega_L + \delta$. The identification of this peak is complicated by the unresolved presence of Rayleigh scattering and a filtering effect in the Na vapor. The filtering occurs because the Na vapor has a spatially nonuniform refractive index at all frequencies near the D lines, except at the dispersion-free point (at ~19,966 cm^{-1}). As a result light at this frequency propagates through the system, whereas light at surrounding frequencies is strongly dispersed and appears to be much weaker. Indeed, in spectra obtained for laser frequencies blue-shifted up to ~5 cm^{-1} from the D_1 resonance, the peak at ω_d appeared to be stationary.

The region near the dispersion free point also exhibits a strong on-axis emission when the laser is tuned to the blue side of D_2. This feature can be eliminated by cleaning the dye laser from the ASE present in the beam. The exact nature of the interaction of the ASE with the Na vapor is still unclear, but it appears that the Na is not merely an optical filter for eliminating (by absorption or dispersion) a part of the ASE spectrum, but rather that certain frequencies present in the ASE undergo amplification.

In Fig. 3 we show the stimulated emission frequencies observed in the forward direction as a function of laser frequency. The experimental points were obtained from many stimulated emission spectra and at various angles for each laser frequency. The solid lines indicate the expected positions of the various TPS and ESR processes at zero-field intensity. The lines labeled A and B are ESR scattering from the $P_{1/2}$ and $P_{3/2}$ states, respectively, and lines C and D are TPS to the $P_{3/2}$ and $P_{1/2}$ final states, respectively (see Fig. 2). The line labeled R corresponds to Rayleigh scattering. Also indicated in the figure are the positions of the D_1 and D_2 atomic resonances. Not shown in Fig. 3 are the observed Rayleigh-scattered peaks, which were always present, often with a pronounced asymmetric broadening, which we tentatively ascribe to self-phase modulation.

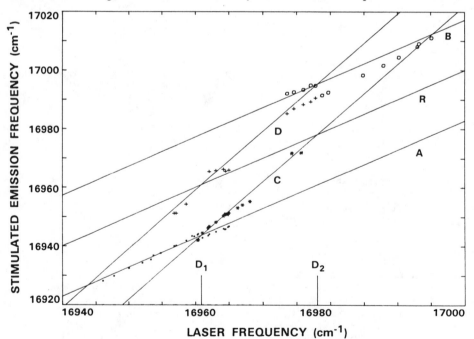

Fig. 3. Stimulated-emission frequencies versus ω_L. Solid lines are zero-field calculations of A, ESR, from $P_{1/2}$, B, ESR from $P_{3/2}$, C, TPS to $P_{3/2}$, D, TPS to $P_{1/2}$, and R, Rayleigh scattering.

The dots and squares in Fig. 3 correspond to peaks attributed to ESR from the $P_{1/2}$ and $P_{3/2}$ states, respectively. Asterisks correspond to TPS to the $P_{3/2}$ state with strong resonance enhancement for ω_L close to the D_1 line. The two TPS data points at ~16,975 cm^{-1} actually correspond to strong emission peaks with an asymmetric broadening toward low frequencies and as yet unexplained structure. The crosses correspond to TPS to the $P_{1/2}$ state with resonance enhancement for ω_L close to the D_2 line. The laser-frequency-independent TPS at ~16,966 cm^{-1}, close to the Na dispersion-free point, is clearly apparent. The observation of the various components of the spectra in different branches is an indication of the relative populations of the levels as a function of laser frequency as well as the importance of resonance enhancement of the TPS process.

The small deviations between the solid lines and the data points in Fig. 3 are due primarily to light shifts of the three-level system. The data are repeatable with a precision of ~0.25 cm^{-1}, and thus these deviations are significantly above the experimental uncertainty. The direction and the magnitude of the shifts agree with rough calculations of the expected light shifts. However, a detailed calculation of the light shifts for a three-level system in which the strong but unsaturated coupling of the laser is to both of the upper states will have to be performed. The strong light shifts observed for both the ESR and RETPS signals for laser frequencies at or to the blue of the D_2 resonance are due to the stronger oscillator strength (and thus larger Rabi shift) of the D_2 transition and to self-focusing effects in this region, which further increase the power density in the interaction region.

CONCLUSION

We have shown that TPS, RETPS, and ESR scattering are important mechanisms in the production of stimulated radiation in Na vapor. Work is in progress to calculate and measure accurately the angular dependence of the emission and to compute the laser-induced light shifts of a three-level system in which both excited states are strongly coupled to the laser field.

REFERENCES

1. Y. Shevy, M. Rosenbluh, and H. Friedmann, Phys. Rev. A 31, 1209 (1985).
2. Y. Shevy, M. Rosenbluh, and H. Friedmann, Opt. Lett. Jan (1986).
3. A.C. Tam, Phys. Rev. A 19, 1971 (1979).
4. C.H. Skinner and P.D. Kleiber, Phys. Rev. A 21, 151 (1980).
5. G.L. Burdge and C.H. Lee, Appl. Phys. B28, 197 (1982).
6. D.J. Harter and R.W. Boyd, Phys. Rev. A 29, 739 (1984), and references therein.
7. E.A. Chauchard and Y.H. Meyer, Opt. Commun. 52, 141 (1984).
8. I. Golub, G. Erez, and R. Shuker, J. Phys. B, in press (1986).

CARS STUDIES OF NONADIABATIC COLLISION PROCESSES

P. Hering, S.L. Cunha, and K.L. Kompa

Max-Planck-Institut für Quantenoptik
8046 Garching bei München
Federal Republic of Germany

INTRODUCTION

Nonadiabatic collisions between atoms and diatomics have drawn a large amount of attention both in theoretical and experimental studies. In particular, transfer of electronic energy of an atom to vibrational, rotational and translational energy of an diatomic molecule (electronic quenching) can be considered as an important fundamental process and is thus extensivly investigated /1,2/. Our interest in this field is two-fold. We apply a new experimental technique in the field of nonadiabatic processes to obtain a detailed understanding of the quenching process. We use Coherent Anti-Stokes Raman Spectroscopy (CARS) to measure the internal energy distribution of the diatomic molecule. Crucial for the prediction of the internal energy distribution is the knowledge of the crossings between the potential energy surfaces of the ground and ex-cited states. Our second interest is to develop a spectroscopic method to measure the location of these crossings. This can be done by ab-sorption during collision and subsequent analysis of the internal state distribution of the collision partners.

EXPERIMENTAL

The $Na + H_2$ system has been chosen both for theoretical and ex-perimental reasons. Ab initio potential energy surfaces are available /3/ and trajectory calculations are currently performed. CARS has been used to measure the state distribution of H_2 in photolysis /4/ and reaction dynamics /5/. Fig. 1 shows our experimental setup. In a probe cell we produce a gas mixture of Na (550 K, 10^{-2} torr) and H_2 (10 to 1000 torr). A flashlamp-pumped dye laser (FLP) with variable pulselength excites sodium to the 3P-state. A CARS laser system detects the rovi-bronic population of hydrogen. Our present sensitivity is approximately 10^{12} particles per cm^3 and quantum state. With an adjustable time control the CARS laser system can monitor the requested population before, during or after the irradiation of the FLP-laser with a time resolution of 10 ns.

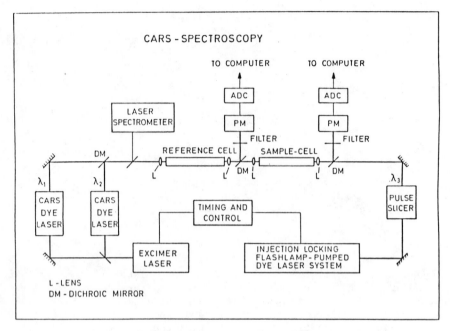

Fig. 1.   Experimental setup

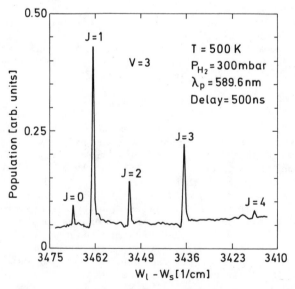

Fig. 2.   Rotational distribution of v=3
           band after quenching

RESULTS

The quenching process produces vibrationally hot H_2 up to v=3. The rotational distribution was determined for different vibrational bands depending on pressure and time after the excitation of Na. Fig. 2 shows an example for v=3. This plot displays the relative population of individual rotational transitions. The corresponding CARS transitions are Δv=+1 and Δj=0 (Q-branch). This distribution was recorded at a fixed time delay between FLP-laser and CARS-laser system.

We also measured the time behavior of the population of individual quantum states of H_2. To that end the wavelength of the CARS probe laser was tuned to the respective transition and locked. We then changed the time delay between FLP-laser and CARS laser. Fig. 3 displays a selection

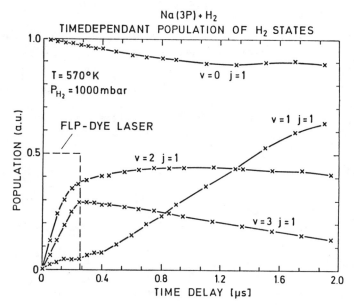

Fig. 3.   Time dependent population of H_2

of different quantum state populations (different v, same j). At time zero only the vibrational ground state is populated. The FLP-laser excites Na (saturating the transition) and energy transfer becomes possible. Different vibrational levels are populated differently. After the production of excited sodium has come to an end we can follow the vibrational relaxation of H_2 (v=1,2,3) in collisions with H_2 (v=0).

We also measured the population dependence of individual H_2 states for far red-detuning of the exciting laser and found, dependent on the excitation wavelength, different line shapes and different populations for different vibrational levels /6/.

Beside the known CARS lines of hydrogen also new, unknown lines were detected. Fig. 4 displays an example. These lines only appear if both excited sodium and hydrogen are present. We also observed an isotope shift when hydrogen was replaced by deuterium or HD. We attribute

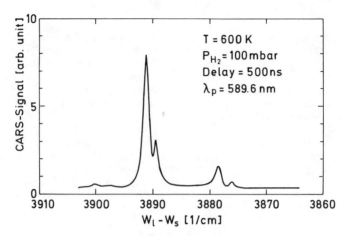

Fig. 4.   Unidentified CARS lines

these unknown lines to Na^+H_2 or NaH_2^+ which is formed during the collision. Further confirmation however is necessary.

   With these first experimental results we show that CARS (or other nonlinear methods) applied to the study of nonadiabatic collision processes gives interesting results. It appears to be a powerful tool to measure the state distribution of many collision partners.

REFERENCES

/1/   F. Rebentrost in "Theoretical Chemistry: Advances and Perspectives", p.1, vol. 6B, Ed.: D. Henderson, Academic Press, N.Y. 1981

/2/   I.V. Hertel, Adv. Chem. Phys. 50 (1982) 475

/3/   P. Botschwina et al., J. Chem. Phys. 75 (1981) 5438

/4/   D. Debarre et al., J. Chem. Phys. 83 (1985) 4476

/5/   D.P. Gerrity and J.J. Valentini, J. Chem. Phys. 82 (1985) 1323

/6/   S.L. Cunha, P. Hering, K.L. Kompa, Opt. Comm. to be published

OPTICAL BISTABILITY OF MOLECULAR SYSTEMS

Meir Orenstein, Jacob Katriel and Shammai Speiser

Department of Chemistry
Technion - Israel Institute of Technology
Haifa 32000, Israel

Recently we developed[1,2] the non-linear-complex-eikonal approximation. Our goal was to set a standard mathematical treatment for the analysis of the propagation of light waves through non-linear media. The linear eikonal approximation is a cornerstone for optical systems design and it is hoped that the non-linear eikonal method will gain some of the universality of its linear counterpart.

The non-linear eikonal approximation results in an integral equation for the complex accumualted phase. This integral equation can be applied to a variety of non-linear media and boundary conditions. After demonstrating the applicability of the non-linear eikonal approach to the non-linear Fabry Perot and ring resonators[1,2], we have used it for analyzing complicated configurations, especially two coupled non-linear resonators[2]. The results, which indicate a complicated multistable pattern, were applied to the design of several switching schemes and to optical pulse reshaping.

Further applications of the non-linear eikonal approximation are discussed here. We are interested in absorptive non-linearity whose sources are molecular photoexcitatons. Propagation effects of an intense light field in organic molecular media are discussed and their connection to molecular parameters is emphasized.

The complex non-linear eikonal approximation is written as

$$\phi(x) = \frac{2\pi}{\lambda_o} \int_o^x n[I(x')]dx' \tag{1}$$

where $\phi(x)$ is the complex accumulated phase along the distance of propagation x through a medium whose index of refraction n depends on the local intensity $I(x)$. When the main source of non-linearity is the imaginary (absorptive type) part of n, eq(1) reads

$$\phi(x) = \int_o^x \alpha[I(x')]dx' \tag{2}$$

where α is the intensity dependent absorption coefficient. This integral equation can be solved analytically for many interesting cases[3] and some results concerning propagation effects in molecular media are presented here.

Propagation effects are analyzed by following the complex electrical field amplitude E and assuming self consistency at steady state[3]. The local intensity of the propagating light is thus given by

$$I(x) = K\exp[-\phi_A(x)] \tag{3}$$

where K for some simple cases, is given by

$$K = I_{in}/n_o \qquad\qquad \text{for free propagation}$$

$$K = I_{in}(I-R)/\{n_o[1-R^2\exp(-\phi_A)]\} \qquad\qquad \text{for a ring resonator (Fig.1) and incoherent light source.}$$

$$K = I_{in}\ (1-R)/\{n_o[1-2R\exp(-\phi_A/2) \qquad\qquad \text{for a matched ring resonator and coherent}$$
$$+R^2\exp(-\phi_A)]\} \qquad\qquad \text{light source}$$

where I_{in} is the input intensity, $\phi_A \equiv \phi_A(L)$ and R is the resonator mirrors intensity reflectivity.

By combining eqs.(2) and (3) and differentiating we obtain the following differential equation

$$d\phi_A(x)/\alpha\{I[\phi_A(x)]\} = dx \tag{4}$$

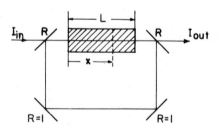

Fig. 1.   A ring resonator.

In media consisting of large organic molecules the main source of nonlinearity is the large $\alpha(I)$ resulting from population trapping and excited state absorption[4]. Using the rate equation approximation, which describes the kinetics of the general scheme of Fig. 2, we can obtain the steady state $\alpha(I)$ [5]

$$\alpha(I) = (A+BI)/(C+DI+EI^2) \tag{5}$$

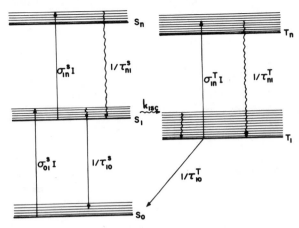

Fig. 2. Schematic level diagram for a large molecule showing singlet (S_i) and triplet (T_i) manifolds, radiative ($\longrightarrow$) and nonradiative transitions ($\leadsto$). Absorption cross-sections are σ_{ij}, τ_{ji} are the level lifetime and k_{ISC} denotes the intersystem crossing rate.

where A,B,C,D, and E are constants consisting of various combinations of molecular parameters (Fig. 2) which depend on the particular molecular absorption process[4],[5]. By substituting $\alpha(I)$ in eq. (4) and integrating we obtain[5]

$$\phi_A = AL/C + (EA/BC)[\exp(-\phi_A)-1]$$

$$-[1+EA^2/CB^2-DA/BC]\ln\{[A+BK\exp(-\phi_A)]/(A+BK)\} \equiv F[\phi_A(L)] \qquad (6)$$

Equation 6 which is the general solution to a variety of exitation schemes, can be solved graphically for $\phi_A(L)$. This is shown here for two cases. The solutions for a medium exhibiting $S_0 \longrightarrow S_1 \leadsto T_1$ transition (Fig. 2) incorporated in a matched ring resonator. The results for a coherent field pumping are depicted in Fig. 3. The results for ϕ_A are used to calculate the output intensity as a function of Iin (Fig. 4). The bistable behavior of this system (lower branch due to almost linear absorption and upper branch due to saturated absorption) can be also demonstrated in many other cases.[5] The solutions for reverse saturable absorber - a medium exhibiting $S_0 \longrightarrow$ $S_1 \leadsto T_1 \longrightarrow T_n$ ($\sigma_{1n} \gg \sigma_{on}$) transitions - in the same optical resonator are depicted in Fig. 5. The bistable behavior here is of a reverse absorption saturation (lower branch) and an absorption saturation (upper branch). Absorption experiments are performed in matched resonators because such devices are capable of accumulating energy and thus sustain absorption saturation. This cannot be accomplished in situations of free propagation. It is worth noting, however, that bistability of molecular systems exhibiting reverse saturation can be obtained even for an incoherent light source and without a resonator[5]. The general features of the system such as threshold switching intensities in a given optical configuration, depend on the various molecular parameters. Thus, optical propagation effects and in particular optical bistability can be used for probing molecular photophysical properties[5].

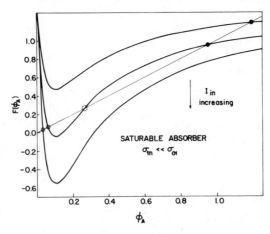

Fig. 3. Graphical solution of eq. (6) for a saturable absorber, excited by a coherent light source, incorporated in a ring resonator. ● linear absorption branch ○ unstable solution o saturated absorption branch.

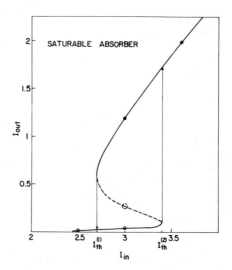

Fig. 4. Optical bistability obtained for the saturable absorber of Fig. 3.

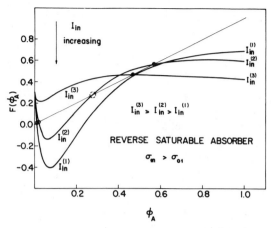

Fig. 5. Same as Fig. 3 for a reverse saturatable absorber.
• reverse saturated absorption branch, ○ unstable
solution, ○ saturated absorption branch.

References

1. M. Orenstein, S. Speiser and J. Katriel, Opt. Commun. 48: 367
   (1984).
2. M. Orenstein, S. Speiser and J. Katriel, IEEE. J. Quant.
   Electron 21:1513 (1985).
3. M. Orenstein, J. Katriel and S. Speiser: "The nonlinear
   complex eikonal approximation. I: Optical bistability
   in absorbing media", (to be published).
4. S. Speiser and N. Shakkour, Appl. Phys. B38:191 (1985).
5. M. Orenstein, J. Katriel and S. Speiser: "The nonlinear
   complex eikonal approximation. II: Photophysical and optical
   aspects of optical bistability in molecular media",
   (to be published).

ELECTROMAGNETIC FIELD INTERACTION WITH FAST-MOVING DIPOLES

Gershon Kurizki*, M. Strauss & J. Oreg[†], and A. Ben-Reuven**

*Department of Chemical Physics
The Weizmann Institute of Science
Rehovot 76100, Israel

[†]The Negev Nuclear Research Center, Israel

**School of Chemistry
Tel-Aviv University, Israel

The two conventional Hamiltonians of quantum electrodynamics, namely, the minimal-coupling ($\vec{p}\cdot\vec{A}$) and the multipolar ($\vec{\mu}\cdot\vec{E}$) Hamiltonians are known to yield the same rates of resonance transitions (absorption or emission) for <u>stationary</u> atoms. These two Hamiltonians, related by the Power-Zienau canonical transformation, are shown to produce identical rates for <u>moving</u> dipoles, in the nonrelativistic limit ($v/c \ll 1$) only if the <u>magnetic moment</u> associated with the moving dipole is included in the multipolar interaction. The effect of the motion is then, regardless of the chosen Hamiltonian, a reduction of the dipole-field coupling constant by a factor of $1-\vec{v}\cdot\hat{q}/c$, where $\hat{q}$ is the unit vector in the direction of the emitted or absorbed radiation. In the case of dipoles associated with relativistic spin $-1/2$ particles (e.g. electrons or positrons channeled in crystals), the coupling constant obtained from the Dirac Hamiltonian differs from that of the non-relativistic limit by a factor of $2\gamma/(\gamma+1)$, where $\gamma=(1-v^2/c^2)^{-\frac{1}{2}}$. The wave equation and generalized N-particle Bloch equations are written in terms of electric-field and polarization operators of the active modes, with the motionally-reduced field-dipole coupling constants, thus providing the framework for the analysis of spontaneous emission, superfluorescence and lasing from ensembles of fast-moving dipoles. The first-order perturbative result for the γ-dependence of gain obtainable by single-mode field stimulation of channeling radiation from relativistic electrons is retrieved from this analysis in the small - T_2, steady-state, small-signal regime. Substantial modifications of this γ-dependence are predicted in other regimes.

The rate of radiation from dipolar emitters moving with high velocities is appreciably reduced compared to the rate of ordinary electric-dipole radiation at Doppler-shifted frequencies. This reduction is due to the magnetization arising in a dipolar emitter moving relative to the detector, which is revealed by a nonrelativistic treatment. Additional corrections to the emission rate are encountered at relativistic velocities. The nature of all these corrections is discussed below. They are then included in the analysis of spontaneous and stimulated emission from relativistic channeled particles (electrons and positrons) in a crystal, as a model example. In this system, the particle propagates relativistically in the

crystal along a channel formed by adjacent atomic planes or rows, while
the dipolar emission is associated with its undulatory motion across the
channel[1,2].

Within the framework of nonrelativistic quantum electrodynamics, the
emission in electric-dipole transitions can be treated using two alternative Hamiltonians for field-matter interaction, i.e. a multipolar Hamiltonian and a minimal-coupling ($\vec{p} \cdot \vec{A}$) Hamiltonian, since the two are related by
a canonical transformation[3,4]. In what follows, the results concerning
motional effects on the emission will be discussed and checked by showing
that they are obtainable from both Hamiltonians.

The field-matter interaction in the minimal-coupling form (ignoring
diamagnetic effects) is given by

$$H_{int.}^{min.coupl.} = \sum_i \frac{e_i}{m_i c} \vec{p}_i \cdot \vec{A}(\vec{r}_i) \tag{1}$$

where $\vec{r}_i$ and $\vec{p}_i$ are the canonical coordinates and momenta of the constituents of the system, with charges e_i and masses m_i and A is the vector
potential, satisfying here the Coulomb gauge. The equivalent interaction
in the multipolar Hamiltonian is

$$H_{int.}^{mult.} = -\int d\vec{r}\, \vec{P}(\vec{r}) \cdot \vec{E}(\vec{r}) - \int d\vec{r}\, \vec{M}(\vec{r}) \cdot \vec{B}(\vec{r}) \tag{2}$$

where $\vec{E}$ and $\vec{B}$ are the transverse electric and magnetic radiation fields,
and $\vec{P}$ and $\vec{M}$ are the electric polarization and the magnetization of the
system. In the dipole approximation, one usually replaces $\vec{A}(\vec{r}_i)$ by $\vec{A}(\vec{R})$,
and $\vec{P}(\vec{r})$ by $\sum_i \vec{\mu}_i \delta(\vec{r}-\vec{R})$, where $\vec{\mu}_i$ are the dipole moments and $\vec{R}$ is the center-
of-mass coordinate, neglecting the magnetic multipole contributions. This
approximation is no longer valid, as shown below, when dealing with
emission from a system whose velocity is non-negligible compared to c.

The emission rates into a single radiation mode for a two-level resonance transition e→g in a system moving with constant velocity $\vec{v}$ are given
to lowest order (neglecting self-energy corrections) by the familiar Golden
rule

$$W_{\vec{q}\lambda} = (2\pi/\hbar)|V_{\vec{q}\lambda}|^2 (n_{\vec{q}\lambda}+1)\delta(\omega_{eg}-\omega_{\vec{q}}-\vec{q}\cdot\vec{v}) \quad . \tag{3}$$

Here $\vec{q}$ is the photon wavevector and $\hat{e}_\lambda$ its polarization. The initial photon
occupation number $n_{\vec{q}\lambda}$ and the matrix element for vacuum excitation (coupling
element) $V_{\vec{q}\lambda}$ determine the emission rate at the Doppler-shifted frequency
$\omega_{\vec{q}}$. The imperative question is what is the appropriate coupling element
to be used in Eq. (3)?

In the electric-dipole approximation, the interactions (1) and (2)
yield coupling elements of different form:

$$V_{\vec{q}\lambda}^{el.dip.} = -\sum_i \vec{\mu}_i <1|\vec{E}_{\vec{q}\lambda}|0> = (\omega_{\vec{q}}/\omega_{eg})(V_{\vec{q}\lambda})^{min.coup.} \tag{4}$$

For emitters at rest ($\vec{v}=0$), the δ-function in Eq. (3) renders the two forms
identical. However, for a finite velocity of the emitter, the electric-
dipole coupling element differs from its minimal-coupling counterpart by
the factor

$$\chi = \omega_{\vec{q}}/\omega_{eg} = 1/(1-\hat{q}\cdot\vec{v}/c) \tag{5}$$

$\hat{q}$ being a unit vector in the direction of emission.

The equivalence of the multipolar and the minimal-coupling forms of

$V_{\vec{q}\lambda}$ is restored on adding to the electric-dipole term a magnetic-dipole term, known as the Röntgen magnetization. The addition of this term is required by the canonical transformation between the two forms in the case of center-of-mass motion[5,6]. To lowest order accuracy, this term is given by

$$\vec{M}_R(\vec{r}) = \frac{1}{2}\,[\vec{P}(\vec{r})\times\dot{\vec{R}} - \dot{\vec{R}}\times\vec{P}(\vec{r})]/c \tag{6}$$

where $\dot{\vec{R}}=\vec{v}$ in our case. Since $\vec{B}_{\vec{q}\lambda}=\hat{q}\times\vec{E}_{\vec{q}\lambda}$, the addition of this magnetic correction to the $\vec{q}\lambda$-mode interaction in Eq. (2) yields

$$V_{\vec{q}\lambda} = -[1-(\hat{q}\cdot\vec{v}/c)](\sum_i\vec{\mu}_i<1|E_{\vec{q}\lambda}|0>)-\sum_i(\vec{\mu}_i\cdot\hat{q})(<1|\vec{E}_{\vec{q}\lambda}|>\vec{v}/c). \tag{7}$$

The first term in Eq. (7) contains the factor χ^{-1} (see (5)), i.e. the inverse of the difference factor in (4), and is therefore <u>independent</u> of whether the minimal-coupling or the multipolar Hamiltonian is used. The second term represents a velocity effect on emission with $\hat{q}||\vec{\mu}_i$. This effect is rather curious since such emission can be <u>perpendicular</u> to $\vec{v}$, i.e. its frequency can be Doppler-free. This term can also be obtained directly from the minimal-coupling Hamiltonian, if $\vec{A}(\vec{r}_i)$ is expanded to first order in $\vec{r}_i-\vec{R}$.

The Röntgen-magnetization correction $\hat{q}\cdot\vec{v}/c$ can become nearly unity as v attains relativistic values. For spin-½ systems (such as a channeled electron) it then becomes necessary to replace the nonrelativistic inter-action (1) by the Dirac-Hamiltonian interaction

$$H_{int.}^{rel.} = -\sum_i e_i\,\vec{\alpha}_i\cdot\vec{A}(\vec{r}_i) \quad ; \quad \vec{\alpha}_i = \begin{pmatrix} 0 & \vec{\sigma}_i \\ \vec{\sigma}_i & 0 \end{pmatrix} \tag{8}$$

expressed in terms of the Pauli matrices $\vec{\sigma}_i$.

Considering the Golden-rule energy conservation and tracing over spin variables, one gets the following correction factor to (7):

$$V_{\vec{q}\lambda}^{rel.} = 2\gamma/(1+\gamma)V_{\vec{q}\lambda} \quad ; \quad \gamma=(1-v^2/c^2)^{-\frac{1}{2}} . \tag{9}$$

The motional effects described above strongly influence the radiation from fast charged particles (usually electrons or positrons) channeled in crystals. Channeling[1,2] occurs when the particles are incident on a crystal nearly along a crystalline axis or plane. Then their transverse motion, i.e. the motion perpendicular to this axis or plane, is sufficiently slow to be confined by the crystal potential to a region smaller than a lattice unit cell (a channel) within which it undulates. In contrast, the longitudinal motion (i.e. the motion along the axis or plane) is nearly free (unaffected by the crystal potential) with a velocity $v_z \simeq v$. Consistently with this picture, the main type of radiation emitted by a channeled parti-cle (channeling radiation) is describable as resulting from the transverse undulation with a dipole moment $\vec{\mu}$ about an "average" longitudinal path[7]. Most experiments in this field are performed with highly relativistic particles ($10^3 \gtrsim \gamma \gtrsim 10^2$). Then channeling radiation is strongly concentrated about the forward direction $\hat{z}$ and consists of spectral peaks whose frequen-cies (usually in the x- or γ-range) are the dipole transition frequencies ω_{eg} multiplied by the huge Doppler shift $(1-\hat{q}\cdot\vec{v}/c)^{-1}\sim 2\gamma^2$.[2]

This system in which the longitudinal and transverse motions of the emitter are separated to a good approximation, provides a convenient example for the consideration of the motional effects discussed above. As shown by Healy[4], the canonical transformation from the minimal-coupling to the multipolar Hamiltonian has the same form for <u>any</u> convenient reference-point $\vec{R}$ (not necessarily the center-of-mass) relative to which the polari-zations are defined. This arbitrariness, which amounts to a gauge transformation (compatible with the Coulomb gauge), allows to identify $\vec{R}$ with the longitudinal coordinate z for channeled particles. The relativistic coupling element (9) is thus appropriate for channeling radiation. Its scaling with γ must allow for the γ-dependence of the dipole transition frequencies ω_{eg} derivable from a transverse force-constant, due to the relativistic mass upshift of the particle. Thus $\omega_{eq} \propto \gamma^{-\alpha}$, where $\alpha \approx \frac{1}{2}$ for positrons channeled along planes which oscillate nearly harmonically, whereas for electrons channeled along planes which oscillate in an anharmonic potential $\alpha \approx 1/3$ [1]. We therefore get finally for channeling radiation (with a polarization perpendicular to z):

$$V_{\vec{q}\lambda}^{rel.} = \frac{2\gamma^{1-\alpha}}{1+\gamma} \chi^{-1} V_{\vec{q}\lambda}^{el.dip.} \equiv \eta \, V_{\vec{q}\lambda}^{el.dip} \tag{10}$$

For $\gamma \gg 1$ and nearly-forward emission, $\eta \sim \gamma^{-(2+\alpha)}$.

As a potentially appealing application of these ideas, we proceed to consider the possibilities for lasing action in a field mode on-resonance with channeling radiation, propagating alongside a beam of channeled electrons. Due to the anharmonicity of the channeling potential for electrons, the ω_{eg} of different transitions are well-separated and the two-level model for emission applies. Define a cooperative Bloch-vector density for this system

$$\underset{\sim}{R}(\vec{r}) = (R_1, R_2, R_3) = \sum_j \varrho j \delta(\vec{r} - \vec{R}_j) \tag{11}$$

where ϱ is the j-electron Bloch-vector operator (represented by Pauli matrices). The corresponding polarization density can be written as $\vec{P}^\pm = \frac{1}{2}\vec{\mu}(R_1 \mp iR_2)$. The Heisenberg ("generalized Bloch") equations for the system[8] obtained on using (10) are (omitting $\vec{q}\lambda$-subscripts for brevity)

$$(\frac{\partial}{\partial t} + v\frac{\partial}{\partial z})\vec{P}^\pm = [\mp i(\omega_{eq} + \vec{v}\cdot\vec{q}) - 1/T_2]\vec{P}^\pm \pm \frac{i}{\hbar}\eta\mu^2\vec{E}^+(R_3 + \frac{1}{2}n_0) ;$$

$$(\frac{\partial}{\partial t} + v\frac{\partial}{\partial z})R_3 = \sum_\pm (\pm i/\hbar)\eta\vec{P}^\pm \cdot \vec{E}^\pm \tag{12}$$

where n_0 is the mean number density of the emitting electrons, T_2 is the relaxation time of the polarization and η (defined in (10)) incorporates all motional corrections. The corresponding Maxwell equation for the electric-field operator $E^\pm$ of the mode $(\vec{q}||z,\lambda)$, driven by $\vec{P}$ and $\vec{M}_R$ (see Eq. (6)), can be reduced to the form

$$(\frac{\partial}{\partial t} + \frac{1}{c}\frac{\partial}{\partial z} + \frac{1}{2}\delta)\varepsilon^\pm = \mp i2\pi q(1-v/c)\wp^+ \tag{13}$$

by introducing the envelopes $\varepsilon, \wp$:

$$E^\pm = \varepsilon^\pm e^{\pm i(qz - \omega_q t)} ; \quad P^\pm = \wp^\pm e^{\pm i(qz - \omega_q t)} \tag{14}$$

Here δ is a term describing the propagation losses of the field.

In order to study amplification (lasing action), it is appropriate to adopt a semiclassical description of the field and assume as boundary conditions a forward-travelling wave with a given initial value $\varepsilon(0,0)$ of the field amplitude[9]. Removing the arbitrary initial phase of the field, it is possible to eliminate the Bloch vector from the combined equations (12) (13). Under steady-state conditions, as ε becomes t-independent, one can introduce a reduced amplitude

$$b(z) = e^{\frac{1}{2}\delta z}\varepsilon(z)/\varepsilon(0) \tag{15}$$

which obeys the ordinary differential equation

$$b'' = -\frac{1}{vT_2}b' + K^2 b \tag{16}$$

with

$$K^2 = (1 - \frac{v}{c})\eta(v_{q\lambda}^{el.dip.})^2 \frac{1}{vc} <\rho_z> \qquad (17)$$

where $<\rho_z>$ is the steady-state population inversion between the excited and ground state of the dipole transition.

The travelling-wave solution can be written

$$b(z) = e^{\lambda z} . \qquad (18)$$

Here is one of the two eigenvalues

$$\lambda_{1,2} = -\frac{1}{2vT_2} \pm \frac{1}{2} \sqrt{(\frac{1}{vT_2})^2 + 4K^2} \qquad (19)$$

of which only the λ_1 (positive) eigenvalue produces gain if it exceeds the loss ($\lambda_1 > \frac{1}{2}\delta$).

Two extreme cases are worth considering to demonstrate the implications of Eq. (19). Case (A) - strong dephasing, namely

$$\frac{1}{vT_2} \gg 2K \quad ; \quad \lambda_1 \simeq K^2 vT_2 , \qquad (20)$$

implying that the gain scales as $\gamma^{-(2+\alpha)}$ for $\gamma \gg 1$. This result coincides with a previous perturbation-theory result[10]. Case (B) - weak dephasing

$$\frac{1}{vT_2} \ll 2K \quad ; \quad \lambda_1 \simeq K \qquad (21)$$

Now the gain scales as $\gamma^{-(1+\alpha/2)}$ for $\gamma \gg 1$, and therefore decreases much more slowly with γ. This result is thus more promising than the previous one as regards the prospects for channeling-radiation lasing at short wavelengths. The derivation of this result demonstrates the need to allow properly for propagation effects (such as the second-derivative term in Eq. (16) which is absent in a perturbative treatment[10]) as well as for motional corrections in the field-dipole coupling.

References

1. J. U. Andersen and E. Bonderup, in "Annual Review of Nuclear Particle Science", J. B. Jackson, H. E. Gove and R. Y. Schwitters, eds., Annual Reviews, Palo Alto (1983).
2. V. V. Beloshitsky and F. F. Komarov, Phys. Rep. 93:117 (1982).
3. E. A. Power "Introductory Quantum Electrodynamics", Longmans, London (1964).
4. W. P. Healy "Nonrelativistic Quantum Electrodynamics", Academic Press, London and New York (1982).
5. B. U. Felderhof and D. Adu-Gyamfi, Physica 71:399 (1974).
6. W. P. Healy, J. Phys.A 10:279 (1977).
7. The limitations of this picture are discussed by G. Kurizki and J.K. McIver, in Phys. Rev. B 32:4358 (1985), where a unified quantal treatment of all types of radiation from fast particles in crystals is presented.
8. M. S. Feld and J. C. MacGillivray, in "Coherent Nonlinear Optics", M. S. Feld and V. S. Letokhov, eds.,Springer, Berlin (1981).
9. H. Haken "Laser Theory", Springer, Berlin (1983).
10. V. V. Beloshitskii and M. A. Kumakhov, Sov. Phys. JETP 47:652 (1978)

GENERATION AND IMPORTANCE OF LINKED AND IRREDUCIBLE MOMENT

DIAGRAMS IN THE RECURSIVE RESIDUE GENERATION METHOD*

Israel Schek and Robert E. Wyatt

Department of Chemistry
Institute for Theoretical Chemistry
The University of Texas
Austin, Texas 78712-1167 U.S.A.

ABSTRACT

Molecular multiphoton processes are treated in the Recursive Residue Generation Method [A. Nauts and R. E. Wyatt, Phys. Rev. Lett **51**, 2238 (1983)] by converting the molecular-field Hamiltonian matrix into tridiagonal form, using the Lanczos equations. In this study, the self-energies (diagonal) and linking (off-diagaonal) terms in the tridiagonal matrix are obtained by comparing linked moment diagrams in both representations. The dynamics of the source state is introduced and computed in terms of the linked and the irreducible moments.

I. INTRODUCTION

The theory of molecular multiphoton excitation (MPE) induced by an intense laser field, including the role played by the interaction of an active mode with optically inactive (background) modes, has been thoroughly treated in the past few years (1-3). In a strong laser field, myriads of molecular levels can be excited and must be taken into consideration in theoretical analysis of the molecular dynamics. The need for inclusion of high energy discrete levels as well as congested regions (quasicontinuum) in the theoretical-computational models, led to the development of a new approach, the Recursive Residue Generation Method (RRGM) (4,5), in which thousands of dressed molecule-field states can be considered. The RRGM recursively converges a very large Hamiltonian matrix into a tridiagonal matrix, employing the Lanczos algorithm (6,7), and then calculates time-dependent transition amplitudes between any two dressed states. This is done by recursive generation of poles and residues of the associated Green operator, thus avoiding the calculation and storage of a myriad of eigenvectors of the huge Hamiltonian matrix. This approach was mainly inspired by studies of the spectra of disordered solids by Haydock et. al. (8-10), who computed the local density of electronic or vibrational states about a particular site. In Sec. II the method of calculation is briefly represented and in Sec. III the numerical results are shown and discussed.

II. METHOD

In order to tridiagonalize the Hamiltonian matrix, a source state (a linear combination of the initial and final states) is defined, and additional recursion states are born one-by-one out of the previous two recursion states, using the Lanczos procedure. The diagonal (a_k) and off-diagonal (β_k) elements of the tridiagonal matrix are interpreted (2,11) as masses and linking terms, respectively, of quasiparticle members of a linear mechanical chain. In our previous work (2,11-13), a diagrammatic construction of the moments of the Hamiltonian in both the zero-order and tridiagonal representations was developed, where both radiative and intramolecular coupling terms were included. This permits the description of energy transfer between a laser active mode and a manifold of optically inert modes. The competition bertween the two channels was emphasized by treating several limiting cases. However, power moments, even for relatively low order, are numerically unstable due to propagation of roundoff errors, and therefore do not supply a credible tool to calculate transition amplitudes. More concise forms of linked moments (12) (where the dynamical system is refrained from returning to the source state in intermediate steps, before completing the full number of interactions needed for the calculation) and irreducible (13) (where the system is refrained also from higher order recursive states in intermediate steps) moments were described, which minimize redundant information, already included in previous lower moments, and by which the tridiagonal elements can be expressed in terms of the matrix elements of the zero-order Hamiltonian. Equivalently, the Green's function for the surviving probability can be represented by the handy continued fraction (M is the number of recursion steps)

$$G(E)_{00} = 1/E - a_0 - \beta_1^2/E - a_1 - \beta_2^2/\ldots - \beta_M^2/E - a_M \tag{1}$$

which were similarly obtained by Haydock et. al. (8-10). The method of subtracting the lowest paths of the full moments, in order to remove redundant information, is shown in Fig. 1 for the third order moment, describing radiative coupling among states of an anharmonic oscillator. Slant (horizontal) links describe radiative interactions (zero-order energies). Consequently, the number of products and additions needed to calculate the relevant contribution to the tridiagonal chain elements decreases enormously as one proceeds to higher order moments.

III. RESULTS AND DISCUSSION

The model system contains a radiatively active mode, harmonically coupled to an inert mode. This intermode coupling is described by the operator $V_{int} = c(ab^\dagger + a^\dagger b)$, where a and b are the destruction operators of the active and inert modes, respectively, and c is the coupling parameter. Other physically pertinent parameters are the field amplitude ε, the molecular dipole moment d, the laser-molecule detuning parameter Δ, and the mutual mode-mode frequency shift s. The tridiagonal matrix elements are calculated and the eigenvalues and elements in the first row of the eigenvector matrix are then determined by a modification of the QL algorithm that was introduced by Wyatt and Scott (14). The survival probability of the ground molecular dressed state is then determined:

$$P_{00}(t) = |<0|\exp(-iHt)|0>|^2 = |\sum_{j=1}^{M} R_j \exp(-iE_j t)|^2 \tag{2}$$

where $\{E_j\}$ are the eigenvalues, $\{R_j\}$ the residues, and M is again the number of recursion steps. The time parameter t is expressed in natural units of (energy)$^{-1}$.

348

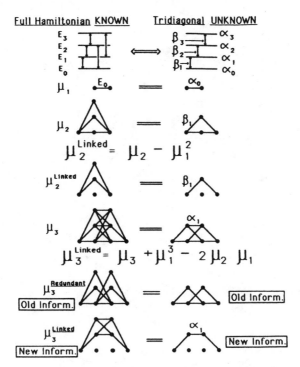

Fig. 1. Equivalence of moments in the zero-order and tridiagonal representation for an anharmonic oscillator excited by a laser field.

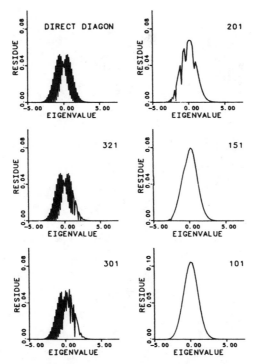

Fig. 2. Convergence of residues of the model system with 301 zero-order states, as compared to direct diagonalization. The number of recursive steps: $101 \leqslant M \leqslant 321$.

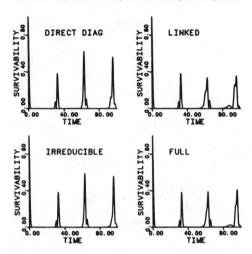

Fig. 3. The ground state survival probability calculated for the model system for various degrees of reduction of redundant information, compared to a direct diagonalization approach.

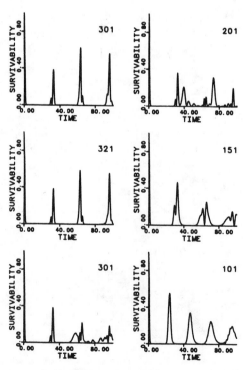

Fig. 4. Convergence of the ground state survival probability for the model system for various number of recursion steps, as in Fig. 2.

350

In Fig. 2, we show the convergence of residues associated with each eigenvalue for a system with N = 301 zero-order states, for various numbers of recursion steps (M) as compared to a direct diagonalization of the Hamiltonian. For the present analysis we choose the strong laser case $(d\varepsilon/\Delta):(c/s) = 10$, so that the high lying zero-order states are effectively populated. That is, the radiative ('vertical') channel is favored over the intermode ('horizontal') channel. From this case, it is clear that only if $M \geq N$ are the fine details of the spectrum reproduced, whereas for $M \ll N$ only the overall envelope is being sketched.

In Fig. 3, we compare time development of the ground state survivability obtained for the hierarchy of reduction of irrelevant information for the present system (the number of recursion steps M exceeds the number N of zero-order states, in order to extend the spectrum of non-ghosts eigenvalues (5,13)). These results are compared with our reference obtained from direct diagonalization. Obviously, the irreducible (and partly the linked) moment calculations do reproduce the results obtained from the reference calculations.

Finally, in Fig. 4 convergence of the time evolution of the ground state survivability is shown with respect to the number of recursion steps for the calculation employing irreducible moments. There is no question that for faithful reproduction of the fine features one needs to include many recursion steps if the strong coupling regime prevails, where high zero-order states are effectively populated, and consequently do contribute to the dynamics. At least for long times (t > 40) one needs $M \approx N$. However, for a weak laser field, just a few recursion steps are required.

It has been shown in this study how the elimination of irrelevant paths from moment diagrams contributes to an enhanced understanding of the pertinent terms (first the chain parameters, then the eigenvalues and residues) which determine the dynamics. Due to conservation of the orthogonality of the recursive states with respect to the source state, assured by the present diagrammatic moment procedure, satisfactory convergence of the numerical results is shown and the contribution of numerical round-off errors is suppressed.

REFERENCES

1. See the following, and references therein: Schultz, P. A., Sudbφ, Aa. S., Grant, E. R., Shen, Y. R., and Lee, Y. T., 1980, J. Chem. Phys., 72:4985 .
2. Schek, I., and Wyatt, R. E., 1985, J. Chem. Phys., 83: (6) 3028.
3. Schek, I., and Jortner, J., 1986, Theoretica Chemica Acta, in press.
4. Nauts, A., and Wyatt, R. E., 1983, Phys. Rev. Lett., 51:2238.
5. Nauts, A., and Wyatt, R. E., 1984, Phys. Rev. A, 30:872 .
6. Lanczos, C., J., 1950, Res. Nat. Bur. Stand., 45:255 .
7. Paige, C. C., 1972, J. Ins. Math. Appl., 10:373 .
8. Haydock, R., 1980, Comp. Phys. Commun., 20:11.
9. Haydock, R., 1980, in "Solid State Physics," H. Ehrenreich, F. Seitz, and D. Turnbull, eds., Academic, New York.
10. Haydock, R., 1981, in "Excitations in Disordered Systems," M. F. Thorpe, ed., Plenum, New York.
11. Schek, I., and Wyatt, R. E., 1985, J. Chem. Phys., 83:(9)4650.
12. Schek, I., and Wyatt, R. E., J. Chem. Phys., in press.
13. Schek, I., and Wyatt, R. E., submitted for publication.
14. Wyatt, R. E., and Scott, D., 1986, in "Large Eigenvalue Problems," J. Cullum and R. A. Willoughby, eds., North-Holland, Amsterdam.

*Supported in part by the National Science Foundation and the Robert A. Welch Foundation.

LASER STIMULATION AND OBSERVATION OF ELEMENTARY REACTIONS IN THE GAS PHASE

Jürgen Wolfrum

Physikalisch-Chemisches Institut der Universität Heidelberg
Im Neuenheimer Feld 253, D-6900 Heidelberg, W. Germany

INTRODUCTION

It has been known, since the first use of fire by mankind, that the rates of chemical reactions in the gas phase depend strongly on the energy of the reactants. Traditionally, the energy dependence of the chemical reaction rate is studied under conditions in which the rate of reaction is slow compared to that of collisional energy transfer. Under these conditions, the energy of the reactants is characterized by a temperature. The temperature variation of the reaction rate can then often be expressed with sufficient accuracy by the Arrhenius equation. The Arrhenius parameters obtained in this way, however, contain no direct information on how the various degrees of freedom of the reacting molecules and in the "activated complex" contribute to the potential pathways of product formation in the chemical reaction. Investigations on the chemical reactivity under a wide range of conditions such as pressure variation and specific excitation of the reactants give important insights into the microscopic dynamics of the chemical reaction. This information on one hand can be compared with the results of theoretical predictions using potential energy surfaces for chemical reactions obtained by ab initio methods, and is also of basic interest to improve the kinetic data used in detailed chemical kinetic modelling. The experimental possibilities to study elementary chemical reactions in detail have expanded quite dramatically in recent years as a result of the development of various laser sources.

Beside such microscopic details laser spectroscopic methods with their high temporal and spatial resolution are especially important for nonintrusive measurements in practical systems in which elementary chemical reactions couple stongly with various transport processes. Data gained from such experiments yield the basis for comparison with detailed mathematical modelling of situtations which require coupling between elementary reactions and transport phenomena such as chemical processes in laminar and turbulent flows including heat and species transport and the influence of walls.

In the present paper three examples are presented. The first part describes the investigations on the specific effect of translational excitation in the reaction of hydrogen atoms with oxygen molecules and the spectroscopy of stable and unstable triatomic molecules using LIF spec-

troscopy. The second part describes temperature measurements in a stationary laminar counterflow diffusion flame using CARS spectroscopy. In the
last part CO_2-laser radiation is used to stimulate instationary ignition
processes in ozone.

SELECTIVE TRANSLATIONAL EXCITATION OF REACTANTS

Laser light sources with their high quantum flux within a narrow
spectral region are able to excite a large number of chemically reactive
substances in specific degrees of freedom and after reaction to investigate product excitation state-selectively. In this way the relative
velocity of reaction partners can be selected precisely over a wide
range, molecules allowed to rotate slower or faster, and atoms in a
molecule drawn out variably and vibrations excited. With polarized lasers
the mutual orientation of the particles during the reaction can be established and even short-lived "transition states" excited specifically and
observed.

The technique of flash or atomic resonance line photolysis for the
production of hot atoms in conjunction with time-integrated product
detection methods has long been used to obtain information about relative
reaction cross sections and excitation functions as a function of the
translational energy of the reactants. In these experiments, the measured
reaction yields could only be used to calculate reaction probabilities
or cross sections, with estimations of the collisional cooling process of
the hot particles. The use of pulsed lasers to produce translationally hot
atoms or radicals by photofragmentation of small molecules in combination
with fast time-resolved product detection techniques such as laser induced
fluorescence (LIF), multiphoton ionization (MPI) or coherent anti-Stokes
Raman spectroscopy (CARS) allows product detection under approximately
single collision conditions. Detailed product state distributions and
absolute total reaction cross sections can be obtained in this way. These
microscopic data can be compared with dynamic calculations using ab initio
potential energy surfaces.

The experimental apparatus used here has been described in detail elsewhere[1,2]. Two antiparallel laser beams are directed coaxially through a
flow reactor equipped with a baffle system to reduce the scattered light
from the excimer laser photolysis pulse and from the excimer laser pumped
dye laser analysis pulse.

The observed rotational energy distributions for the reaction

$$H + O_2 \longrightarrow OH + O \tag{1}$$

are shown in Fig. 1. The surprisal analysis of the rotational distributions in OH (v"=0) with H-atoms from HBr (2.6 eV) and HI photolysis
(E=1.9 and 1.= eV) give straight lines with θ_R = +1.2, -0.7, and -1.2 at E
= 1.0, 1.9 and 2.6 eV respectively. All three slopes are quite small
compared to those of other "nonstatistical" reactions. Spin-orbit and
orbital-rotation interactions in the OH radical cause fine-structure
splittings for each rotational level. Each of these fine structure levels
can be probed by different rotational sub-bands. The two OH spin states
$^2\Pi_{1/2}$ and $^2\Pi_{3/2}$ are, within experimental error, equally populated.
However, the Λ-doublet fine structure states show a clear preference for
the lower energy Π^+ component. The experimental result shows that break-up
of the reaction complex generates forces in a plane containing the bond to
be broken. The OH radical rotates in that plane and J_{OH} is perpendicular
to it and to the broken bond. This picture is consistent only with a
preferential planar exit channel in these reactions. This could also be
directly demonstrated using polarized photolysis and analysis laser beams.

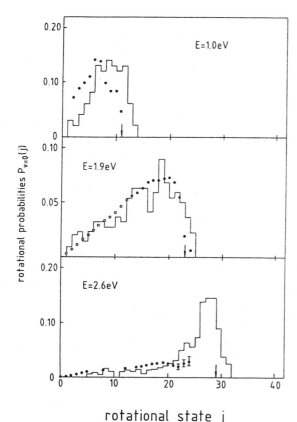

Fig. 1 Deconvoluted OH (v=0) rotational distributions. The open circles in (b) are from the surprisal analysis. This histograms are obtained from classical trajectory calculations on the Melius-Blint surface. The arrows show the maximum OH energies a vailable in reaction (1) at E.

The polarization experiments are based on measuring the distribution of orientations of the OH angular momentum vector J by using polarized dissociation and analysis lasers. The OH - radical fluorescence intensity is observed with the electric vectors of both lasers E_D and E_E parallel and perpendicular to each other. The cases (a) $E_E \parallel E_D \parallel$ Z and (b) $E_E \perp E_D$ $\parallel$ Z, with the lasers propagating along X and the phototube along Y, are used. The polarization ratio R is defined as the ratio of the fluorescence intensity in case (a) to that in case (b). R is measured as function of J for Q branch transitions, because here μ_{OH} lies along J_{OH} and larger polarization effects are expected. For P or R transitions, μ_{OH} is perpendicular to J_{OH} and rotates in the OH plane of rotation. The experimental apparatus is shown in Fig. 2. Here, both lasers were linearly polarized (around 95 % polarization) by using 10 Brewster quartz plates respectively (rack-polarizer). Both light beams are then directed through λ/2 plates so that the electric vectors of both lasers can be adjusted to any desired angle independently.

Dissociation of HBr at 193 nm to H + Br ($^2P_{3/2}$) is induced by a perpendicular transition, so that the H atom flight direction is aligned with a $\sin^2\theta$ distribution along E_D, i.e. preferentially $v_H \perp E_D$. Interesting information on H-O_2 reaction dynamics can be obtained by investigating possible memory effects of OH product alignment on reactant alignment. Fig. 3. shows the variation of the OH-$Q_1$16 (v"=0) fluorescence intensity with polarization of the dissociation laser E_D relative to analysis laser E_E. We assume the observed preference $J_{OH} \parallel E_D \perp v_H$ to be due to restrictions in the possible reaction geometries at high collision

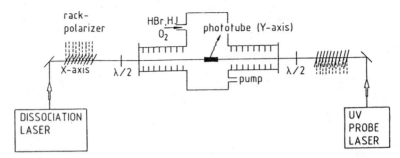

Fig. 2. Experimental arrangement for the polarization experiment. Both lasers are linearly polarized and their vectors can be adjusted to any desired angle.

energies. Λ-doublet measurements and trajectory calculations suggest that the $H + O_2$ reaction occurs essentially in a plane at high collision energies.[2] From that we expect $J_{OH} \perp v_H$ for randomly oriented O_2 molecules[5]. The transition moment μ_E of Q-lines is perpendicular to the OH rotation plane ($\parallel J_{OH}$) for high OH rotational states. Thus we obtain the maximum OH excitation probability $|E_E \cdot \mu_E|^2$ for $\mu_Q \parallel E_E \parallel J_{OH} \perp v_H$ resulting in higher fluorescence intensity for $E_B \parallel E_D$ than for $E_E \perp E_D$. The model is quantitatively described elsewhere[2].

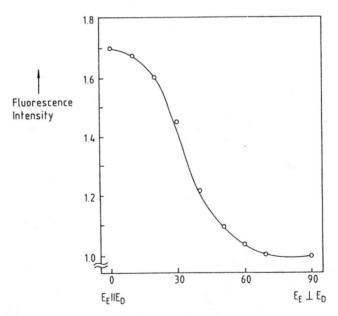

Fig. 3 Variation of the OH-Q16 (v''=0) fluorescence intensity with polarization of the dissociation lasr E_D relative to analysis laser E_E.

SPECTROSCOPY OF DISSOCIATING MOLECULES

Reaction (1) is knwon to take place adiabatically on the ground state potential surface of the HO_2 ($^2A''$)-radical. Trajectory calculations[3]-[4] (s.Fig. 1) on an ab initio surface (Melius-Blint[4]) are in agreement with calculated OH rotational distributions from the phase space theory[5] for low relative translational energy. With increasing relative translational energy the OH rotational distribution becomes considerably hotter than the statistical one, and no long living HO_2-complex exists during the reaction. An total reaction cross section of 0.42 ± 0.2 [Å]^2 at E = 2.6 eV is obtained for reaction (1) in the experiments described here. The theoretical cross section obtained under this condition by quasi classical trajectory calculations[3,6] on the Melius-Blint surface is 0.38 [Å]^2 . These numbers cannot be compared directly, because the multiplicity of the $^2A"$-surface at infinite $H-O_2$ seperation is not taken into account. Miller[6] uses a multiplicity factor of 1/3. This would yield a theoretical cross section of 0.13 [Å]^2 which is significantly outside the experimental range. The observed discrepancies could be attributed to a reduction of calculated reaction cross sections due to a 'rigid' character and a barrier of 8 kJ/ mol in the Melius-Blint surface for dissociation of the HO_2 in reaction (-1a)[7]. Calculations by Dunning et al.[8] reduce this barrier to less than 1.7 kJ/mol. Also for reaction (-1) the Melius-Blint surface apparently overestimates the long range O-OH attraction[9] while the Quack-Troe interpolation scheme[10] leads to better agreement with the experimental values at low temperatures. However, more work should be done on the potential energy surface used for this system. Direct information on the potential energy surface of the HO_2 intermediate can be obtained experimentally by using the laser induced fluorescence signal[11] generated by transitions to higher repulsive potential energy surfaces. As depicted in Fig. 4, photoemission occurs when the HO_2-wave packet, in the course

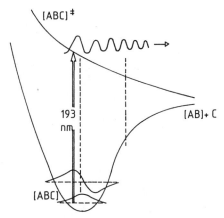

Fig. 4 LIF (Resonance-Raman)-Spectroscopy of HO_2 radials.

of its movement on the upper repulsive potential surface, favorably overlaps the HO_2 vibrational levels in the lower (bound) electronic state. Here, the emission to higher vibrational overtones and therefore at longer wavelengths corresponds with later times in the course of photodissociation. Fig. 5 shows the experimental arrangement for such measurements, which can give data for the vibrational frequencies of the HO_2 electronic ground state up to the dissociation limit. HO_2 radicals are generated by H – atom addition to O_2 at high pressures using ArF–laser photolysis of NH_3 as hydrogen atom source. A part of the laser pulse for photolysis is delayed and the frequency shift of the fluorescence lines relative to the excitation wavelength is recorded with a gated multiplier and boxcar system. HO_2 concentrations measured by UV-absorption spectroscopy.

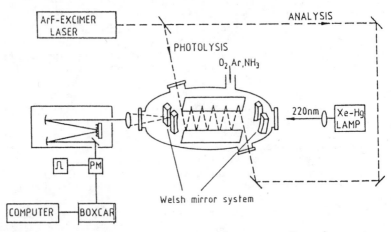

Fig. 5 Experimental arrangement for LIF (Resonance-Raman) spectroscopy of HO_2.

First experimental results obtained with this arrangement for the stable triatomic molecule H_2S are depicted in Fig. 6 show the long progression to high vibrational states of the electronic ground state.

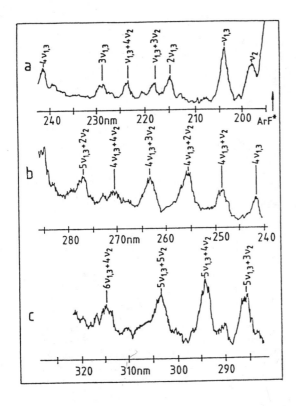

Fig. 6 LIF-(Resonance-Raman)-Spectrum of H_2S [11].

CARS SPECTROSCOPY IN LAMINAR COUNTERFLOW DIFFUSION FLAMES

Diffusion flame fronts are basic elements of turbulent (technical) diffusion flames. Therefore, exact information on the detailed structure of diffusion flame fronts is essential for the understanding of such flames using flamelet models, enabling the inclusion of full detailed chemistry for the pediction e.g. of the formation of pollutants like soot and nitrogen oxides and of realistic heat release in technical combustion processes.

Figure 7 shows the schematic drawing of the burner. An airflow, regulated by Tylan flow controllers, is blown against the cylindrical burner head. The fuel gas methane emanates through the lower half-shell consisting of water cooled porous sintered brass cylinder. A flame of 1 mm thickness burns around the cylinder with high temporal and spatial stability. Two Schott glass filter windows in the stainless steel housing allow the optical access to the flame. The housing is fixed on a mounting system, so that the flame can be moved with a precision of 0.01 mm relative to the laser beams.

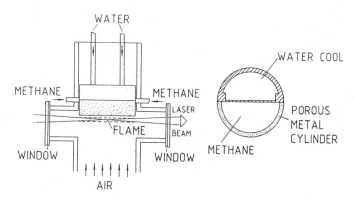

Fig. 7   Cross section of the laminar counterflow diffusion burner

The CARS apparatus is depicted in Fig. 8. The pump beam is produced by a frequency-doubled Nd:YAG laser (Quanta Ray Model DCR 1A; 10 ns pulse length). The linewidth amounts to 0.3 cm^{-1} at 532 nm. Because of the unstable resonator of the cavity, the laser is running in a doughnut mode. A part of the laser intensity is used to pump a dyelaser (Quanta Ray POL-1) operating at the Stokes shifted wavelengths. The polarization of both laser- beams can be adjusted with the help of polarizers and λ_2-plates. So the polarization technique is used to suppress the nonresonant CARS background.To improve the spatial resolution of the CARS apparatus the pump and the Stokes beams are adjusted into so called "USED CARS" technique. By using this phase matching technique the spatial resolution is about 4 mm on the beam axis and 50 - 100 μm perpendicular to the axis. The anti-Stokes beams from a reference-cell filled with a nonresonant gas and the burner are adjusted in a 1m Monochromator (Mc-Pherson) and detected by two photomultipliers. Both signals are sampled by boxcar integrators and digitalized. In a computer the resonant signal is rationed by the nonresonant CARS signal to get an improved S/N-ratio and for normalization.

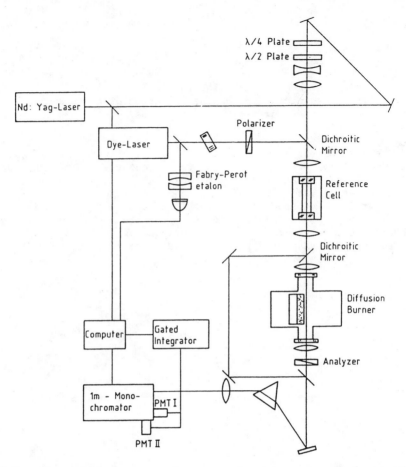

Fig. 8 Experimental arrangement for measuring temperature and concentration profiles in a counterflow diffusion flame with USED-CARS.

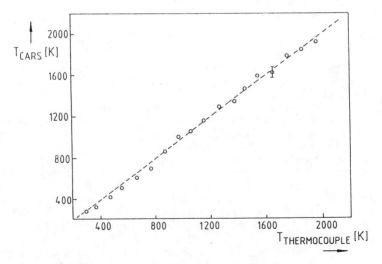

Fig. 9 Comparision of the CARS and thermocouple readings an the same location in the furnace. The error bar gives the temperature uncertainly in the CARS measurements.

To determine the accuracy of the temperature and concentration
measurements N_2-CARS spectra up to 2000 K were recorded in a furnace. The
CARS intensities are divided by the nonresonant reference signals for
normalization purposes and improving the S/N ratio. A least-squares
fitting routine, developed and kindly supplied by W. Kreutner[12] compares
the measured data points with computer spectral shapes. Fig. 9 displays
the temperature T_{CARS} as a function of the Pt-Rh/Pt thermocouple readings
in the furnace. The discrepancy amounts to less than 40 K over the whole
covered temperature range (300 - 2000 K). The dependence of the peak
intensity of the CARS spectra must be known, to get concentrations from
peak height analysis. To reduce measurement errors caused by intensity
fluctuations of the CARS signals at the maxima of the spectra, they were
convoluted with a triangular slit function of 2.5 cm^{-1} FWHM before the
peak-value was determined. The dependence of the peak intensity from the
concentration was assumed to be quadratic. This was checked at room
temperature and 1200 K.

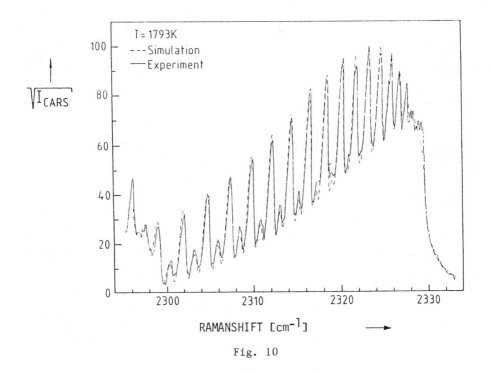

Fig. 10

In Fig. 10 an experimental rotation-vibrational CARS-spectrum
(Q-branch of N_2) from the burner is shown as a solid line curve. The best
fit for T=1793 K is plotted as a dashed line. To get a flame profile the
height of the burner relative to the laser beam axis was varied in steps
of 0.25 mm. Each CARS spectrum takes 10 - 15 min. In Fig. 11 the measured
temperature points are presented as circles. The theroretical curve is
shown as dashed line.

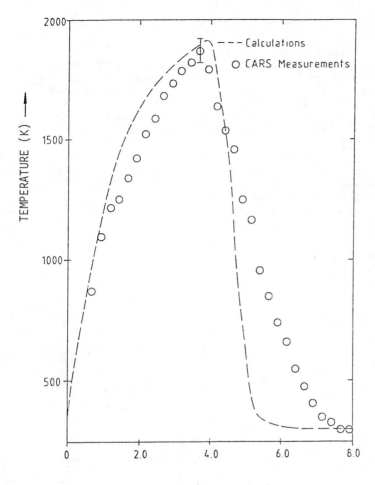

Fig. 11 Temperature profile through the flame front of the counterflow diffusion burner with (ϕ_{air} = 80 1/min and ϕ_{CH4} = 2,6 1/min). Circles are the measured temperature points deduced from CARS signatures. The dashed line gives the temperature profile as computed from a diffusion flame model.

DISTANCE FROM BURNER-HEAD (mm) ⟶

The error of the CARS temperature measurements is expected to be largest at the peak of the temperature profile, where movements of the flame front relative to the laser beam focus during the measurement period (10 - 15 min) can bias the CARS temperature to lower values since signals from cooler regions are weighted stronger in comparison to those from hotter gas regions. Also flame radiation corrections not included in the calculations might partially explain the lower peak temperature obtained in the CARS measurements. Larger discrepancies observed between reported thermocouple measurements and the calculations[13] are attributed to the omission of radiation corrections in the thermocouple measurements[14]. While similar deviations to lower temperatures from the calculated profile are observed in thermocouple and CARS measurements on the fuel side of the temperature distribution significant higher temperature are measured by CARS at the oxygen side of the diffusion flame. In the present experiments slow oxidizer flow velocities are applied in order to simulate conditions important in "flamelets" of technical turbulent diffusion flames. Under such conditions diffusion effects in the boundary layer between oxidizer and fuel gases may not be neglected. Since boundary layer approximations are used in the calculations[13] this can be lead to serious distortions between measurements and calculations. The extended temperature profile towards larger distances from the burner head will also cause deviations in the calculated and measured N_2 profiles. Further calculations should therefore replace the use of boundary layer apapproximations by true two-dimensional representations of the flow field.

CO_2 LASER INDUCED IGNITION OF O_3–O_2 GAS MIXTURES

The detailed description of the ignition and deflagration of homogeneous gas mixtures contained in closed vessels requires the solution of
the conservation equations for mass, momentum, energy, and the mass
fractions of each of the chemical species. The solution of this problem is
done by the method of lines approach, because of its complexity in one
spatial dimension and for simple gas mixtures. Of importance is the
specification of the source term that is done well in the case of ignition
of sensitized gas mixtures by irradiation with suitable laser light. A
simple test system is the ignition of O_2/O_3 mixtures by irradiation of CO_2
laser pulses along the axis of a cylindrical vessel.

The experimental set up is shown in Fig. 12. Attenuated light pulses
of a TEA-CO_2 laser (9P20 line) are focussed down to a beam waist with
fairly constant diameter (FWHM) along the length of the cell and op-
tionally reflected back into the cell. A part of the laser radiation is
coupled out and measured by two energy detectors in order to determine the
absorbed energy. The course of the deflagration is followed by means of uv
absorption of ozone on different light paths through the cell parallel and
perpendicular to the axis.

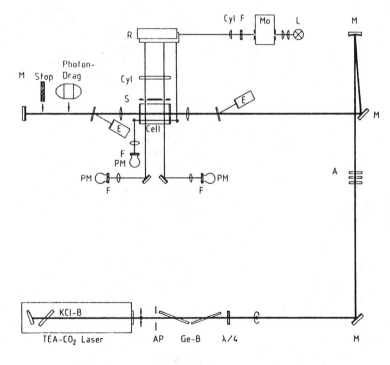

Fig. 12 Experimen-
tal setup; B :
Brewester windows,
AP: circular
aperature, λ/4:
(reflective)
quaterwave plate,
M: mirror, A:
absorbers, E:
pyroelectric
energy detector,
L: HgXe lamp. Mo:
monochromator, F:
uv transmitting
filter, R: rotates
slit image to
horizontal plane,
Cyl: cylindrical
lens, S: slits,
PM: photomulti-
plier

The ozon uv absorption coefficient strongly depends on temperature
except in a region at about 280 nm. Therefore, at 280 nm variations of the
absorbance are given by the variing O_3 concentration and the onset of the
reaction should be observed. Figure 13 gives an example of the absorbance
at a short time scale, at   312.6 nm where the O_3 absorption increases s-
trongly with rising temperature. During the laser pulse the transmitted
intensity reaches its minimum depending on the order of non-equilibrium
excitation and then increases back reflecting the relaxation of the O_3
molecules to thermal equilibrium. Provided that from measurements at

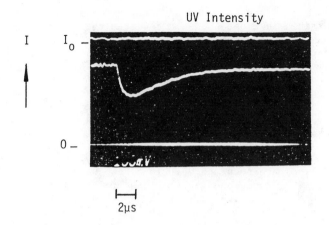

UV Intensity

I

$I_0 -$

$0 -$

⊢—⊣
2μs

Fig 13: Time dependence of incoming laser intensity and of uv intensity (= 312.6 nm) transmitted at the diameter l= 5.2 cm for a O_2/O_3 mixture (p_{tot} = 0.40.bar, 21% O_3 by a laser pulse of F_{in} = 1.50 J/cm^2 (E_{abs}=0.114 J/cm^3)

280 nm the onset of the reaction may be excluded at this time scale the local temperature reached by absorption of the laser light and the duration of the energy source, i.e. the relaxation process may be observed, both being important parameters for the numerical simulation. First experiments dealing with the laser ignition of O_2/O_3 mixtures were performed investigating the radial moving of the deflagration and the ignition limit of various mixtures [15]. The results for the flame propagation indicated a nearly radial moving for higher incoming fluences. The investigation of the ignition limit was done by slowly rising the incoming fluence for a given O_2/O_3 mixture and inspecting the gas mixture whether reaction occured or not.[3] The experimental values are averaged over the whole irradiated gas volume and should be taken as a lower limit since the deflagration just above the ignition limit starts near the entrance window of the cell in the region of highest absorbed energy density.

Mathematical simulation of the ignition process is done by solving the corresponding system of conservation equations. For one-dimensional geometries (infinite slab, infinite cylinder or sphere), after transformation into Lagrangian coordinates and using the uniform pressure assumption. Spatial discretization using finite differences leads to a system of ordinary differential and algebraic equations which can be solved numerically. Due to the large ratio of vessel diameter to flame front thickness and to diameter of the artificial energy source adaptive gridding has to be used in case of ignition by artificial energy sources. Determination of the grid point density is done by equipartitioning the integral of a mesh function and inverse interpolation, the mesh function given by a weighted norm of temperature gradient and curvature. Piecewise monotone cubic Hermite interpolation [16] is used for the static regridding. The resulting systems of ordinary differential/algebraic equations can be solved using the computer codes DASSL [17] or LIMEX [18]. These solvers lead to the same results and need comparable calculation times.

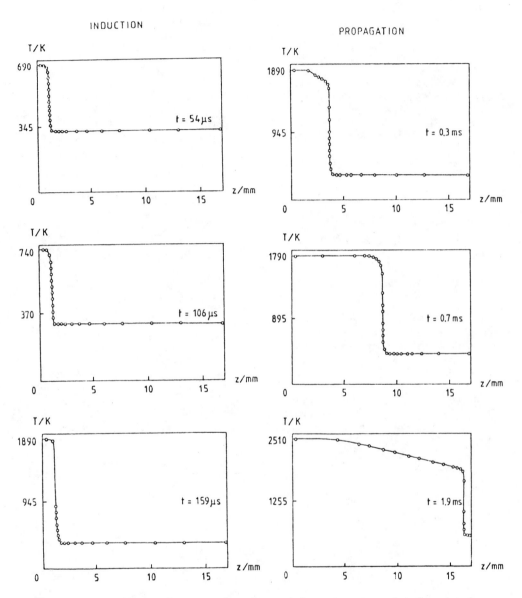

INDUCTION

PROPAGATION

Fig. 14 Modelling of temporal and spatial temperature development in infrared laser-induced ozon decomposition.

ACKNOWLEDGEMENT

The financial support by the "Deutsche Forschungsgemeinschaft", the "Stiftung Volkswagenwerk", the "Fonds der Chemischen Industrie", and (within the frame of the TECFLAM projekt) the BMFT is cordially acknowledged.

REFERENCES

1. K. Kleinermanns and J. Wolfrum, J. Chem. Phys. 80, 1446 (1984)
2. K. Kleinermanns and J. Wolfrum, Appl. Phys. B 34, 5 (1984)
3. K. Kleinermanns and R. Schinke, J. Chem. Phys. 80, 1440 (1984)
4. C. F. Melius, and R. J. Blint, Chem. Phys. Letters 64, 183 (1979)
5. P. Pechukas, J. C. Light, and C. Rankin, J. Chem. Phys. 44, 794 (1966)
6. J. A. Miller, J. Chem. Phys. 74, 5120 (1981)
7. C. Cobos, H. Hippler, and J. Troe, J. Phys. Chem. 89, 342 (1985)
8. T. H. Dunning jr., S. P. Walch, and M. M. Goodgame, J. Chem. Phys. 74, 3482 (1981)
9. S. N. Rai and D. G. Truhlar, J. Chem. Phys. 79, 6046 (1983)
10. M. Quack and J. Troe, Chem. Soc. Spec. Period. Report Gas Kinet. Energy Transfer 2, 175 (1977)
11. D. G. Imre, J. L. Kinsey, R. W. Field and D. H. Katayama, J. Phys. Chem. 86, 2564 (1982).
    K. Kleinermanns, R. Suntz (to be published).
12. Kreutner, W., Stricker, W., Just, T.: Ber. Bunsenges. Phys. Chem. 87, 1045 (1983).
13. Dixon-Lewis, G. Fukutani, G., Miller, J.A., Peters, N., Warnatz, J: Twentieth Symposium (International) on Combustion, p. 1893. The Combustion Institute, 1985.
14. Tsuji, H.: Progr. Energy Combust. Sci. 8, 93 (1982)
15. Raffel, B., Warnatz, J., Wolfrum, J: Appl. Phys. B 37, 189 (1985)
16. Fritsch, F.N., Butland, J.: SIAM J. Sci. Stat. Comput. 5, 300 (1984)
17. Petzold, L.R.: A Description of DASSL: A Differential/ Algebraic System Solver, Sandia Report SAND82-8637, Sandia National Laboratories, Livermore (1982); IMACS World Congress, Montreal 1982.
18. Hairer, E., Zugck, J., Deufelhard, P.: Publication in preparation.

UV LASER IONIZATION SPECTROSCOPY

AND ION PHOTOCHEMISTRY

M. Braun, J.Y. Fan*, W. Fuß, K.L. Kompa
G. Müller, and W.E. Schmid

Max-Planck-Institut für Quantenoptik
8046 Garching, Germany

ABSTRACT

This paper discusses two aspects of UV-laser induced photo-ionization of polyatomic molecules, taking benzene as an example. First some work is reported concerning the ionization spectroscopy in the energy range from threshold to 2000 cm^{-1} excess energy. Vibrational structure has been extensively analyzed and autoionization resonances could be tentatively assigned. In the second paragraph the use of benzene photoions in practical photochemistry is considered. It is shown that this concept indeed opens up new ways for chemical synthesis.

I. INTRODUCTION AND BACKGROUND

UV laser induced 2-photon ionization is a theme with variations. It involves the transformation of a polyatomic molecule - often an aromatic compound - to the corresponding radical cation going through a resonant intermediate state, usually the first excited singlet state of the molecule. This is typified in fig.1. In historical perspective starting with the year 1979 the following steps in the development of the concept may be distinguished.

_____

* permanent address: Institute of Optics and Fine Mechanics,
                     Academia Sinica, Shanghai, China

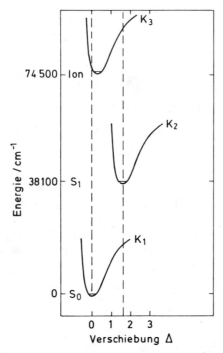

Fig. 1: General scheme of 2-step photoionization of a molecule with a
resonant intermediate state. UV (excimer or doubled dye) lasers
can conveniently used in this scheme /1/. As is typical for a
molecule like benzene the electron which is excited becomes auto-
bonding in the first step $S_o \rightarrow S_1$ and obviously nonbonding in the
second $S_1 \rightarrow X$. Corresponding shifts in the potentials along certain nuclear coordinates (e.g. υ_1) are indicated.

1) Studies of yield and selectivity with excimer and tunable UV lasers.

2) Secondary ion photodissociation and further fragmentation.

3) Spectroscopy and relaxation of intermediate states.

4) Mechanisms of ionization.

5) Use of photoions
   a) ion spectroscopy
   b) mass spectrometry
   c) photochemistry
   d) ion molecule reactions

The present paper deals with points (4), (5a) and (5c) out of this catalog and uses the benzene molecule (plus some benzene derivatives as far as (5c) is concernd) as the model example.

We first report on a two-laser photoionization experiment of benzene where the stepwise excitation was accomplished by frequency doubled dye lasers. The first laser pumped the molecule to a selected vibronic level of its first excited siglet state $^1B_{2u}$ from where it was ionized by the time-delayed pulse of a second laser. The ion yield depends on the chosen intermediate state as well as on the wavelength of the ionizing laser. From the structure and intensities of the recorded ion spectra vibrational frequencies and molecular parameters of the ground electronic state of the ion were derived. The contributions of autoionizing Rydberg levels to the ionization cross section can dearly be distinguished from direct ionization. Some of these resonace peaks could be correlated with vibrations within these Rydberg states /2/.

In the context of this work we found it interesting to make the transition from ion spectroscopy in the gas phase to ion photochemistry in liquid solution. While the basic processes are still the same the different environment drastically modifies the results known from gas phase studies. The mobility of ions and electrons in solution is greatly different, the energy of the spectroscopie states is changed, and reactive interaction with the solvent or other added species becomes possible. A very first step in this direction of preparative ionic photochemistry was reported before from our laboratory treating the benzene molecule as an example. More results have now been obtained relating to the mechanism of ion formation and its yield, the nature of the intermediates, and the conditions favouring product yield. As a first case the conversion of benzene to phenol and biphenyl in aqueous solution was studied followed by the study of reactions of benzene derivatives in water and other solvents /3/.

In the last paragraph of this report some brief comments are made concerning the usefulness of resonant multiphoton ionization for the construction of an cold ion beam source and the diagnostic use of photoionization in collisional energy transfer studies /4/.

II. IONIZATION SPECTROSCOPY OF BENZENE

Benzene exhibits a lowest ionization potential of 9.241 eV in the gas phase and an electronic level $^1B_{2u}$ a little more than halfway between (4.72 eV = 38086 cm^{-1}) originating from the excitation of an e_{1g} π-electron to an e_{2u} π-orbital. If the spectral structures of the REMPI spectrum are compared to the LIF spectra observing the fluorescence from the intermediate state back to the ground state there is complete conformity of the spectral positions and no new peaks are found (Fig. 2). However, the ion yields of some excitation bands show considerable variations (>50%) which can be assigned to structures in the ionizing step.

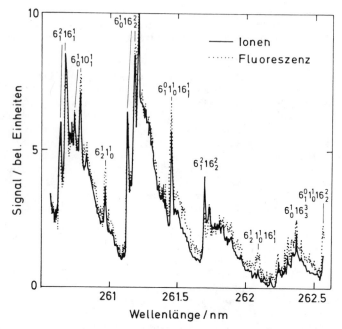

Fig. 2: Part of the ion versus fluorescence ($S_1 \rightarrow S_0$) of spectrum of benzene exhibiting the same spectral peaks in both cases but different peak heights /2/.

The excess photon energy of the order up to a few thousand wavenumbers will end up partly as internal energy of the ion. The resulting vibrational populations depend on the amount of excess energy, the selection rules for transitions from the intermediate vibronic states and the corresponding Franck-Condon factors.

One limiting case of the process is the direct ionization describing an immediate coupling to the ionization continuum. The other limiting case is that of auto-ionization and is superimposed on the signals originating from direct ionization. The excitations to Rydberg levels are resonant processes, more or less broadened due to their individual state lifetimes. This then manifests itself by added peaklike structures in the vicinity of Rydberg resonances.

Smalley et al. /5/ found considerable auto-ionization in their thre-shold ionization spectrum of benzene exhibiting unexpected positions of several of the peaks. Conversely Reilly et al. /6/ showed from photo-electron spectra that the features observed could be well understood as vibrational energy levels split by Jahn-Teller forces acting upon the degenerate electronic state of the benzene ion. This paper repeats the two-color ionization experiment on benzene under optimized conditions and provides a rather complete interpretation of the observed ion spectrum.

Experimentally the UV sources are provided by two dye lasers giving a minimum of 2 x 250 µJ UV-light of 10 ns pulse duration in the range 260 - 285 nm. The benzene pressure was between 10 µbar and 200 µbar. The electrodes were biased with 16 V and the ion current following laser ir-radiation was collected by a gated integrator. Details can be found in Fig. 3. In this way a complete ionization spectrum for several selected intermediate ($^1B_{2u}$) vibronic levels could be recorded from slightly ne-gative excess energies up to excess energies of 2000 cm^{-1} with a resolu-tion of 1.5 cm^{-1}. A computer-based subtraction method was used to eli-minate contributions from one laser only and from higher order proces-ses. Care had to be taken to avoid recombination losses for the charged particles.

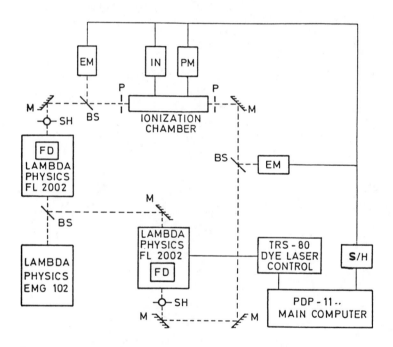

S/H : SAMPLE AND HOLD   M : MIRROR

EM : ENERGY METER, R$_j$ 7100   BS: BEAMSPLITTER

IN : INTEGRATOR   FD: FREQUENCY DOUBLING

PM : PRESSURE METER   SH : SHUTTER

                             P : PINHOLE

Fig.: 3: Experimental diagram of 2-color laser photoionization experiment /2,7/.

Figure 4 shows a representitive 2-laser ionization spectrum from the vibrationless groundstate of S_1 formed by the transition 6_1^0, i.e. $\upsilon=6$, $v=1$ in the groundstate and $v=0$ in the excited state at 266.5 nm. A sudden first step at zero excess energy (9.241 eV) occurs and is followed by other steps which are likely to be due to direct ionization going into different vibrational levels of the ion. The most prominent steps whose exact positions may be determined by numerically differentiating the spectra are at 0 cm^{-1} and 980 cm^{-1} due to excitation of the ionic levels zero and $\upsilon_1=1$, the totally symmetric breathing vibration of both transitions. The ratio of both transition intensities appears to be close to 1 indicating a strong displacement for the equilibrium positions for this mode in both the states connected by the transition. From a numerical Franck-Condon calculation we determined a displacement of $\Delta = 1.4$, which is almost the same shift tabulated for the υ_1-potentials of groundstate and S_1-state. This result indicates, that the equilibrium bond lengths along the υ-1 coordinate nearly correspond for S_0 and the electronic groundstate of the benzeneion. This is in agreement with photoelectron spectroscopy in the vacuum UV. A quite similar result has also been derived from one-color two-photon ionization spectra of benzene by comparing ionization cross sections of transitions with and without excited υ_1-quanta in the intermediate S_1-state. This argument can be used to determine the bonding characteristics of the photoelectron in the S_0 and S_1 electronic states (with respect to mode υ_1). A more detailed discussion can be found elsewhere /7/.

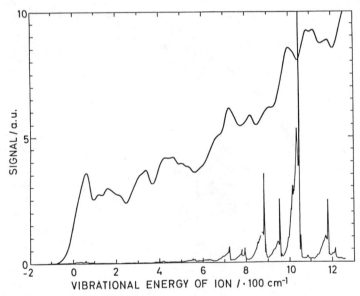

Fig. 4 Ionization spectrum for 6_1^0 excitation of C_6H_6 /7/.

So far the two most prominent steps of the spectrum are given by totally symmetric (D_{6h}) transitions from B_{2u}-state S_1 to the e_{1g}-ion core. However, one further clearly distinguished step centered at 650 cm^{-1} can not be assigned to any totally symmetric vibration, because the only other totally symmetric vibration (υ_2) has a mode, which is mainly determined by stretching of the C-H-bonds. Its high frequency of 3073 cm^{-1} is caused by strong bonding forces between the C and H atoms, which will therefore hardly change upon excitation of a $2p_z$-π-electron of a carbon atom. This can be seen from the very similar frequencies, that the υ_2-mode has in S_0 and S_1. The expected ion fundamental frequency does certainly not fall in the low energy range investigated here.

However, the e_{1g}-electronic ground state of the benzene ion is degenerate and its wave function can be perturbed by certain vibrations of the molecule. This leads to Jahn-Teller active modes which are distinguished by angular momenta $j = \pm 1/2$ and $j = \pm 3/2$. This will allow further transitions from the vibrationless groundstate of S_1 to vibrational levels of $j = \pm 1/2$ in the ion.

The Jahn-Teller-active vibration υ_6, has a groundstate frequency of 608 cm^{-1}, whereas the frequencies of the other modes lie more or less above 1100 cm^{-1}. So the assignment of the step at 650 cm^{-1} to the 1/2-component of υ_6 is rather clear.

Reilly et al. /6/ did not assign the 1/2 component of $v_6 = 1$ of the 6_1^0 transition, but did so for several other transitions. They got an averaged value of 83.2 meV (671 cm^{-1}) with an error of 0.4 meV (3 cm^{-1}). From a theoretical calculation they determined a value of 82.8 meV (668 cm^{-1}). We estimated an error of about $\pm$ 15 cm^{-1} to locate the center of our signal step at 650 cm^{-1}, which is broadened by the rotational structure of the thermally populated vibrational levels.

Figure 5 shows an ionization sepctrum from the S_1-state $v_{16} = 1$ excited by the transitions 6_1^0 16_1^1 at 267.95 nm. The structure of this spectrum is very similar to the one before discussed, except that it is shifted by 295 cm^{-1}. The shift is caused by the excitation of one υ_{16}-quantum in the ion, to obtain a totally symmetric transition from the state $v_{16} = 1$ of S_1 to the ion.

In addition to the steplike features of the spectra in fig. 4 and 5 the distinct resonace peaks have to be discussed. As already mentioned these peaks are expected to be autoionizing contributions of vibronic Rydberg states, (vibrationally) excited above the ionization potential of 9.241 eV, which can be optically excited from the electronic groundstate and therefore must have symmetries a_{2u} or e_{1u}. Additionally, there is a close lying e_{1u} (n = 3)-Rydberg level at 9.16 eV (-653 cm^{-1}) belonging to an excitation of an e_{2g}-σ-electron with a series limit of 11.49 eV.

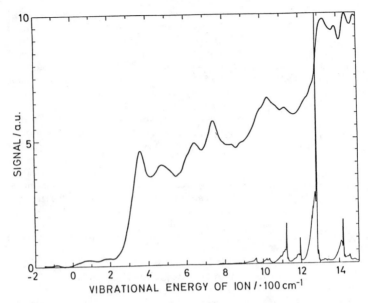

Fig. 5: Ionization spectrum for 6_1^0 16_1^1 excitation of C_6H_6 /7/.

One result of our measurements is the ovservation, that the appearence of prominent auto-ionizing resonance peaks decreases with increasing energy of the intermediate vibrational levels of S_1. The reason is probably a stronger coupling of the Rydberg levels to the ionization continuum at higher vibrational energies, which will smear out transitions to these levels and make them less distinguishable from direct ionization.

Many other ionization spectra have been taken from vibrationally excited levels of the S_1-state of benzene, which cannot be discussed here in full detail. The energies derived from these spectra for the vibrational states of the benzene ion in its electronic groundstate are given in ref. /7/.

Multiphoton ionization experiments mostly do not take care of structures arising from the ionization process itself. We have shown in this experiment, that the ionization cross section can vary rather strongly at low excess energies of the ionizing electron, caused by the vibrational structure of the generated ions and superimposed Rydberg structure of the parent molecule.

The two color ionization experiment povides a valuable technique to investigate a large number of vibrational levels of an ion, that are not accessible by excitations from the electronic groundstate, and to determine the behaviour of the ionization cross section in the vicinity of the ionization energy. Contributions from auto-ionizing processes can be distinguished from ordinary non-resonant ionization spectra. A super-sonic molecular beam with greatly reduced rotational population would therefore improve the error in determining vibrational frequencies to a very few wave numbers.

III. LASER GENERATION AND REACTIONS OF BENZENE RADICAL CATIONS IN SOLUTION

The high ion yield of benzene and other aromatic molecules found in low pressure gas phase studies suggests the use of these ions in prepar-ative phtochemistry in solution. The absorption spectrum of benzene in water as a solvent shows a weak band around 250 nm ($E_{max} \cong 200$ l mole^{-1} cm^{-1}) corresponding to the $S_1 \leftarrow S_0$ transition (see section II). The pho-tochemistry fo this molecule exhibits a great manifold of processes. Pure benzene can be transformed by irradiation in this spectral region into various ring isomers, open chain valence isomers and other minor products. In aqueous solution benzene may be photochemically oxidized to formyl-cyclopentadiene-1,3. This oxidation is believed to proceed via the formation of benzvalene in the first step which is rapidly oxi-dized - with and without oxygen, eqn. (1) - to the aldehyde

(1)

If the same aqueous benzene solution is irradiated by a UV laser (KrF, λ = 248 nm) one finds at low light intensity the known 1-photon photoproduct (1) while at higher intensity the UV absorption of the solution changes, gas bubbles are evolving and two new photoproducts, phenol and biphenyl are formed according to (2)

(2)

Identification is provided by liquid phase chromatography (HPLC). Following the laser pulse transient electric conductivity can be mea-sured. In view of the intensity dependence a 2-photon process is sug-gested with the wellknown benzene ion as the intermediate. The observed

375

$$(3)$$

Photoionization in condensed phases, both liquids and ordered solids, of benzene has received attention before, e.g. by Albrecht et al. /8/. The Wavelengths, however, as well as other conditions were different $\lambda > 355$ nm) and no direct 2-photon ionization could be observed. Also in the context of our work no absolute claim is made that the number of photons be always restricted to 2. The question of the actual energy needed for ionization in aqueous solution was adressed in the following way: It is clear that it is lower than in the gas phase by the solution energy of the ions P_+ and the electrons V_o

$$IP_{liq} = IP_{gas} + P_+ + V_o$$

The value of IP_{liq} may be estimated to be 6 ± 0.5 eV (see fig. 6). An interesting possibility to test our mechanism is then offered if the laser energy is lowered to a value where the 2- (or more) photon excitation can still supply the ionization energy but is insufficient to induce on-photon photochemistry. The excitation under these conditions is not resonance-enhanced by S_1 but proceeds through a virtual state. Light of 308 nm (XeCl laser) was chosen for this test which is below the long wavelength absorption of benzene. The multiphoton products could still be identified, however, as expected, any one-photon products were totally absent. We thus believe that eqn. (4) holds also under these conditions (B standing for benzene).

$$B^{**} \rightarrow [B^+ e^-]_{solv.\ cage} \rightarrow B^+_{solv} + e^-_{solv} \rightarrow Products \qquad (4)$$

One further point should be noted in relation to eqn. (3). The hydroxy cyclohexadienyl radical formulated there is an intermediate known also from the radiochemistry of benzene where it is often generated by addition of a hydroxyl radical to neutral benzene.

$$(5)$$

This idea opens up new possibilities for qualitative checks on the mechanism by adding radical scavengers, electron scavengers and varying

the p$_H$-value of the solution. All these tests gave additional plausibi-
lity to the mechanism of eqn. (3). Details may be found elsewhere /9/.If
addition of a solvent molecule (H$_2$O in the first case) to the radical
cation is the first srep in the product formation this can be tested by
trying other solvents. Consistently, the same reaction in methanol gives
anisol.

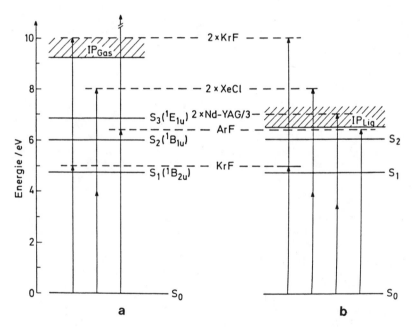

In a further investigation of this concept of ionic photochemistry a
more extensive study of substituted benzenes was conducted in order to
prove the validity of this mechanism to practical photochemistry. This
will be reported in a forthcoming paper /10/.

Fig. 6: Comparison of benzene ionization in the gas phase (a) and in
aqueous solution (b) showing some of the common excimer laser
photon energies /3,9/.

IV. CONCLUSION AND OUTLOOK

As was already suggested in the introduction, once the mechanism as well as the experimental parameters of laser induced multiple photon ionization have been investigated these ions may be used in spectroscopic and ion-molecule interaction studies. For two cases such an application has now been demonstrated in this laboratory by Proch, Trickl and Sha /11,12/. For practical experimental reasons the two diatomic molecules N_2 and CO were chosen to address questions of cold-ion beam formation and E to E energy transfer. Nitrogen molecular ions N_2^+ $^2\Sigma^+$ were made in selected vibrational states by going through the following sequence of excitation steps.

$$N_2(x\ ^1\Sigma_g^+) \xrightarrow{\text{2 photons}} N_2(a\ ^1\Pi_g) \xrightarrow[\text{photons}]{\text{1 or 2}} N_2^+\ (X\ ^2\Sigma_g^+) \qquad (7)$$

$$\downarrow \text{1 Photon}$$

$$N_2^+\ (B\ ^2\Sigma_u^+)$$

The intermediate $N_2(a)$ state, alternatively to being ionized, could also be used, although not yet under beam conditions, to transfer energy to the CO molecule to form $CO(A'\Pi, v = 0...5)$. For analysis the excited CO molecules were probed by 2-step photoionization going through CO(B) to the $CO^+(X)$ level. Both electronic and vibrotational energy is exchanged in the collision which can very clearly be analyzed by this technique. The brief mentioning of this work is only meant to indicate a fruitful direction of future research taking laser photoionization into the realm of ion-molecule scattering.

REFERENCES

/ 1/ S.D. Rockwood, J.P. Reilly, K. Hohla, K.L. Kompa, UV Laser Induced Molecular Multiphoton Ionization and Fragmentation, Opt.Comm. 28, 175 (1979).
J.P. Reilly, K.L. Kompa, Laser Induced Multiphoton Ionization Mass Spectrum of Benzene, J.Chem.Phys. 73, 5468 (1980).
/ 2/ Ph.D. Dissertation G. Müller, Universität München 1985, Lab. Report MPG 93, 1985.
G. Müller, K.L. Kompa, J.L. Lyman, W.E. Schmid, S. Trushin, 2-Step Photoionization of Benzene: Mechanism and Spectroscopy in Multiphoton Processes, P. Lambropoulos, S.J. Smith, Eds. Springer Series on Atoms and Plasmas 1984.
/ 3/ Ph.D. Dissertation M. Braun, Universität München 1986.
/ 4/ D. Proch, G.H. Sha, T. Trickl, K.L. Kompa, to be published.
/ 5/ M.A. Duncan, T.G. Dietz, R.E. Smalley, J.Chem.Phys. 75, 2118 (1981).
/ 6/ S.R. Long, I.T. Week, J.P. Reilly, J.Chem.Phys. 79, 3206 (1983).
/ 7/ G. Müller, J.Y. Fan, J.L. Lyman, W.E. Schmid, K.L. Kompa, submitted J.Chem.Phys.
/ 8/ T.W. Scott, H.C. Albrecht in Adv.Laser Spectr. I, B.A. Garetz, J.R. Lombardi, Eds., Heyden and Sons, London 1982.
/ 9/ M. Braun, W. Fuß, K.L. Kompa, to be published.
/10/ M. Braun, W. Fuß, K.L. Kompa, to be published.
/11/ D. Proch, T. Trickl, to be published.
/12/ G.H. Sha, D. Proch, to be published.

SURFACE EFFECTS IN VIBRATIONALLY EXCITED

MOLECULAR BEAM SCATTERING

J. Misewich, P.A. Roland and M.M.T. Loy

IBM Thomas Watson Research Center
Yorktown Heights, N.Y.

INTRODUCTION

Recently we have combined laser spectroscopic techniques with UHV technology to develope an apparatus for studying vibrationally excited molecular beam scattering.[1-5] In this apparatus, a tunable infrared laser is used to vibrationally excite a supersonic molecular beam prior to scattering from a single crystal surface in a UHV scattering chamber. Scattered molecules are then detected in a state-specific manner through multiphoton ionization using a tunable ultraviolet laser. In this manner, state-to-state surface scattering information is obtained. We have applied this technique to the scattering of nitric oxide. The pump laser excites a single quantum state $NO(v=1, J=3/2, \Omega=1/2)$. Then angular, velocity, rotational and spin-orbit distributions are obtained after scattering from a variety of surfaces, including air-cleaved and vacuum-cleaved $LiF(100)$[3], $Ag(111)$[4,5] and $Ag(110)$[4]. From this collection of experiments, the effects of conduction electrons, surface corrugation and surfaces roughening on vibrational, rotational and translational energy transfer during molecule-surface collisions has been studied.

Earlier work on vibrational energy transfer during gas-surface collisions has involved experimental setups in which careful characterization of the surface using conventional LEED and Auger techniques was not possible. Rosenblatt and coworkers have developed a vibrating surface technique in which internal energy accommodation coefficients can be extracted.[6] This work provided one of the earliest estimates of internal energy accommodation, but could not directly determine vibrational energy transfer information. Vibrational energy exchange from a hot surface to an incoming beam of vibrationally cold molecules has been studied by Mantell et. al. using Fourier transform emission spectroscopy.[7] Vibrational accommodation coefficients were obtained for carbon monoxide and carbon dioxide scattered from hot platinum surfaces. In order to study vibrational relaxation, Houston et. al. have used a laser-induced fluorescence technique.[8] In that experiment, a static gas sample is excited with a pulsed infrared laser, then the decay of the fluorescence signal is related to vibrationally inelastic collisions with the walls of the gas container. Vibrational energy deactivation probabilities were obtained for carbon monoxide and carbon dioxide deactivation by evaporated silver walls. In contrast, we study vibrational energy transfer in scattering with a well-defined surface from a single quantum state using laser spectroscopic techniques.

Laser spectroscopic techniques have recently been employed to study rotational energy distributions. Cavanagh and King have determined the rotational distributions for molecules leaving a surface during a thermal desorption.[9,10] For the case of NO desorbing from Ru(001), they have observed a rotational temperature $T_{rot} \sim 0.5 \ T_{sur}$, but for NO desorption from an oxidized surface they observed $T_{rot} \sim 0.7 \ T_{sur}$. In molecular beam scattering experiments, NO rotational distributions have been obtained by several groups.[11-14] As in the case of thermal desorption, very different behavior has been observed depending on surface properties. In our work, we hope to extend the scope of these studies to include vibrational energy. In this paper, the effects of scattering from different surfaces on internal energy distributions, particularly the effect on the vibrational energy survival probability, will be examined.

EXPERIMENTAL

Shown schematically in figure 1 is the experimental apparatus which consists of an ultra-high vacuum molecular beam scattering chamber, a tunable infrared excitation laser and a tunable ultraviolet detection laser. Turbo and titanium sublimation pumps establish a base pressure of 3×10^{-10} torr in the stainless steel UHV scattering chamber which is equipped with a cylindrical mirror analyzer (CMA) for Auger electron spectroscopy and LEED optics in order to verify surface cleanliness and order. Silver surfaces were obtained from a high purity rod after orientation to $\pm$ 0.5 degrees using Laue x-ray diffraction. Cycles of gentle ion bombardment (1 keV, 5 μA, 15 minutes) and annealing (750 K, 30 minutes) were used to prepare clean and well-ordered silver surfaces after introduction into the scattering chamber. For experiments with LiF(100) a cleaver was installed in the chamber which enabled us to cleave fresh samples at 5×10^{-10} torr.

Supersonic expansion of NO is used to produce a rotationally cold molecular beam. For all experiments described here, a seeding mixture of 10% NO

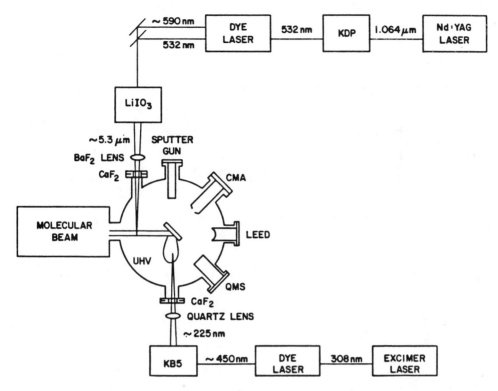

Figure 1: Experimental Apparatus

in He at a total backing pressure of 450 torr was used, which produces a rotational distribution that can be approximately described by a temperature of 8.5 K.[15] Thus most of the NO is in the (v=0,J=1/2,Ω=1/2) state. Tunable infrared laser radiation is produced by difference frequency mixing in a lithium iodate crystal of ~ 590 nm output from a Nd:YAG pumped dye laser and part of the 532 nm second harmonic of the YAG. The infrared laser is tuned to the $R_1(J''=1/2)$ line at 1876.076 cm⁻¹ to populate a single rotational state NO (v=1,J=3/2,Ω=1/2). Infrared pulse energies were approximately 5-10 microjoules in a bandwidth of 0.4 cm⁻¹.

Resonance enhanced two-photon ionization via the $^2\Pi$ A <-- $^2\Sigma$ X γ(0-0) and γ(1-1) bands was used for state-specific detection of NO. Frequency doubling the output of a XeCl excimer pumped dye laser in a potassium pentaborate crystal produced tunable ultraviolet radiation for the ionization with UV pulse energies of approximately 30 microjoules in a bandwidth of about 0.4 cm⁻¹. The focussed UV beam crossed the molecular beam at right angles and could be moved over a variety of radii and angles about the scattering sample surface. Ionized NO molecules were detected using a Johnston MM-1 multiplier.

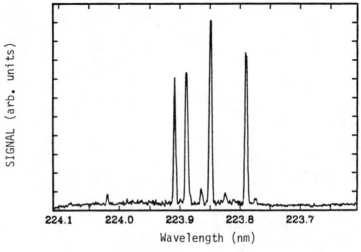

Figure 2: Incident Beam Ultraviolet Multiphoton Ionization Spectrum

The high spectral, spatial and temporal resolution of the laser excitation and detection system is utilized to perform three types of measurements: rotational distributions, angular distributions and velocity distributions. Rotational and electronic (spin-orbit) distributions are obtained by scanning the ultraviolet laser frequency. State-specific angular distributions are obtained by fixing the ultraviolet laser frequency to detect a single well-defined final state and then moving the position of the probe laser beam with respect to the surface. Finally, velocity information is obtained for both the incident and scattered beams by varying the delay time between the firing of the infrared pump and the ultraviolet probe lasers. This laser spectroscopic time-of-flight technique is the state-specific analog of chopped beam/mass spectroscopic velocity analysis. In addition to being state-specific, this technique also offers the advantage of a radially movable as well as rotatable detector.

INCIDENT BEAM CHARACTERIZATION

Although various surfaces will be discussed, all experiments reported here have a similar incident molecular beam in common. The infrared laser pulse excites a small part of the incident beam 230 microseconds after the start of the 600 microsecond molecular beam pulse, exciting only molecules initially in the $v=0$, $J=1/2$ level of the $^2\Pi_{1/2}$ electronic state to produce molecules in the state $NO(v=1, J=3/2, \Omega=1/2)$. An incident beam ultraviolet scan in the region of the $\gamma(1\text{-}1)$ band is shown in figure 2. Four prominent lines are observed representing the four branches: P_{11}, $(Q_{11}+P_{21})$, $(R_{11}+Q_{21})$ and R_{21}

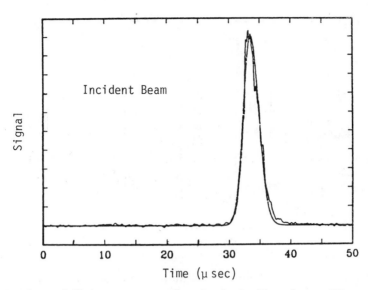

Figure 3: Time-of-flight spectrum taken in the incident beam. The probe laser is tuned to detect NO ($v=1, J=3/2, \Omega=1/2$).

all originating from the ($v=1, J=3/2, \Omega=1/2$) state. Since there are no collisions in the molecular beam there are no lines from other rotational levels. The minor lines present in figure 2 are from high rotational states of $v=0$. As discussed in detail in reference 15, the rotational distribution in a supersonic expansion cannot be completely characterized by a Boltzman distribution; there is a tail extending out to high J. It is important to note that lines originating from background NO molecules rather than from molecules which were vibrationally excited by the infrared laser pulse can be distinguished by their very different time scales, as shown below.

To determine the incident beam velocity distribution, a time-of-flight spectrum shown in figure 3 was taken with the ultraviolet probe laser positioned in the incident beam. Since the infrared laser fires for a few nanoseconds during the 600 μsec duration of the molecular beam pulse, a short pulse of NO ($v=1, J=3/2, \Omega=1/2$) is produced on top of a 600 microsecond background pulse of NO $v=0$ distributed over the lowest few rotational states. The molecular beam flux-velocity distribution, $I(v)dv$ = number of particles/area/sec with velocities in the range $v \rightarrow v+dv$, can be characterized by:

$$I(v) = Av^3 \exp [-(v-v_f)^2/\alpha^2]$$

where v_f is the flow velocity of the beam and α is a measure of the width of the velocity distribution along the direction of the incident beam. After transformation of the flux-velocity distribution from velocity space to time

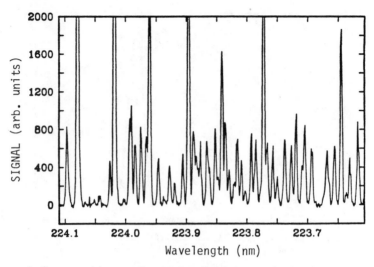

Figure 4: Scattered beam ultraviolet multiphoton ionization spectrum.

space for density dependent detection, the observed incident beam time-of-flight spectra are fit using a nonlinear least squares fitting routine.[5] Such a fit is also illustrated in figure 3, which indicated an incident beam flow velocity of $v_{f,i} = 1.48 \times 10^5$ cm/sec and width of $\alpha_i = 7.7 \times 10^3$ cm/sec. From the velocity distribution an incident beam translational energy, $E_i = 340$ meV, was determined. Since the incident angle is $\theta_i = 45$ degrees, the normal energy, $E_n = E_i \cos^2\theta$, is then 170 meV.

SCATTERED BEAM DISTRIBUTIONS

Rotational, electronic (spin-orbit), angular and velocity distributions have been obtained for scattering from a variety of surfaces. These distributions in conjunction with the incident beam data allow us to determine a probability for survival of vibrational energy after the collision, S.P., by integrating over the scattered beam distributions and comparing with the amount of NO(v=1) pumped by the IR laser in the incident beam.

A scattered beam ultraviolet multiphoton ionization spectrum covering the same wavelength range of figure 2 is illustrated in figure 4. This particular spectrum was taken near the specular angle for scattering from LiF(100). Spectra taken with the IR laser blocked so that only the background appears are used to determine the rotational and electronic distributions for scattered NO that was pumped by the IR laser.[2] Rotational distributions were found to be reasonably well described by Boltzman distributions. The corresponding rotational temperatures for scattering from the various surfaces are summarized in table 1. Comparison of ultraviolet multiphoton ionization spectral lines originating in the $^2\Pi_{1/2}$ spin-orbit electronic state and lines originating

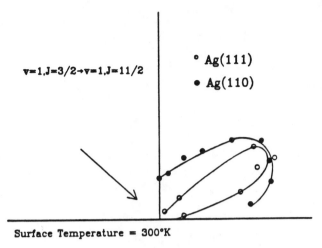

Figure 5: Angular distribution for scattering into a final state of $(v=1, J=11/2, \Omega=1/2)$ from Ag(111) and Ag(110).

in the $^2\Pi_{3/2}$ state allowed electronic temperatures to be determined which are also given in table 1.

Angular distributions for scattering into a specific final quantum state, NO$(v=1, J=11/2, \Omega=1/2)$, are illustrated in figure 5 for scattering from Ag(111) and Ag(110). The full width at half maximum (in degrees) of the angular distributions, λ_f, are reported for the surfaces studied in table 1.

Laser spectroscopic time-of-flight distributions were obtained for scattering into a single final quantum state at a well-defined scattering angle and distance from the surface. An example is shown in figure 6 for scattering into the NO$(v=1, J=11/2, \Omega=1/2)$ quantum state at a scattering angle of $\theta_f = 60$ degrees from Ag(111). Time-of-flight spectra were observed to be largely independent of final J for $J \leq 41/2$ in scattering from all the surfaces studied here. Also shown is the fit to the data obtained by a nonlinear least squares fit of the data to a convolution of the density dependent time-frame distribution function and the experimentally determined arrival function.[5] From the velocity distributions, an average final translational energy was determined for scattering into the most probable angle. In scattering from all surfaces except sputter damaged LiF, average final translational energies much greater than expected from accommodation to the surface temperature were obtained. This in conjunction with the angular distributions establishes that the scattering was direct-inelastic (as opposed to trapping-desorption in which case isotropic angular distributions and accommodation of translational energy to the surface temperature are expected).

Table 1: Scattering properties from various surfaces. T_{rot} is the rotational temperature in K ± 40 K, T_{ele} is the electronic (spin-orbit) temperature in K ± 50 K and λ_f is the full width at half maximum of the angular distribution of the scattered molecules, in degrees. S.P. (survival probability) is the probability for an incident $v=1$ molecule to remain in $v=1$ after the collision with the surface.

	T_{rot}	T_{ele}	λ_f	S.P.
Ag(111)	380	250	40	0.9 ± 0.1
Ag(110)	340	250	65	0.85 ± 0.15
Vacuum or air-cleaved LiF(100)	340	200	40	0.9 ± 0.1
Ion damaged or polished LiF	Below detection threshold			

Experiments with air-cleaved LiF(100) and vacuum-cleaved LiF(100) gave identical results to within our uncertainties. In order to determine the effect of surface disorder on the scattering, an argon sputter ion gun was used to damage a cleaved LiF surface in situ. First, a good LiF(100) sample was obtained by cleaving in vacuum and the normal $NO(v=1) \rightarrow NO(v=1)$ signal was obtained. Then without moving the surface, the ion gun was turned on. The $NO(v=1) \rightarrow NO(v=1)$ signal was observed to decrease with increasing ion bombardment until the signal fell below our detection threshold. In an earlier experiment, polished LiF was used for scattering.[1] As in the scattering experiment with ion bombardment damaged LiF, the $NO(v=1) \rightarrow NO(v=1)$ signal was below our detection threshold.

DISCUSSION AND CONCLUSIONS

Table 1 summarizes the main observations in scattering from a variety of surfaces. The most conspicuous effect is the absence of an observable $NO(v=1) \rightarrow NO(v=1)$ signal when the LiF surface was damaged by ion bombardment. Surface damage greatly increases the microscopic roughness and significantly alters the gas-surface interaction potential. In addition, microscopic roughness leads to multiple collisions and an increased probability for trapping. If the residence time is longer than the vibrational lifetime on the surface, the vibrational energy becomes accommodated to the surface temperature. This experiment and the experiment on polished LiF (which is

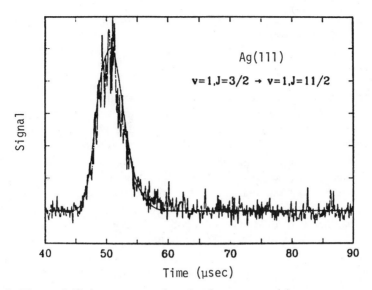

Figure 6: Time-of-flight spectra taken in the scattered beam at a scattering of angle of θ_f = 60 degrees. The probe laser is tuned to detect NO(v=1,J=11/2,Ω=1/2).

also characterized by great surface roughness) demonstrate the need for well-ordered surfaces to observe direct-inelastic scattering of NO(v=1) --> NO(v=1).

Scattering results from the well-ordered surfaces have certain features in common. In all these cases, direct-inelastic scattering of NO(v=1) --> NO(v=1) was observed, as characterized by angular distributions peaked near the specular direction and translational energies much higher than expected from accommodation to the surface temperature.

A comparison of the survival probabilities in scattering from LiF(100) and Ag(111) and Ag(110) reveals an interesting similarity. A high survival probability is observed from both the insulating and the metal surfaces, despite the fact that there are conduction electrons in Ag, which opens up another vibrational energy deactivation channel. In calculations of the vibrational lifetime near a metal surface, Persson and Persson have found very short lifetimes due to excitation of electron-hole pairs in the metal surfaces.[16] Our observations suggest that this mechanism is inefficient in deactivating NO(v=1) in direct-inelastic scattering.

Surface corrugation effects are examined in experiments from Ag(111) and Ag(110). Both surfaces are found to have a very large survival probability; however, the details of the scattering are very different. As shown in figure 5, the angular distribution in scattering from Ag(110) is quite a bit broader than that from Ag(111). Also, there is a greater degree of translational energy exchange with the Ag(110) surface. However, the average transla- tional energies in scattering from Ag(110), although smaller than the translational energies in scattering from Ag(111), are still much larger than expected from thermal accommodation. This is also true for distributions taken at the surface normal, so even though there is scattered flux at the surface normal, the scattering is direct-inelastic.

In conclusion, scattering from a single quantum state $NO(v=1, J=3/2, \Omega=1/2)$ has been studied for a variety of surfaces. In scattering from well-ordered LiF(100), Ag(111) and Ag(110) surfaces direct-inelastic scattering with a large probability for survival of vibrational energy was observed. Conduction electrons were shown to have little effect on vibrational energy survival probability for the systems studied. Surface corrugation was shown to change the angular and velocity distributions, but had little effect on the vibrational energy survival probability. In contrast, the survival probability is dramatically changed when the surface is microscopically rough, falling below our detection threshold.

This work is partially supported by the Office of Naval Research.

REFERENCES

1. H. Zacharias, M.M.T. Loy and P.A. Roland, Phys. Rev. Lett. *49*, 1790 (1982).
2. J. Misewich, H. Zacharias and M.M.T. Loy, J. Vac. Sci. Technol. B, *3*, 1474 (1985).
3. J. Misewich, H. Zacharias and M.M.T. Loy, Phys. Rev. Lett. *55*, 1919 (1985).
4. J. Misewich and M.M.T. Loy, J. Chem. Phys., in press.
5. J. Misewich, P.A. Roland and M.M.T. Loy, submitted for publication.
6. G. Rosenblatt, Acc. Chem. Res. *14*, 42 (1981).
7. D.A. Mantell, S.B. Ryali, G.L. Haller and J.B. Fenn, J. Chem. Phys. *78*, 4250 (1983).
8. J. Misewich, P. L. Houston and R. P.Merrill, J. Chem. Phys. *82*, 1577 (1985).
9. R.R. Cavanagh and D.S. King, Phys. Rev. Lett. *47*, 1829 (1981).
10. R.R. Cavanagh and D.S. King, J. Vac. Sci. Technol. A2, 1036 (19 84).
11. G.D. Kubiak et.al., J. Chem. Phys. *79*, 5163 (1983).
12. A.W. Kleyn, A.C. Luntz and D.J. Auerbach, Surf. Sci. *117*, 33 (1982).
13. J. Segner et.al., Surf. Sci. *131*, 273 (1983).
14. M. Asscher et.al., J. Chem. Phys. *78*, 6992 (1983).
15. H. Zacharias et. al., J. Chem. Phys. *81*, 3148 (1984).
16. B. N. J. Persson and M. Persson, Surf. Sci. *97*, 609 (1980).

PERTURBATION FACILITATED OODR RESOLVED FLUORESCENCE SPECTROSCOPY OF THE

$a^3\Sigma_u^+$ STATE OF Li_2 AND Na_2

Steven F. Rice and Robert W. Field

Department of Chemistry
Massachusetts Institute of Technology
Cambridge, Massachusetts

In the past, perturbations in diatomic molecules have been used to gain information about the positions of potential energy curves and electronic properties of states not directly seen in a simple fluorescence or absorption spectrum. This is typically done through the analysis of shifts in the position of main lines and the assignment of "extra" lines.[1] Recent work in our laboratory has focused on taking advantage of perturbations in a somewhat new way. Levels that are superpositions of singlet and triplet electronic states due to spin-orbit mixing are used as "stepping stones" in an optical-optical double resonance scheme to access triplet manifolds of states that cannot be directly excited from the molecule's singlet ground state. Two-photon excitation spectroscopy of triplet Rydberg states in Na_2 and Li_2, using this scheme, has already resulted in the identification of many high lying states,[2-4] providing rigorous experimental tests of recent ab initio calculations.

Double resonance excitation in these molecules is detected as fluorescence to the low lying $a^3\Sigma_u^+$ triplet state originating from the ground atomic limit, $^2S + ^2S$, and to the $b^3\Pi_u$ state originating from the $^2S + ^2P$ excited atomic limit. By resolving these fluorescence spectra, we have been able to determine accurately the shape and position of the $a^3\Sigma_u^+$ potential curve in Na_2 and Li_2 over a large range of internuclear distance. The experimental $a^3\Sigma_u^+$ potential curves provide tests of theoretical calculations of the fundamental bonding interaction between two spin unpaired ^2S centers.

The results from Perturbation Facilitated Optical-Optical Double Resonance (PFOODR) resolved fluorescence spectroscopy on the weakly bound region of the Na_2 $a^3\Sigma_u^+$ state will be presented here and shown to provide a model independent experimental determination of long range induced multipole interactions and the shorter range two-center electronic exchange interaction when combined with results for the $X^1\Sigma_g^+$ state.[5] We will also present resolved fluorescence from PFOODR excitation of Li_2 to the $b^3\Pi_u$ state which has resulted in the determination of the $a^3\Sigma_u^+$ curve in the region that is significant for molecular predissociation, 2.3 -2.6Å.[6] This curve has been determined in a unique way by examining the dissociative $a^3\Sigma_u^+$ state's effect on the linewidths and lifetimes of twelve vibrational levels of $b^3\Pi_u$. The analysis of the $b^3\Pi_u$ linewidth dependence on vibrational energy results in a complete description of the predissociation of the $A^1\Sigma_u^+$ and $b^3\Pi_u$ states in Li_2, determining both the position of this dissociative curve relative to these other states and the electronic matrix elements connecting them.

The experimental approach is fairly simple. Two c.w. tunable dye lasers (one for exciting from the ground state to the intermediate perturbed level and the other for exciting from the intermediate to final level (Pump and Probe respective-ly), are combined at a 50% beam splitter and copropagated through a hole in a collection mirror into a heat pipe containing a partial pressure of the metal dimer being studied. Visible and u.v. fluorescence is collected "on axis" by the mirror and focused onto the slits of a scanning monochro-mator, operated in 2nd order. The maximum attainable resolution of the fluorescence spectrum varied with signal intensity but in some instances was as high as 0.2 cm^{-1}. Absolute frequency measurements were accurate to ± 0.3 cm^{-1}.

Figure 1 shows the excitation scheme used to obtain fluorescence from $2\,^3\Pi_g$ and $3\,^3\Pi_g$ in Na$_2$ to the bound and continuum regions of the a$^3\Sigma_u^+$ state.

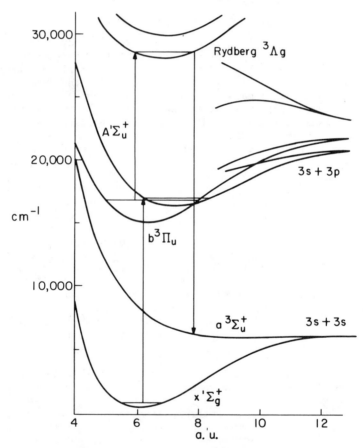

Fig. 1. Potential energy curves and excitation scheme for PFOODR experiments on Na$_2$. The Pump laser excites from the diatomic ground state, X$^1\Sigma_g^+$, to a superposition of b$^3\Pi_u$ and A$^1\Sigma_u^+$. The probe laser excites from this level to $^3\Lambda_g$ Rydberg states. Near u.v. fluorescence to a$^3\Sigma_u^+$ is resolved with a monochromator and detected with a PMT.

The Pump laser was tuned to the $b^3\Pi_{1u} - X^1\Sigma_g^+$ (21,0) P(17) transition at 16525.954 cm^{-1}. This intermediate state, $\tilde{b}^3\Pi_{1u}$ (v=21, J=16) at T_{vJ} = 16651.88 cm^{-1} is perturbed through spin-orbit coupling by an $A^1\Sigma_u^+$ level (v=17, J=16) at T_{vJ} = 16652.490 cm^{-1}, resulting in some singlet character in the nominal triplet level. This admixed singlet character provides transition strength from the ground state. Figure 2 shows the fluorescence spectrum to the bound region of the $a^3\Sigma_u^+$ state when the Probe excitation laser was tuned to excite the $3^3\Pi_{0g}$ (v=17, J=15) level at T_{vJ} = 33856.262 cm^{-1}. The spectrum shows a vibrational progression of two rotational lines, one strong and one weak, along with still weaker features due to collisional relaxation of the upper level. The strong feature is "R" = $[Q_{11}(15) + R_{12}(14)]$ and the weaker feature is "P" = $[Q_{13}(15) + P_{12}(16)]$. The intensity ratio of $^R/_P$ = 5.8 is predicted by intensity formulas for an intermediate Hund's case(a) – case(b) ($3^3\Pi_g$) to a pure Hund's case(b) ($a^3\Sigma_u^+$) transition where $A_v(3^3\Pi_g)$ = 2.5 cm^{-1} and $B_v(3^3\Pi_g)$ = 0.0855 cm^{-1}. This ratio is in good agreement with the spectrum shown in Figure 2. The term values for v = 0-12 (N= 14,16) of the $a^3\Sigma_u^+$ state are then used to determine the rotationless potential curve through an RKR inversion procedure.

The potential curves[8] for the $X^1\Sigma_g^+(-)$ state and the $a^3\Sigma_u^+(+)$ state are represented by

$$V(R) = \pm\, V_{EX}(R) - C_6R^{-6} - C_8R^{-8} - \cdots - C_{2N}R^{-2N}\ .$$

C_6R^{-6} and C_8R^{-8} are the first two terms in the long range multipolar

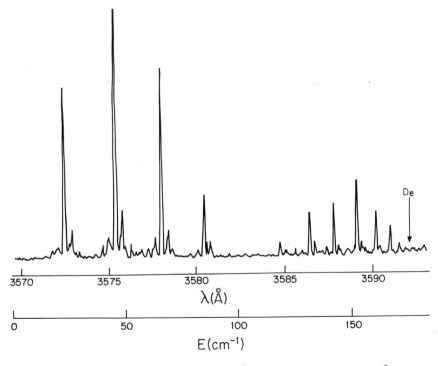

Fig. 2. Resolved Fluorescence from $3^3\Pi_{0g}$ (v=17, J=15) to $a^3\Sigma_u^+$ (v=0-12) with ~1 cm^{-1} resolution. Abscissas are the mono-chomator wavelength and the energy from the minimum of the $a^3\Sigma_u^+$ (J=0) potential curve.

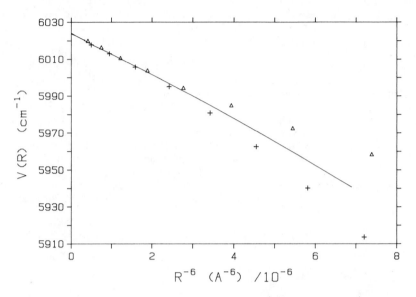

Fig. 3. Potential energy curves of $a^3\Sigma_u^+$ (Δ) and $X^1\Sigma_g^+$ (+)
at the RKR turning points for the experimentally ob-
served vibrational levels and the average of these
two curves (solid line) plotted vs. R^{-6}.

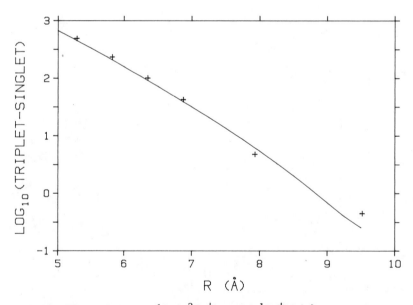

Fig 4. Plot of $\log_{10} \{[V(a^3\Sigma_u^+) - V(X^1\Sigma_g^+)]/2\}$ vs. R (solid
line). The crosses are the same function obtained
pointwise from the <u>ab initio</u> calculation in Ref. 11.

expansion (induced dipole–induced dipole and induced dipole–induced quadrupole) of two polarizable ^2S centers. In the limit where the interaction between the two centers is viewed as the simple interaction between two nonoverlapping 3s orbitals, V_{EX} is the valence bond exchange integral[9]

$$J(R) = \int \phi_{3s}^A (1) \; \phi_{3s}^B (1) \; V_I \; \phi_{3s}^A (2) \; \phi_{3s}^B (2) \; d\tau$$

where A and B refer to the Na atoms 1 and 2 refer to the electrons, and

$$V_I = e^2 \left(- \frac{1}{r_{B,1}} - \frac{1}{r_{A,2}} + \frac{1}{r12} + \frac{1}{R_{AB}} \right).$$

Recent work by Barrow et al.[10] has determined the $X^1\Sigma_g^+$ potential curve to within several wavenumbers of the Na_2 dissociation limit. The results on the $a^3\Sigma_u^+$ state determine its curve (which shares this limit) to a comparable energy. The sum and difference between these two experimental curves allow the multipolar part of the two electronic potentials to be separated from the exchange part, regardless of the exchange term's functional form or the rate of convergence of the multipolar expansion.

Figure 3 shows the average of these two curves plotted along with the individual outer turning points of the experimentally observed levels vs. R^{-6}. This curve can be fitted to the function

$$V_{AVG} = D_e - C_6 R^{-6} - C_8 R^{-8}$$

to determine the dissociation limit of the $X^1\Sigma_g^+$ and $a^3\Sigma_u^+$ states (relative to the minimum of the $X^1\Sigma_g^+$ curve) along with the first two multipole terms. The fit yields D_e = 6022.868 (96) cm^{-1}, C_6 = 7.89 (13) x 10^6 cm^{-1} Å^6, and C_8 = 2.119 (64) x 10^8 cm^{-1} Å^8. From the accurate determination of $D_e(^1\Sigma_g^+)$, the $D_e(^3\Sigma_u^+)$ is determined to be 174.45 cm^{-1} measured from the bottom of its potential well at 5.0911 Å.

Figure 4 shows the difference curve

$$V_{DIFF} = V_R(^1\Sigma_u^+) - V_R(^1\Sigma_g^+)$$

(divided by two) plotted logarithmically. The crosses are the values for the difference between the $a^3\Sigma_u^+$ and $X^1\Sigma_g^+$ curves determined by Konowalow and Rosenkrantz.[11] The solid line is an experimental determination of the molecular exchange interaction regardless of its functional form. As has been suggested,[8] it appears as though an exponential is a reasonable approximation to the true functional form of this interaction over three orders of magnitude in its absolute value.

PFOODR resolved fluorescence spectroscopy of the Na_2 $a^3\Sigma_u^+$ state has produced an accurate determination of this low-lying triplet's potential energy curve in the weakly bound region of internuclear distance. The combination of these results with the $X^1\Sigma_g^+$ potential curve has yielded an accurate determination of the relative magnitudes of the different electronic effects, exchange and electric multipolar, that cause the distinction between the triplet and singlet molecular states.

The interaction between the $A^1\Sigma_u^+$ and $b^3\Pi_u$ states in Li_2 is much weaker than in Na_2 because the atomic spin–orbit interaction is much smaller for the lighter atoms. However, recent measurements of the lifetimes of a large number of rotation–vibration levels of $A^1\Sigma_u^+$ have identified numerous predissociations and perturbations between these two states.[12] Several of these lend themselves to be used as intermediate levels in a PFOODR scheme similar to that used for studying Na_2. PFOODR excitation spectroscopy on the triplets of Li_2 has resulted in the assignment of many vibrational levels of several high lying $^3\Lambda_g$ states.[3]

Figure 5 shows a monochromator scan of the fluorescence from the $2^3\Pi_g$ state in Li_2 to the $a^3\Sigma_u{}^+$ and $b^3\Pi_u$ (v=0-2) states. Note the three bound vibrational levels of the $a^3\Sigma_u{}^+$ state on the short wavelength side of the distorted "reflection" of the excited state v=13 wavefunction onto the continuum of $a^3\Sigma_u{}^+$.[13] These bound levels are analogous to those studied in the previous section in Na_2. The sharp features to the long wavelength side of the scan are transitions to the first few vibrational levels of $b^3\Pi_u$.

Figure 6 shows a higher resolution spectrum of the fluorescence from another excited state, $1^3\Lambda_g$ (N=31, F_1) to the v=0 level of $b^3\Pi_u$. Because these are both good Hund's case(b) states, due to small spin-orbit coupling, large rotational constants, and high J, the spectrum exhibits a P,Q,R pattern. The weaker features are due to collisional relaxation in the upper state, as in Na_2 (Figure 1). The Q(N=31) line in this fluorescence scan is broader than the instrumental resolution of about 0.8 cm^{-1} due to predissociation by the $a^3\Sigma_u{}^+$ vibrational continuum. The R(30) and P(32) lines are sharp. However, the collisionally relaxed Q(30) and Q(32) lines are sharp and the R and P lines adjacent to R(30) and P(32) are broad. The reason for this selectivity in the predissociation of the $b^3\Pi_u$ rotational levels is explained by examining Figure 7.

The Pump laser (14367.935 cm^{-1}) excites the mixed levels (e-parity only) of the $A^1\Sigma_u{}^+$ (N = J = 33) $\sim$ $b^3\Pi_u$ (F_1; N = 32, J = 33) perturbation complex. The f-parity levels of $b^3\Pi_u$ cannot interact with $A^1\Sigma_u{}^+$ because the spin-orbit perturbation preserves parity and $A^1\Sigma_u{}^+$ contains only e-parity levels. Excitation by the Probe laser on the $P_{11}(33)$ line at 16614.301 cm^{-1} prepares the F_1 (N = 31, J = 32) e-parity level of the $1^3\Lambda_g$ state due to the $\Delta N=\Delta J$ case(b) electric dipole transition selection rule and the general $\Delta J = \pm 1$ e$\leftrightarrow$e, $\Delta J = 0$ e$\leftrightarrow$f electric dipole selection rule. Collisionally unrelaxed P,Q, and R fluorescence to any vibrational level of $b^3\Pi_u$ is to the F_1 spin comments of N = 30 (e-parity), N = 31 (f-parity), and N = 32 (e-parity).

Interaction of the bound $b^3\Pi_u$ levels with the $a^3\Sigma_u{}^+$ continuum can originate from the spin-orbit and/or the L-uncoupling, $B(R)\underset{\sim}{N}\cdot\underset{\sim}{L}$, terms in

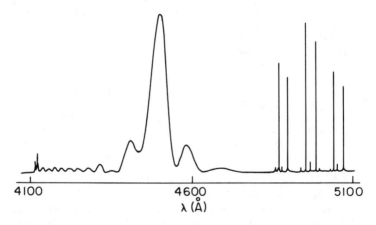

Fig. 5. Long range monochromator scan of fluorescence to $a^3\Sigma_u{}^+$ (bound and continuum levels) and $b^3\Pi_u$ (v=0-2) from $2^3\Pi_g$ (v=13, N=33).

the Hamiltonian. The N·L operator has the rigorous selection rules of $\Delta J = 0$, (e↔e, f↔f), and $\Delta \tilde{N} = 0$. Only one of the parity components of a given Λ-doubled level of $b^3\Pi_u$ has an $a^3\Sigma_u^+$ state in the continuum of the same J, N and parity. Neglecting spin-orbit effects, only the e-component of the $b^3\Pi_u$ F_2 level and the f components of F_1 and F_3 levels can be predissociated. Because F_1 e-parity levels are the lower levels of the P and R transitions of the OODR fluorescence when the second laser excites $1^3\Delta_g$ by a P or R line, these fluorescence lines should be sharp. The Q line (f-parity component) should be broadened by predissociation. Conversely, if the $Q_{11}(33)$ line is used to excite $1^3\Delta_g$, the P and R fluorescence lines are observed to be broadened and the Q line is sharp. Referring to Figure 6, the collisionally populated levels in $1^3\Delta_g$, $N = 32$, $F_1(f)$ and $N = 30$ $F_1(f)$ fluoresce via Q line fluorescence at 5059.6 and 5066.9 Å which are the first two sharp lines to the left and right of the Q(31) line. Examining the P and R branches, the first two features on either side of the sharp unrelaxed lines $\big(P(31)$ and $P(33)$ or $R(29)$ and $R(31)\big)$ are broadened.

The spin-orbit interaction between two case(b) triplet states follows a

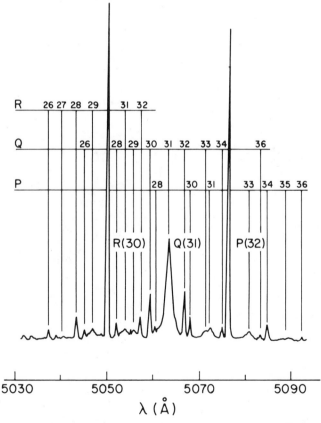

Fig. 6. PFOODR fluorescence spectrum from $1^3\Delta g$ ($v=4$, F_1, $N=31$, e) to $b^3\Pi_u$ ($v=0$). The weak features surrounding the main lines originate from collisional relaxation within $1^3\Delta_g$.

$\Delta N = \pm 1, 0$ selection rule. Because of this selection rule the data, exhibiting the alternating sharp and broad lines predicted by the L-uncoupling predissociative interaction, indicate that the predissociation of $b^3\Pi_u$ at $N \approx 31$ is due almost entirely to the L-uncoupling $b^3\Pi_u \sim a^3\Sigma_u^+$ interaction. If the magnitude of the spin-orbit interaction were competitive with that due to L-uncoupling in causing the broadening of the $b^3\Pi_u$ levels, the data in Figure 6 would should broad P and R lines as well as a broad Q line.

The spectrum of the fluorescence to $b^3\Pi_u$ shows a progression from v=0 to v=11. The linewidth of the Q(31) line is a sensitive function of v because of the variation in the vibrational integral, Bv_av_b, in the expression for the linewidth

$$\Gamma = 2\pi |\langle b^3\Pi_u,v,J| \; \underset{\sim}{H} \; |^3\Sigma_u^+,\varepsilon,J\rangle|^2$$

where $\underset{\sim}{H} = B(R)(N^+L^- + N^-L^+)$, $B(R) = \hbar^2[2\mu R^2]^{-1}$. This simplifies to

$$\Gamma = 4\pi |\langle \underset{\sim}{L}^+\rangle|^2 |Bv_av_b|^2 N(N+1)$$

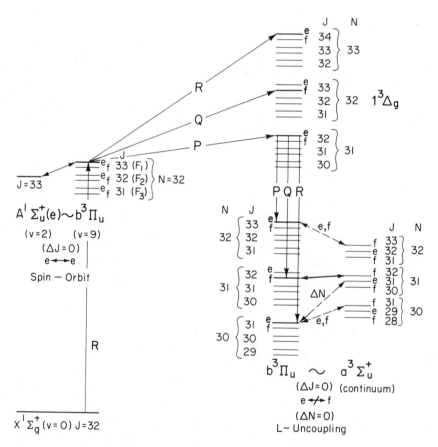

Fig. 7. Perturbation-Facilitated OODR fluorescence scheme used to obtain $b^3\Pi_u$ predissociation data. Perturbation and optical selection rules are indicated. Dashed lines between $b^3\Pi_u$ and $a^3\Sigma_u^+$ levels depict interactions which are forbidden by the selection rules cited adjacently to them.

where $\langle L^+\rangle = \langle b^3\Pi_u, F_1, \Lambda=+1|L^+| a^3\Sigma_u^+, F_1, \Lambda=0\rangle$ and $Bv_av_b = \langle v_\Pi|B(R)|\varepsilon_\Sigma\rangle$. The position and shape of the $a^3\Sigma_u^+$ potential curve relative to the bound vibrational wavefunctions of $b^3\Pi_u$, along with the absolute value of the electronic matrix element, $\langle L^+\rangle$, determine the overall linewidth of an individual level.

Because the $b^3\Pi_u$ state has been well studied, the vibrational wavefunctions can be calculated precisely. The experimental linewidths can then be fit to several parameters describing the $a^3\Sigma_u^+$ curve and the electronic matrix element. Figure 8 shows the results of this fit when just two parameters α and A are used to describe the $a^3\Sigma_u^+$ potential as a simple exponential, $V(R) = \Delta E(^3\Pi_u -^3\Sigma_u^+) + A\exp(-\alpha R)$. $\Delta E(^3\Pi_u -^3\Sigma_u^+)$ is the known experimental difference between the $a^3\Sigma_u^+$ dissociation limit and the minimum of the $b^3\Pi_u$ potential curve, 2724.3 cm^{-1}. The agreement between the calculated and observed linewidth is within the accuracy of the measurements. The $a^3\Sigma_u^+$ curve is accurate over the energy range sampled by the v=0 to 11 $b^3\Pi_u$ vibrational wavefunctions, 11200 cm^{-1} to 14900 cm^{-1} relative to the minimum of the ground state potential.

The agreement between a recent <u>ab initio</u> calculation[14] and our results is excellent. The $a^3\Sigma_u^+$ potential curves agree with our fitted standard deviation of 81 cm^{-1} over the entire energy range sampled and the electronic matrix element from the fit $\langle L^+\rangle = 1.216$ is nearly identical to the <u>ab initio</u> value of 1.228. This suggests that the <u>ab initio</u> electronic wavefunctions for both the $a^3\Sigma_u^+$ state and $b^3\Pi_u$ state are very accurate and can be used with confidence in calculating other electronic properties of these states. These results provide for the determination of the predissociation rate of any vibrational level (up to about v=25 by extrapolating the $a^3\Sigma_u^+$ curve) of the $b^3\Pi_u$ state and, when combined with recent studies in our laboratory[15] regarding the details of the spin-orbit

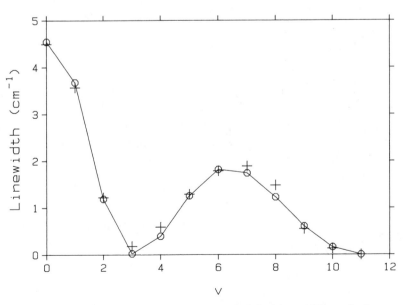

Fig. 8. Observed (+) and calculated (o) linewidths of the $b^3\Pi_u$ (N=31, F$_1$, v=0-11) levels in the PFOODR resolved fluorescence spectrum from $1^3\Delta_g$ (N=31, F$_1$, v=4).

interaction of the $A^1\Sigma_u^+$ state with the $b^3\Pi_u$ state, generate a complete quantitative experimental description of the entire $A^1\Sigma_u^+ \sim b^3\Pi_u \sim a^3\Sigma_u^+$ predissociation complex.

ACKNOWLEDGMENTS

We would like to thank Dr. Li Li and Dr. Xing-bin Xie for their contributions to this work, Prof. W. Meyer for sending us the results in Ref. 14 prior to publication and Dr. R.F. Barrow for useful discussions and criticism. This research was supported by a grant from the National Science Foundation (CHE81-12966).

REFERENCES

1. H. Lefebvre-Brion and R.W. Field, "Perturbations in the Spectra of Diatomic Molecules", Academic Press, Orlando (1986).
2. Li Li and R.W. Field, Direct Observation of High-Lying $^3\Pi_g$ States of the Na_2 Molecule by Optical-Optical Double Resonance, J. Phys. Chem. 87:3020 (1983).
3. X. Xie and R.W. Field, Perturbation Facilitated Optical-Optical Double Resonance Spectroscopy of the $^6Li_2$ $3^3\Sigma_g^+$, $2^3\Pi_g$. $1^3\Delta_g$, $b^3\Pi_u$, and $a^3\Sigma_u^+$ States, J. Mol. Spectrosc. submitted for publication.
4. Li Li, S.F. Rice, and R.W. Field, Observation of the v = 0 – 13 Levels of Na_2 $b^3\Pi_u$ by OODR $^3\Delta_g$ – $b^3\Pi_u$ Fluorescence Spectroscopy, J. Mol. Spectrosc. 105:344 (1984).
5. Li Li, S.F. Rice, and R.W. Field, The Na_2 $a^3\Sigma_u^+$ State. Rotationally Resolved OODR $^3\Pi_g$ – $a^3\Sigma_u^+$ Fluorescence Spectroscopy, J. Chem. Phys. 82:1178 (1985).
6. S.F. Rice, X. Xie, and R.W. Field, The Predissociation of Li_2 $b^3\Pi_u$ by $a^3\Sigma_u^+$, Chem. Phys., submitted for publication.
7. I. Kovacs, "Rotational Structure in the Spectra of Diatomic Molecules, American Elsevier, New York (1969).
8. E.A. Mason and L. Monchick, Determination of Intermolecula Forces in: "Advances in Chemical Physics XII", J.O. Hirschfelder, ed., Wiley, New York (1967).
9. C.A. Coulson, "Valence", Oxford University Press, Oxford (1961).
10. R.F. Barrow, J. Verges, C. Effintin, R. Hussein, and J. D'Incan, "Long Range Potentials for the $X^1\Sigma_g^+$ and $(1)^1\Pi_g$ States and the Dissociation Energy of Na_2", Chem. Phys. Lett. 104:179 (1984).
11. D.D. Konowalow and M.E. Rosenkrantz, Long Range Interactions of Na (3s ^2S) with Na (3s ^2S) or Na (2p ^2P), J. Phys. Chem. 86:1099 (1982).
12. W. Preuss and G. Baumgartner, Time Spectroscopy in the $A^1\Sigma_u^+$ State of Li_2. Perturbations by the $a^3\Pi_u$ State, Z. Phys. A. 320:125 (1985).
13. J. Tellinghuisen, The Franck-Condon Principle in Bound-Free Transitions in: "Photodissociation and Photoionization", K.P. Lawley, ed., Wiley, New York (1985).
14. I. Schmidt-Mink and W. Meyer, Predissociation Lifetimes of the $b^3\Pi_u$ State of Li_2 and the Accidental Predissociation of its $A^1\Sigma_u^+$ State, Chem. Phys., submitted for publication.
15. X. Xie and R.W. Field, The $^6Li_2$ $A^1\Sigma_u^+ \sim b^3\Pi_u$ Spin-Orbit Perturbations: Sub-Doppler Spectra and Steady State Kinetic Lineshape Model, Chem. Phys. Submitted for publication.

THE Cs(7P)+H$_2$ PHOTOCHEMICAL REACTION

R. Vetter

Lab. Aimé Cotton, CNRS II
Bât. 505
91405 Orsay Cedex, France

ABSTRACT

A crossed-beam experiment has been devoted to study the photochemical reaction at threshold : Cs(7P)+H$_2$ → CsH+H. It is shown that use of lasers (excitation of Cs atoms, detection of CsH products) and calculation of diabatic potential energy surfaces are able to provide a good understanding of this particular reactive process.

It is now well established that the use of sophisticated experimental techniques and the development of Quantum Chemistry calculations are providing a better understanding of the different mechanisms which lead to the formation and the rupture of chemical bonds. The necessary key for comparison between theory and experiment is the choice of a simple system where all parameters which determine the path of the reactive collision can be well defined and well controlled before and after (if possible, during !) it takes place. Experimentally, this is achieved by use of supersonic molecular beams which permit a good definition of the energy and the mutual orientation of particles. In this context, use of lasers is of considerable importance since they allow state selection of reagents, orientation of excited atoms and molecules, creation of unstable species (radicals), high resolution determination of product states by laser-induced fluorescence techniques. An other interesting feature is that electronic excitation of atoms and molecules is able to compensate the endoergiticity of certain reactions, thus to open the wide field of photochemical studies.

In the case where electronic excitation of reagents takes place, new problems are raised to the theorist : first, calculation of excited state potential energy surfaces is a difficult task in itself but, second, the existence of many underlying surfaces (often leading to quenching processes) hardly complicates the determination of the reaction path. From the entrance valley of excited reagents to the exit valley of products (often in the ground state) the system has to cross a number of intermediate surfaces : under these conditions, is a single collision process able to yield the products or is it necessary to invoke more complicated mechanisms ?

The reactive system which is discussed here is the Cs/H$_2$ one. From the dissociation energy of H$_2$ [1] and CsH [2], it is easy to determine the endoenergiticity of the reaction in the ground state : ΔH = 2.7 eV. However, as was shown earlier in cell experiments [3][4], excitation of Cs atoms to the 7P state by laser beams of suitable wavelength can compensate the energy defect and promote the reaction :

$$Cs(7P)+H_2 \rightarrow CsH+H$$

Actually, this photochemical process is isoenergetic since the potential energy which is available above threshold is drastically small : 0.0016 eV and 0.024 eV for Cs(7P$_{1/2}$) and Cs(7P$_{3/2}$) excitation respectively : only remains kinetic energy. An explanation for the efficient formation of CsH molecules which is observed in cell experiments can be found in multiple collision processes whose energy balance is more favorable or in high energy encounters in the tail of the velocity distribution [5].

A crossed-beam experiment has been devoted to study this photochemical reaction at threshold ; in principle, it is able to clear up the situation and provide unambiguous results since the observation of a signal relative to CsH molecules is the signature of single collisions of well determined geometry and energy.

THE EXPERIMENT

It is a classical crossed-beam experiment where a supersonic beam of molecular hydrogen (technique developed by R. Campargue [6]) crosses a supersonic beam of Cs atoms at right angle in a collision chamber (residual pressure $\sim$ 10^{-7} Torr). Typical densities at the collision volume are 3.10^{12} mol./cm^3 and 10^{10} at./cm^3 for H$_2$ and Cs respectively. The supersonic expansion in the H$_2$ beam leads to an efficient cooling of the different degrees of freedom of the molecules, up to several Kelvin : only the first vibrational level and the two or three first rotational ones are populated, and the velocity spread is reduced.

A first laser beam enters the collision chamber perpendicularly to the plane of collisions and excites Cs atoms from 6S$_{1/2}$ to 7P$_{1/2}$ or 7P$_{3/2}$, according to its wavelength : λ = 459.3 nm or λ = 455.5 nm respectively. It is delivered by a single mode C.W. jet stream dye laser whose frequency is locked to the desired transition through a servo-controlled system alimented by the fluorescence light emitted by Cs atoms excited at the collision volume.

A second laser beam, collinear to the first one, is used to characterize the CsH molecules which are produced at the collision volume, by exciting them from the X$^1\Sigma^+$(v"=0,j") levels* to the A$^1\Sigma^+$(v'=5,j') ones from which they fluoresce. It is provided by a single mode C.W. ring laser whose frequency is locked to the desired transition by use of a sigmameter [7]. A preliminary study of the CsH spectrum has been made by comparing the CsH absorption lines to the I$_2$ ones [8][9].

The fluorescence light due to CsH molecules is collected by a para-
boloid mirror (efficiency ∿ 30 % of the full solid angle) and received
onto the cathode of a low noise photomultiplier through a series of
filters. The corresponding signal is discriminated and stored for short
time intervals (typically 1 second). The role of the filters is to eliminate stray light at the laser wavelengths and the fluorescence light emitted by Cs atoms (on the 7P-6S, 6P-6S and "forbidden" 5D-6S transitions).
A compromise must be found between an efficient rejection of this undesi-
red light and the spectral width of the detection window : under appropriate conditions, the "signal" is of the order of 10^2 cs/s, superimposed to
a background of the same order of magnitude. A better contrast is obtained
by further reducing the detection window. Then, by slowly scanning the
frequency of the second laser, one obtains the spectral "profile" of the
signal, the width of which is typically 3 or 4.10^{-3} cm^{-1} (∿ 100 MHz)
(figure 1).

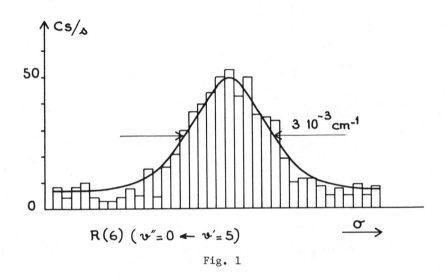

Fig. 1

---

* Due to energy considerations, only the $X^1\Sigma^+(v''=0)$ levels are populated
in the experiment.

EXPERIMENTAL RESULTS

Three principal results have been gathered up to now :

- The reaction proceeds through a single collision process :
$Cs(7P)+H_2 \rightarrow CsH+H$. Signals due to CsH in the $X^1\Sigma^+(v''=0)$ level have been
recorded in the range $J'' = 1$ to $J'' = 16$, in agreement with the energy
balance of the reaction, when Cs atoms are excited to the $7P_{1/2}$ level. In
accordance to the estimated values of the particle densities in the two
beams, to the efficiency of the fluorescence collection and to the different parameters of the CsH molecule, they correspond to a reactive cross-
section of the order of 10^{-16} cm^2. This is only a rough estimate but it is
enough to ascertain that a two collision process cannot be observed in
this experiment since it would require cross sections of 10^{-13} cm^2 for
each collision to yield the same signal ; such cross sections are highly
unlikely [10].

- The reactive cross section for $Cs(7P)_{1/2}$ excitation is roughly
ten times larger than the one for $Cs(7P)_{3/2}$ excitation. Such a "fine
structure" effect was already observed for a limited number of reactions
[11][12].

- In the $X^1\Sigma^+(v''=0)$ level, the population of the product molecule
varies with j" values according to figure 2 ($Cs(7P_{1/2})$ excitation). The
number of counts per second is obtained for a constant value of the second laser power and is reported to the number of counts relative to the
background (mainly proportional to the $Cs(7P_{1/2})$ population). The kinetic
energy which is available in the center of mass is indicated on figure 2,
with error bars corresponding to the velocity spread of the hydrogen beam.
Although still preliminary, these measurements indicate a linear variation
for j" values up to 12, 13, followed by a rapid decrease for higher
values. It has not been possible to observe a signal for j" = 17 and on
molecular transitions originating from the $X^1\Sigma^+(v''=1,j''=4)$ level.

DISCUSSION

The first experimental result - the reaction proceeds through a unique collision - is in itself an important result which deserves discussion. The problem comes out due to the fact that the reagent system lies,
in the most general case of an undetermined geometry of approach, on the
11th potential energy surface from the ground state and has no potential
energy to lose - only relative kinetic one. Two collision processes, as
discussed in [5] and [10] should offer a much better energy balance and
probably occur in cell experiments [3][4][13], in addition to the single
collision process ; for example :

$$\begin{cases} Cs(7P)+H_2(v''=0) \rightarrow Cs(6S)+H_2(v''=6) \\ Cs(6S)+H_2(v''=6) \rightarrow CsH(X^1\Sigma^+)+H \end{cases}$$

or
$$\begin{cases} Cs(7P)+H_2(v''=0) \rightarrow [CsH_2]^{\neq *} \\ [CsH_2]^{\neq *}+Cs(6S) \rightarrow 2CsH(X^1\Sigma^+) \end{cases}$$

This last mechanism could explain the nucleation of CsH molecules giving
rise to "laser snow" formation [3].

A first confirmation of the single collision process is provided by calculation of potential surfaces in the frame of the adiabatic approximation : as shown in [5], there is no activation barrier on the ground state surface of the products, in the case where the collision proceeds through a collinear approach ; thus the reaction is at least possible.

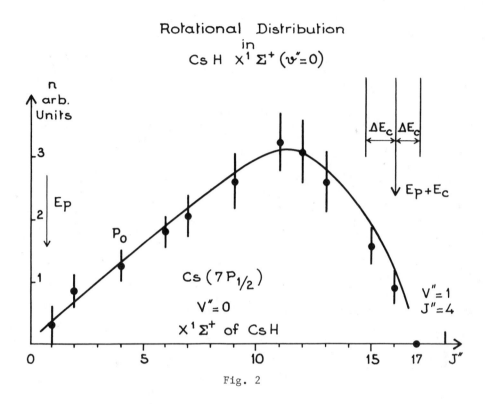

Fig. 2

More precise results can be derived, still in the case of a collinear approach, from the recently calculated quasidiatatic surfaces including : $Cs(6S)+H_2$, $Cs(6P)+H_2$, $Cs(5D)+H_2$, $Cs(7S)+H_2$, $Cs(7P)+H_2$ and $Cs^+ + H_2^-$ [14][15] ; it is shown that the ionic diabatic surface crosses all the neutral ones, the $Cs(7P)+H_2$ one in the entrance valley ($R_{Cs-H_1} \sim 6.5$ as) and the others for smaller $Cs-H_1$ and larger H_1-H_2 distances. The first crossing ($Cs(7P)+H_2$ and $Cs^+ + H_2^-$) occurs at an energy of the order of 300 cm^{-1} above the asymptotic value and is classically allowed since the kinetic energy available in the experiment is of the order of 800 cm^{-1}. The other intersections are below the asymptotic value.

These results confirm the harpooning mechanism which was suggested in [10] ; the Cs(7P)+H_2 system transfers the Cs optical electron to the molecular hydrogen at large Cs-H_1 distance and the newly formed Cs^++H_2^- slowly transforms into Cs^+H^-+H as the distance between the two hydrogen atoms increases. The efficiency of the process depends on the probability of staying on the ionic surface compared with the probability of jumping back to the lower neutral "quenching" channels. To estimate the corresponding losses, the electronic couplings between neutral surfaces and the ionic one have been calculated ; results indicate that the Cs(7P)+H_2 → $Cs^+H_2^-$ transfer is efficient whereas the deexcitation mainly occurs through Cs(7S) and Cs(5D) states.

Given these accurate quasidiabatic surfaces, it should be now of interest to introduce the effect of the spin-orbit interaction to determine if one can give account of the difference of reactivity $Q(7P_{1/2})/Q(7P_{3/2})$ which has been experimentally observed. Work is in progress in this direction.

The last experimental observation which deserves discussion is the rotational distribution of products in the v" = 0, $X^1\Sigma^+$ levels. Although still preliminary, the measurements indicate a quasilinear variation of the reaction probability with J" values, followed by a rapid decrease for J" > 12,13 (cf. figure 2). It is interesting to note that J" = 12 corresponds to an equal partitionning of the incident kinetic energy into rotational energy and kinetic energy of the products : then, does the reaction probability decrease for J" values in connection with the reactive mechanism itself or is it due to the decrease of energy caused by the velocity spread in the hydrogen beam ?

A statistical approach (the "Prior" Distribution [16]) indicates that the reaction probability should vary as $(2J"+1)(E-BJ"(J"+1))^{1/2}$ where E is the total available (here kinetic) energy of reagents and B is the rotational constant of the CsH molecule. It assumes that there is no constraint in the reactive system, but conservation of energy ; then, all quantum states are equally probable. The corresponding curve is shown on figure 2 for comparison with experimental data.

On an other hand, a very simple model of collision can be proposed where there is simultaneously breaking of the hydrogen bond and formation of the CsH one ; half the kinetic energy of H_2 is converted into rotational energy of CsH. By writing the conservation of angular momenta, $\hbar\sqrt{j"(j"+1)}$ = $\mu V_r b$ where μ is the reduced mass of CsH, V_r is the relative velocity and b the impact parameter, one gets b $\sim$ 5 a_0 for J" = 12 ; this value of b is to be compared with the value of the Cs-H_1 distance at which occurs the jump from the Cs(7P)+H_2 surface onto the $Cs^+H_2^-$ one (R_{Cs-H_1} $\sim$ 6,5 a_0). In this last case, a linear variation of the reaction probability should be observed as soon as the "opacity function" [17] is constant with b values.

One of the conclusions which can be drawn from this discussion is in the necessity of having a much better accuracy in the determination of the reaction probability, especially for large J" values. This is related to a much better control of the overlapping of the four beams in the collision chamber. Future developments include variations of the kinetic energy which is available to the reactive system : for instance, it should be of interest to cool the hydrogen beam nozzle to liquid nitrogen temperature, hence to reduce the kinetic energy to roughly 300 cm^{-1}, the height of the barrier calculated in the quasidiabatic approximation. Other experiments - linked to the use of laser beams - one planned : rotation of the Cs(7P) orbitals with the direction of the laser beam polarization, excitation of the Cs atom towards the Rydberg states, determination of the differential cross-sections. There is no doubt that in the future, the interfacing of molecular beams and laser beams will provide deeper insights into reactive mechanisms, in particular those related to more complicated systems.

REFERENCES

[1] G. Herzberg, Molecular Spectra and Molecular Structure. I. Spectra of Diatomic Molecules. Van Nostrand (1950).

[2] C. Crépin, J. Vergès, C. Amiot, Chem. Phys. Lett., $\underline{112}$, 10 (1984).

[3] A. Tam, G. Moe, W. Happer, Phys. Rev. Lett., $\underline{35}$, 1630 (1975).

[4] J.L. Picqué, J. Vergès, R. Vetter, J. de Physique, Lett., $\underline{41}$, 305 (1980).

[5] F.X. Gadea, G.H. Jeung, M. Pelissier, J.P. Malrieu, J.L. Picqué, G. Rahmat, J. Vergès, R. Vetter, Laser Chem., $\underline{2}$, 361 (1983).

[6] R. Campargue, A. Lebéhot, J.C. Lemonnier, Rarefied Gas Dynamics, Progress in Astronautics and Aeronautics, $\underline{51}$, 1033 (1976).

[7] P. Juncar, J. Pinard, Opt. Comm., $\underline{14}$, 438 (1975).

[8] C. Crépin, Thèse de 3ème Cycle, Université Paris XI (1984).

[9] S. Gerstenkorn, P. Luc, Atlas du Spectre d'Absorption de la Molécule d'Iode, Edition du C.N.R.S., Paris (1978).

[10] C. Crépin, J.L. Picqué, G. Rahmat, J. Vergès, R. Vetter, F.X. Gadea, M. Pelissier, F. Spiegelmann, J.P. Malrieu, Chem. Phys. Lett., $\underline{110}$, 395 (1984).

[11] T.D. Dreiling, D.W. Setser, J. Chem. Phys., $\underline{79}$, 5423 (1983).

[12] H.J. Yuh, P.J. Dagdigian, J. Chem. Phys., $\underline{79}$, 2086 (1983) ; M.L. Campbell, N. Furio, P.J. Dagdigian, Laser Chem., to be published (1986).

[13] B. Sayer, . Ferray, J. Lozingot, J. Berlande, J. Chem. Phys., $\underline{75}$, 3894 (1981) ; M. Ferray, J.P. Visticot, B. Sayer, J. Chem. Phys., 81, 3009 (1984).

[14] G. Rahmat, J. Vergès, R. Vetter, F.X. Gadea, M. Pelissier, F. Spiegelmann, Recent Advances in Molecular Reaction Dynamics, Edited by R. Vetter and J. Vigué, Editions du C.N.R.S., Paris (1986).

[15] F.X. Gadea, F. Spiegelmann, M. Pelissier, J.P. Malrieu, J. Chem. Phys., to be published (1986).

[16] A. Ben-Shaul, Y. Haas, K.L. Kompa, R.D. Levine, Lasers and Chemical Change, Springer-Verlag, Berlin (1981).

[17] J.S. Mac Killop, C. Noda, M.A. Johnson, J.R. Waldeck, R.N. Zare, to be published (1985).

ACKNOWLEDGMENTS

The author wishes to thank his colleagues who have participated to the experiment and the interpretation, namely G. Rahmat, J. Vergès, F.X. Gadea, J.P. Malrieu and F. Spiegelmann.

TIME RESOLVED SPECTROSCOPY OF JET-COOLED ACETONE

O. Anner, G.D. Greenblatt, Y. Haas and H. Zuckerman

Department of Physical Chemistry, The Hebrew University of
Jerusalem 91904, Israel

INTRODUCTION

Acetone occupies a unique position in organic photochemistry. It has
been one of the most intensively studied molecules for over fifty years,
both in liquid solution and in the gas phase [1]. Yet, the details of its
photodecomposition process, particularly of the isolated molecule, remain
practically unknown.

We have recently shown [2] that acetone exhibits relatively strong
fluorescence at low pressures (10^{-4}-10^{-3} Torr), with decay times of about
10^{-6} sec for the protonated molecule and almost 10^{-5} sec for the deuterated one. This result holds for excitation near the origin of the $n\pi^*$
electronic transition. Very large quenching cross sections were measured
for practically any added gas, that of acetone itself being about an order
of magnitude larger than the gas-kinetic one. This result accounts for the
much shorter decay times reported previously [1], of the order of 10^{-9} sec.
They were measured at pressures exceeding 1 Torr, and were thus totally
collision controlled.

A detailed study of the evolution of electronically excited acetone
molecule was hampered by the severe spectral congestion prevailing at room
temperature even at very low pressures. In order to define properly the initial energy, supersonic cooling must be used for gaseous samples. The fluorescence excitation spectrum of jet-cooled acetone was first published by
Hanazaki and coworkers [3] who also provided an assignment for the vibronic
bands observed. We have recently reported the study of time resolved emission from single vibronic levels (SVL's) of jet-cooled acetone [4]. In this
paper a more detailed description of experiments conducted to date is given.

EXPERIMENTAL

Acetone was seeded in helium or argon, and expanded from a high pressure region to a vacuum chamber maintained at 10^{-5}-10^{-4} Torr. A pulsed conical nozzle was used, with a diameter of 0.35 mm. The unskimmed molecule
beam was crossed 30-40 nozzle diameters downstream by a pulsed UV laser
beam, that could be continuously tuned with a spectral resolution of about
1.5 cm^{-1}, and a pulse width of $\sim$7 nsec. The pulse energy was usually 0.1-
0.5 mJ, and the laser beam was collimated to a diameter of about 2 mm. Both

devices (the laser and nozzle) were operated at a repetition rate of 10 pps, and were synchronized to yield a proper signal as described below.

Fluorescence was observed at right angles to both the molecular and light beams, passed through a cutoff filter (λ >400 nm) and detected by a fast photomultiplier (RCA 8850, risetime 3 nsec). The signal was digitized by a 500 MHz digitizer (Tektronix 7912) and averaged until a desired signal to noise was obtained. Data were usually recorded as decay curves. Excitation spectra were obtained by scanning the UV laser and measuring the emission intensity about 100-200 nsec after the excitation pulse.

In the early experiments, reproducibility was observed to be poor. This turned out to be due to the extreme sensitivity of the signal's intensity to any change in the experimental conditions: Total pressure, gas composition, valve opening time, the time of delay between the valve opening and the laser pulse and of course exact tuning and laser power.

As an example, Figure 1 shows the signal intensity as the laser pulse is scanned across the molecular pulse. The latter was about 600 µsec long. It is seen that at low acetone pressures a fairly smooth curve is obtained with a broad maximum at about the center. However, as the acetone partial pressure is increased (with the overall pressure being kept constant), a distinct dip in the fluorescence intensity is obtained in the middle. We assign this decrease in intensity as due to cluster formation: as the valve opens up, cooling becomes more efficient, and acetone molecules tend to form clusters. This tendency increases with acetone pressure. Electronically excited acetone, when adjacent to one or more ground state molecules, is expected to fluoresce weakly, due to the high quenching efficiency of acetone [3].

Decay times of up to about 3 µsec could be measured by the method described above. Longer decay times were distorted by the fact that the fluorescent molecules were leaving the field of view. These were measured by setting the photomultiplier and the collecting lens on the axis of the molecular beam, facing the nozzle at a distance of about 50 cm. In this way lifetimes up to 100 µs could be measured.

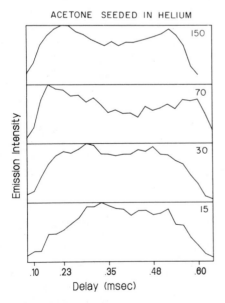

ACETONE SEEDED IN HELIUM

Fig. 1: The fluorescence signal of acetone seeded in helium, as a function of the delay time between the beginning of the molecular beam pulse and the laser pulse. The decrease observed as the valve reaches its maximum opening at high acetone pressures probably indicates cluster formation. The pressure of acetone (Torr) in the expansion mixture, is shown in each panel.

RESULTS

As indicated in the experimental section, cluster formation is deduced from the reduction in the fluorescence signal intensity at high acetone pressure. We tried to detect clusters (of acetone with itself, or with argon) by the appearance of new excitation bands. These attempts were unsuccessful. In the following discussion we shall use their numbering to designate individual lines. Counting from the origin, only the first 15 lines (in acetone-h_6 are well separated from each other. At higher excitation energies, the spectrum shows structure, but emission results upon excitation by any laser frequency. Within our resolution, a quasi-continuum thus prevails. This phenomenon, which is yet more pronounced in acetone-d_6, hampers the observation of new weak peaks. However, even in the discrete part of the spectrum, no added lines were observed upon increasing the acetone pressure. We conclude that clusters, if present, are weakly fluorescent upon excitation in this region. Further evidence for their existence is obtained from the fact that the signal in argon is weaker than in helium.

Decay times were approximately, but not strictly, exponential. The numbers given below were obtained from a fit for about two 1/e lifetimes. For any given line, the decay in argon was somewhat longer than in helium. The decay time at the origin was about 1.0 µsec for acetone-h_6 and 3 µsec for acetone-d_6 [4]. Upon excitation into successively higher levels, it gradually increased up to about 2.65 µs for acetone-h_6 in argon (Fig. 2). The same qualitative trend was observed in helium, although the increase was less pronounced, up to 1000 cm^{-1}, a gradual decrease is obtained. In the range of 1000-2000 cm^{-1} the decay rate is roughly constant, and beyond that there is an increase.

A very dramatic deuterium effect was observed, exemplified by Fig. 3. The initial lifetimes were, for an excess energy of 1606 cm^{-1} ((CH_3)$_2$CO) and 1761 cm^{-1} ((CD_3)$_2$CO), 0.9 µs and 3.3 µs, respectively. In the deuterated samples, one could clearly observe a decay component exceeding 10 µsec. This component was revealed upon looking straight at the jet, and although of small amplitude, was distinctly above the noise level.

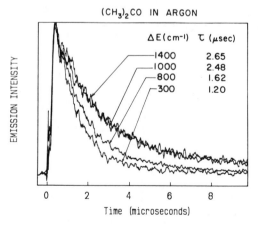

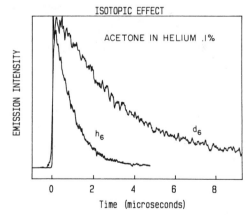

Fig. 2: Representative decay curves of acetone seeded in argon. ∆E is the vibrational energy above the zero point level of S_1, and τ the observed decay time.

Fig. 3: Decay curves of acetone seeded in helium, upon excitation to about the same energy increment (∆E) above the origin for acetone-h_6 (bottom curve, ∆E=1761 cm^{-1}) and acetone-d_6 (top curve, E=1606 cm^{-1})

DISCUSSION

1. The decays observed are not exponential. In particular, a long-lived emission component appears to be present, and is more pronounced in acetone-d_6 than in acetone-h_6.

2. The decay times measured in argon are longer than those observed with helium as a carrier gas.

3. The isotopic effect is quite large—decay times in the deuterated molecule are about three times longer than in the protonated one.

4. The decay rate decreases as the excitation energy is increased up to 1000 cm^{-1} excess energy, remained roughly constant from that energy up to 2000 cm^{-1}, then rose again somewhat. Further increase in energy does not affect the decay rate, but the intensity appears to reach a maximum at an excess energy of about 3000 cm^{-1} and then decreases.

The multiexponential decay can be understood in terms of the preparation of a variety of different species upon optical excitation. According to the analysis of Hanazaki et al., single vibronic levels can be prepared. However, our laser cannot resolve rotational structure, leading to the preparation of many J,K states, most of which are highly degenerate. Work on formaldehyde showed that radiationless transitions depend strongly on the rotational state, leading to a large change in decay rates. This phenomenon could well be observed in acetone as well. However, attempts to excite different parts of the rotational envelope of a given vibrational band, failed to reveal different decay kinetics. The observation of a long decay time, larger than the one calculated for the radiative lifetime of acetone, indicates triplet state formation by a non radiative process. The large isotope effect also points in that direction.

Another possible source for species inhomogeneinty is cluster formation. Our experiments strongly indicate cluster formation, and suggest that clusters are more weakly fluorescent than monomers. The appearance of fluorescence at excess energies as high as 4500 cm^{-1} is somewhat puzzling, since the dissociation threshold for acetone is much lower. The fact that the decay rate constant (in helium) is as low as 10^6 sec^{-1} at these high energies is also surprising for an isolated molecule.

Dimer formation may account for these observations: Upon light absorption, the dimer falls apart, and leaves behind an electronically vibrational excited acetone molecule, with only about half the nomial excess energy (the other half goes to the other, groundstate, monomer). Aggregation may also account for the effect of argon: argon is heavier than helium, and may enhance the spin orbit coupling. Upon dissociation of a cluster containing an argon atom, the contribution from a triplet state is more pronounced, leading to a longer decay rate. It should be mentioned that acetone clusters have been abundantly observed in supersonic jets using mass spectrometry. In our experiments, the signal intensity is linear with the acetone pressure, as expected for monomer. The contribution of dimers is thus still not clear.

REFERENCES

1. E.K.C. Lee and R.S. Lewis, Adv. Photochem. 12, 1 (1981); R.B. Cundall and A.S. Davies, Proc. Roy. Soc. A290, 563 (1966).
2. G.D. Greenblatt, S. Ruhman and Y. Haas, Chem. Phys. Lett. 112, 200 (1984).
3. M. Baba and I. Hanazaki, Chem. Phys. Lett. 103, 93 (1983); M. Baba, I. Hanazaki and U. Nagashima, J. Chem. Phys. 82, 3928 (1985).
4. O. Anner, H. Zuckermann and Y. Haas, J. Phys. Chem. 89, 1336 (1985).

EXPERIMENTAL DETERMINATION OF THE ClO_2- $\tilde{A}(^2A_2)$ (0,0,0) PREDISSOCIATION LIFETIME

K.J. Brockmann and D. Haaks
Universität G.H. Wuppertal

FB 9, Physikalische Chemie
Gaußstraße 20
5600 Wuppertal 1 West Germany

Abstract

The time resolved ClO_2 $\tilde{A}(^2A_2)$(0,0,0) $\longrightarrow \tilde{X}(^2B_1)$(0,0,0)) fluorescence excited with the 4765 Å single mode argon- ion laser line leads to two distinct lifetimes corresponding to the spin states Fl(N=J+1/2) and F2(N=J-1/2) with tl=(56 ± 20) ps and t2=(336±27) ps respectivily. Our tl- value is in good agreement with the tl- value calculated from the two recent rotational linewidth measurements /1,2/. Contrary to this finding t2 is prolonged by more than a factor of three to the linewidth results. We suggest a strong spin- rotation coupling of $\tilde{A}(^2A_2)$ with a state of higher density of states, possibly $\tilde{A}(^2A_1)$ and a feed back to $\tilde{A}(^2A_2)$ in order to explain the pronounced differences in the Fl- and F2- lifetimes and the differences in the results of the distinct experimental methods.

Introduction

The first observations of the faint emission spectra of ClO_2 induced by various argon- ion laser lines have been reported by Sakurai, Clark and Broida /3/ . Their attempt to measure the lifetime of the excited ClO_2 with a pulsed dye laser was not successfull due to the 10 ns width of the laser pulse. The fluorescence decayed almost within the excitation pulse. Later Curl et al. /4/ assigned the 4765 Å excitation savely to the ((0,0,0) --- (0,0,0))band of the (A $\longleftarrow$ X)- transition. From the fact that the emission intensity is very faint the authors concluded, that most of the excited molecules do not fluoresce but undergo an intersystem crossing to a nonradiating, possibly predissociative electronic state. Using Fabry-Perot spectrometry McDonald and Innes /1/ calculated the lifetimes of various vibrational levels from their measurements of the rotational linewidth of several absorption profiles. These measurements lead to the result, that the two spin components Fl and F2 have distinct lifetimes which are indipendent on the rotational quantum number N but slightly dependent on the vibrational quantum. The lifetime of the Fl component varied between 46 and 95 ps and that of the F2 component from 53 to

220 ps for vibrational excitation of the symmetric stretching mode to a maximum of v'= 5 . For the ratio of the F2- to the F1- lifetime a factor near two was observed. The short life- times were interpreted as being due to rapid predissociation by spin- orbit interaction of the A - state with a lower elec- tronic state of A and/or B symmetry and not to the radiative decay. In a more recent work Michielsen et al. /2/ reinvesti- gated linewidth measurements with the Fabry- Perot inter- ferometry methode. Their average t1- and t2- values calculated from the individual linewidths are not dependent on v' until v' = 3 and increase slightly for higher vibrational quantum numbers v'. In accordance with McDonald and Innes, within the experimental error limits no dependence of the lifetimes on the rotational quantum number N was observable too. The ratio of t2/t1 was found to be 1.6 . Michielsen et al. assumed that the difference in the lifetimes of the two spin components is due to spin-rotation coupling of the above mentioned states and they calculated a t2/t1 ratio of 1.2 from measured spin- rotation coupling constants. The goal of the present work was the direct determination of the t1- and t2- lifetimes with a mode locked single mode argon- ion laser.

Experimental

Chlorine dioxide was prepared in vacuum from potassium chlorate and oxalic acid by the method of Bray /5/. After carefull trap to trap distillation ClO_2 was stored in the dark at -78 C on silicagel. Samples were tested for impurities with an FTIR spectrometer. Throughout our measurements ClO_2- pres- sures between 800 mTorr and a maximum of 45 Torr were used. Lightpulses with an average power of several milliwatts at 4765 Å and a pulse width of 300 ps were generated from a modelocked single mode argon laser. The fluorescence of CLO_2 was monitored via a SPEX 75 cm Czerny-Turner monochromator and a very fast channel plate photomultiplier (HAMAMATSU Typ. 164U-01). Decay curves were measured with the delayed coin- cidence technique and were stored in a LSI 11 micro computer. The delayed coincidence was triggered by a fast avalange photo diode via a calibrated variable delay line. This delay line served also for calibration purposes. Throughout all our ex- periments a calibration constant of k = (4.962 ± 0.05) ps/ channel has been used. A great deal of effort was investi- gated in the carefull design of the fluorescence cell and the pass of the fluorescence within the cell. The entrance and exit windows for the argon laser beam were placed at Brewster angle and the fluorescence was collected with a concave mirror and focussed onto the entrance slit of the monochromator. In order to minimize problems arrising from different transit times of the fluorescence, the direct fluorescence was sup- pressed by a set of spatial filters. The fluorescence signal was deconvoluted with the Raman scattered signal observed at atmospheric pressure by using a computer programm written by Striker /6/. This deconvolution was only successfull in the case of using two exponential fit parameters t1 and t2. The Raman signal was also used as a monitor for the alignment of the argon laser within the cell, by comparing this signal with the signal from the photo diode monitored with a fast Sampling Oscilloscope.

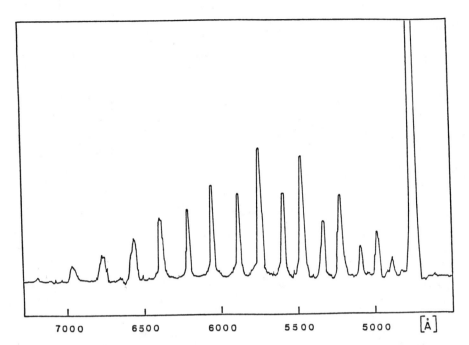

Fig.1 A typical $ClO_2(A \rightarrow X)$- spectrum excited at 4765 Å.

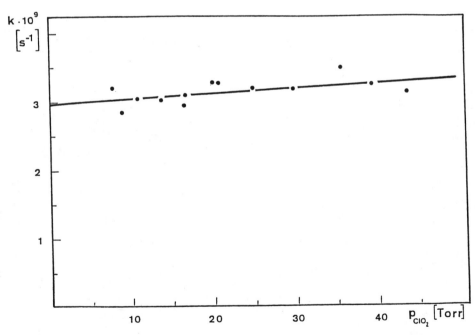

Fig. 2 The dependence of the decay rate K2 observed in the ClO_2- fluorescence excited at 4765 Å as a funktion of the ClO_2- pressure.

Results

In fig. 1 a typical fluorescence spectrum of ClO_2 excited at 4765 Å and monitored with low spectral resolution is shown. All lifetime measurements were carried out at the observation wavelength of 5780 Å. For example the decay rate K2 is plotted versus the ClO_2- pressure in fig. 2. The extrapolation of K1 and K2 to zero pressure leads to two distinct lifetimes t1 = (56±20) ps and t2 = (336±27) ps. The short lifetime t1 is within the experimental error limits in good agreement with the results obtained from the linewidth measurements /1,2/, but t2 is prolonged by more that 300%. For comparison see table I.

Table I

	this work	McDonald et al. /1/	Michielsen et al. /2/
t1 (F1)	(56±20) ps	(47±10) ps	(67±14) ps
t2 (F2)	(337±27) ps	(101±10) ps	(90±18) ps

Discussion

In accordance with Michielsen et al. /2/ the difference in the predissociation rates of the two spin components F1 and F2 is interpreted by a strong spin rotation coupling of $\tilde{A}$ ($^2A_2$) with an intermediate state of a longer lifetime and a higher density of states, see fig. 3. This state is possibly $\tilde{A}$ ($^2A_1$)/7/.

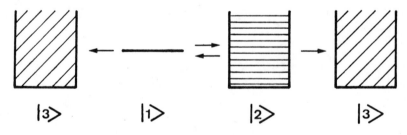

$$|3\rangle \qquad |1\rangle \qquad |2\rangle \qquad |3\rangle$$

Fig.3

The difference between the directly measured lifetimes and the values calculated from rotational line width measurements can be explained by assuming that the decomposition of the intermediate state $|2\rangle$ to the continuum state $|3\rangle$ is weaker than the coupling of $|1\rangle$ to the continuum. Fermi's golden rule shows that the transition rate from a state $|1\rangle$ to a state $|2\rangle$ is given by the product of the final state density n (E) and the transition matrix element.

$$P12 = \frac{2\pi}{\hbar} |\langle 1 | H_{12} | 2\rangle|^2 \, n_2(E)$$

We suggest that the state $|2\rangle$ feeds back to state $|1\rangle$ with the rate

$$P21 = \frac{2\pi}{\hbar} |{<}2 | H_{21} | 1{>}|^2 \, n_1(E)$$

Assuming $|{<}1 | H_{12} | 2{>}|^2 = |{<}2 | H_{21} | 1{>}|^2$ the feed back transition rate P_{21} is given by:

$$P21 = P12 \, \frac{n_1(E)}{n_2(E)}$$

State $|2{>}$ is expected to be well below state $|1{>}$ /7/ with a larger density of states $n_2(E)$ in the energy region of state $|1{>}$. Therefore this feed back transition rate $P12$ should be smaller than $P21$ leading to a prolonged lifetime of the radiative state $|1{>}$ compared with the values calculated from the integrated absorption coefficient /8/. We therefore expect a stronger influence of this effect on the weaker coupled doublet component F2 than on F1 in agreement with the experimental results.

Acknowledgement

We thank the Deutsche Forschungsgemeinschaft (SFB 42) and the Fonds der Chemischen Industrie for the financial support of this work. We also thank J.Schaab for the assistance in the evaluation of the data.

References

1. P.A. McDonald and K.K. Innes
   Chem. Phys. Lett. _59_, 562 (1978)
2. S. Michielsen, A.J. Merer, S.A. Rice, F.A. Novak, K.F. Freed and Y. Hamada
   J. Chem. Phys. _75_, 3029 (1981)
3. K. Sakurai, J. Clark and H.P. Broida
   J. Chem. Phys. _54_ 1217 (1971)
4. R.F. Curl, K. Abe, J. Bissinger, C. Bennet and K.F. Tittle
   J. Mol. Spectr. _48_, 72 (1973)
5. W. Bray
   Z. Physik. Chemie _54_, 569 (1906)
6. G. Striker
   Priv. Commun.
7. J.L. Gole
   J. Chem. Phys. _84_, 1333 (1980)
8. A.E. Douglas
   J. Chem. Phys. _45_, 1007 (1966)

LASER SPECTROSCOPY OF PROTON-TRANSFER IN MICROSOLVENT CLUSTERS

Ori Cheshnovsky[a] and Samuel Leutwyler[b]

[a]School of Chemistry, Tel-Aviv University
Ramat-Aviv, Tel-Aviv 69978, Israel

[b]Institut fur anorganische Chemie, Universitat Bern
Freiestrasse 3, 3012 Bern, Switzerland

INTRODUCTION

The effect of stepwise solvation of molecules and ions on their acidity has been the subject of extensive investigation in the last decade [1]. Here we report [2] the observation of proton transfer reaction in isolated ultracold <u>neutral</u> clusters of an acid molecule HA with n base molecules B:

$$HA \cdot B_n \rightleftharpoons A^- \cdot H^+ B_n$$

where HA is the aromatic excited state acid α-Naphtol and B is NH_3. In aqueous solution α-Naphtol is known to change its acidity in its excited state. A weak base in its ground state S_0 ($pK_a=9.4$) it becomes a strong acid in its S_1 excited state ($pK_a*=0.5\pm0.2$). This phenomenon, manifested in proton transfer to the solvent by the excited molecule, has been widely investigated and reviewed [3]. The spectroscopic fingerprints of proton transfer in aqueous solution are the appearance of a broad emission band of the excited α-Naphto peaked at 465 nm, which is $\sim$7000 cm^{-1} red shifted from the emission band of the undissociated excited α-Naphtol.

In the molecular beam, species selective absorption spectroscopy of α-Naphtol $(NH_3)_n$ clusters was performed by one-colour resonant two-photon-ionization (R2PI). Based on the identification of the spectral features of each complex, the laser induced fluorescence (LIF) emission spectra were then measured as a function of the cluster size. The search for proton transfer in these clusters involved the monitoring of fluorescence emission from the undissociated α-Naphtol* and/or the Naphtolate anion.

EXPERIMENTAL

Microsolvent complexes of α-Naphtol $(NH_3)_n$ were synthesized in pulsed adiabatic expansions. α-Naphtol was seeded into a gas mixture of the composition: NH_3 (0.5%)/noble carrier gas. The total backing pressures were in the range of 1.2-1.6 bars. The pulsed heatable valve delivered gas pulses of about 200 μsec duration. Three types of experiments were performed:
1. Laser induced fluorescence using quartz collection optics and an unfiltered RCA 7265 photomultiplier.

2. Resolved emission using the same optics and a Jobin-Ynon H25 mono-chromator. The optical bandpass employed was $\Delta\lambda = 7$ Å.
3. Resonant two-photon ionization (R2PI): The skimmed core of the expan-sion was passed into a second differentially pumped vacuum chamber $(p<5\cdot10^{-8})$, which housed a 44 cm long space-focusing time of flight mass spectrometer.

Optical excitation and ionization were performed with the frequency doubled output of a Lambda Physik FL2002 dye laser pumped by the second harmonic of a Quanta-Ray DCR 2A Nd: Yag Laser.

LIF AND R2PI SPECTRA

Following the opening of the pulsed valve, the beam density increases to a maximum plateau value. Simultaneously, the beam temperatures drop, and cluster nucleation and growth shifts the average cluster size to higher values. We have utilized the variable delay between the valve opening and the laser pulse to monitor different size distribution regions: At short delays ($\tau_D<80$ μsec) only microclusters of n=0,3 were observed whereas for longer delays (up to 250 μsec) a wide range of medium size clusters (n=1,20) was produced. Spectra in the frequency range 29800-32500 cm^{-1} were obtained for the α-Naphtol NH_3 clusters by both the LIF and R2PI techniques. The mass resolved R2PI spectra make two distinct groups:
1. Well resolved, vibrationally and rotationally cold spectra are found for microsolvent clusters with n=1,3 (see Fig.1).
2. By contrast, the R2PI spectra of medium-size solvent clusters with n>3 are completely broadened, although a vibrational structure, with spectral features, hundreds of wave numbers wide, is still discernible.

EMISSION SPECTROSCOPY

At short delay times, when clusters with $n\geqslant4$ are absent from the cluster distribution, excitation at the various bands of the n=0,3 species yields discrete, well structured emission spectra with maximum intensity in the near UV ($\nu_{max}\sim31000\ cm^{-1}$). All of those spectra are very similar to the emission spectrum of the bare α-Naphtol (Fig.2a). When $\tau_D>100$ μsec a non structured emission band is observed in the blue-violet ($\nu_{max}\sim24000\ cm^{-1}$). The intensity of this feature increases with increasing τ (Fig.2b and 2c). The excitation spectrum of this "blue" emission is very broad resembling the R2PI spectra of medium size clusters ($n\geqslant4$).

PROTON TRANSFER IN MICROSOLVENT CLUSTERS

We conclude that the excitation of the large microsolvent clusters to the S_1 state gives rise to the broad emission characteristic of the α-Naphtolate anion. The appearance of the anion fluorescence marks the onset of intracluster ion-pair formation which can be described as proton transfer from the excited α-Naphtol acid to the microsolvent cluster. The critical size of the solvent cluster for detectable proton transfer is placed at n=4 as: (a) Smaller species do not exhibit the broad emission of the α-Naphtolate anion. (b) The appearance of this species within the duration of the gas pulse coincides with the appearance of the α-Naphtolate emission. (c) This is the smallest species which does not exhibit a structured absorption.

No substantial change in the features of the broad emission spectra was observed with experimental parameters that lead to increase of the average cluster size. Due to their continuous overlapping absorption spectra, as monitored by R2PI, species-selective excitation of solvent clusters with $n\geqslant4$ is not feasible.

Note the large red-shift ($\sim7000\ cm^{-1}$) in the emission spectrum of the solvated molecule. In comparison the spectral shift in the absorption spectrum is modest. This observation can be explained in terms of intra-molecular vibrational redistribution of the excited cluster to the more stable proton-transfer form. In the S_0 ground state the proton transfer form of the cluster is higher in energy than the neutral form, resulting in the large red shift in the emission.

The assignment of the blue-violet fluorescence to α-Naphtolate emission following intra-cluster proton transfer was supported by the following experiments:

1. When employing a solvent with lower gas-phase proton affinity, such as H_2O no Naphtolate emission was observed from α-Naphtol $(H_2O)_n$ clusters up to n=20.

2. When employing excited state acid with lower pK_a^* such as β-Naphtol $(pK_a^*=2.8\pm0.3)$ no β-Naphtolate emission was observed from β-Naphtol·(NH_3) clusters up to n∿10.

3. The α-Naphtolate emission was also found to increase with higher backing pressures, which lead to increase in the average cluster size.

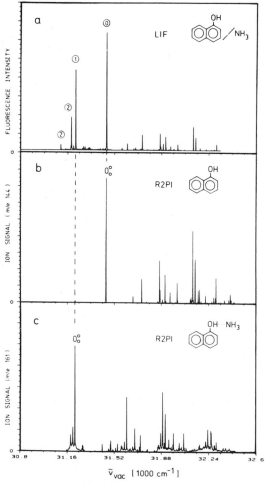

Fig. 1. LIF and R2PI spectra at short delay time ($\tau_D=75\mu sec$). (a) LIF spectrum of α-Naphtol/helium expansion ($p_o=1.3$ bar) showing the overlapping spectra of α-Naphtol·$(NH_3)_n$ with n=0.2. n is indicated above the corresponding electronic origin; (b,c) mass resolved R2PI spectra of α-Naphtol and α-Naphtol NH_3.

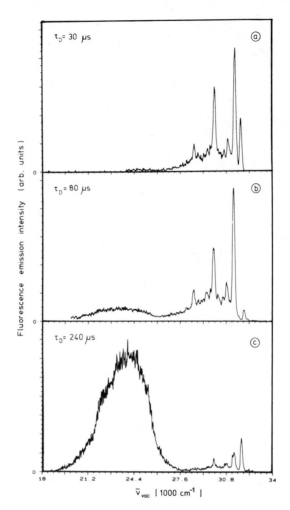

Fig. 2. Dependence of the fluorescence emission spectra on delay time τ_D. Excitation at ν_{vac}=31622.6 cm^{-1} is resonant with the ν=388 cm^{-1} intramolecular excitation of the α-Naphtol·$(NH_3)_2$ complex. The spectra demonstrate the crossover from pure -Naphtol* structured emission (a), to predominant α-Naphtolate* blue unstructured emission (b,c) due to increasing concentration of large α-Naphtol·$(NH_3)_n$ clusters which are excited at the same laser frequency.

REFERENCES

1. P. Kaberle, Ann. Rev. Phys. Chem. 28:445 (1977); C. R. Moylen and J. I. Brauman, Ann. Rev. Phys. Chem. 34:187 (1983).
2. O. Cheshnovsky and S. Leutwyler, Chem. Phys. Lett. 121:1 (1985).
3. J. F. Ireland and P. A. H. Wyatt, Advan. Phys. Org. Chem. 12 (1976); D. Huppert, M. Gutman and K. J. Kaufman in: Advances in Chemical Physics, Vol. 47, Part II, eds. J. Jortner, R. D. Levine and S. A. Rice (Wiley-Interscience, New York, 1981).

GIANT INTENSITY IR-ACTIVE VIBRATIONS PHOTOINDUCED IN CONDUCTING POLYMERS

E. Ehrenfreund, Z. Vardeny and O. Brafman

Physics Department and Solid State Institute
Technion, Haifa, Israel

It is now well known that charged defects in a conjugated polymer chain show infra-red active vibrations (IRAV) with huge oscillator strength (1,2). The strong optical absorption of these modes are the result of very efficient coupling between the backbone modes of the polymer and the π-electrons, giving rise to localized charge oscillations in response to the lattice vibrations (3).

These localized IRAV can be separated into two categories: "translational" modes which are associated with the center of mass motion of the charge(4), and (weaker) "bound" modes which are sensitive to the topology of the defect(5). By considering a rigid motion of the charge, it has been shown (4) that the translational modes (T-modes) are independent of the intrinsic structure of the charge distribution. Their intensity is inversely proportional to the kinetic mass of the defect and their frequency depends on the pinning of the charged defect(4); the stronger the pinning the higher the frequency. The T-modes can therefore be used to study the dynamics of the charge carriers. The bound modes (B-modes), on the other hand, are associated with the internal degrees of freedom of the charged defects and are therefore sensitive to their microscopic structure(5,6,7). Both the intensity of the B-modes and their frequency depend upon the exact structure of the charged defect and its boundaries and therefore can, in principle, be used to study the nature of the defects. This is particularly important in the case of polyacetylene [$(CH)_x$] where solitons as well as one-dimensional polarons may be photogenerated.

In this work we report the observation of IRAV modes in partially isomerized $(CH)_x$. We pay special attention to the differences in the IRAV spectra of predominantly trans-$(CH)_x$ and predominantly cis-$(CH)_x$ samples. The latter shows newly discovered enhanced B-mode ir activity coexisting with the previously observed T-mode ir activity. It appears that charge confinement in the partially isomerized samples strengthen the B-mode ir activity.

The samples were thin films ($\sim 1000 A^o$ thick) of $(CH)_x$ and $(CD)_x$ polymerized on KBr substrates. The as grown films, denoted t(c), contained 80% cis - 20% trans. Measurements on predominantly trans-$(CH)_x$ were carried out on the same films after isomerization of ten minutes at 170^oC. The experimental setup for steady state photoinduced absorption

(PA) measurements in the pump and probe technique is described elsewhere (8). The carriers were photogenerated by an Ar^+ laser at 2.7eV chopped at 140 Hz thus optimizing the measurement to carrier lifetime of order $\lesssim$1msec. The changes ΔT of the transmission T were measured via a probe beam that was provided by broad band CW lamps followed by a monochromator.

The photoinduced absorption $\Delta\alpha(\sim -\Delta T/T)$ of t(c) and of predominantly trans-$(CH)_x$ samples at 20K in the energy range up to 1eV is shown in Fig. 1. In both samples the response $\Delta\alpha$ originates from the trans chains only since the cis isomer does not have any ir activity(2). The spectra in both samples are composed of a broad electronic band (LE), three intense translational IRAV lines (t_1, t_2 and t_3) and two weaker IRAV lines (b_1 and b_3) which are more intense in the t(c) samples. The LE band arises from the electronic transitions between the conduction band (valence band) and the charged soliton ($S^\pm$) defect level in the gap(2,9). The T-modes correlate with the LE band(2) and appear to be weaker in t(c) [Fig. 1] shifting towards higher frequencies as seen in Table 1. The higher T-mode frequencies in t(c) indicate stronger pinning(4) of the photogenerated charged solitons. Their weaker intensity in t(c) means larger soliton kinetic mass(4,5), indicating a shorter defect length compared to that of predominantly trans-$(CH)_x$. As the laser intensity increases the LE band as well as the T-modes in t(c) saturate more easily than the respective PA bands in the predominantly trans-$(CH)_x$ sample, implying thus lower density of trap centers (possibly neutral solitons, S^0)(10) in t(c). Also, as the temperature increases from $\sim$20K to 280K the PA bands in t(c) exhibit considerably weaker decrease in intensity, indicating reduced mobility of the charge solitons ($S^\pm$) in t(c), possibly due to less mobile S^0 defects or due to relative isolation of trans chains in cis-rich samples.

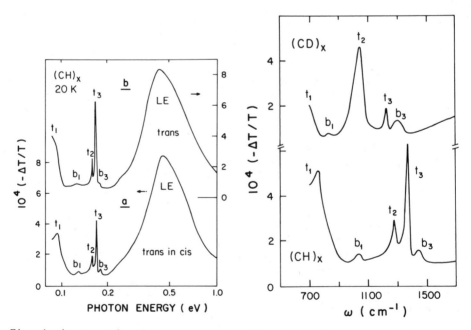

Fig. 1. $\Delta\alpha$ vs probe photon energy in $(CH)_x$. (a) 20% trans in cis, (b) predominantly trans. t_1, t_2 and t_3 denote T-modes and b_1 and b_2 denote B-modes. LE is the $S^\pm$ electronic transition.

Fig. 2. photoinduced IRAV of t(c) at 20K.

Table 1. The IRAV frequencies (in cm^{-1}) of trans chains in predominantly trans polyacetylene and t(c). (a) G.B. Blanchet et.al., Phys. Rev. Lett. 50, 1938 (1983). (b) Only the high frequency tail is observed by us since our spectrometer is limited to ∿720 cm^{-1}. (c) calculated using $D_0(\omega)$, Ref. 11.

		T-modes			B-modes		
		t_1	t_2	t_3	b_1	b_2	b_3
$(CH)_x$	trans	∿500[a]	1280	1365			
	t(c)	750	1280	1372	1035	(1260)[c]	1440
$(CD)_x$	trans	∿400[a]	1036	1223			
	t(c)	(700)[b,c]	1045	1225	835	(1192)[c]	1300

The principal observation of this study is the appearance of the additional ir modes (b_1 and b_3 in Fig. 1) in the spectra. In this respect the main difference between t(c) and predominantly trans-$(CH)_x$ samples is the strength of these B-modes: their intensity in t(c) is about 4-5 times larger. In Fig. 2 we show on an enlarged scale the full photoinduced IRAV spectra for both isotopes of t(c). The frequencies of the various modes are given in Table 1. In addition, we have found the B-modes and the T-modes to have similar laser excitation dependence implying that the same charge defect is responsible for both. We therefore identify the two pairs b_1 and b_3 in Fig. 2 as the bound modes associated with t_1 and t_3 respectively. The intermediate bound mode (b_2), which is associated with t_2, is expected to be very weak and is therefore not observed. Using the amplitude mode formalism(4) with the vibrational response function $D_0(\omega)$ determined earlier from the Raman modes(11), we can account in detail for the frequencies and relative intensities of the B-modes in both isotopes. The calculated frequency for the missing b_2 line is given in Table 1 for both $(CH)_x$ and $(CD)_x$. The observed frequencies agree roughly with the recent calculations(5,6,7) of B-modes ir activity of either one-dimensional polarons or soliton defects. However, due to the sensitivity of these calculations to the boundary conditions (12,13) it is impractical at this stage to carry a detailed comparison with the experimental results.

The fact that the B-mode activity in t(c) is more pronounced may be correlated with a higher degree of charge confinement in t(c). The stronger confinement in t(c) is manifested in the T-mode properties and in the photocarriers recombination kinetics. Previously, we have already shown(14) via Raman scattering in t(c) that the two-fold ground state degeneracy of the trans backbone structure is lifted in t(c), resulting in a finite confinement parameter. Consequently, S^0 defects are more strongly confined and so are $S^{\pm}$, thus giving rise to appreciable B-mode activity. On the other hand, there exists only a weak charge confinement in predominantly trans-$(CH)_x$ due perhaps to interchain coupling(15) or chain ends(16), leading to only minor B-mode ir intensity. The correlation of the confinement with the B-modes is also supported by recent calculations(12,13) showing more intense B-modes for more confined charged defects.

Finally, it is important to note that in our steady-state experiment we detect mainly the long lived ($\sim$1 msec) excitations. At such long times the initially photogenerated carriers end up either as a single charged soliton (S^+ or S^-)(10) on a chain or as a pair of charged solitons $S^+ S^+$ or $S^- S^-$)(17) held close together by the relatively strong confinement in t(c). Experimentally, these two possibilities could not be distinguished by our results and the determination of the exact nature of the photoinduced charged defect must await for further experimental and theoretical studies.

Acknowledgements

We thank Prof. J. Tanaka for the polyacetylene films. This work was supported by the US-Israel BSF and by the Israel Academy for Basic Research, Jerusalem, Israel.

References

1. C.R. Fincher, M. Ozaki, A.J. Heeger and A.G. MacDiarmid, Phys. Rev. B19, 4140 (1979).
2. Z. Vardeny, J. Orenstein and G.L. Baker, J. Phys. (Paris) 44, C3-325 (1983).
3. E.J. Mele, and M.J. Rice, Phys. Rev. Lett. 45, 926 (1980).
4. B. Horovitz, Solid State Commun. 41, 729 (1982).
5. E.J. Mele and J.C. Hicks, Phys. Rev. B32, 2703 (1985).
6. H. Ito, A. Terai, Y. Ono and Y. Wada, J. Phys. Soc. Jpn. 53, 3519 (1984).
7. A. Terai, Y. Ono and Y. Wada, J. Phys. Soc. Jpn. 54, 196 (1985).
8. Z. Vardeny, J. Tanaka, H. Fujimoto and M. Tanaka , Solid State Commun. 50, 937 (1984).
9. Z. Vardeny, E. Ehrenfreund and O. Brafman, Mol. Cryst. Liq. Cryst. 117, 245 (1985).
10. J. Orenstein, Z. Vardeny, G.L. Baker, G. Eagle and S. Etemad, Phys. Rev. B30, 786 (1984).
11. Z. Vardeny, E. Ehrenfreund, O. Brafman and B. Horovitz, Phys. Rev. Lett. 51, 2326 (1983).
12. E.J. Mele and J.C. Hicks, private communications.
13. A. Terai, H. Ito, Y. Ono and Y. Wada, preprint.
14. Z. Vardeny, E. Ehrenfreund, O. Brafman and B. Horovitz, Phys. Rev. Lett. 54, 75 (1985).
15. D. Baeriswyl and K. Maki, Phys. Rev. B28, 2068 (1983).
16. H.W. Gibson et.al., Phys. Rev. B31, 2338 (1985).
17. F. Moraes, Y.W. Park and A.J. Heeger, Synthetic Metals (in press).

EXCITED STATE INTRAMOLECULAR PROTON TRANSFER

IN SUPERSONIC JETS

Niko P. Ernsting and Andrzej Mordzinski

Max-Planck-Institut für biophysikalische Chemie
Abteilung Laserphysik
D-3400 Göttingen, FRG

The development of picosecond laser techniques has led to
a renewed interest in the spectroscopy and kinetics of aromatic
molecules which may undergo excited state intramolecular proton
transfer (ESIPT, fig. 1). The ESIPT reaction is evidenced by a
large Stokes shift for fluorescence from the proton-transferred
molecule.[1]

The current model for ESIPT holds that optical excitation
leads directly to a vibrational level of some adiabatic excited
state. This state is the result of strong interactions between
the initial and proton-transferred forms of the excited molecule.
Its fluorescence may, in principle, have dual character. A weak
component of normal Stokes shift should reflect the relative
weight of the primary form for the emitting vibronic states,
while the stronger fluorescence with large Stokes shift is
attributed to the dominance of the final form. The reaction
enthalpy in the excited state controls the duality of fluorescence . For a non-zero reaction enthalpy, the reaction occurs
without a significant energy barrier.

Weak emission with normal Stokes shift may also be caused
by molecules having no suitably prepared H-bonds in the ground
state, either because of specific H-bonding solvent impurities[2]
or because of unfavourable molecular conformations[3]. It is
therefore necessary to characterize the diverse populations by

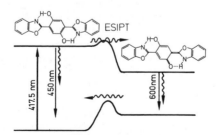

Fig. 1. Photochemical cycle
involving ESIPT.

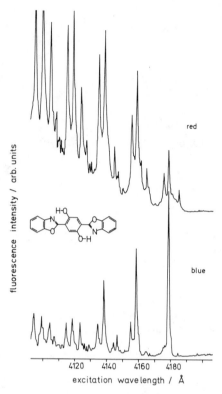

Fig. 2. Fluorescence excitation
spectra of jet-cooled BBXHQ

vibronic resolution of the optical spectra, and to selectively
excite the species with intramolecular H-bonds. High resolution
studies at low temperatures have been reported, among other
cases, for methyl salicylate[3-5], tropolone and similar
compounds[6-9], and 1,5-dihydroxyanthraquinone[10], in noble gas
matrices, Shpol'skii matrices, and supersonic jets. The detailed
vibronic structure of jet-cooled molecules with ESIPT may
already provide useful information on the excited state
potential.

Dual fluorescence has also been reported for isolated, jet-
cooled 2,5-bis(2-benzoxazolyl)hydroquinone[11] (BBXHQ, fig. 2).
The observation agrees with the current ESIPT model, as the
reaction enthalpy is rather small (0.5 kcal/mole in a hydrocarbon solvent)[12]. Figure 2 shows the excitation spectra of dual
fluorescence from the isolated molecule in a supersonic argon
jet. The balance of red to blue emission within a central, main
progression of $\Delta \nu' = 115$ cm^{-1} shows a strong change with excess
vibrational energy. This unusual effect does not conform with
the barrierless adiabatic model mentioned above. Instead it may
indicate the existance of a barrier to ESIPT in this case.

Complexation with a single noble gas atom effectively
enhances the red fluorescence at the cost of the primary
emission. Deuteration leads to two successive blue shifts of the
progressional origin, by 57 cm^{-1} each. This proves the existance
of two equivalent N...HO hydrogen bonds, so that the conformer
shown in fig. 2 must be responsible for the spectra.

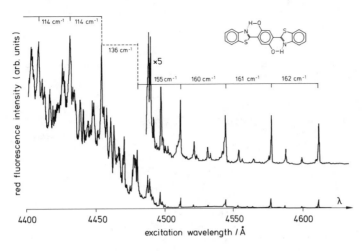

Fig. 3. Fluorescence excitation spectrum
of jet-cooled BBTHQ.

In this paper we report novel observations on the analogue
thiazole compound BBTHQ (fig. 3). In a supersonic neon jet, only
red fluorescence associated with the proton-transferred form was
detected. Figure 3 shows the excitation spectrum of red fluorescence for BBTHQ. The spectrum has two distinct regions. At wavelengths below 450 nm, the spectrum is relatively intense and
congested; however, the bands form progressions with $\Delta\nu' = 114$
cm^{-1} in close similarity to the parent oxazole compound. At
wavelengths longer than 450 nm, a weak set of bands is organized
in progressions with $\Delta\nu' = 161$ cm^{-1}. The electronic origin occurs
at 461.5 nm; it has a hot band shifted by $- 123$ cm^{-1}.

Partial deuteration again gives information on the H-bonds
in the excited state. Two new origin bands are observed upon
partial deuteration. A band shifted by 44 cm^{-1} to the blue from
the d_0-origin may be assigned to d_1-BBTHQ with one intramolecular N...DO hydrogen bond. The other, weaker band has a shift of
158 cm^{-1}. It can be concluded that the transition leads to an
excited state which does no longer have two equivalent hydrogen
bonds. We suggest here that this state corresponds to the excited proton-transferred form of the molecule.

In summary, the spectroscopy of jet-cooled BBXHQ, and of
its thiazole analogue BBTHQ, likely reveals two kinds of
potentials for the excited state intramolecular proton transfer
reaction. The first case is characterized by a small reaction
enthalpy and by a small but significant energy barrier for the
reaction. In the second case, the reaction enthalpy exceeds
1.5 kcal/mole; there is no energy barrier and the excited
proton-transferred molecule may be directly reached by optical
excitation, albeit with small Franck-Condon factors for
non-vertical transitions.

427

References

1. D. Huppert, M. Gutman, and K. J. Kaufmann,
   Laser studies of proton transfer, Advan.Chem.Phys.
   47:643 (1981)

2. D. McMorrow and M. Kasha, Intramolecular excited state
   proton transfer in 3-hydroxyflavone. Hydrogen-bonding
   solvent perturbations, J.Phys.Chem., 88:2235 (1984).

3. L. A. Heimbrook, J. E. Kenny, B. E. Kohler, and
   G. W. Scott, Lowest excited singlet state of hydrogen-
   -bonded methyl salicylate, J.Phys.Chem.,87:280 (1983).

4. J. Goodman and L. E. Brus, Proton transfer and
   tautomerism in an excited state of methyl salicylate,
   J.Am.Chem.Soc., 100:7472 (1978).

5. P. M. Felker, Wm. R. Lambert, and A. H. Zewail,
   Picosecond excitation of jet-cooled hydrogen-bonded
   systems: Dispersed fluorescence and time-resolved
   studies of methyl salicylate, J.Chem.Phys., 77:1603
   (1982).

6. R. Rossetti and L. E. Brus, Proton tunneling dynamics
   and an isotopically dependent equilibrium geometry in
   the lowest excited $\pi - \pi^*$ singlet state of tropolone,
   J.Chem.Phys., 73:1546 (1980).

7. R. Rossetti, R. C. Haddon, and L. E. Brus,
   Intramolecular proton tunnelling in the ground and
   lowest excited singlet states of 9-hydroxyphenalenone,
   J.Am.Chem.Soc., 102:6913 (1980).

8. Y. Tomioka, M. Ito, and N. Mikami, Electronic spectra
   of tropolone in a supersonic free jet. Proton tunneling
   in the S_1 state, J.Phys.Chem., 87:4401 (1983).

9. V. E. Bondybey, R. C. Haddon, and P. M. Rentzepis,
   Spectroscopy and dynamics of 9-hydroxyphenalone
   and of its 5-methyl derivative in solid neon: Effect
   of methyl group upon vibrational relaxation,
   J.Am.Chem.Soc., 106:5969 (1984).

10. M. H. Van Benthem and G. D. Gillispie, Intramolecular
    hydrogen bonding. 4. Dual fluorescence and excited
    state proton transfer in 1,5-dihydroxyanthraquinone,
    J.Phys.Chem., 88:2954 (1984).

11. N. P. Ernsting, Dual fluorescence and excited-state
    intramolecular proton transfer in jet-cooled
    2,5-bis(2-benzoxazolyl)hydroquinone, J.Phys.Chem.,
    89:4932 (1985).

12. A. Mordzinski, A. Grabowska, and K. Teuchner,
    Mechanism of excited-state proton transfer in "double"
    benzoxazoles: bis-2,5-(2-benzoxazolyl)hydroquinone,
    Chem.Phys.Lett., 111:383 (1984).

ANGULAR MOMENTUM-VELOCITY CORRELATION OF

OCS PHOTODISSOCIATION PRODUCTS

G.E. Hall[*], N. Sivakumar[*], P.L. Houston[*] and I. Burak[†]

[*]Department of Chemistry, Cornell University, Ithaca, N.Y.
N.Y. 14853, U.S.A.

[†]School of Chemistry, Tel-Aviv University, 69 978 Tel-Aviv
ISRAEL

INTRODUCTION

The photodissociation of molecules by polarized light is characterized
by a nonisotropic angular distribution of the photofragments. For a dissociation characterized by a sharp momentum distribution, the momentum orientation in space has a cylindrical symmetry with respect to an axis coin-
ciding with the polarization vector of the dissociation beam and is
characterized by the anisotropy parameter β through:[1,2]

$$D(\theta')=1/4\pi\{1+\beta P_2(\cos\theta')\}, \tag{1}$$

Here θ' is the angle between the polarized electric field of the dissociation beam and the velocity vector of the photofragment. The (angular
and magnitude) velocity distribution of the photofragments produced in a
specific internal state may be investigated through the use of Doppler
spectroscopy. In this technique the Doppler profile of a transition
involving an internal state of one of the fragments is obtained with a
monitoring laser beam using LIF state selective detection. Let's define
a coordinate system in which the Z axis coincides with the propagation
vector of the monitoring beam. For a sharp velocity distribution v_0 of the
photofragments, the fragments with velocities oriented with an angle θ with
respect to the Z axis, are monitored with a frequency characterized by a
Doppler frequency shift of:

$$\Delta\nu=(\nu_0/c)v_0\cos\theta \tag{2}$$

429

When averaged over the azimutal angle ϕ the angular velocity distribution is given as a function of θ by:[3]

$$D(\theta) = 1/4\pi\{1 + \beta'P_2(\cos\theta)\} \qquad \beta' = \beta P_2(\cos\alpha) \qquad (3)$$

here α is the angle between the polarization vector of the dissociating beam and the propagation vector of the monitoring beam. $P_2(x)$ is the second degree associated Legendre function. If the interaction of the monitoring beam with the studied fragment is independent of its velocity orientation then the Doppler profile follows the $D(\cos\theta)$ function where $\cos\theta = v/v_0$. v is the velocity component along the propagation vector of the monitoring beam. Doppler profiles of H atoms dissociated from the diatomic HI[4] or CN fragments dissociated from thermalized ICN samples[5] agree very well with eq.3. In this paper we challenge the assumption previously made that the interaction of the photons of the monitoring beam with the photofragments is independent of their velocity orientation. The Doppler profiles of CO molecules obtained from the dissociation of rotationally cold OCS molecules suggest that the internal angular momentum of the CO fragments is aligned perpendicularly to the recoil velocity vector. As a result CO fragements with different orientation of their velocity vectors have different distributions of rotational M quantum numbers and will have therefore a specific interaction with the monitoring beam.

The experimental results are described in Section II. The shapes of the Doppler profiles are accounted in Sec. III in terms of the specific alignment of the rotational magnetic quantum numbers and the angular distribution of the CO photofragments.

II.   EXPERIMENTAL RESULTS

Rotationally cold OCS molecules are ejected from a pulsed supersonic jet. The molecular beam is dissociated by a pulsed polarized laser beam at 222 nm. The internal states of the CO fragments are monitored by a circular polarized tunable VUV beam aligned perpendicularly to the dissociation beam. The VUV beam monitors the various vibrational rotational states through the CO's X $^1\Sigma \rightarrow$ A $^1\pi$ transition. It is shown[6] that the CO photofragments are produced in the vibrationless ground state with a highly excited rotational distribution sharply peaked at J = 56.

The CO's rotational state are shown to be associated with the process:

$$OCS \rightarrow CO(v = 0, \ J = 40 - 60) + S(^1D). \qquad (4)$$

The recoil velocity v_0, of the fragments produced in a quantum number J is given in terms of the energy of the dissociating photon, D.E., the dissociation energy of the parent molecule and the rotational energy:

$$1/2 \mu v_0^2 = h\nu - D.E. - BJ(J+1). \tag{5}$$

The Doppler profiles of various transitions have been taken at two excitation modes: the parallel mode corresponding to an angle $\alpha=0$ and the perpendicular mode $\alpha=\pi/2$. The results may be summarized as follows:

1) Doppler profiles corresponding to the same P.R or Q transitions are different. (see fig.1)

2) Doppler profiles of the same transition, obtained for the parallel and perpendicular excitation modes, correspond to two values of β when fitted with eq.3.

III. DISCUSSION

The deviations of the Doppler profiles from the behavior predicted by eq.3 are explained in terms of a correlation between the translational and angular momenta of the recoil fragments. If J_i,L and J are the angular momentum vectors of the parent molecule, the orbital motion of the two fragments and the internal rotational motion of the diatomic fragment respectively, then conservation of angular momentum implies: $J_i = L + J$. If the parent molecules are initially cooled then $J_i \sim 0$ or $L = -J$. This implies that the photodissociation process prepares the molecular fragment with an angular momentum perpendicular to the recoil velocity. The magnetic quantum number along an axis parallel to the recoil velocity vector is: $M_{v(\theta,\phi)} = 0$. Here the angles θ and ϕ specify the orientation of the velocity vector with respect to the z axis (see eq.3). The state: $|J,M_{v(\theta,\phi)} = 0 \rangle$ is expanded in terms of a basis set $|Jm\rangle$ defined with respect to the z axis:

$$|J,M_{v(\theta,\phi)} = 0 \rangle = \Sigma_m D_{m,o}(\phi,\theta,0)|Jm\rangle. \tag{6}$$

$D_{m,o}(\phi,\theta,0)$ is the rotation matrix element by the three Euler angles.[7] In order to evaluate the Doppler signal one defines[8] the Absorption and Fluorescence matrices: A and F respectively. The elements of A and F are given by:

$$A_{m'm} = 1/4\pi\{1+\beta P_2(\cos\theta'')\}\Sigma_{nn'} \langle J'm|\mu \cdot E_D|Jn'\rangle \rho_{n'n} \langle Jn|\mu \cdot E_D|J'm\rangle, \tag{7}$$

$$F_{M'M} = \Sigma_m \langle J'M'|\mu \cdot E_f|J''m\rangle\langle J''m|\mu \cdot E_f|JM\rangle, \tag{8}$$

here J, J' and J" correspond to the initial intermediate and final rotational states respectively, μ is the transition dipole moment, E_D and E_f are the polarization vectors of the dissociation and fluorescence photons respectively. $\rho_{n'n}$ is the density matrix element given by:

$\rho_{n'n} = D^J_{n'o}(\phi,\theta) D^{*J}_{no}(\phi,\theta)$. The value of $\cos\theta''$ is equal to $\cos\theta$ for the parallel excitation and $\sin\theta\sin\phi$ for the perpendicular excitation mode. The LIF signal is expressed as a function of $\cos\theta$ (corresponding to molecules with the same Doppler shift) and is given by:

$$\text{Signal}(\cos\theta) = \text{Tr}\underline{A}\underline{F}, \tag{9}$$

the elements of the matrices $\underline{A}$ and $\underline{F}$ are the elements of A and F averaged over ϕ. It is shown[9] that after averaging, the contribution to the signal of terms with nondiagonal density matrix elements is negligible. As a result, result, the Doppler signal is given by:

$$\text{Signal}(\cos\theta)=(4\pi)^{-1}\times\{1+\beta'P_2(\cos\theta)\}\times I(\cos\theta)=D\times I, \tag{10}$$

The $D(\theta)$ function is expressed in eq.3. The function $I(\theta)$ is given by:

$$I(\theta)=\text{Tr}A'F$$

$$A'_{m'm}=\Sigma_n <J'm'|\mu\cdot E_D|Jn>\rho_{nn}(\theta)<Jn|\mu\cdot E_D|J'm> , \tag{11}$$

$$\rho_{nn}(\theta) = |D^J_{no}(0,\theta,0)|^2$$

The calculated and experimental results are displayed in fig.1.

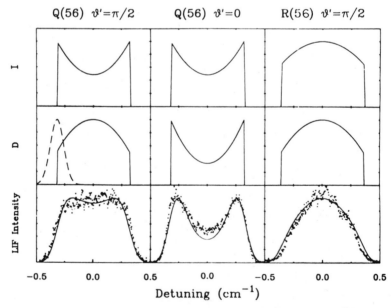

Fig. 1. Experimental and calculated data for the Q(56) and R(56) lines of CO. First row (left to right): Calculated $I(\cos\theta)$ functions. Second row: Calculated $D(\cos\theta)$ functions (solid curves); the Gaussian laser line width is shown in the dashed curve. Third row: Experimental (dots) and calculated (solid curve) Doppler profiles. The curves are synthesized using the values: $\beta=0.8$, FWHM laser linewidth=0.14 cm^{-1}. The magnitude of the CO's velocity is calculated: $V_0=1430$m/s. The Doppler profiles are computed for a circular polarized monitoring beam.

The left panels correspond to the Q(56) transition with a perpendicular excitation mode: $\alpha=90°$, the middle panels are for the Q(56) with a parallel excitation mode: $\alpha=0$. The right panels correspond to a perpendicular excitation of the R(56) transition. The upper row displays the calculated correction function $I(\cos\theta)$. The middle panels display the angular distribution function $D(\cos\theta)$. These curves should describe the Doppler profiles in the absence of any v-J correlation. A comparison between experimental and calculated Doppler profiles is shown in the bottom panels. Only one parameter namely: $\beta=0.8$ is used in calculating the three curves. It is clear from the figure that the line shapes in the absence of a v-J correlation will be qualitatively different from those when there is no correlation.

The simplified model described in this paper should be compared with a formal and exact quantum mechanical treatment of the photofragmentation of triatomic molecules.[10] According to the theory the helicity λ (corresponding to M_v) is restricted by $\lambda=\min(J_i,J)$. This is consistent with the assignment $M_v=0$ in our experiment as a result of the specific conditions: $J>J_i$ and $J_i\sim 0$.

References

1. R.N. Zare and D.R. Herschbach, Proc. IEEE 51, 173 (1963).
2. S. Young and R. Behrson, J. Chem. Phys. 61:4400 (1974).
3. K.H. Welge and R. Schmidel in Adv. in Chem. Phys. Vol.47 (Wiley N.Y. 1981).
4. R. Schmidel, H. Dugan, W. Meier, and K.H. Welge, Z. Phys. A403:137 (1982).
5. I. Nadler, D. Mahgerefteh, H. Reisler and K. Wittig, J. Chem. Phys. 82:3885 (1985).
6. N. Sivakumar, I. Burak, W.Y. Cheung and P.L. Houston, J. Phys. Chem. 89:3609 (1985).
7. A.R. Edmonds, Angular Momentum in Quantum Mechanics (Princeton Univ. Press, 1957).
8. R.N. Zare, J. Chem. Phys. 45:4510 (1966).
9. G.E. Hall, N. Sivakumar, P.L. Houston and I. Burak, to be published.
10. G.G. Balint-Kurti and M. Shapiro, Chem. Phys. 61:137 (1981).

VIBRATIONAL-ENERGY REDISTRIBUTION IN HIGHLY EXCITED MOLECULES

Gabriel Hose

Chemical Physics Department
Weizmann Institute of Science
76100 Rehovot, Israel

INTRODUCTION

Considerable amount of evidence indicates that spectra involving highly excited vibrational states of polyatomic molecules are generally structured. Overtone progressions of stretching vibrations extending as high as ten quanta have been observed with the use of CW lasers in the visible [1], in energy-loss measurements of fast-ion collisions [2], and also in short-duration fluorescence (or enhanced Raman) experiments [3]. Infra-red emission of molecules initially excited to high energies by step-wise multiphoton absorption shows broad profiles red shifted from stretching fundamentals of the high resolution spectrum [4]. Combination bands involving stretching motions were obtained in double-resonance multiphoton experiments [5].

Recently we have advanced a theory [6] that associates the broad structures in the high-energy spectra with bands of mode-localized extreme-motion (EM) vibrational states. Here localized extreme motion means high excitation concentrated in one stretching mode of motion. A wave packet created from such a band, say by overtone absorption, would initially (at least) describe a large-amplitude coherent vibration of the nuclei. This excess energy may in time spread over other modes of motion, and we shall discuss below a mechanism for vibrational-energy redistribution (VER) among bands of extreme motion in different modes. This mechanism is essentially the same as anharmonic (Fermi) resonances [7], and it should be valid in the presence of collisions or under collisionless conditions.

MODE-LOCALIZED BANDS

The high-energy vibrational-state manifold of polyatomic molecules can be understood [6] starting from an anharmonic-normal-modes (ANM) description. To zeroth order in mode-mode coupling this approach successfully explains the positions of lines in the C-H overtone spectrum of molecules like methane [8]. Theoretical considerations based on numerical studies [9] of model coupled-oscillator systems, show that the influence of inter-mode anharmonicities need not altogether destroy this picture at high energies. We have found that generically [6,9] high-energy ANM configurations having most of the excited vibrational quanta in one high-frequency mode of motion are relatively immune to mode-mode coupling. In contradistinction, high-energy configurations in which the excitation localizes in a low-frequency vibration or alternatively spreads over many modes, mix thoroughly. In the low density-of-states (small molecule) limit we therefore expect to find high-energy vibrational states that are dominated (in the probability density sense) by EM configurations. This should generally lead to narrow spectral features associated with various stretching modes [6,10].

In case of large density of states as is found in many-atom molecules the EM configurations are still feeling relatively weak inter-mode couplings but with much more nearby non-extreme-motion (NEM) ANM configurations. In analogy [6] with resonances described as bound states adjacent to continuum, the resultant situation is of bands comprising vibrational states centered around the EM configuration. To see this imagine a hypothetical molecule with one high-frequency stretching mode and numerous low-frequency (e.g., bending) vibrations. Denoting n to be the quantum number for the stretching motion, and letting m represent the collective of low-frequency quantum numbers, the ANM configurations (zero-order states) are written as [n,m]. We shall partition the ANM basis into two subspaces. The first is the set { [n,0]; n⩾0} which contains the ground state, the pure-stretch low-lying excited states that are not extreme motion, and all the EM configurations of the stretching mode. We shall refer to it as the EM subspace and denote the corresponding projection operator by Q. We stress that a combination of high excitation of the stretching mode with one or more low-frequency fundamentals may also be an EM configuration in the sense of immunity to mode-mode coupling. However, for simplicity we shall ignore this and assume that the remainder of the basis, i.e., the set {[n,m]; n⩾0,m>0} with the projection operator N=1-Q, includes only NEM configurations.

Consider the matrix of the vibrational Hamiltonian H in the ANM basis. The off-diagonal elements are precisely the mode-mode couplings between the ANM configurations. Obviously QHQ is a diagonal matrix whose elements could be taken to obey a Birge-Sponer relationship,

$$\langle n,0|H|n,0\rangle = -An^2 + Bn + C; \quad B > A > 0.$$

The matrix NHN however is not diagonal, and its eigenstates are linear superpositions of the NEM ANM configurations. At high energies each one of the latter strongly mixes with many other such configurations. The resulting high-energy spectrum of NHN is thus a dense quasi-continuum of states which are linear superpositions of many small weight NEM configurations. Now transform the full Hamiltonian matrix to a new basis comprising the subspace of EM configurations and the eigenstates of NHN. The two submatrices QHQ and NHN, which are diagonal in this basis, are coupled by the "off-diagonal" rectangular matrices QHN and NHQ. Generally these are weak couplings because the eigenstates of QHQ are the immune EM configurations [6], but nevertheless, they increase with the energy. Note that starting from the stretching fundamental the spectra of QHQ and NHN overlap. Hence at high energies, where the density of NHN eigenstates is large, each EM configuration will generally behave like an isolated resonance adjacent to continuuum and is thus broadened [6]. The width is expected to increase with the energy.

Optical transitions into bands are generally stronger than those involving high-energy vibrational states lying between bands. The latter are linear combinations of many NEM configurations each entering the superposition with small weight and arbitrary sign. Since the dipole selection rules for the zeroth-order ANM configurations are sharp the combined transition dipole must be small. With band states the situation is different because one EM configuration enters the superposition with a weight much larger than all the rest; leading to relatively strong dipole transitions. This explains [6,10] the origin of the high-energy spectral structure. There is an analogy here with bound-resonance or resonance-resonance transitions as compared to bound-free or free-free transitions in atomic physics. The former are much stronger than the latter.

ANHARMONIC RESONANCES

To describe a realistic many-atom molecule we have to add several high-frequency modes (e.g., symmetric and anti-symmetric stretches) to the model of the former section. Also, the EM subspace should be extended to include EM configurations that were previously ignored, i.e., those involving some low excitation in a mode other than the extremely excited one. The net result is that QHQ is no longer a diagonal matrix. Its off-diagonal elements consist of inter-mode couplings between EM configurations. These couplings are high-order mode-mode product terms in the Taylor-series expansion of the full anharmonicity of the potential [6]. Because there are large quantum-number differences between EM configurations in different modes, the corresponding off-diagonal elements of QHQ should be small. Slightly larger values are expected between configurations that are extreme motion in the same mode but differ in the low excitation of other modes.

Because the off-diagonal terms of QHQ are small, we may neglect them to first order. In this case the EM configurations couple indirectly via the NEM eigenstates of NHN. This coupling is negligible so that essentially we have an extension of the previous situation of one high-frequency mode. Neglecting direct couplings, the high-energy vibrational spectrum is composed of progressions of EM bands of different high-frequency modes. For example, assuming two stretching modes and many low-frequency ones, and labelling bands by their EM ANM configuration, we have several possible progressions: [n,0,0], [0,m,0], [n,1,0], [n,0,1], [1,m,0], etc. Here the third quantum number stands for all the low-frequency motions.

Employing the Birge-Sponer relationship with coefficients taken from the high resolution data we can estimate center-of-band progressions. Because of the width it is clear that EM bands from different progressions may overlap. The small off-diagonal elements of QHQ, which can hardly affect isolated bands, may significantly mix overlapping EM bands. In the sparse low-energy range the mixing of nearly degenerate states is referred to as anharmonic (Fermi) resonances. It is an efficient mechanism for energy transfer among the vibrational modes of motion [7]. The very idea of EM bands that (like resonances) describe motion localized in modes is suggestive of such a mechanism for VER in the high-energy range.

First consider the submatrix QHQ alone, i.e., the undressed EM configurations. If two such states are nearly degenerate, then the small inter-mode couplings will cause further splitting and mix the EM configurations. That is, each of the corresponding eigenstates of QHQ will now describe a combined EM in two vibrational modes. Classically energy is being transferred back and forth between the modes. The closer in energy are the EM configurations, the stronger is the mixing. Now dress the EM configurations so that we have bands. The broadening may wipe the splitting altogether; but this is not always the case. For example, Fermi-resonance induced splitting of high-overtone profiles were observed in benzene [1]. The net result of overlapping EM bands is therefore a joint band that is a superposition of extreme motions in several modes. Any wave packet created within a joint band will extend over a number of modes among which the corresponding classical motion displays energy transfer.

Unlike isolated EM bands, joint bands can be optically active rather intensely at several wavelengths. This is one important consequence of VER implied by the existence of joint bands. If the system is been excited into such a band, say via step-wise multiphoton absorption in one progression, it can nevertheless emit IR photons at other wavelengths [10]. This energy transfer is instantaneous (on a vibrational time scale) because the joint band extends over several modes. Observe that the joint-band (anharmonic-resonance) mechanism for VER does not require collisions. Nevertheless the latter may promote VER by relaxing an excited system into joint bands. In the high-energy dense quasi-continuum this is expected to be a prominent process induced by collisions.

SUMMARY

Our overall view of the high-energy vibrational-state manifold of the polyatomic molecule is that of a "network" of EM bands embedded in an optically inactive quasi-continuum of evenly-mixed states. The bands are organized in vertical columns (i.e., progressions) corresponding to the high-frequency modes. Usually these are all the stretching vibrations. Bands within the same column are optically (IR or Raman) active. That is, transitions between consecutive bands are strong whereas overtone-like transitions are weaker. The ground state is common to all the pure EM progressions. In addition there are other progressions that begin with some combination band. A band in a progression may be isolated and describe an extreme motion in the relevant mode. It may also be (due to anharmonic resonances) joint with other progressions. These joint bands are the horizontal connections in the network. They are responsible for VER monitored by optical means.

REFERENCES

1. K.V. Reddy, D.F. Heller and M.J. Berry, J. Chem. Phys. 76, 2814 (1982).
2. T. Ellenbroek, U. Gierz, M. Noll, and J.P. Toennies, J. Phys. Chem. 86, 1153 (1982).
3. D.G. Imre, J.L. Kinsey, R.W. Field and D.H. Katayama, J. Chem. Phys. 76, 2564 (1982).
4. I. Glatt and A. Yogev, Chem. Phys. Letters 77, 228 (1981).
5. E. Borsella, R. Fantoni, A. Giardini-Guidoni and C.D. Cantrell, Chem. Phys. Letters 87, 284 (1982).
6. G. Hose and H.S. Taylor, Chem. Phys. 84, 375 (1983).
7. E.L. Sibert, W.P. Reinhardt, and J.T. Hynes, Chem. Phys. Lett. 92, 455 (1982).
8. I. Abram, A. de Martino, and R. Frey, J. Chem. Phys. 76, 5727 (1982).
9. G. Hose and H.S. Taylor, J. Chem. Phys. 76, 5356 (1982).
10. G. Hose, A. Yogev, and R.M.J. Ben Mair, Chem. Phys. to be published.

RECENT ADVANCES IN INTRAMOLECULAR ELECTRONIC ENERGY TRANSFER

Shammai Speiser

Department of Chemistry
Technion - Israel Institute of Technology
Haifa 32000, ISRAEL

A study of intramolecular energy transfer (Intra-ET) in a specifically designed[1] series of bichromophoric molecules consisting of cyclic α-diketones incorporating an ortho-, meta-, or para substituted benzene ring (Fig. 1) is reported. Most spectroscopic properties of these molecules are described by a superposition of those of their constituent chromophores. Unique for the bichromophoric molecule is the fact that, depending on the molecular geometry, energy absorbed by the aromatic chromophore is transferred in part to the α-diketone and both chromophores emit their characteristic fluorescence spectra[2]. An extensive study was made of the intramolecular electronic energy transfer process in solution as a function of temperature[3]. The results (Fig. 2) indicate that the transfer efficiency is strongly structure dependent suggesting that Dexter type exchange interaction is responsible for Intra-ET between close chromophores in a bichromophoric molecule. The thermal dependence observed in some cases is atributed to conformational factors.

$$P_{nn} \qquad M_{nn} \qquad O_{nn}$$

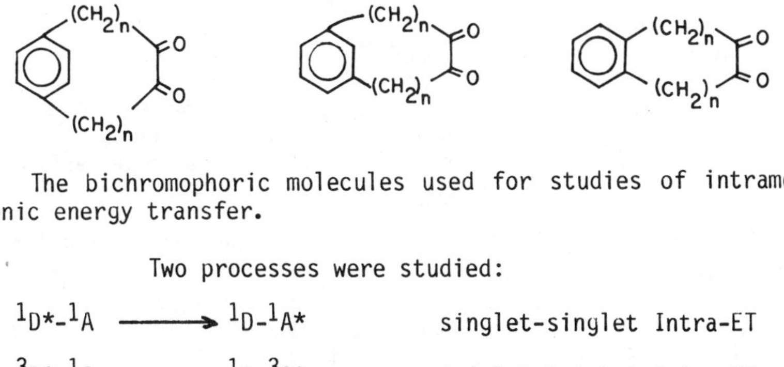

Fig. 1: The bichromophoric molecules used for studies of intramolecular electronic energy transfer.

Two processes were studied:

$$^{1}D*-^{1}A \longrightarrow {}^{1}D-^{1}A* \qquad \text{singlet-singlet Intra-ET}$$

$$^{3}D*-^{1}A \longrightarrow {}^{1}D-^{3}A* \qquad \text{triplet-triplet intra-ET}$$

where D and A denote donor and acceptor chromophores, respectively.

A general theoretical analysis of Intra-ET in bichromophoric molecules D - A provides expressions for donor fluorescence (phosphorescence) decay and for its fluorescence (phosphorescence) quantum yield in terms of the average distance between donor and acceptor moieties and the flexibility of the chains connecting donor and acceptor[4]. Comparison with the present experimental data supports the predictions of this analysis. It is concluded that singlet-singlet (S-S) Intra-ET and triplet-triplet (T-T) Intra-ET in bichromophoric molecules are indeed governed by short range exchange interaction (Dexter mechanism).

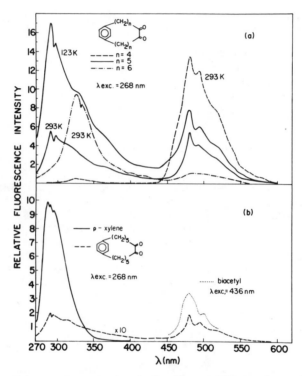

Fig. 2: Temperature dependent dual fluorescence from O_{nn} molecules indicating a strong molecular structure dependence of intramolecular electronic singlet-singlet energy transfer. Similar results were obtained with the P_{nn} and M_{nn} series.

Using time resolved laser induced luminescence spectroscopy the relative yields of T-T Intra-ET and S-S Intra-ET were determined. Typical examples are shown in Figs. 3-6 where comparison is made between the relative quantum yields of fluorescence and phosphorescence of the α-diketone chromophore upon direct excitation of the α-diketone chromophore at 430 nm to that obtained by exciting the aromatic moiety at 266nm followed by S-S and T-T Intra-ET. The results indicate competition between T-T Intra-ET and S-S Intra-ET leading to a more efficient T-T process whenever S-S Intra-ET becomes less efficient.

Using the results for direct excitation of the α-diketone at 430 nm the fluorescence to phosphorescence quantum yield ratio ρ_d' can be determined. This ratio can then be used as a reference for the corresponding ratio obtained for the α-diketone emissions upon excitation of the aromatic chromophore at 266 nm followed by intra-ET. The parameter ξ given by

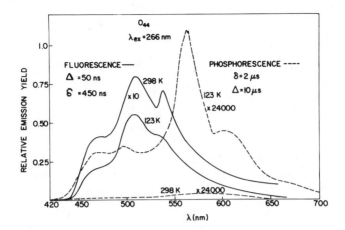

Fig. 3 : Relative temperature dependent time resolved fluorescence and phosphorescence yields for the α-diketone moiety of O_{44} upon excitation of the benzene cromophore at 266nm. The resulting emission is a manifestation of S-S and T-T intra-ET. The boxcar overager aperture duration Δ and aperture delay δ used for the time resolved spectroscopy are shown.

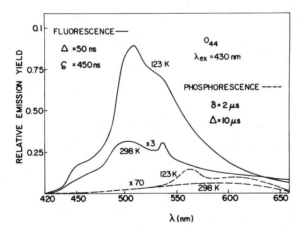

Fig. 4 : Relative temperature dependent time resolved fluorescence and phosphorescence yields for the α-diketone moiety of O_{44} upon direct excitation of the α-diketone chromophore at 430nm.

$$\xi = 1 + (k_{ISC}/\tau_A k_{ISC} k_{ET})Q_{ET} \tag{1}$$

where k_{ISC}'s are the intersystem crossing rates of the donor (aromatic) and acceptor (α-diketone) moieties τ_A is the α-diketone fluorescence lifetime, k_{ET} is the S-S intra-ET rate and Q_{ET} is the T-T intra-ET efficiency.

The enhancement in the α-diketone phosphorescence is mainfested in $\xi > 1$ as exemplified here for P_{55} and O_{44}. A detailed discussion is given elsewhere[5].

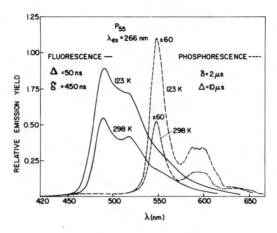

Fig. 5: Same as Fig. 3 for P$_{55}$.

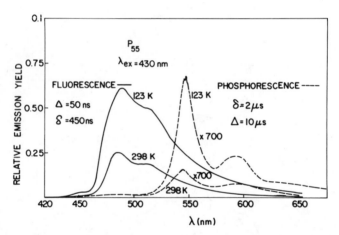

Fig. 6: Same as Fig. 4 for P$_{55}$.

References

1.   M.B. Rubin, S. Migdal, S. Speiser and M. Kaftory, Isr. J. CHem. 25, 66 (1985).
2.   S. Hassoon, H. Lustig, M.B. Rubin and S. Speiser, Chem. Phys. Lett. 98, 345, (1983).
3.   S. Hassoon, H. Lustig, M.B. Rubin and S. Speiser, J. Phys. Chem. 88, 6367 (1984).
4.   J. Katriel and S. Speiser, Chem. Phys. Lett. 102, 88, 1983.
5.   S. Speiser, S. Hassoon and M.B. Rubin (to be published).

TIME-RESOLVED PHOTOFRAGMENT FLUORESCENCE AS A PROBE OF LASER-PULSE MOLECULAR

DISSOCIATION

G. Kurizki[*] and A. Ben-Reuven[†]

[*]Department of Chemical Physics
Weizmann Institute of Science
Rehovot 76100, Israel
[†]School of Chemistry
Tel-Aviv University, Tel-Aviv 69978, Israel

   Recently we have carried out theoretical studies of time-resolved
fluorescence from isolated systems of fragments of a molecular reaction
(or collision), in which the fragments are coupled by long-range interactions. These interactions result from radiative exchange between fragments
whose dipole transitions are resonant or near-resonant[1,2], or from Stark-
mixing of levels in a fragment by the Coulomb fields of its ionic partners[3].
These studies, which are based on a quantal master-equation approach, indicate that fluorescence from such systems exhibits in many cases cooperative
features, until the fragments reach large separations - thousands of Å for
visible fluorescence. The cooperative features include ringing of the
emission rate as a function of time and nonexponential decay of the envelope of the ringing pattern. They are detectable provided the required
temporal resolution, ranging in different systems from hundreds to several
psec, is available. A pioneering experiment in this field, which was
performed by Grangier, Aspect and Vigué[4], demonstrated the observability
of ringing in the time-resolved fluorescence from photofragments (of Ca_2,
in this particular experiment).

   In this note, the cooperative fluorescence (CF) features will be
illustrated for systems of neutral and ionic fragments, emphasizing their
main interest - the strong dependence of these features on the electronic
angular momentum and orbital symmetry of the state of the molecular complex
as the fragments emerge from the quasi-molecular region, i.e. the domain
of short-range ($\lesssim$few nm) molecular forces. These attributes of the
molecular complex (the "parent molecule") can thus be diagnosed on observing CF features and fitting them to the theoretical expressions.

   Such diagnostic information is often required for electronically-
excited bound or quasi-bound molecular states formed, e.g. in charge-
exchange reactions of molecular beams with other species[5] or with surfaces[6]. A short (subnanosecond) laser pulse dissociating the molecule
from such a state can allow to monitor the CF with sufficient temporal
resolution, in order to probe the properties of this state. Another fore-
seeable application of CF effects is the study of molecular symmetry
(related to the geometrical configuration) of clusters in laser-excited
quasi-bound or dissociative electronic states.

A simple system where CF features are important is a <u>homonuclear diatom</u> dissociating into a singly-excited pair of atoms:

$$X_2^* \begin{array}{c} \nearrow X^{(A)} + X^{(B)*} \searrow \\ \searrow X^{(A)*} + X^{(B)} \nearrow \end{array} X^{(A)} + X^{(B)} + \hbar\omega_{\vec{k}}$$
(1)

where A and B label the atoms, $\omega_{\vec{k}}$ is the frequency and $\vec{k}$ the wavevector of the emitted fluorescence photon. The two channels contributing to the fluorescence in (1) interfere with each other, since the dissociative state of X_2^* correlates to a superposition of the states $|e_A g_B\rangle$ and $|g_A e_B\rangle$, e and g being the relevant upper and lower atomic states.

When space-degeneracy effects play no role and the atomic states have definite parities (e.g. S and P states of alkaline and alkaline-earth elements), the superpositions allowed by the electronic inversion symmetry of X_2^* are the symmetric (constructively interfering) and antisymmetric (destructively interfering) $\{|e_A g_B\rangle \pm |g_A e_B\rangle\}/\sqrt{2}$, corresponding to ungerade and gerade X_2^* states, respectively.

The emission rate $\dot{P}_{\hat{k}}(t)$ into a unit solid angle in the direction of unit vector $\hat{k}$ is obtained by averaging the temporal interference pattern corresponding to the ungerade or gerade state of the diatom over all possible orientations of the internuclear axis relative to $\hat{k}$ and summing over polarizations[1]. In deriving this rate the axial-recoil approximation is made, i.e. the rotational velocity of the diatom is neglected compared to the interfragment separation velocity. The resulting expression[1] is then independent of $\hat{k}$ (isotropic):

$$\dot{P}_{\hat{k}}(t) = \frac{1}{4\pi} \dot{P}_{total} = [\gamma \pm \Gamma(t)]\exp[-\gamma t \mp \int_0^t \Gamma(t')dt']$$
(2)

the upper (lower) sign corresponding to the ungerade (gerade) state. Here γ is the Einstein A-coefficient for emission in the e→g transition of an independent atom and $\Gamma(t)$ is the rate of photon exchange between the two atoms. The time-dependence of $\Gamma(t)$ is due to its dependence on the inter-atomic separation $R(t)$. At t=0 (which can be chosen after a short delay from the onset of the dissociating laser pulse, so that R(0) is outside the quasi-molecular region), the product of R(0) with the emission wave-vector k is small ($\xi(0) \equiv kR(0) \ll 1$). In this limit, $\Gamma \approx \gamma$ and the emission rate in an ungerade state is <u>doubled</u> compared to the normal rate, whereas in a gerade state it is totally suppressed (as required by the molecular selection rule g↛g). These are elementary examples of superradiance and subradiance, respectively. As R(t) increases, the first factor in (2) exhibits a ringing behavior in time, due to the oscillation of $\Gamma(\xi(t))$ (provided the ensemble-distribution of separation velocities is sharp enough - a condition satisfied in the experiment of Grangier et al.[4]). Concurrently, the second factor causes the nonexponential decay of the ringing envelope. For the ungerade (superradiant) state the decay is sped up, whereas for the gerade (subradiant) state it is slowed down. The oscillations of Γ as a function of $\xi(t)=kR(t)$ depend on the projection of the electronic angular momentum Λ^* of the diatom on the R-axis in the excited state, which decays radiatively to the $\Sigma(\Lambda=0)$ ground state $|g_A g_B\rangle$. On substituting Γ_{Λ^*} into (2), one obtains the combined effects of both factors on $\dot{P}$ for ungerade and gerade states with $|\Lambda^*|=|\Delta\Lambda|=1$ or 0 ($\Pi^* \to \Sigma$ or $\Sigma^* \to \Sigma$ emission). These effects are demonstrated in Figures 1 and 2 for different ratios $\gamma/\dot{\xi}$ (i.e. different separation velocities). It is obvious from these figures that by observing the evolution of $\dot{P}$, Λ^* and the inversion symmetry of the dissociative state of the diatom are easily recognized.

The simplest system of <u>nonidentical fragments</u> which exhibits CF

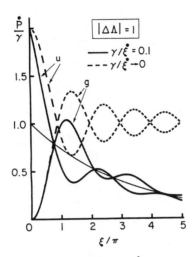

Figure 1. The cooperative emission rate ($\dot{P}$) as a function of time (in units of $\xi = K\dot{R}t$) for a diatom in a $\bar{\Pi}_u^*$ or $\bar{\Pi}_g^*$ state dissociating into two atoms in an even and an odd parity states (e.g. s and p), calculated for two values of $\chi = \gamma/K\dot{R}$: Solid line —$\chi = 0.1$, showing ringing accompanied by nonexponential decay; broken line —$\chi \rightarrow 0$, showing ringing only. Thin line shows for comparison the independent-atom emission rate $\gamma \exp(-\gamma t)$, the emission wavelength being $2\pi/K$.

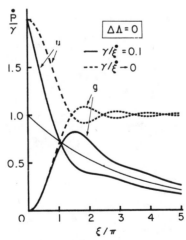

Figure 2. As in Figure 1, but with diatom prepared in a Σ_u^* or a Σ_g^* state.

445

features is a singly-excited <u>heteronuclear diatom</u> (AB)[*] in which the near-resonant transitions $e_A \rightarrow g_A$ and $e_B \rightarrow g_B$ do not involve space-degeneracy effects (e.g. same types of transitions as the ones discussed above). As in a homonuclear diatom, the essence of CF features in this system is the interference between the contributions of the correlated emitting channels $|a> \equiv |e_A g_B>$ and $b> \equiv |g_A e_B>$. The amplitudes and the relative phase of $|a>$ and $|b>$ as they emerge from the quasi-molecular region are not prescribed by symmetry in this case. The main diagnostic purpose of CF studies here is to determine these amplitudes and phase and thus gain insight into the dissociation dynamics.

Taking, for illustration purposes, the oscillator strength of the two near-resonant transitions corresponding to $|a>$ and $|b>$ to be nearly equal and denoting their frequencies by ω_a and ω_b, we distinguish between two generic cases:

<u>Case A - small detuning</u>: $\Delta \equiv |\omega_a - \omega_b| << |\Omega(R(0))|$, where $\hbar\Omega$ is the retarded dipole-dipole interaction (radiative coupling) which varies as ξ^{-3} for $\xi << 1$. This inequality is satisfied by all diatoms wherein A and B are isotopic species, as well as by systems such as $Na(5p \rightarrow 3s) + Mg(^1S_0 \rightarrow ^1P_1)$.

The deviation from exponential decay is now given by the exponent of the time-integral

$$\beta(t) \equiv \int_0^t [\Gamma(1 - \Delta^2/2\Omega^2) - \Delta^2 \dot{\Omega}/2\Omega^3] dt' . \qquad (3)$$

The term containing $\dot{\Omega} \propto k\dot{R}$ in (3) is <u>positive</u> and therefore, for sufficiently large separation velocities $\dot{R}$, it compensates for the reduction of $\beta(t)$ by other Δ-dependent term. Thus, fast changes in the coupling Ω can drive the system into resonance. The analog of (2) in the domain $|\Omega(t)| \gtrsim \Delta$ has the form[2]

$$\dot{P}_{\hat{k}}(t) = \frac{1}{4\pi} e^{-\gamma t} |C(t) - u(0)D(t)| \qquad (4)$$

where

$$C(t) = -\gamma\cosh\beta + \Gamma\sinh\beta - (\Delta^2/2\Omega^2)(\Gamma + \dot{\Omega}/\Omega)\sinh\beta;$$

$$D(t) = -\gamma\sinh\beta + \Gamma\cosh\beta - (\Delta^2/2\Omega^2)\{2(\Gamma + \dot{\Omega}/\Omega)\cosh\beta \qquad (5)$$

$$- \gamma\Omega^2 \int_0^t dt'[\cosh\beta(\Gamma + \dot{\Omega}/\Omega)/\Omega^2]\}$$

and $u(0) = \rho_{ab}(0) + \rho_{ba}(0)$ is the sum of the cross-products of the amplitudes of $|a>$ and $|b>$ at $R(0)$. The observation of ringing (caused by the $\Gamma(t)$ terms) and of nonexponential decay (caused by the β terms) and their curve-fitting to (4) and (5) can yield $u(0)$ and thus provide information on the correlations between the fragments as they emerge from the quasi-molecular region.

<u>Case B - large detuning</u>: $\Delta >> |\Omega(t)|$. In this case, the parameter of interest is the "inversion" $w(0) = \rho_{aa}(0) - \rho_{bb}(0)$, measuring the sharing of excitation between the populations of the energetically-unequal states $|a>$ and $|b>$. Nonexponential decay in this case is governed by the exponent of

$$\beta(t) \equiv \int_0^t [(\Omega/\Delta)\Gamma] dt' \qquad (6)$$

where Ω/Δ expresses the off-resonance reduction of the cooperative photon exchange. The corresponding emission rate[2]

$$\dot{P}_{\hat{k}}(t) \simeq \frac{1}{4\pi} e^{-\gamma t} |-\gamma(\cosh\beta + w(0)\sinh\beta + (\Omega/\Delta)\Gamma(\sinh\beta + w(0)\cosh\beta)| \quad (7)$$

allows to infer $w(0)$ from observations. Numerous systems are describable by this case, such as $Li(2s \rightarrow 2p) + H(3p \rightarrow 2s)$ or $B_e(^1S_0 \rightarrow ^1P_1) + C_d(^1P_1 \rightarrow ^1S_0)$.

Analogous features to the ones described above should exist in the fluorescence from excited fragments perturbed by the Stark fields of their ionic partners[3]. These features are most salient for a diatom dissociating into an ion and a hydrogenic emitting fragment (H, He^+, Li^{2+} etc.), whose nearly-degenerate fine-structure states are strongly mixed by the ionic field. The immediate diagnostic information obtainable on observing such features in this system is that the emission corresponds to a $\Delta\Lambda = 0$ transition. This conclusion stems from the fact that at $R(t)$ much larger than the impact parameter, the field of the ionic fragment points along the separation axis, whence the projection Λ on this axis is conserved. Thus $\overline{\Pi}^* \rightarrow \Sigma$ emission decays exponentially, whereas $\Sigma^* \rightarrow \Sigma$ emission is affected by the ionic Stark field.

To illustrate the evolution of $\Sigma^* \rightarrow \Sigma$ emission in this system, consider the case of Ly-α radiation ($2p \rightarrow 1s$) from the hydrogenic emitter. For $R(t) \lesssim Z^{\frac{1}{2}} 500 \mathring{A}$ (Z being the ion charge), the Stark interaction $|\Omega_S| = 6Zea_o/\hbar R^2$ (where a_o is the Bohr radius) is much larger than the fine-structure splitting Δ_F between $2p_{3/2}$ and $2s_{1/2}$. In this domain (the strong Stark regime), the total Ly-α emission rate is given by[3]

$$\dot{P} = e^{-\gamma t/2} \frac{\gamma}{2} \sum_{\pm} \{\rho_{\pm} \exp[\mp(\gamma/3) \int_0^t (\Delta_F/|\Omega_S|)dt'](1 \pm 2\Delta_F/3|\Omega_S|)\} \quad (8)$$

+ ringing terms.

Here γ is the Einstein A-coefficient of the 2p state and $\rho_{\pm}$ are the separation-dependent (i.e. time-dependent) populations of the Stark states (i.e. the states diagonalizing the Stark perturbation). In the strong Stark regime, these states are given by the $m_j(=\pm\frac{1}{2})$-dependent expressions $|\pm\rangle \simeq 2^{-\frac{1}{2}}|2s_{\frac{1}{2}}\rangle \mp 6^{-\frac{1}{2}}sgn(m_j\Omega_S)|2p_{\frac{1}{2}}\rangle \pm 3^{\frac{1}{2}}|2p_{3/2}\rangle$. They are energetically correlated to the following diatom states in homonuclear systems (e.g. H_2^+) $|+\rangle \leftrightarrow |2s\sigma_g\rangle, |3p\sigma_u\rangle$; $|-\rangle \leftrightarrow |3d\sigma_g\rangle$, $|4f\sigma_u\rangle$ whereas in a heteronuclear systems such as $(HeH)^{2+}$ $|+\rangle \leftrightarrow |2s\sigma\rangle$; $|-\rangle \leftrightarrow |3d\sigma\rangle$.

As seen from Eq. (8), the $|+\rangle$ state contributes to a superradiant nonexponential deviation from the average emission rate $(\gamma/2)\exp(-\gamma t/2)$ of an equal mixture of $|2s\rangle$ and $|2p\rangle$, whereas the $|-\rangle$ state contributes to a subradiant deviation of the same magnitude. The ringing terms in (8) oscillate with the phase $_0\int^t |\Omega_S| dt'$, and their magnitude depends on the coherences between Stark states with specific magnetic quantum numbers $m_j = \pm\frac{1}{2}$. The observability of ringing in this system would indicate that the relative phases of such states are well-defined in the quasi-molecular region. Whenever this is not the case, ensemble averaging eliminates the ringing terms, and comparison of the observed emission with (8) then yields directly the sharing of excitation between the $|+\rangle$ and $|-\rangle$ states.

As a final example of the diagnostic applications of CF, we consider the emission from electronically-excited quasi-bound or dissociating clusters composed of N>2 identical neutral fragments. The commonly used basis for the treatment of CF from N identical two-level emitters is that of the pseudospin Dicke states[7] $|s, s_z, \alpha\rangle$, where s is the cooperation number, $s_z = N_e - N/2$ (N_e being the number of excited emitters) and α a degeneracy index. The drawback entailed by the use of Dicke states is that states with different s are mixed by long-range dipole-dipole interactions, unless all fragments are geometrically equivalent (as in an equilateral triangle or a tetrahedron).

Therefore, we propose to describe CF in such systems using states adapted to the molecular symmetry group of the cluster, which automatically diagonalize these dipole-dipole interactions. Such states are easily constructed from products of single-fragment states by the projection formula of group theory. They are labeled $|\delta,\Delta,\underset{\sim}{n}>$, where δ is a component of a degenerate basis for the irreducible representation Δ of the cluster symmetry group, and $\underset{\sim}{n}$ are all other quantum numbers. Young-tableau techniques yield readily the superpositions of Dicke states (which, for given s and s_z transform as the basis for the permutation group $\mathscr{S}_N$) corresponding to each $|\delta,\Delta,\underset{\sim}{n}>$. This correspondence leads to the following result for the emission rate at $t<<\gamma^{-1}$ from a cluster much smaller than a wavelength, which is prepared by the excitation process in a state with symmetry Δ and N_e excited fragments:

$$\dot{P} \simeq \gamma \sum_{\delta} \rho_{\delta\delta}^{(\Delta)}(0) \sum_{s,\alpha} (s+s_z)(s-s_z+1)|<\delta,\Delta,\underset{\sim}{n}|s,s_z,\alpha>|^2 \ . \qquad (10)$$

The measurement of $\dot{P}$ at $t<<\gamma^{-1}$ can thus reveal the symmetry and the initial populations $\rho_{\delta\delta}^{(\Delta)}(0)$. For example, in a doubly-excited N=3 linear chain with a center of inversion ($D_{\infty h}$ symmetry), (10) yields zero for ungerade states and γ for gerade states.

References

1. G. Kurizki and A. Ben-Reuven, Phys. Rev. A 32:2560 (1985).
2. G. Kurizki and A. Ben-Reuven, Phys. Rev. A (to be published).
3. G. Kurizki and A. Ben-Reuven, Nucl. Instrum. Methods B (to be published).
4. P. Grangier, A. Aspect and J. Vigué, Phys. Rev. Lett. 54:418 (1985).
5. V. Sidis and D. P. de Bruijn, Chem. Phys. 85:201,215 and 233 (1984).
6. B. Willerding, W. Heiland and K. J. Snowdon, Phys. Rev. Lett. 53:2031 (1984).
7. G. S. Agarwal,"Quantum Statistical Theories of Spontaneous Emission" in Springer Tracts in Modern Physics 70, Springer, Berlin (1974).

COVERAGE EFFECT ON THE WORK FUNCTION OF A METAL FILMS COVERED WITH

ORGANIZED ORGANIC SPACER

A. Petrank, R. Naaman*, and J. Sagiv

Department of Isotope Research
Weizmann Institute of Science
Rehovot, Israel

The effect of adsorbed molecules or atoms on the surface's work function (wf) raised interest already in the beginning of the century. Langmuir established[1] that molecules or atoms which have an ionization potential lower than the surface wf, tend to reduce the effective wf as they are adsorbed. Later Gomer and co-workers made quantitative measurements both on metal surfaces covered with alkalies[2] and for adsorption of rare gases[3,4]. It was found that while the first monolayer cause a decrease, the second either has no effect, or may cause an increase in the wf.

The focus of this work was on the effect of spacers on the reduction of the work function due to adsorbed molecules. The surfaces studied were evaporized metal films (Al,Ag), and the spacer consisted of organized organic monolayers of long chain amphiphiles with varying length. They were chemically bound to a metal film[5] with the techniques developed by Sagiv and co-workers.[5,6] The organic molecules are chemisorbed with a Si-O bond to the surface. For comparison two types of spacers were investigated - A long organic chain that contains eighteen carbons (OTS=trichloroctadecyl-silan), and short chain which contain only a single methyl group (methyl silane=Me-Si).

The experimental setup was described in detail before[7], however it is important to note that the experiment was carried out under low vacuum conditions (10^{-3} to 10^{-6} torr). Hence the bare metal surfaces, when used were always covered by one or more monolayers of dirt. As organized organic monolayer covered the surface, it became oelphobic and hydrophobic, and because of the low sticking coefficient of water or of organic material on this type of surface it remained clean.

The photoemission signal dependence on the laser energy for aluminum or silver metal films coated with OTS or Me-Si could be fitted to Fowler's formula.[8] The wf obtained for aluminum covered either with OTS or Me-Si was the same. For silver the wf with Me-Si is lower than with OTS.

In figure 1 the photoemission signal is shown as function of adsorption time for naphthalene on a full OTS monolayer on Ag (full dots). For comparison the same experiment was performed with partial OTS monolayer (open dots). The naphthalene pressure in the cell during the experiment was 4×10^{-4} torr. A theoretical curve could be fitted to the data with the full OTS monolayer in figure 1, based on the Topping equation.[9]

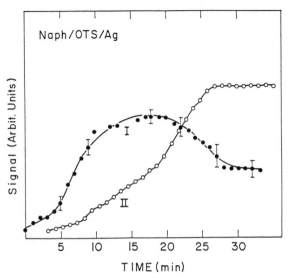

Figure 1: The photoemission signal as function of
coverage for naphthalene on full OTS monolayer (full
dots), and on partial OTS monolayer (open dots).

The model assumes that the adsorbed molecules create dipole moment
on the surface, while due to the polarizability of the adsorbance there
is a depolarization effect as the coverage increases. Hence the dependence
of the work function (ϕ) on the coverage (n) follows the equation:

$$\Delta\phi = c_1 n/(1+c_2 n^{3/2}) \qquad (1)$$

Based on the Fowler formula the change in the work function can be
converted to the change in the measured current (Δj) so that:

$$\Delta j = k n^2/(1+\ell n^{3/2})^2 \qquad (2)$$

From the parameter ℓ obtained, the polarizability of the adsorbed
monolayer can be deduced, and for naphthalene adsorbed on OTS it was found
to be 80×10^{-25} cm^3. The change in the current corresponds to a change of
about 0.003 V in the work function. From independent measurements it was
found that the OTS thickness is 27+3A, this means that even at such distances
the work function is effected and the reduction can be observed.

In figure 2 the signal dependence on the bias voltage is plotted for
naphthalene adsorbed on three type of substrates: OTS/Ag (I), Ag (II), and
Md-Si/Ag (III).

The signal observed with the OTS as spacer is smaller by about an order
of magnitude than in the other two cases. It is important to note that
the similarity between the uncovered silver surface and the one with Me-Si
is probably due to our poor vacuum. The silver film is covered by impurities
(water, oxygen etc.), hence also in this case a "spacer" exists although not
a controlled one. The similarity in the signal between this dirty surface
and the one covered with Me-Si indicates that the dirt has an effective thickness which is equivalent to the methyl silane length.

Also important is the fact that the aluminum film used always oxidized,
even before the organic monolayer was put on it. Thus the fact that the
work function of the surface covered with aluminum does not vary between
OTS and Me-Si coverage, may result from the existence of this oxide layer.

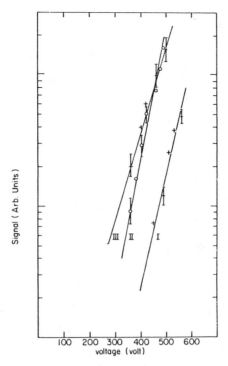

Figure 2: The photoelectron signal dependence on the bias voltage for naphthalene adsorbed on OTS/Ag (I), Ag (II), and Me-Si/Ag (III).

In order to check that the effect observed is due to the spacer and is not a result of "holes" in the spacer, a partial monolayer was prepared. In this case "holes" are created in a controlled manner. We found here that the current dependence on the coverage does not show the decrease at high coverage. This result can be rationalized by the small size of the "holes" which does not allow the cooperative polarizability effect. Hence a dipole is created which cannot be canceled even partially by the polarizability. As it is well known more monolayers have no effect on the work function, therefore a plato is observed, following the first monolayer.[4]

REFERENCES

1. I. Langmuir, J. Am Chem. Soc. 54:2798 (1932).
2. L.D. Schmidt and R. Gomer, J. Chem. Soc. 45:1605 (1966).
3. R. Gomer, Australian J. Phys. 45:1605 (1966).
4. C. Wang and R. Gomer, Surface Sci. 91:533 (1980).
5. J. Gun, R. Iscovici and J. Sagiv, J. Colloid and Interfaces Sci. 101:201 (1984).
6. R. Maoz and J. Sagiv, J. Colloid and Interface Sci. 100:465 (1984).
7. R. Naaman, A. Petrank and D.M. Lubman, J. Chem. Phys. 79:4608 (1983).
8. A.V. Sokolov, Optical Properties of Metals, ed. O.S. Hearens, London (1967) p. 399.
9. Solid Surface Physics, ed. G. Hohler, New York (1979) Vol. 85 p. 29.
*   Incumbent of the C.S. Koshland Career Development Chair

Acknowledgements

This work was supported partially by the Fund for Basic Research administrated by the Israel Academy of Sciences.

PHOTOIONIZATION AND DISSOCIATION OF THE H_2 MOLECULE NEAR THE IONIZATION THRESHOLD

H. Rottke and K. H. Welge

Universität Bielefeld

D - 4800 Bielefeld 1, FRG

INTRODUCTION

The investigation into molecular hydrogen Rydberg states has gained new interest in the last few years, since laser excitation into these states has become possible. The first detailed spectroscopic studies on highly excited H_2 Rydberg states were done by Herzberg and Jungen[1] for the singlet np-series using classical spectroscopy methods, and the dynamical processes autoionization and predissociation in the np-Rydberg series were for example investigated by Dehmer and Chupka[2]. A laser experiment using three photon excitation of the singlet np-Rydberg series from the $X^1\Sigma_g^+$ ground state was reported recently by Helm et al.[3]. Several other groups deal presently with laser spectroscopy and the study of autoionization and predissociation of triplet nd-Rydberg series excitable from the meta-stable $c^3\Pi_u^-$ state[4,5,6].

We carried out experiments on resonant two photon excitation of high singlet gerade ns- and nd-Rydberg series. As resonant intermediate state we used $B^1\Sigma_u^+$ (v = 0) excited with tunable vuv laser radiation. The second step, excitation from the B-state to the Rydberg states, was done by uv laser radiation. Below the H_2 ionization threshold we observed predissotiation of the highly excited states and above threshold autoionization of Rydberg series converging at rotationally and vibrationally excited levels of the H_2^+ molecular ion was detected.

EXPERIMENTAL

Fig. 1 shows schematically the experimental arrangement. Hydrogen gas was introduced in an effusive beam through a capillary into the excitation region between two parallel metal plates. The counterpropagating vuv and

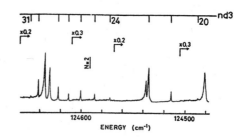

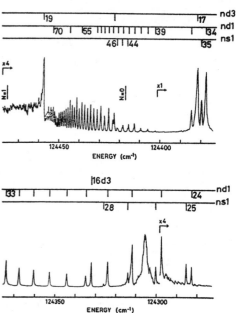

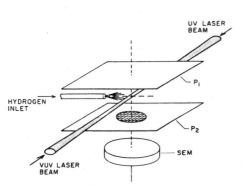

Fig.1 Experimental arrangement

Fig.2 Section of ionization disso-ciation spectrum with initial state $B^1\Sigma_u^+$ (v = 0, J = 0). N = 0,1,2 are thresholds for H_2^+ ($X^2\Sigma_g^+$, v = 0,N) + e.

uv laser beams crossed the H_2 molecular beam at right angles. Photo-ionization of the hydrogen molecules was monitored by detecting ions with a secondary electron multiplier (SEM). The ions were extracted from the excitation region by a pulsed electric field which was applied to the metal plates after excitation. For the results it was important that during the excitation process no electric field was present in the interaction region to avoid the Stark effect. Besides ion extraction the pulsed elec-tric field (178.6 V/cm) served to field ionize bound states with high principal quantum numbers.

The vacuum ultraviolet laser radiation near 111 nm, necessary to excite the (0,0) band of the $B^1\Sigma_u^+ \longleftarrow X^1\Sigma_g^+$ transition, was generated by frequency tripling the radiation of an excimer laser pumped uv dye laser near 333 nm in the noble gas Krypton. The uv radiation exciting from $B^1\Sigma_u^+$ (v = 0) to the Rydberg states (tuning range ∿270 nm – ∿298 nm) was generated by frquency doubling the output of a second dye laser operating in the visible ($\sim$540 nm – $\sim$596 nm). The bandwidth of the vuv laser radiation was 0.95 cm^{-1} and of the uv radiation 0.45 cm^{-1}. The wavelength calibration in the spectra was done absolutely using Neon hollow cathode lines and relatively by means of a Fabry-Perot etalon with a free spectral range of 1.6931 cm^{-1} $\pm$ 0.0009 cm^{-1}. This allowed the absolute term energy of observed lines to be determined to within 0.7 cm^{-1}.

RESULTS

Fig. 2 shows as an example a section of a photoionization spectrum obtained near the ionization threshold of the H_2 molecule at 124416.1 cm^{-1}. It was obtained by holding the vuv laser wavelength fixed at the P(1) line of the (0,0) band of the $B^1\Sigma_u^+ \longleftarrow X^1\Sigma_g^+$ transition, thus exciting the state $B^1\Sigma_u^+$ (v = 0, J = 0). This choice of intermediate state allows the uv laser to excite Rydberg states on the R(0) transition only thus reaching states with total angular momentum J = 1. Taking into account the united atom configuration of $B^1\Sigma_u^+$, which is ($1\sigma_g$, $2p\sigma$), the united atom configuration of the excitable Rydberg states is determined to be ($1\sigma_g$, nsσ), ($1\sigma_g$, ndσ), ($1\sigma_g$, ndπ) with the highly excited electron either in an s- or d-orbital. The desciption of the high Rydberg states in terms of the Born-Oppenheimer approximation and therefore Hund's case (a),(b) coupling of angular momenta fails, a more adequate description is in terms of Hund's case (d) coupling. This coupling scheme views the highly excited molecule as consisting of an electron in an s- or d-orbital coupled loosely to the H_2^+ ion core in a definite electronic ($X^2\Sigma_g^+$) vibrational (v^+) and rotational (N^+) state, leading to the state notation (nlN^+) used in the spectrum (fig. 2, n = principal quantum number). All identified lines belong to states having the H_2^+ ion core in its vibrational ground state v^+ = 0, the resonances not identified have the H_2^+ ion core in a vibrationally excited state. Due to the total angular momentum J = 1 of the Rydberg states only the three identified (ns1), (nd1) and (nd3) Rydberg series can be excited for each vibrational state v^+ of the ion. An analysis of the spectra can be done using the multichannel quantum defect theory (MQDT) developed for molecules by Fano et al.[7,8] and Jungen and Atabek[9]. We used MQDT to calculate Rydberg state term values theoretically by starting with quantum defect curves $\mu_{l\lambda}$(R) (R = internuclear distance) of the $H\bar{H}^1\Sigma_g^+$($1\sigma_g$, 3sσ), $GK^1\Sigma_g^+$($1\sigma_g$, 3dσ), $I^1\Pi_g$($1\sigma_g$, 3dπ) and $J^1\Delta_g$($1\sigma_g$, 3dδ) states. Despite this crude approximation quite good agreement of the calculation with the measured line positions could be achieved.

Different processes lead to the ion signal observed in the spectrum in fig. 2. The lowest H_2 ionization threshold, where a Rydberg series excitable on a R(0) transition from $B^1\Sigma_u^+$ converges, is the threshold with the H_2^+ ion core in $v^+ = 0$ and $N^+ = 1$ (indicated by N = 1 in fig. 2). Below this threshold down to about 124400 cm^{-1} the observed ion signal is due to electric field ionization by the applied pulsed electric field (178.6 V/cm) for all members of the (nd1), (ns1) Rydberg series. Below 124400 cm^{-1} electric field ionization is no longer possible. In this range the excited Rydberg states predissociate into either one of the following two energetically possible channels:

$$H(1s) + H(1s) \text{ or } H(1s) + H(2s,2p).$$

The channel $H(1s) + H(2s,2p)$ is monitored in our experiment below 124400 cm^{-1} through photoionization of the excited H atom $H(2s,2p)$ by absorption of one uv laser photon. The predissociation of the high n Rydberg states is probably due to coupling to the dissociation continuum of the doubly excited state $F^1\Sigma_g^+((1\sigma_u)^2)$ which adiabatically correlates to the separated atoms state $(1s,2s)_g$. The (nd3) lines appearing in the spectrum above the ionization threshold can be due to two processes, autoionization and the just mentioned predissociation. Autoionization is possible through coupling to the (ϵs1) or (ϵd1) ionization continua. As can be seen from fig. 2 (nd3) lines appear only near perturber states with the H_2^+ ion core vibrationally excited. Thus one can conclude that direct autoionization via rotation-electron coupling is very weak but autoionization through vibration-electron coupling occurs quite strongly. Using Fano's l-uncoupling treatment[7,8] we determined the eigen quantum defects $\mu_{d\sigma}$, $\mu_{d\pi}$ for the (nd1), (nd3) Rydberg series from our spectrum to be very small. This fact confirms that rotation-electron coupling is actually very small at least within the (nd1), (nd3) channels.

Besides the small section in fig. 2 we took and analysed spectra starting at $B^1\Sigma_u^+$ (v = 0, J = 0) up to the threshold where H_2^+ can be produced in its first vibrationally excited state ($v^+ = 1$) and spectra near the ionization threshold starting at $B^1\Sigma_u^+$ (v = 0, J = 1).

REFERENCES

1) G. Herzberg and Ch. Jungen, Jour. Molec. Spectrosc., 41:425 (1972)

2) P. M. Dehmer and W. A. Chupka, Jour. Chem. Phys., 65:2243 (1976)

3) N. Bjerre, P. Kachru and H. Helm, Phys. Rev. A, 31:1206 (1985)

4) H. Helm, D. P. de Bruijn and J. Los, Phys. Rev. Lett., 53:1642 (1984)

5) R. Kachru and H. Helm, Phys. Rev. Lett., 55:1575 (1985)

6) R. D. Knight and L. Wang, Phys. Rev. Lett., 55:1571 (1985)

7) U. Fano, Phys. Rev. A, 2:353 (1970)

8) E. S. Chang and U. Fano, Phys. Rev. A, 6:173 (1972)

9) Ch. Jungen and O. Atabek, Jour. Chem. Phys., 66:5584 (1977)

PHOTODISSOCIATION DYNAMICS OF TERT-BUTYL-NITRITE

D. Schwartz-Lavi, I. Bar and S. Rosenwaks

Department of Physics
Ben-Gurion University of the Negev
Beer-Sheva 84105, Israel

INTRODUCTION

The dynamic aspects of the photofragmentation of the t-butyl-nitrite $(CH_3)_3CONO$ (TBN) molecule were studied by determining the NO photoproduct internal states distributions for different initial states of the excited parent molecule.

The TBN electronic transitions can be compared to those of the closely related molecule HONO. Studies on this molecule show that its first electronic excited state is planar with an A'' symmetry. The CONO part of TBN is also expected to be planar and its first electronic transition $\tilde{X}^1A' \rightarrow \tilde{A}^1A''$, like that of HONO, corresponds to the excitation of one electron of the lone pair of oxygen to the π^* antiboding orbital. The absorption band associated with the $\tilde{X}^1A' \rightarrow \tilde{A}^1A''$ transition consists of a vibrational progression assigned to the excitation of the -N=O stretching mode.

The photofragmentation of TBN was carried out at several wavelengths: 339, 351.8, 365.8 and 381.1nm corresponding to the specific NO vibrational stretching level of the parent molecule in its $n\pi^*$ excited state. The quantum states of the NO fragment were determined for each dissociation wavelength. The relevant energy levels of TBN and NO are presented in Fig. 1.

EXPERIMENTAL

The photodissociation of TBN was performed in a flow cell at room temperature at a pressure of 0.1 Torr. Two antiparallel propagating laser beams were used: A frequency doubled, Nd:YAG pumped dye laser beam to photodissociate the TBN and a frequency doubled Excimer pumped dye laser beam to probe the NO fragment via a single photon laser induced fluorescence (LIF). An external oscillator and a delay generator were used in order to synchronise the lasers so that the probe beam will be delayed by ~50ns with respect to the photolysing one. The LIF from NO was collected by a solar blind photomultiplier, averaged and normalized to the probe lasers intensity by a boxcar, and recorded on a strip-chart recorder. In some of the experiments the relative polarizations of the two lasers were of importance, as will be explained later. For

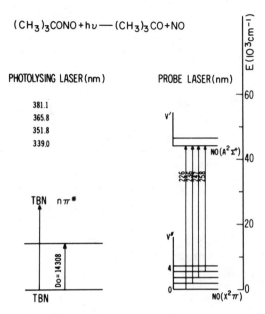

$(CH_3)_3CONO + h\nu \longrightarrow (CH_3)_3CO + NO$

Fig. 1.   Diagram of energy levels of TBN and NO
          and of the photolysing laser wavelengths.

these experiments a polarization rotator was used in order to rotate the
polarization of the photolysing beam.

RESULTS AND DISCUSSION

Rotational Distribution

     A detailed analysis of the rotational state distribution for the
different branches of all the spectra collected indicates a preference
in populating specific rotational levels:

     1.   The population of the rotational levels of the F_1 $(X^2\Pi_{\frac{1}{2}})$ spin
state probed by the $Q_{11} + P_{21}$, P_{11}, $R_{11} + Q_{21}$ and R_{21} branches is
preferred over the population of the rotational levels of the F_2 $(X^2\Pi_{3/2})$
spin state probed by the P_{12}, $Q_{12} + P_{22}$, $R_{12} + Q_{22}$ and R_{22} branches.

     2.   The population of one of the Λ doublet levels probed by the
$Q_{11} + P_{21}$ and $Q_{22} + R_{12}$ branches is preferred over the population of
the other Λ doublet levels probed by the R_{22}, $P_{22} + Q_{12}$, P_{11} and
$R_{11} + Q_{21}$ branches. The Λ doublets population will be discussed later.

     In order to compare the rotational distribution in the different
vibrational levels at each dissociating wavelength we have selected the
Q_{22} branch which contains enough resolved lines to build a distribution.

     Our analysis indicates high rotational non-Boltzmann distribution
peaking in high rotational levels ($J'' \sim 31\frac{1}{2}$). The different vibrational
levles of the NO fragment show similar rotational distributions. Identical
distributions were derived for each dissociating wavelength showing that
the rotational state distribution is unaffected by the different initial

states of the excited parent molecule. A typical excitation spectrum of NO following TBN photolysis is presented in Fig. 2.

Vibrational Distribution

The vibrational state distribution indicates population of higher vibrational levels of the $NO(X^2\Pi)$ fragment upon dissociating the TBN at shorter wavelengths. Since the rotational population is unaffected by the dissociating wavelength, it seems that the vibrational energy deposited in the -N=O stretching mode of the TBN $n\pi^*$ state is retained in the vibration of the NO fragment. Retainment of the vibrational energy has previously been reported for the photodissociation of CH_3CONO.[1] In order to gain further understanding regarding this point, an analysis of the total vibrational energy content of the NO fragment for each dissocia- tion wavelength, compared with the excess vibrational energy deposited in the -N=O stretching mode of the excited parent molecule must be performed. This analysis will be done in the near future.

Λ Doublets Population

The two Λ doublet states of $NO(X^2\Pi)$ are nondegenerate due to the coupling of electronic and nuclear motion. In the classical high J limit, one Λ doublet (Π^+) corresponds to the π lobes of the molecular orbital being in the plane of rotation perpendicular to the angular momentum J and to the internuclear axis. The other Λ doublet (Π^-) refers to the π lobes being perpendicular to the plane of rotation

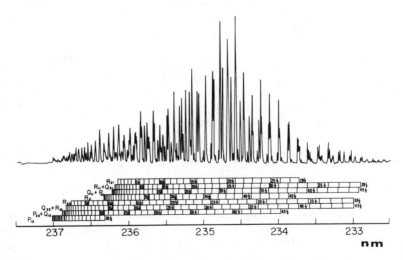

Fig. 2.   NO excitation spectrum ($v'' = 1 \rightarrow v' = 0$) following TBN photolysis at 381.1nm.

and to the internuclear axis but parallel to J. The electronic wavefunction of the Π^- Λ doublet state is antisymmetric with respect to reflection at the plane of rotation whereas the electronic wavefunction of the Π^+ Λ doublet is symmetric with respect to the same reflection.

The TBN excited state is of A'' symmetry where the π lobes are perpendicular to the molecule rotation plane. The conservation of symmetry would cause the antisymmetric Λ doublet component of the NO fragment to be highly populated relative to the symmetric Λ doublet component provided that the dissociation process is fast and planar. An indication to the planarity of the TBN photofragmentation process is found from the alignment factor derived in the next section. It has been shown[2] that in a $^2\Pi$ state the Q lines probe the Π^- Λ doublet levels while the P and R lines probe the Π^+ Λ doublet levels. Accordingly Λ doublets population ratio can be deduced from the relative intensities of the Q and P and R branches. Our measurements and calculations of the Λ doublets population indicate preferential population of the Π^- antisymmetric Λ doublet levels.

However, it has been shown that the internal state distributions are not accurately and directly deducible from the experimental line intensities due to the alignment of the fragment.[3] In the limit of fast fragmentation an experimental geometry where the dissociating and probe electric vectors are parallel, $\hat{\varepsilon}_d || \hat{\varepsilon}_p$, favors the detection of Q transition whereas an experimental geometry where $\hat{\varepsilon}_d \perp \hat{\varepsilon}_p$ favors P and R transitions. One has to correct the intensities for the alignment of the fragment. This can be done following the procedure of Greene and Zare.[4]

These calculations are yet to be performed. We can only present a lower limit to the relative population of the Λ doublets levels. They are deduced from the spectra collected with the $\hat{\varepsilon}_d \perp \hat{\varepsilon}_p$ geometry which favors as noted above the P and R transitions over Q. Our experimental observations indicate a Π^-/Π^+ population ratio > 1.8 for $J'' > 30\frac{1}{2}$.

$\bar{J}_{NO}$ Alignment

In a completely planar fragmentation, the plane defined by the fragment rotation is identical to the parent molecule plane. In order to check for the planarity of the photodissociation process of TBN, we examined the spectra collected in the two different experimental geometries mentioned above. ($\varepsilon_p \perp \varepsilon_d$ and $\varepsilon_p || \varepsilon_d$).

We analysed the polarization data for alignment of $\bar{J}_{NO}$ with the aid of a classical dipole model presented by Andresen et al.[5] According to this model, one calculates a joint probability for: 1) a specific orientation of $\bar{J}_{NO}$ after dissociation, 2) exciting that NO with a particular transition, and 3) finding the component of each P, R or Q emission dipole on a particular Cartesian axis.

Relying on our experimental data and on the limited calculations performed to date, we deduce an alignment factor $A_0^{(2)}$ of 0.32 for the high J limit.

The value of $A_0^{(2)}$ for total alignment is 0.8. Our value of $A_0^{(2)}$ indicates there is partial planarity in the photofragmentation of TBN. Deviation from complete planarity could arise from out of plane rotation or torsional motion of the parent molecule.

Further measurements of the vibrational and rotational distributions are now in progress and will be published in the near future.

References

1. O. B. D'azy, F. Lahmani, C. Lardeux and D. Solgadi, State-Selective Photochemistry: Energy Distribution in the NO Fragment After Photo-dissociation of the CH_3ONO $n\pi^*$ State, Chem. Phys. 94:247 (1985).

2. M. H. Alexander and P. J. Dagdigian, Clarification of the Electronic Asymmetry in π-State Λ Doublets with Some Implications for Molecular Collisions, J. Chem. Phys. 80:4325 (1984).

3. R. Vasudev, R. N. Zare and R. N. Dixon, State-Selected Photo-dissociation Dynamics: Complete Characterization of the OH Fragment Ejected by the HONO Ã State, J. Chem. Phys. 80:4863 (1984).

4. C. H. Greene and R. N. Zare, Determination of Product Population and Alignment Using Laser-Induced Fluorescence, J. Chem. Phys. 78:6741 (1983).

5. P. Andresen, G. S. Ondrey, B. Titze and E. W. Rothe, Nuclear and Electron Dynamics in the Photodissociation of Water, J. Chem. Phys. 80:2548 (1984).

THREE PHOTON SPECTROSCOPY OF THE LOWEST RYDBERG- AND VALENCE

STATES OF THE CHLORINE MOLECULE

M. Swertz   and D. Haaks

Universität G.H. Wuppertal
FB 9 Physikalische Chemie
Gaußstr. 20
5600 Wuppertal 1   West Germany

Abstract

In the last years the electronically excited states of the chlorine molecule have attracted considerable experimental /1,2/ and theoretical /3/ attention due to the observation of laser action in Cl_2 /4/ and the important role of Cl_2^+ in the rare gas chloride excimer formation processes /5/.
The results of our spectroscopic studies on the first Rydberg-series of the chlorine molecule in the energy range between 70000 and 80000 cm^{-1} using three photon excitation are reported in the present work.

Experimental

The chlorine was excited by focussing UV laser light into a fluorescence cell. The UV radiation at wavelengths between 4000 and 4120 A was obtained by frequency mixing of the radiation of a pulsed dye laser (Quantel TDL IV) with the fundamental wave of a Nd-YAG laser. The typical laser power at these wavelengths was 3 - 8 mJ with a pulse duration of 10 ns. The fluorescence was monitored between 1200 and 2800 Å via a 1 m-VUV-monocromator (McPherson Mod. 225) and a solar blind photomultiplier (EMR 542F-08-18).

Results and Discussion

In order to figure out the complexity of the potentials of the excited states, selected potential curves are depicted in fig.1 /3/.
In the energy range from 70000 to 80000 cm^{-1} the $1^1\Sigma_u^+$, $2^1\pi_u$ and $2^3\pi_u$ Rydberg states can be excited. In fig. 2 a Cl_2 excitation spectrum in the energy range of $\tilde{\nu}$ =73000 - 74800 cm^{-1} is shown. According to Douglas /1/ and Möller et al. /2/ the bands were assigned to transitions from the $X^1\Sigma_g^+$ ground state to the $1^1\Sigma_u^+-$, $2^1\pi_u-$, and $2^3\pi(0_u^+,1_u)$-Rydberg states. In contrast to the cited investigations two strong blue shaded bands with band heads at 74500 and 74675 cm^{-1} were

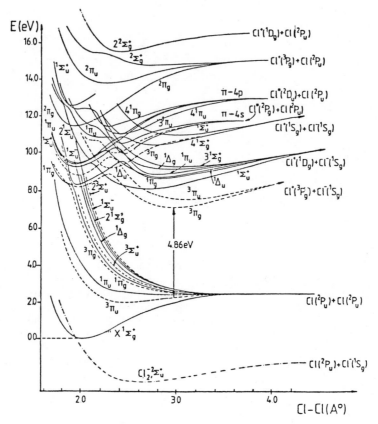

Fig.1   Potential Diagram of the Cl₂ Molecule /3/

464

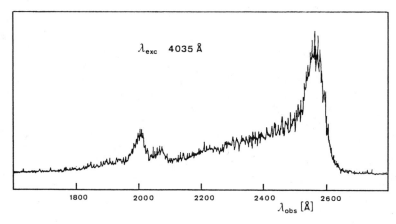

Fig.4   Fluorescence Spectrum of Cl_2

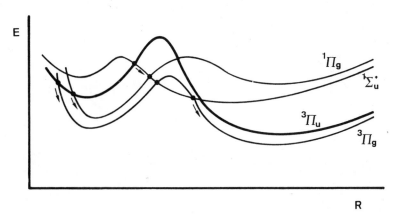

Fig.5   Mechanism for the Collisionally Induced
             Decay of Rydberg Cl_2 Molecules

Acknowledgements

We thank the Deutsche Forschungsgemeinschaft (SFB42) and the
Fonds der Chemischen Industrie for the financial support of
this work.

References

1. A.E. Douglas
        Can. J. Phys. $\underline{59}$, 835 (1981)
2. T. Möller, B. Jordan, P. Gürtler, G. Zimmerer,
    D. Haaks, J. LeCalvé and M.C. Castex
        Chem. Phys. $\underline{76}$, 295 (1983)
3. S.D. Peyerimhoff and R.J. Buenker
        Chem. Phys. $\underline{57}$, 279 (1981)
4. A.K. Hays
        Opt. Commun. $\underline{28}$, 209 (1979)

5. J.LeCalvé, M.C. Castex, D. Haaks, B. Jordan,
   G. Zimmerer
       Il Nuovo Cimento 63, 265 (1981)
   Ch.A. Brau
       in Excimer Lasers Topics in Applied Phys. 30
       ed. by Ch.K. Rhodes 1979 p.87
6. T. Ishiwata and I. Tanaka
       Chem. Phys. Lett. 107, 434 (1984)

UNEXPECTED RESULTS ON NO_2 AND

AN UNCONVENTIONAL EXPLANATION

H.G.Weber *

Physikalisches Institut, Universität Heidelberg

Heidelberg, Fed.Rep.of Germany

Several experiments on NO_2 reveal unexpected results, for which we have no conventional explanation[1-6] . The conditions in these experiments are to our knowledge analogous to those of experiments on atoms, namely preparation of a free molecule into a well defined excited state and detection of the radiative decay of this state. This well defined experimental situation is clearly demonstrated by a number of results, which indicate that laser light induces a transition between two isolated and well defined fine structure levels of the ground and electronically excited state. However, under the same experimental conditions there are simultaneously also other observations with unexpected results, which are in disagreement with the conventional description of the prepared experimental situation. The two classes of experimental results appear consistently in different experiments: in level-crossing experiments, in optical radio-frequency double resonance experiments, in "laser scanning" experiments, in experiments under molecular beam conditions and under static gas conditions, and with use of different lasers in different spectral regions of the NO_2 absorption spectrum. The experiments show that two characteristic times, the "intramolecular decay time" $\tau_0 \approx 3$ μs and the radiative decay time $\tau_r \approx 35$ μs have to be associated with each excited state, which according to other experiments is an isolated quantum state. We come to the conclusion that a consistent description of all experimental results is possible with the assumption that underlying the conventionally expected quantum states of NO_2 (within an excitation width of 0.2 MHz) is a substructure which evolves irreversibly in time [3,4]. As an example of our investigations we describe here a "laser scanning" experiment.

The experimental situation is indicated in Fig.1. A light beam of a single mode, cw dye laser is split into two beams L_1 and L_2, which both cross a beam of freely propagating NO_2 molecules (molecular beam axis is z-axis, the average velocity of molecules along the z-axis is u=610±25 ms^{-1}). It is possible to vary the angle α and the gap width s (travelling time T_s = s/u) between L_1 and L_2, and the aperture width d which determines the transit time T_L=d/u through L_1 or L_2. In the experiments the laser light (polarization parallel to z-axis) is tuned with α=0 to a molecular transition near 593 nm and the fluorescence light is detected either versus the angle α or versus a magnetic field B (parallel to z-axis). The detection of fluorescence light can be along the molecular beam axis or perpendicular to it.

* present address
Heinrich-Hertz-Institut, Einsteinufer 37, D-1000 Berlin 10, FRG

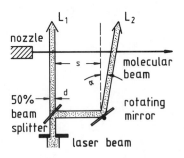

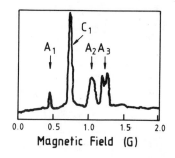

Fig.1: Experimental arrangement. Not indicated are the magn.fields and the fluorescence detection.

Fig.2: Magnetic resonance spectrum of an optical transition between two fine structure levels.

In a first experiment α and d are fixed at $\alpha=0$ and d=1 mm. We apply a radio frequency field (frequency ν', polarization perpendicular to z-axis) and sweep the static magnetic field B through a value B' which satisfies the magntic resonance condition $2\pi\hbar\nu' = g\mu_B B'$. Fig.2 shows an experimental result. There is a residual Doppler width of about 50 MHz in the molecular beam. The laser light depopulates all hyperfine components (hyperfine splitting >5 MHz) of the lower state fine structure level $< N=0, J=1/2 |$ and populates all hyperfine components of the upper state fine structure level $< N=1, J=3/2 |$. Here J=N+S, where N and S are the rotational and electron spin quantum numbers respectively. There appears an increase of the total fluorescence intensity whenever the magnetic resonance condition is met for an upper state (the resonances A_1, A_2, A_3 for the hyperfine components F=1/2, 3/2,5/2, the nuclear spin is I=1), or a lower state (the resonance C_1, the two hyperfine components F=I ±J with J=1/2 have the same value $|g|$). There are no additional resonances at higher magnetic fields. Measurements with similar results are performed on 10 different absorption lines of NO_2. The required light intensity is in general >10 mW/mm^2. The measured g-factors (precision $\approx 1\%$) are in agreement with optical rf-double resonance experiments and with theoretical expectations for the ground state. The width (FWHM) of the resonances for low rf-field power is given (in the frequency domain) by $\Delta\nu'=(0.86\pm0.05)/(T_s+T_L)$ where T_s+T_L is the travelling time from the center of L_1 to the center of L_2. This indicates pure transit time broadening. From the amplitudes of the resonances versus the gap width s we find that branching back of fluorescence light to the initial state and intrabeam collisional perturbations are negligible, and we find a value for the excited state lifetime $\tau_r \approx 35$ µs which is in good agreement with results of radiative decay and Hanlé measurements. The resonances can be explained by optical pumping and laser saturation effects.

The resonance spectrum in Fig.2 disappears (for d≤1 mm) if we use only one light beam. In this case (with use of polarization sensitive detection) we see optical rf-double resonance signals at A_2 and A_3 having a width determined by the radiative lifetime $\tau_r \approx 35$ µs (see perturbation at A_3 in Fig. 2). However, for d>1 mm a resonance spectrum with broad resonances at A_1, A_2, A_3 and C_1 appears also with one light beam only (see Fig.2 of Ref.2). These "broad magnetic resonances" (the width is independent of the light intensity) have to be associated with a lifetime $\tau_o \approx 3$ µs. Similar "broad magnetic resonances" (additionally to the much narrower Hanle signals) are also observed in level crossing experiments. Both experiments[2,5,6] reveal the existence of two characteristic times τ_r and τ_o associated with each excited state of NO_2. We have no conventional explanation for the "broad magnetic resonances" and for the time τ_o associated with them. τ_o is

470

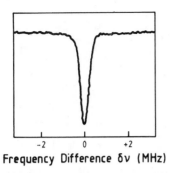

Frequency Difference δν (MHz)

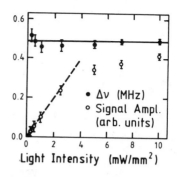

- Δν (MHz)
∘ Signal Ampl.
(arb. units)

Light Intensity (mW/mm^2)

Fig.3: Resonance signal in the fluorescence intensity versus the frequency difference $\delta\nu=\nu\alpha(u/c)$.

Fig.4: The width $\Delta\nu$ and the amplitude A_ν of the resonance signal shown in Fig.3 vs.the light intens.

however in good agreement with the lifetime evaluated from the integrated absorption coefficient of NO_2. Also the experiment described in the following reveals the existence of both lifetimes.

In the "laser scanning" experiment[4] we use no rf-field and no static field B (compensation of earth magnetic-field). The fluorescence light is detected versus the angle α. If we vary α we change the relative frequency difference $\delta\nu = \nu\alpha(u/c)$ of L_1 and L_2 as seen by the molecules (ν light frequency, c light velocity). As Fig.3 shows there appears a resonance at $\alpha=0$ indicating a minimum (up to 20%) of the total fluorescence intensity at $\alpha=0$. Similar resonances are obtained on all investigated absorption lines of NO_2 in the red, green and blue part of the absorption spectrum. We investigate the width $\Delta\nu$ (FWHM) and the amplitude A_ν of this resonance versus the light intensity (Fig. 4), versus the aperture width d (Fig. 5) and versus the gap width s (Fig. 6). The amplitude A_ν is defined as $A_\nu=(P-P_0)/P_0$ where P is the fluorescence intensity as seen by the photomultiplier and $P=P_0$ if α is far off resonance. We find that $\Delta\nu$ is independent of the light intensity but that A_ν shows a strong saturation behaviour (Fig. 4). $\Delta\nu$ is given by $\Delta\nu=\Delta\nu_0+\Delta\nu_t+\Delta\nu_i$ with $\Delta\nu_0=0.13\pm0.05$ MHz, $\Delta\nu_t=(0.87\pm0.05)/T_L$ and $\Delta\nu_i=(7.6\pm1.0)10^9 (T_s+T_L)(Hz)^2$. Here $\Delta\nu_t$ is in good agreement with the expected transit time broadening and $\Delta\nu_i$ describes the frequency jitter in the laser light. $\Delta\nu_0$ is solely a property of the molecule. The uncertainty of $\Delta\nu_0$ includes also the velocity distribution in the molecular beam. We find $\Delta\nu_0=(\pi\tau_0)^{-1}$ in good agreement with the value τ_0 measured before. Fig. 6 shows (for corrections see Ref.4) that A_ν depends on the distance $s=uT_s$ between L_1 and L_2 as $\exp(-T_s/\tau_r)$. That is, it depends only on the excited state radiative lifetime τ_r. There is obviously no contribution from the ground state.

The investigations of NO_2 reveal two classes of results. On the one hand we find a confirmation of the prepared experimental situation. We obtain expected results. On the other hand nearly all experiments reveal more "structure" than expected. The unexpected experimental results are in disagreement with the conventional description of the prepared experimental situation. For instance in the "laser scanning" experiment we expect a saturation resonance whose width depends on the light intensity and approaches the value $(2\pi\tau_r)^{-1}\approx5$ kHz for vanishing light intensity. The amplitude of this resonance is expected to depend primarily on the population hole in the lower state. However, the experimental results are very different from these expectations. Experimental errors can be excluded. The same experiments are also performed on I_2 with the same experimental apparatus,

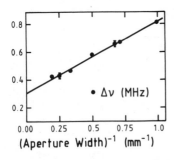

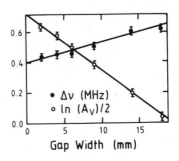

Fig.5: The width $\Delta\nu$ of the resonance signal shown in Fig.3
versus $(d)^{-1}$ (d=aperture width).

Fig.6: The width $\Delta\nu$ and the quan-
ty $\ln(A_\nu)/2$ versus the gap
width s. A_ν is signal amplitude.

and good agreement with theoretical expectations is obtained in this case.
Moreover, the unexpected results in the "laser scanning" experiment are in
agreement with the unexpected results in the magnetic resonance experiments. The lifetime τ_0 extracted from the "broad magnetic resonances"
appears also in the "laser scanning" experiment. The lifetime τ_0 was interpreted as characterizing the time scale of an irreversible intramolecular
evolution in the excited state [2,3]. Obviously (for vanishing transit time
broadening) the intramolecular evolution and not the radiative decay limits
the absorption process and determines therefore the absorption width $\Delta\nu_0$.
We obtain $\Delta\nu_0 = (\pi\tau_0)^{-1}$ if we use the analogy of the change of the absorption
process in a collision broadened absorption line. The most surprising observation of the "laser scanning" experiment is the result on the amplitude
A_ν. This result is consistent with all other unexpected observations, if an
irreversible intramolecular evolution similar to the excited state (but
much faster) is also assumed for the ground state [4].

In conclusion, we find consistency of the unexpected results in very
different experimental situations. The consistency is the strongest indication for a "structure" underlying the conventionally expected quantum
states of this molecule. We have no explanation for this "structure".

References

1. H.G.Weber, F.Bylicki and G.Miksch: Inversion of polarization by light-
   induced stabilization, Phys.Rev.A30:270 (1984)
2. H.G.Weber and G.Miksch: Magnetic resonances verified in agreement with
   an assumed time-asymmetric evolution, Phys.Rev.A31:1477 (1985)
3. H.G.Weber: Explanation of unexpected results with the use of a time-
   asymmetric evolution in a free molecule, Phys.Rev.A31:1488 (1985)
4. H.G.Weber: Anomolous laser saturation resonances in NO_2, Z.Phys.D-Atoms
   Molecules and Clusters 1 (1986)
5. H.G.Weber and G.Miksch: Unexpected properties of a quantum mechanical
   system, Phys.Lett. 106A:239 (1984)
6. H.G.Weber: Unexpected results in level crossing spectroscopy on NO_2,
   submitted for publication.

PARTICIPANTS

AHARON Y.
Dept. of Chemistry
Ben Gurion University
Beer Sheva
Israel

ALBECK D.
Dept. of Physics
Bar Ilan University
Ramat Gan
Israel

ANNER O.
Dept.of Physical Chemistry
The Hebrew University
Jerusalem
Israel

ASIDA N.
Dept. of Physics
Bar Ilan University
Ramat Gan
Israel

BAND Y.
Dept.of Chemistry
Ben Gurion University
Beer Sheva, Israel
(Bitnet:GEAQ100 AT BGUNOS)

BAR I.
Dept. of Physics
Ben Gurion University
Beer Sheva.
Israel

BAR-JOSEPH I.
Dept. of Electronics
Weizmann Institute of Science
Rehovot 76100
Israel

BAVLI R.
Dept. of Chemistry
Ben Gurion University
Beer Sheva
Israel

BEN-REUVEN A.
School of Chemistry
Tel-Aviv University
Tel-Aviv
Israel

BREWER R.G.
IBM Research Lab
5600 Cottle Rd.
San Jose, Ca. 95193
USA

BRONSTEIN S.
Department of Physics
Technion
Haifa 32000
Israel

BURAK I.
School of Chemistry
Tel-Aviv University
Tel-Aviv,Israel
(Bitnet: F62 AT TAUNOS)

CHANG R.K.
Yale University
Applied Physics
P.O.B. 2157, Yale Station
New Haven, CT. 06520, U.S.A.

CHESHNOVSKY O.
School of Chemistry
Tel Aviv University
Ramat Aviv, Tel Aviv 69978
Israel

CHU S.
AT&T Bell Laboratories
Room 4F433
Crawfords Corner Road
Holmdel, N.J. 07733, U.S.A.

COHEN E.
Department of Physics
Technion
Haifa 32000, Israel
(Bitnet: PHR1OEC AT TECHNION)

COHEN S.
Dept. of Isotopes
W.I.S.
Rehovot 76100
Israel

COHEN-TANNOUDJI C.
E.N.S
24 Rue Lhomond
75231 Paris, Cedex 05
France

DOTAN M.
Dept. of Physics
Bar Ilan University
Ramat Gan
Israel

DUCLOY M.
Laser Physics Lab
Universite Paris Nord
Avenue J.B.Clement
93430 Villetaneuse, France

ECKBRETH A.C.
United Technologies
Research Center, M.S. 90
Silver Lane
E. Hartford, CT. 06108, U.S.A.

EHRENFREUND E.
Dept. of Physics
Technion
Haifa 32000
Israel

EREZ G.
NRCN
P.O.Box 9001
Beer Sheva
Israel

ERNST W.E.
Inst. fur Molekulphysik
Freie Univ. Berlin
Arnimallee 14
D-1000, Berlin 33, W.Germany

ERNSTING N.
MPI fur Biophysikalische Chemie
Abt. Laserphysik
Goettingen
W.Germany

FREED C.
Lincoln Lab.
M.I.T.
Lexington, Ma. 02173-0073
U.S.A.

FREEDHOFF H.
Dept. of Physics
Technion
Haifa, Israel
(Bitnet: PRQO8HF AT TECHNION)

FRIEDMAN H.
Dept. of chemistry
Bar Ilan University
Ramat Gan 52100
Israel

FRIESM A.A.
Dept. of Electronics
Weizmann Institute of Science
Rehovot 76100
Israel

GERSHONI D.
Solid State Institute
Technion
Haifa 32000
Israel

GOELMAN G
Dept. of Isotope
Weizmann Institute of science
Rehovot 76100
Israel

GOLUB I.
Dept. of Physics
Ben Gurion University
Beer Sheva 84105
Israel

GREENBLATT G.
Dept. of Chemistry
Hebrew University
Jerusalem
Israel

GRISCHKOWSKY D.
IBM T.J.Watson Research Center
P.O.Box 218
Yorktown Heights, NY 10598
U.S.A.

HAAKS D.
Univ. G.H.Wuppertal
Gaussstr. 20
D-5600 Wuppertal 1
W.Germany

HAHN E.L.
Department of physics
Univ. of California
Berkeley, Ca. 94720
U.S.A.

HANSCH T.
Physics Department
Stanford University
Stanford, Ca. 94305
U.S.A.

HARDE H.
Univ. de Bundeswehr
Postfach 700822
D-2000 Hamburg 70
West Germany

HAROCHE S.
E.N.S.
24 Rue Lhomond
75231 Paris, Cedex 05
France.

HARTMAN S.R.
Physics Department
Columbia University
New York, N.Y. 10027
U.S.A.

HAAS Y.
Dept. of Chemistry
Hebrew University
Jerusalem
Israel

HERING P.
Max Planck Inst. fur Quantenoptik
Postfach 1513
D-8046 Garching, West Germany
(Bitnet: PRH AT DGAIPP1S)

HORESH N.
Dept. of Chemical Physics
Weizmann Institute of Science
Rehovot 76100
Israel

HOSE G.
Dept. of Chemical Physics
Weizmann Institute of Science
Rehovot 76100
Israel

HUPERT D.
School of Chemistry
Tel Aviv University
Ramat Aviv, Tel Aviv
Israel

IVRI J.
NRCN
P.O.Box 9001
Beer Sheva
Israel

JAFFE H.
Dept. of Electronics
Weizmann Institute of Science
Rehovot 76100
Israel

JORTNER J.
School of chemistry
Tel Aviv University
Ramat Aviv, Tel-Aviv 69978
Israel

KAISER W.
Fakultat fur Physik E11
Technische Universitat Munchen
Arcisstrasse 21
8000 Munchen 2, W. Germany

KARNI Z.
Laser Industries Inc.
Atidim Industrial Park
Tel-Aviv
Israel

KATZENELLENBOGEN N.
Dept. of Chemical Physics
Weizmann Institute of Science
Rehovot 76100, Israel
(Bitnet: CFKATZEN AT WEIZMANN)

KIMEL S.
Department of Chemistry
Technion
Haifa 32000, Israel
(Bitnet:CHR3OAR AT TECHNION)

KOFMAN A.
Dept. of Chemical Physics
Weizmann Institute of Science
Rehovot 76100
Israel

KOMPA K.
Max Planck Inst. fur Quantenoptik
Postfach 1513
D-8046 Garching
W.Germany

KRUTMAN I.
Dept. of Physics
Ben Gurion University
Beer Sheva 84105
Israel

KURIZKI G.
Dept. of Chemical Physics
Weizmann Institute of Science
Rehovot 76100
Israel

LAUBEREAU A.
Physical Institute
University of Bayreuth
P.O.B. 3008
D-8580 Bayreuth, W.Germany

LAVI S.
NRCN, P.O.Box 9001
Lasers division
Beer Sheva
Israel

LAVI R.
Dept. of Physics
Ben Gurion University
Beer Sheva
Israel

LEHMITZ H.
Fachbereich Elektrotechnik
Univ. der Bundeswehr
Postfach 700822, D-2000
Hamburg 70, W.Germany

LEVINE A.
NRCN
P.O.Box 9001
Beer Sheva
Israel

LEVINGER A.
Dept. of Chemical Physics
Weizmann Institute of Science
Rehovot 76100
Israel

LICHTMAN E.
Dept. of Electronics
Weizmann Institute of Science
Rehovot 76100
Israel

LOY M.M.T.
IBM T.J.Watson Research Center
P.O.Box 218
Yorktown Heights, N.Y. 10598
USA

MARCUS P.
Chemistry Division
NRCN, POB 9001
Beer-Sheva 84190
Israel

MENAT M.
DO Group
Applied Physics
P.O.B. 5046
26009A Delft, Holland

MIRON E.
Laser Department
NRCN, P.O.Box 9001
Beer Sheva 84190
Israel

METCALF   H.
Department of Physics
State Univ.of N.Y.
Stony Brook,
N.Y. 11794, USA

MICHEL-BEYERLE M.E.
 Univ. Munchen,
 Lichtenbergstrasse 4
 8046 Garching Bei Munchen
 W. Germany

MLYNEK J.
 Inst.fur Quantenoptik
 Universitat Hannover
 Welfengarten 1
 D- 3000 Hannover, W.Germany

MOORADIAN A.
 M.I.T. Lincoln Labs.
 P.O.Box 73
 Lexington, Mass. 73-0073
 U.S.A.

NAAMAN R.
 Dept. of Isotopes
 Weizmann Institute of Science
 Rehovot 76100
 Israel

NAFCHA Y.
 Dept. of Physics
 Bar Ilan University
 Ramat Gan
 Israel

ORENSTEIN M.
 Dept. of Chemistry
 Technion
 Haifa 32000
 Israel

PETRANK A.
 Dept. of Isotopes
 Weizmann Institute of Science
 Rehovot 76100
 Israel

PRIOR Y.
 Dept. of Chemical Physics
 Weizmann Institute of Science
 Rehovot 76100, Israel
 (Bitnet: CFPRIOR AT WEIZMANN)

RABIN I.
 Dept. of Chemical Physics
 Weizmann Institute of Science
 Rehovot 76100
 Israel

RICE S.
 M.I.T. Room 2-025
 77 Massachusetts Ave.
 Cambridge, MA. 02139
 U.S.A.

RON A.
 Dept. of Chemistry
 Technion
 Haifa 32000
 Israel

ROSENBLUH M.
 Dept. of Chemistry
 Bar-Ilan University
 Ramat Gan, Israel
 (Bitnet: F67341 AT BARILAN)

ROSENWAKS Z.
 Department of Physics
 Ben Gurion University
 Beer Sheva, Israel
 (Bitnet: GDBB100 AT BGUNOS)

ROTTKE H.
 Fak. fur Physik
 Universitat Bielefeld
 D-4800 Bielefeld 1
 W.Germany

ROZEN I.
 Dept. of Isotopes
 Weizmann Institute of Science
 Rehovot 76100
 Israel

SCHEK I.
 Dept. of Chemistry
 University of Texas
 Austin, Texas 78712-1167
 U.S.A

SCHREIBER W.
 Dept. of Appl. Sciences
 College of Staten Island
 130 Stuyvesant Pl.
 Staten Island, NY. U.S.A.

SCHWARTZ-LAVI D.
 Dept. of Physics
 Ben Gurion University
 Beer Sheva 84105
 Israel

SEIDMAN T.
 Dept. of Chemical Physics
 Weizmann Institute of Science
 Rehovot 76100
 Israel

SHANK C.V.
 Electron. Research Lab.
 ATT Bell Lab.
 Holmdel, NJ. 07733
 U.S.A.

SHAPIRO M.
 Dept.of Chemical Physics
 Weizmann Institute of Science
 Rehovot 76100, Israel
 (Bitnet:CFSHAPIR AT WEIZMANN)

SHEVY Y.
 Dept. of Physics
 Bar Ilan University
 Ramat Gan
 Israel

SHOSHAN I.
 Laser Division
 EL-OP, P.O.Box 1165
 Rehovot 76110
 Israel

SHUKER R.
 Department of Physics
 Ben Gurion University
 Beer Sheva
 Israel

SOEP B.
 Lab. de Photophysique Moleculaire
 Univ. de Paris-Sud
 91405 - Orsay Cedex
 France

SPEISER S.
 Dept. of Chemistry
 Technion
 Haifa 32000
 Israel

VALENTINI J.
 Dept. of Chemistry
 University of California
 Irvine, Calif.
 U.S.A.

VARDANI Z.
 Dept. of Physics
 Technion
 Haifa 32000
 Israel

VEGA S.
 Isotope Department
 Weizmann Institute of Science
 Rehovot 76100
 Israel

VETTER R.
 Laboratoire Aime Cotton
 CNRS II
 91405 Orsay Cedex
 France

WAARTS R.
 Dept. of Electronics
 Weizmann Institute of Science
 Rehovot 76100
 Israel

WALDMAN M.
 Landseas Corporation
 38 King George st.
 Tel Aviv
 Israel

WALTHER H.
 Univ. Der Munchen
 Am. Coulombwall 1
 8046 Garching
 West Germany

WEBER H.G.
 Univ. of Heidelberg
 Philosophenweg 12
 D-6900 Heidelberg
 W.Germany

WELGE K.H.
 Fakultat f. Physik
 Universitat Bielefeld
 D-4800 Bielefeld 1
 W. Germany

WILSON-GORDON A.
 Dept. of Chemistry
 Bar-Ilan University
 Ramat Gan 52100
 Israel

478

WOLFRUM J.
 Physik. Chemisch Institut
 Im Neuenheimer Feld
 D-6900 Heidelberg 1
 West Germany

YAJIMA T.
 Inst. Solid State Physics
 University of Tokyo
 Roppongi Minato-Ku
 Tokyo 106, Japan

YARIV A.
 CALTECH
 1201 California Ave
 Pasadena, Ca. 91125
 U.S.A.

YATSIV S.
 Racah Institute of Physics
 Hebrew University
 Jerusalem
 Israel

YOGEV A.
 Dept. of Isotopes
 Weizmann Institute of Science
 Rehovot 76100
 Israel

ZAX D.
 Isotope Department
 Weizmann Institute of Science
 Rehovot 76100
 Israel

ZUCKERMAN H.
 Dept. of Chemistry
 Hebrew University
 Jerusalem
 Israel

"FRITZ HABER SYMPOSIUM ON METHODS OF LASER SPECTROSCOPY"

WEIZMANN INSTITUTE OF SCIENCE, REHOVOT, ISRAEL

DECEMBER 16-20, 1985.

1.A. Petrank
2.G. Goelman
3.D. Zax
4.N. Ernsting
5.A.C. Eckbreth
6.Y. Prior
7.T. Seidman
8.J. Ivri
9.H. Yaffe
10.E. Miron
11.R. Bavli
12.R. Chang
13.N. Katzenellenbogen
14.H. Weber
15.K. Shuker
16.I. Rosen
17.G. Erez
18.S. Cohen
19.H. Rottke
20.U. Ayam
21.S. Chu
22.M. Loy
23.H. Lehmitz
24.K. Welge
25.I. Shoshan
26.J. Wolfrum
27.R. Naaman
28.R. Waarts
29.H. Harde
30.O. Cheshnovsky
31.A. Kofman
32.J. Valentini
33.D. Haaks
34.G. Hose
35.S. Rice
36.M. Rosenbluh
37.S. Hartman

38.T. Yajima
39.G. Kurizki
40.K. Kompa
41.N. Asida
42.Y. Band
43.M. Ducloy
44.R. Vetter
45.E. Pines
46.Y. Snevy
47.A. Mooradian
48.B. Reindorf
49.R. Avni
50.D. Albeck
51.H. Walther
52.S. Kimel
53.H. Metcalf
54.C. Shank
55.Y. Nafha
56.A. Laubereau
57.J. Mlynek
58.S. Haroche
59.P. Hering
60.C. Freed
61.D. Grischkowsky
62.C. Cohen-Tannoudji
63.Y. Rabin
64.E.L. Hahn
65.D. Gershoni
66.W. Ernst
67.S. Speiser
68.R. Brewer
69.W. Kaiser
70.A. Ben-Reuven
71.T. Hansch
72.M. Shapiro
73.M. Menat
74.A. Levinger

481

of excited hydrated molecules, 113
  laser control, 239-47
  laser-pulse molecular, 443-48
  polarized, 355-58
Distribution
  rotational, 458-59
  vibrational, 459
Doppler dephasing, 110, 112
Doppler distribution, in RHS, 181-84
Doppler-free laser spectroscopy, 163-72, 183, 191-200
Doppler-free measurements, time-resolved coherence spectroscopy, 103
Doppler-free phase conjugate emission, 279
Doppler-free saturation cell, 42
Doppler shift
  and laser cooling, 33, 34-35
  and magnetic trapping, 37
  in optical molasses, 42
  in picosecond modulation spectroscopy, 94-95
Doppler spectroscopy, 429-33
Double heterojunction diode lasers, 157-58
Double-peak lineshape, 279
Double resonance, 240-47
  optical-optical resolved, 389-98
  optical radio-frequency, 469
Double resonance spectroscopy, 175-85
  laser-microwave, 192-200
Doublets population, 459-60
Douglas, A. E., 463
DPB (1-4-diphenyl butadiene), excited state conformation change, 130-31
Dressed atoms, 4-10
Drexhage, K. H., 15
Droplets, 249-58
Dual fluorescence, 425-26
Dual frequency modulation (DFM), 187-90
Dunning, T. H., Jr., 357
Dye molecules, intramolecular vibrational redistribution, 58

Earnshaw's theorem, 37, 45
Eberly, J. H., 18
8-hydroxy 1,3,6 trisulfonate pyrene
  See HPTS
Eikonal approximations, 335-36
Electro-optic modulator, 42-43
Electrodynamics. See Quantum electrodynamics
Electron diffraction, low-energy, 380

Electronic level inversion, 129
Electronic quenching, 331-34
Electronic resonance enhancement in CARS, 233-34
Electron shelving, 1-10
Electron transfer, excited state, in polar solvents, 131-32
Elliot, D. S., 295
Emission rates of fast-moving dipoles, 341-45
Emission spectroscopy, in micro-solvent clusters, 418-19
Emission wavelength range of TDLs, 157
Endoenergiticity of reaction, 399, 400
Energy distribution, internal, of diatomic molecule, 331
Energy limiters, RSAs as, 119
Energy relaxation times, 84
Energy transfer
  intramolecular, 56-57, 439-42
  molecular, 55-56
Enhanced spontaneous emission, 27
Ertmer, W., 42
ESIPT (excited state intramolecular proton transfer), 425-28
Ethanol droplets, 254-56
Exact semiclassical solution for four-wave mixing, 289-93
Excited state
  conformation change, in DPB, 130-31
  internal charge transfer, 131-32
Excited state intramolecular proton transfer (ESIPT), 425-28
Excited state Raman scattering (ESR), 326-29
Exciton bleaching, 54
Exciton resonances, and optical transitions, 51-54
External cavity lasers, 138-41, 145
Extracavity optical pulse compressor, 117, 123-25
Extra-resonances
  in degenerate four-wave mixing, 301-5
  induced, 295-99
  origin of gain, 307-11
Extreme-motion (EM) vibrational states, mode-localized, 435-38

Fabry-Perot etalon, 455
Fano, U., 313, 315, 455, 456
Fano lineshapes, 315
Fast-moving dipoles, 341-45
Femtosecond optical pulses, 51-54
Femtosecond time resolution, in infrared spectroscopy, 65-66

Fermi golden rule, 6, 414
Field damping, in single atom cavities, 26-27
Field ionization method, 27
Field spectrum of semiconductor lasers, 147-49
Field strengths of Rydberg atoms, 13
Filaments, self-trapped, 317-18
Finesse, of DFM, 189
Fine structure splitting, 105
Flashlamp-pumped dye laser (FLP), 331-33
Flash photolysis, 354
Fleming, M. W., 156
Fluorescence
  of acetone, 407-10
  cooperative, 443-48
  excited state intramolecular proton transfer, 425-27
  laser-induced, 354, 357-58, 400-402
Fluorescence, continued
  in microsolvent clusters, 417-19
  in NO_2, 470-71
  and scattering, 34
  single atom, 1-10
  time-resolved, 441
  and velocity distribution, 34-35
Fluorescence decay, 114-16
Fluorescence sensitive detection, 99
Fluorescence spectroscopy, perturbation facilitated OODR resolved, 389-98
Foot, C. J., 167
Foreign gas broadening, 67-68
  of hyperfine multiplets, 201-4
Forster Cycle, 113
Forward scattering, quantum-beats, 101
Fountain. See Atomic fountain
Four-level system, 301-2
Four-wave mixing, 322, 323
  degenerate, 76-84, 277-81
    induced extraresonances, 301-5
  in droplets, 254-57
  exact semiclassical solution, 289-93
  pressure induced extra-resonances, 295
  stimulated emission, 310, 326-29
  time delayed, 87-95
  transient, 75
Four-wave parametric amplification, 317
Fourier transform coherent Raman spectroscopy (FT CARS), 68-72
Fowler's formula, 449, 450
Franck-Condon principle, 191

Free induction decay (FID), 63, 76, 103
  in cesium, 109-12
  optical measurements, 205-8
Free ion generation, 113-14
Freely falling atoms, 169
Free radicals, 191-200
Frequency-doubled dye lasers, 369
Frequency fluctuations, 205-8
Frequency jitter, in TDLs, 158
Frequency modulation, 187-90
  in SLs, 148
  of TDLs, 152
  Raman heterodyne detection, 177-82
Frequency noise spectral density, 136-37
Frequency offset-locking, 158, 159
  of TDLs, 152
Frequency response of quantum well lasers, 146-49
Frequency stabilization of CO_2 lasers, 152-54
Frequency standards for infrared synthesizers, 151-61
Frequency up-conversion, optical phase conjugation, 285-86
FT CARS (Fourier transform coherent Raman spectroscopy), 68-72
Fujita, T., 136
Fully-saturating diffracting beams, 283
Fundamental constants, 171-72

(GaAl)As lasers, 98, 133-38
  See also AlGaAs lasers
GaAs MQWS, 51, 52
Gabrielse, G., 15
Gain, extra-resonant origin, 307-11
Gas mixtures, laser-induced ignition, 363-64
Geminate recombination, 113-16
Gerade states, 444
G matrix, in CARS, 231-32
Goldman, M., 268
Gomer, R., 449
Goodman, L. S., 195
Goy, P., 15, 27
Gradient traps, 45-47
Grangier, P., 443
Grating-controlled CO isotope lasers, 154
Gratings
  anharmonic, 282, 282
  optically-induced, 278-79
Grischkowsky, D., 108
Gross, M., 15
Ground state transitions, microwave measurement, 191-92

Haar, H. P., 116

Metastable excitation, 270-71
Meter, definition of, 187
Methane, FT-Raman spectroscopy, 69-72
Meystre, P., 41
Michielsen, S., 412, 414
Microsolvent clusters, photon transfer, 417-20
Microwave discharge, CARS spectroscopy, 216
Microwave modulated polarization spectroscopy (MMPS), 194
Microwave optical double resonance spectroscopy (MODR), 193-200
Microwave-optical polarization spectroscopy (MOPS), 193-200
Microwave transitions, 191
Mies, F. H., 315
Miller, J. A., 357
Minogin, V. G., 41
Mobile CARS instrument, 235-36
Mode-localized motion, 435
Mode-locked lasers, 117, 129
  synchronously pumped, 102-3, 106-7
Mode-locking, passive, 123, 125-27
Modulated echoes, 88-94
Modulated pumping, in coherence spectroscopy, 97-100
Modulators
  acousto-optic, 43
  electro-optic, 42-43
Molecular beam
  charge-exchange reactions, 443
  Doppler spectroscopy, 429-33
  of NO_2, 469-70
  supersonic, 399
Molecular-beam laser-mw double resonance spectroscopy, 195-200
Molecular jets, 407-10
Molecular multiphoton excitation (MPE), 347
Molecular predissociation, 389-98
Molecular systems, optical bistability, 335-39
Moller, T., 463
Monoenergetic atomic beams, 33
Monolayers
  organized organic, 449-51
  SHG studies, 260-65
Monomer absorption, 119-21
Mooradian, A., 144, 156, 157
Muller, G., 27-28
Multichannel quantum defect theory, 455
Multiphoton absorption, step-wise, 435, 438
Multiphoton ionization, 354, 384-85
Multiple Quantum Well Structure (MQWS), 51-52

Multiple species measurements, 227-29
  pure rotational CARS, 230
Multiplex CARS, 225-27
Multiwave mixing, degenerate, 280
  in resonant gas media, 281-83
Mutual spin flip interactions, 268

Nearly-free induction decay (NFID), 63, 64
Neutral atom traps, 33-40
Noise
  and energy fluctuations, 29
  in lasers, theories of, 144-45
  technical sources, 155
Noise spectra, in SLs, 135-37
Noise spikes, 87
Nonadiabatic collision processes, CARS studies, 331-34
Non-degenerate wave mixing, 277
  four-wave mixing, 284-86
Non-equilibrium spin distributions, from optical pumping, 267-76
Non-equilibrium state distributions, CARS spectra, 213
Non-linear eikonal approximation, 335-36
Nonlinear optics, 249-58
  in resonant gas media, 277-87
  stochastic fluctuation induced extra-resonances, 295-99
Nonlinear scattering, 325
Nonlinear susceptibility of surfaces, 264
Nonlinear transient spectroscopy, 75-85
Nonresonant cavities, 14-15
Nonthermal population distributions, 51-54
Non-transform-limited light, 78, 84
Noolandi, J., 114-15
NO_2, unexpected experimental results, 469-72

OCS photodissociation products, 429-33
Off-resonant laser-induced ring emission, 317-20
Ohtsu, M., 137
Onsager, L., 114
Optical anisotropy, oscillating, 98
Optical bistability, 335-39
Optical coherences, 110
Optical Earnshaw Theorem, 45
Optical feedback for linewidth reduction, 137-38
Optical heterodyne techniques, 153
Optically thick media, 281
Optical moslasses, 41-46

Optical-optical double resonance
175, 184, 389-98
Optical phase conjugation, 285-86
Optical pulse
circularly polarized, 269
ultrashort, 51-54
Optical pulse compressors, 123-25
Optical pulse train interference
spectroscopy (OPTIS), 97-100
Optical pumping, 36, 175
in magnetic trap, 39
non-equilibrium spin distributions, 267-76
with pulse train, 97-100
velocity selective, 178-79
Optical transitions in semiconductors, 51-54
Orbital symmetry, 443
Oscillator strength, 120
Output resonance, of droplets,
251
Ozone
CARS spectroscopy, 212-17
laser induced ignition, 363-65

Parametric amplification, 317
Parametric up-conversion, 64, 65
Partial deuteration, 427
Paschen-Back states, 109
Passive mode-locking, 123, 125-27
Paul, W., 37
PEMS. *See* Photon Echo Modulation
Spectroscopy
Penning trap, 15
Periodic excitation resonances,
98-99
Persson, M., 387
Perturbation facilitated optical-
optical double resonance
(PFOODR) resolved spectros-
copy, 389-98
Phase-conjugate (PC) processes,
277-87
Phase fluctuations
and laser linewidth, 133
and PIER4, 295
stochastic, 299
Phase-locking
of external cavity lasers, 138,
141
optical cavity to radio frequen-
cy, 187-90
Phase-matched multiwave mixing,
angle-resolved, 282
Phase matching, 224-25, 228, 229-30
in droplets, 254-57
pure rotational CARS, 230
Phase-modulated pulses, 78-79
coherent, 75-76, 80-84
Phase modulation, 188-90
in droplets, 254-55

Phase transitions, of surfaces, 263
Phillips, W. D., 41, 45
Photochemical reactions, crossed-
beam experiment, 399-405
Photodecomposition of acetone, 407
Photodetection signal, darkness
periods, 3-8
Photodissociation, 429, 457-61
CARS spectroscopy, 209-22
Photodynamics, picosecond, 129-32
Photoexcitation, direct, 244
Photofragmentation, 457-60
Photofragments
angular velocity distribution,
429-33
CARS spectroscopy, 210, 213-21
time-resolved fluorescence,
443-48
Photoinduced absorption (PA),
421-23
Photoionization, 313-16
of hydrogen molecules, 454-55
of polyatomic molecules, 367-78
Photolysis
atomic resonance line, 354
of ozone, 212-16
polarized, 354
pulsed-laser, 210
Photomultiplier, 107
Photon antibunching, 1
Photon echo, 75, 76, 87-95
Photon Echo Modulation Spectroscopy
(PEMS), 88-95
Photon statistics, 2-10, 18, 21
Photon-transfer, in microsolvent
clusters, 417-20
Picosecond absorption spectrometer,
129-30
Picosecond excitiation of large
molecules, 57-61
Picosecond experiments, 87-95
Picosecond laser pulses, and FT
CARS, 68-72
Picosecond modulation spectroscopy,
87-95
PIER D4 (pressure-induced extra-
resonances in degenerate
FWM), 301
PIER4 (pressure-induced extrareso-
nances in four wave mixing),
295
Poisson photon statistics, 18, 21
Polarizability
of adsorbed monolayer, 450
of Rydberg states, 13
Polarization
cross-relaxation, 275
of FID, 109
induced, 144, 298
for ring emission, 317
of surfaces, 264

Polarization beats, 108
Polarization selective detection, 99
  of hyperfine quantum beats, 101-4
Polarization spectroscopy, Doppler-free, 191-200
Polarized photolysis, 354
Polar solvents, internal charge transfer, 131
Polyacetylene, 421
Polyatomic molecules
  excited states, 129-32
  highly-excited vibrational states, 435-38
Polymers, conducting, 421-24
Positronium, 164
Pound, R. V., 14
Power-independent linewidth, 136-37
Power limiter, RSAs as, 119
Predissociation, 239
  of hydrogen Rydberg states, 453, 456
  molecular, 389-98
Predissociation lifetimes, 411-15
Pressure broadening, 169
  of hyperfine multiplets, 201-4
Pressure induced extraresonances
    See PIER D4; PIER4
Pressure sensitivity of CARS spectra, 231
Pressure shifts, OPTIS measurement, 99-100
Prior Distribution, 404
Pritchard, D. E., 46
Probe pulse, linearly polarized, 101, 102, 103, 106
Proch, D., 378
Product arrangement channels, 241-43
Programmable infrared synthesizers, 151-61
Propagation, coherent, 66-67
Proton scavengers, 115-16
Proton transfer reactions, 113-16
Pulse compressors, 117, 123-25
Pulsed-laser photolysis, 210
Pulsed lasers, 354
  CARS spectroscopy, 210, 226
Pulses
  circularly polarized, 101, 102
  linearly polarized probe, 101, 102, 103, 106
  ultrashort, 75
Pulse shaping, 118-19, 239
Pulse technology, ultrashort, 75
Pulse trains, picosecond, 97-100
Pump resonance, tuning for, 233-34
Purcell, E. M., 14
Pure rotational CARS, 230-31
Pure transit time broadening, 470

Q factor
  of DFM, 189
  in lead-salt tunable diode lasers, 155
  of resonant cavity, 14, 25, 26-27
Quack, M., 357
Quack-Troe interpolation, 357
Quadrupole atom traps, 37-40
Quantum beats
  in cesium, 101-4
  in forward scattering, 102-4
  in sodium, 105-12
Quantum electrodynamics (QED), 11-23, 170-71
  nonrelativistic, 342
Quantum phase noise limited linewidth, 155
Quantum well lasers, 143-50, 157-58
Quasidiabatic surfaces, 403-4
Quaterphenyl, as saturable absorber, 123-25
Quenching, 331-34
  of acetone, 407, 408

Rabi frequency, 26
Rabi-nutation, 21-22
Radiation force, and scattering, 33
Radiationless processes, 58
  and rotational state, 410
Radiative decay time, 469
Radical cations, 375-77
Radiochemistry of benzene, 376
Radiofrequency (rf) resonances, 176-82
Radiofrequency-optical double resonance spectroscopy, 175-85
Raimond, J. D., 15
Raman heterodyne spectroscopy (RHS), 175-85
Raman scattering, 307, 322, 323
  excited state, 326-29
Raman spectroscopy, 63
  CARS as, 210
  of combustion, 223
  high resolution, 68
Raman waves, in droplets, 250-53
Ramsey, N. F., 182
Ramsey resonances, 17, 31
  collision-induced, 182-85
Ramsey spectroscopy, 169-70
Random walk
  in optical molasses, 42, 43
  of oscillation phase, 155
Rare-earth ions, spin-spin cross-relaxation, 267-76
Rayleigh scattering, 307-11, 328
Rayleigh-type extraresonance, 301
Reaction enthalpy, 425, 427
Reactive cross-sections, 402

Reagents, electronic excitation, 399-400
Reciprocal spectral width, 78, 84
Recombination, of excited hydrated molecules, 113-16
Rectangular cavities, 16
Recursive residue generation method (RRGM), 347-51
Redfield, A. G., 205
Redistribution, vibrational
  in highly excited molecules, 435-38
  intramolecular, 55-61, 419
Refractive effects of combustion, 224-26
Reilly, J. P., 371, 373
Relative-temperature-dependent time-resolved fluorescence, 441
Relaxation, vibrational, 68
Relaxation oscillation
  in SLs, 134, 137
  lasers stabilized against, 117, 123, 125
Relaxation processes, ultrafast, 63
Relaxation rates, 87
  ultrafast, 84
Resolved fluoresence, from PFOODR excitation, 389, 391
Resonance enhancement, 326-29
Resonance fluorescence, 1, 4-10
Resonance lines. See Linewidths
Resonant cavities, 14-16
  collective radiative effects, 28-30
  single-atom effects, 26-28
  single-atom maser, 18-22
Resonant gas media, nonlinear optics, 277-87
Resonant two-photon excitation, 453-56
Resonant two-photon ionization (R2PI), 417-19
Resonant unimolecular reactions, coherent control, 239-47
RETPS (resonance enhanced three-photon scattering), 326-29
Reverse saturable absorbers (RSAs), 117-21, 123-27, 337-38
Rhodamine 6G, 117
Ring emission, off-resonant laser-induced, 317-20
Ringing, cooperative fluorescence, 443, 444, 445
Ring resonator, 336-38
Rontgen magnetization, 343
Rosenblatt, G., 379
Rosenkrantz, M. E., 393
Rotational dephasing, 68

Rotational distributions, 380, 382-84, 458-59
  CARS spectroscopy, 209
  of ozone, 212-15
  in combustion, 231-32
  nonadiabatic collisions, 333
Rotational Raman scattering, 67, 68
Rotation-electron coupling, 456
Rothenberg, J. E., 108
Ro-vibrational spectra, 230
RRGM (Recursive residue generation method), 347-51
R2PI (resonant two-photon ionization), 417-19
Rydberg atoms, 11-23
  and radiation, 25-32
Rydberg constant, 163, 167, 168-71
Rydberg states, 370, 373-74
  chlorine molecule, 463-67
  molecular hydrogen, 453-56
  scaling laws, 12-13
  two-photon excitation spectroscopy, 389, 390

Sagiv, J., 449
Saito, S., 137
Sakurai, K., 411
Samarium vapor, 176-82
Sato, H., 136
Saturable absorbers, 117, 123-27, 338
Saturated absorption process, 280
Saturation
  of degenerate four-wave mixing in resonant gas media, 279-81
  of Rydberg atoms, 13
Saturation cell, Doppler-free, 42
Saturation resonances, 152-53
  in NO_2, 471-72
Saturation spectroscopy, 163-66, 201-2
Scaling laws, for Rydberg atoms, 12
Scattered beam distributions, 384-85
Schawalow, A. L., 42, 143, 155, 191, 201
Schawlow-Townes formula, 155-56
Schmitt-Rink, S., 52
Scott, D., 348
Second harmonic generation, 133
  surface studies, 259-66
Seeding, 380-81
Self-broadening, 67
  of hyperfine multiplets, 201-4
Self focusing, 321
Self-oscillation, phase-conjugate, 281
Self-phase-modulated coherent pulse, 78-79

Self phase modulation, in droplets, 254
Self-trapped filaments, 317-18
Semiclassical exact solution for four-wave mixing, 289-93
Semiconductor lasers, 100
  quantum well, 143-50
  spectral linewidth, 133-42
Semiconductors, optical transitions, 51-52
Sequential decay, 301-5
Sha, G. H., 378
Shortener of pulse, RSAs as, 119
Shot noise, of DFM, 189
Side bands, of semiconductor laser field spectrums, 147-48
Silver electrode, surface studies, 260
Single-atom cavity effects, 26-28
Single-atom masers, 12, 16, 18-22, 25-32
Single-atom spectroscopy, 1-10
Smalley, R. E., 371
Smoother of pulse, RSAs as, 119
Sodium
  atoms
    laser cooling, 35-36
    laser trapping experiments, 41-48
    magnetic trapping, 37-40
  CARS studies, 331-34
  coherent blue-shifted emission, 321-24
  laser induced stimulated emission, 325-29
  off resonant laser induced ring emission, 318-20
  quantum beats, 105
  Stark resonances, 313-16
  vapor
    PC self-oscillation, 281
    picosecond modulation spectroscopy, 87-95
Solitions, 421, 425
Sorem, M. S., 201
Spacers, and work function, 449-51
Spectra, extremely congested, 192-94
Spectral broadening, enhanced, 134
Spectral purity
  of quantum well lasers, 143-46
  of TDLs, 155
Spectroscopy
  of dissociating molecules, 356-58
  semiconductor diode laser as light source, 133
Speed of light, 187
Spencer, W. P., 15
Spin diffusion, 267-71

Spin-orbit coupling, 411-14
Spin-spin cross-relaxation of optically-excited rare-earth ions, 267-76
Splitting frequencies, 101
Spontaneous emission, 11-16, 133, 134, 341
  cavity effects, 26-28
  and frequency modulation, 148
  lead-salt diode lasers, 155
  and magnetic trapping, 37
  of Rydberg atoms, 13, 25-32
  spectral purity, 143-46
Spontaneous Raman scattering, 223
Spontaneous transitions, 6, 7
Sputter ion gun, argon, 386
SRGS (Stimulated Raman Gain spectroscopy), 224
SRS (stimulated Raman scattering), in droplets, 251-54
Stabilized lasers, 152-59, 187-90
Standing wave probe field, 269
Standing-wave saturation resonances, 152-53
Stark field, ionic, 447
Stark resonances, in sodium, 313-16
Stark shift, 47
Stark spectroscopy, 197-99
Stark splitting, 8
State distribution
  modification of density, 145-46
  rotational, 458-59
  vibrational, 459
State-specific angular distributions, 382
Stenholm, S., 41
Stimulated emission, 8, 16-22, 34, 35
  coherent blue-shifted, 321-24
  extraresonant, 309-11
  laser induced, from sodium vapor, 325-29
  off resonant ring emission, 317-20
Stimulated Raman Gain spectroscopy (SRGS), 224
Stimulated Raman scattering (SRS), in droplets, 251-54
Stimulated scattering, 326-29
  Rayleigh scattering, 307-11
Stiumlated transitions, 6, 7
  in Rydberg atoms, 13
Stochastic fluctuation induced extraresonances, 295-99
Stokes lasers, 227
Stokes resonances
  in combustion, 233
  in droplets, 250-54
Striker, G., 412
Strong model, modified Bloch equations, 208